HPE Aruba Networking Certified Professional - Switching

OFFICIAL CERTIFICATION STUDY GUIDE
(EXAM HPE7-A08)

First Edition

Ivannia Alpizar, Henrique Almeida, Mugdha Sathe, Venu Dhanraj Puduchery

HPE Press
660 4th Street, #802
San Francisco, CA 94107

HPE Aruba Networking Certified Professional - Switching
Official Certification Study Guide

Ivannia Alpizar, Henrique Almeida, Mugdha Sathe, Venu Dhanraj Puduchery

Published by:

Hewlett Packard Enterprise Press
660 4th Street, #802
San Francisco, CA 94107

ISBN: 978-1-960058-26-3

Printed in Mexico.

WARNING AND DISCLAIMER

This book provides information about the topics covered in the HPE Aruba Networking Certified Professional - Switching certification exam (HPE7-A08). Every effort has been made to make this book as complete and as accurate as possible, but no warranty or fitness is implied.

The information is provided on an "as is" basis. The author, and Hewlett Packard Enterprise Press, shall have neither liability nor responsibility to any person or entity with respect to any loss or damages arising from the information contained in this book or from the use of the discs or programs that may accompany it.

The opinions expressed in this book belong to the author and are not necessarily those of Hewlett Packard Enterprise Press.

FEEDBACK INFORMATION

At HPE Press, our goal is to create in-depth reference books of the best quality and value. Each book is crafted with care and precision, undergoing rigorous development that involves the expert-ise of members from the professional technical community.

Readers' feedback is a continuation of the process. If you have any comments regarding how we could improve the quality of this book, or otherwise alter it to better suit your needs, you can contact us through email at hpepress@epac.com. Make sure to include the book title and ISBN in your message.

We appreciate your feedback.

Publisher: Hewlett Packard Enterprise Press

HPE Press Program Manager: Michael Bishop

About the Authors

Based in Costa Rica, Ivannia Alpízar is an experienced instructional designer with a passion for bridging learning gaps and fostering global collaboration. With over 8 years of expertise in instructional design, she specializes in crafting engaging and effective technical training materials for diverse global teams.

Henrique Almeida is a Certified HPE Aruba Switching Expert with over 18 years of networking experience. He is based in Brazil and works as an instructional designer and subject matter expert for HPE Aruba Networking Education Services.

Mugdha Sathe is an instructional designer based in India with over 10 years of experience. She has acted as a courseware developer and instructor for technologies related to business integration and analytics and currently develops courseware on networking.

Venu Dhanraj Puduchery is a certified HPE Aruba Mobility and Design expert with over 15 years of networking experience. Based in India, he works as instructional designer and SME for mobility, location services, and network management.

Introduction

The HPE Aruba Networking Certified Professional - Switching study guide will teach you the advanced skills necessary to implement and operate enterprise-level HPE Aruba Networking switching solutions. You will build on the skills you learned at the Associate level to configure and manage modern, open standards-based networking solutions using HPE Aruba Networking CX Switch Operating System (AOS-CX) routing and switching technologies. In this guide, you will learn about AOS-CX switch technologies:

- Securing port access with HPE Aruba Networking Dynamic Segmentation
- Redundancy technologies such as Multiple Spanning Tree Protocol (MSTP)
- Link aggregation techniques, including Link Aggregation Protocol (LACP)
- Switch virtualization with HPE Aruba Networking Virtual Switching eXtension (VSX) and HPE Aruba Networking Virtual Switching Framework (VSF)

Certification and Learning

Hewlett Packard Enterprise Certification and Learning provides end-to-end continuous learning programs and professional certifications that can help you open doors and accelerate your career.

We provide

- **Professional sales and technical training and certifications** to give you the critical skills needed to design, manage and implement the most sought-after IT disciplines; and
- **Continuous learning activities and job-role based learning plans** to help you keep pace with the demands of the dynamic, fast paced IT industry
- **Advanced training** to help you navigate and seize opportunities within the top IT transformation areas that enable business advantage today.

As a Certification and Learning certified member, your skills, knowledge, and real-world experience are recognized and valued in the marketplace. To continue your professional and career growth, you have access to our large HPE community of world-class IT professionals, trend-makers and decision-makers. Share ideas, best practices, business insights, and challenges as you gain professional connections globally.

To learn more about HPE Certification and Learning certifications along with continuous learning programs, visit

http://certification-learning.hpe.com

Audience

This book is designed for those who have experience in advanced level implementation and maintenance of wired solutions. They possess experience assessing and interpreting existing networks and network design documentation. They regularly troubleshoot, resolve issues, and perform ongoing support of large network environments. They have familiarity with wired network security best practices and perform duties independently with at least two years of experience working as a member of the team that supports maintaining multiple campus topologies, edge branches, and data center networks.

Assumed Knowledge

This is a professional plus level guide for senior solution architects. It is assumed that you have a thorough understanding of HPE Aruba Networking switching in small-to-medium businesses (SMBs).

We recommend students to have completed the AOS-CX Switching Fundamentals course or have equivalent on-the-job experience.

Minimum Qualifications

Note that there are no prerequisites for taking the HPE7-A08 exam. However, we recommend having already achieved the Aruba Certified Campus Access Professional exam (HPE7-A01.)

Relevant Certifications

After you pass this exam, your achievement may be applicable toward more than one certification. To determine which certifications can be credited with this achievement, log in to The Learning Center and view the certifications listed on the exam's More Details tab. You might be on your way to achieving additional certifications.

Preparing for Exam HPE7-A08

This self-study guide does not guarantee that you will have all the knowledge you need to pass the exam. It is expected that you will also draw on real-world experience and would benefit from completing the hands-on lab activities provided in the instructor-led training.

Recommended HPE Training

Recommended training to prepare for each exam is accessible from the exam's page in The Learning Center. See the exam attachment, "Supporting courses," to view and register for the courses.

Obtain Hands-on Experience

You are not required to take the recommended, supported courses, and completion of training does not guarantee that you will pass the exam. Hewlett Packard Enterprise strongly recommends a combination of training, thorough review of courseware and additional study references, and sufficient on-the-job experience prior to taking an exam.

Exam Registration

To register for an exam, go to https://certification-learning.hpe.com/tr/learn_more_about_exams.html

CONTENTS

1 Introduction to AOS-CX

We are going to develop an overview of the HPE Aruba Networking CX Switch Operating System (AOS-CX) Switching Portfolio, its characteristics, and its benefits.

In essence, the AOS-CX switching portfolio embodies a blend of innovation, performance, and security, making it a formidable choice for organizations seeking to build resilient, high-performing, and future-ready network infrastructures. Whether it is for a high-density data center or a dispersed campus network, AOS-CX switches deliver the flexibility and intelligence necessary to meet the evolving demands of today's digital enterprises.

OBJECTIVES

After completing this chapter, you should be able to:

- ✓ Compare and contrast the different AOS-CX switching models
- ✓ Describe the important features of the AOS-CX switching product line

---

AOS-CX switch overview

The AOS-CX software forms the heart of HPE's strategy for campus core, access and aggregation switching.

Some benefits of AOS-CX include:

Programmable

AOS-CX is based on open, REST API features that enable external applications to securely view and configure all system components. Developers can use the familiar Python language to script REST API interaction and to create NAE agents for system monitoring and troubleshooting.

Extensible

AOS-CX is extensible, built to support microservices and integrate with other workflow systems and services.

Innovative

AOS-CX takes an innovative approach to high availability and fault tolerance. The software architecture supports resiliency; for example, a process that fails can roll back to a good state using the stateful database. AOS-CX has even been honored by the industry for its innovation. CRN awarded the CX 8400 Series switch its 2017 Product of the Year, in large part due to HPE Aruba Networking Network Analytics Engine (NAE). NAE takes the guesswork out of troubleshooting and optimization and will be a game-changer for many customers.

Secure

Finally, AOS-CX is secure and forms the bedrock for a trusted infrastructure. In addition to standard security features such as access control lists (ACLs), AOS-CX supports a trusted boot process. The software will not boot unless every hardware component checks out as authentic and free from malware and rootkits.

Configuration version rollback may cause issues when going between major AOS-CX software versions.

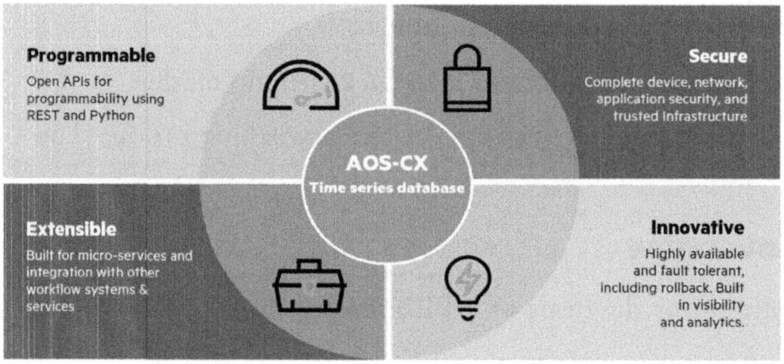

Figure 1-1: Heart of HPE's core, aggregation, and access switching

Customer benefits

By deploying AOS-CX switches in a campus or data center network, you get all the benefits of an advanced operating system. This includes:

- New hardware that brings together Gen 7 HPE ASICs with Virtual Output Queuing (VOQ), Smart Rate ports, and up to Class 8 (IEEE 802.3bt Type 4).
- AOS-CX modern architecture that enables a seamless REST interface, with support for the exclusive NAE

- A set of advanced tools that cover zero-touch or one-touch deployment via the HPE Aruba Networking CX Mobile App, and advanced configuration management with HPE Aruba Networking Fabric Composer.
- Proven technologies like Dynamic Segmentation that unify wired and wireless access, enhance and simplify policy enforcement, and protect IOT and other critical devices.

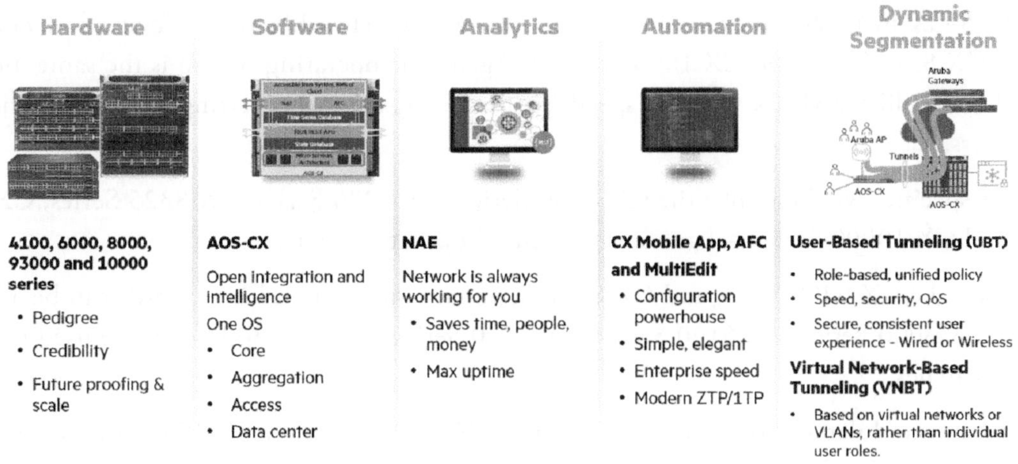

Figure 1-2: Customer benefits

AOS-CX switching portfolio

The AOS-CX switching portfolio represents a cutting-edge suite of network switches designed to meet the demands of modern, dynamic IT environments. Engineered for high performance and scalability, AOS-CX switches leverage a cloud-native operating system—AOS-CX—which offers advanced automation, analytics, and security features. These switches are tailored for a range of use cases, from enterprise data centers to campus networks, providing robust solutions for handling data-intensive applications and complex networking requirements. The portfolio is distinguished by its support for intelligent edge capabilities, simplified management through centralized control, and seamless integration with HPE's broader networking ecosystem, empowering organizations to deliver superior connectivity experiences.

Overall, as we talk about AOS-CX models, it is important to remember that the operating system is the same. If you can configure a CX 6300 Series switch, you will be familiar with the CX 83xx Series or CX 8400 Series. The features are the same, making it so much easier when you are adding products and different AOS-CX switches. This means it is easier to maintain, as there is configuration consistency and a shorter/flatter learning curve. The only change or difference in hardware is capacity: The larger ones can hold more tables, while the smaller ones have a higher capacity

We have modular switches or one rack unit (RU) switches. Some support 24 ports, some 48 ports. Some are just copper. In some instances, you can combine Line Cards.

All AOS-CX switches can open REST API and run a script, which operates like SNMP and gathers information by itself and monitors directly on your switch. They can interact with third-party devices and send it to SNMP and MAP.

Some AOS-CX switches can be configured as a campus switch, while others focus on data centers (like the CX 83xx Series and CX 10000 Series). Again, the operating system is the same, but they vary in features like VXLAN support, speed, performance, and other settings that are applied for data centers.

There are specific switches—like the CX 6400 Series, CX 8320 Series, CX 8325 Series, CX 8400 Series, and CX 10000 Series—that are very focused on data centers.

Remember, the CX 6400 Series and CX 8400 Series are modular and Line Cards can be added. And the CX 6000 Series, CX 6100 Series, CX 6200 Series and CX 6300 Series are used more as campus switches.

We also have modular and fixed options, like the CX 6300M Series and CX 6300F Series (M is modular and F is fixed). A modular chassis is excellent for scalability and flexibility. These are one RU switches with the ability to add modules like fans and power units on the rear of the switch.

HPE Aruba Networking Virtual Switching Framework (VSF) is supported on the CX 6200 Series and CX 6300 Series.

HPE Aruba Networking Virtual Switching eXtension (VSX) is supported on the CX 6400 Series, CX 8xxx Series, CX 9300 Series, and CX 10000 Series. The main difference between VSX and VSF is the separate control. VSX focuses on providing high availability and redundancy for network services in data center environments, while VSF aims to create a high-performance, resilient network fabric by combining multiple switches into a virtualized chassis.

Virtual Output Queuing in the CX 6400 Series

Virtual Output Queuing (VOQ) is a technique used in certain network switch architectures where, rather than keeping all traffic in a single queue, separate queues are maintained for each possible output location. It addresses a common problem known as "head-of-line blocking." One of the most important reasons why use Case VOQ is necessary is due to congestion. If you are facing congestion and you have real-time applications like voice and video, this is mandatory. Hospitals, universities, video editing, whatever you have—all benefit from VOQ.

Always-on POE and quick POE on the CX 4100i Series, CX 6200 Series, CX 6300 Series, and CX 6400 Series

Always-on PoE is a feature that provides the ability to the switch to continue to provide power across a soft reboot. It is applicable only to the interfaces which were connected and delivering before the soft reboot. Also, power will not be delivered if power to the switch is interrupted. By default, always-on PoE is enabled at the switch or the slot level. Always-on PoE is recommended especially for critical things like APs.

Quick PoE is a feature that provides the ability for the switch to provide power to the connected powered device as soon as the switch goes through a cold reboot. When quick PoE is enabled on the subsystem, PoE port disablement and PD demotion is not allowed. Quick PoE enablement is not allowed if any of the ports are disabled on the subsystem. You should not over-subscribe the PoE power when quick PoE is enabled. A saved quick PoE configuration will work irrespective of the configuration change at reboot. By default, quick PoE is disabled for the subsystem.

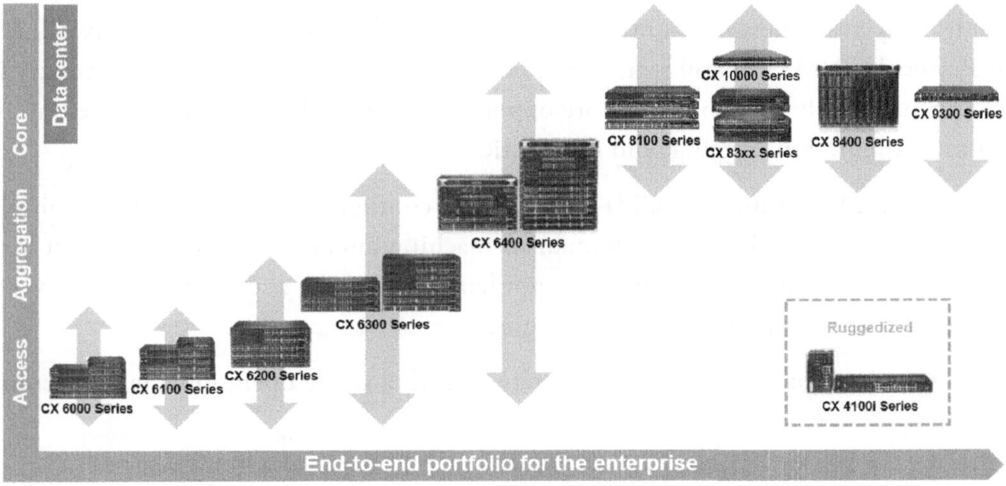

Figure 1-3: AOS-CX switching portfolio

Switch product label

The switch product label is an orange-colored tab on the bottom right side of the switch's front panel. Pull the tab out to view the product label information.

The product label information includes the part number, serial number, and MAC address. Serial numbers and MAC address labels are duplicated on the back or bottom of the product.

Being able to identify this information is important for two main purposes:

- Inventory
- To open a case on the HPE Networking Support Portal for support/warranty.

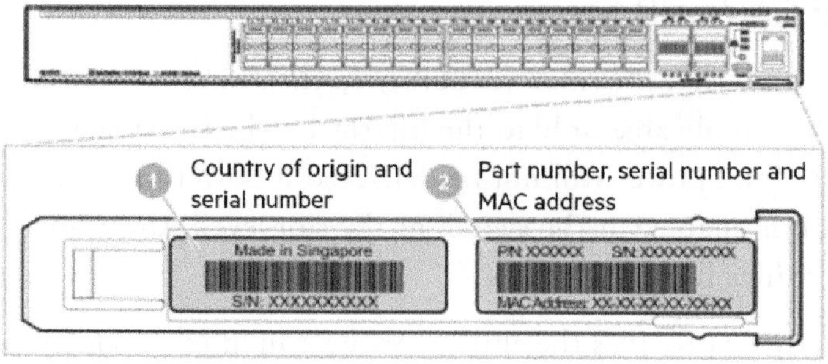

Figure 1-4: Switch product label

Legacy management systems

The following figure shows a typical two-tier network architecture, often duplicated several times in a large campus deployment of dozens or hundreds of switches, multilayer switches, and controllers. For many years, network administrators have relied on the old Simple Network Management Protocol (SNMP). It still works fine, but its "poll driven approach" to telemetry is a bit dated. It remains heavily used, but there are opportunities to find a more streamlined, universal approach to configuration, logging, and diagnostics.

The old Command Line Interface (CLI) with flat-text configuration files is getting a bit old-fashioned as well. The old CLI is great for human-to-machine interaction, but not so great for machine-to-machine interaction. It just does not lend itself to the scalability and ease-of-management requirements of modern campus designs.

Wouldn't it be nice to have a unified infrastructure, with a single operational model across the infrastructure?

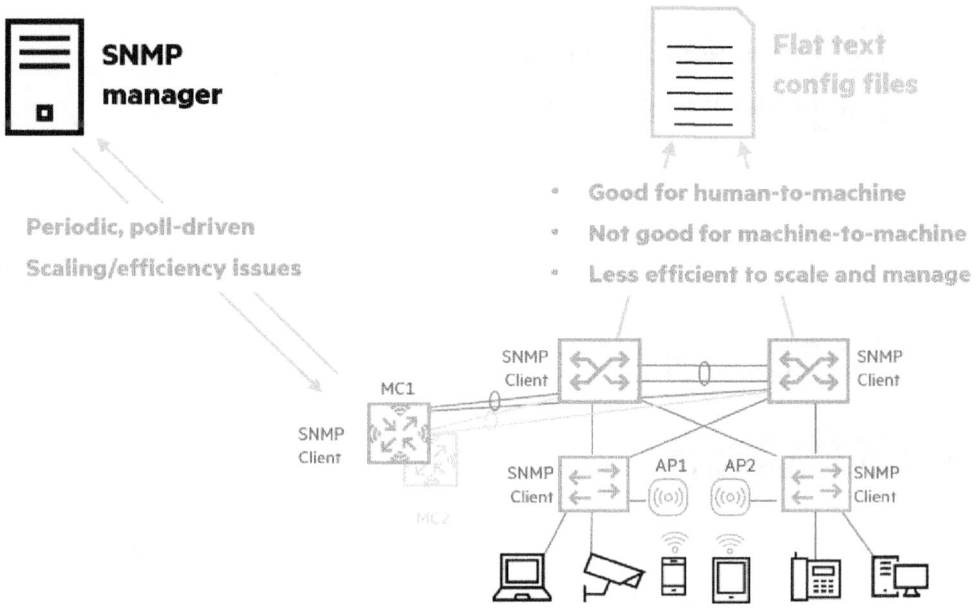

Figure 1-5: Legacy management systems

Modern management approach

AOS-CX switches are built on a microservices-based operating system with an open REST API available for configuration and management functions.

You get easy, secure cloud connectivity, with modern HTTP-based telemetry that uses a more optimized publisher/subscriber model, as opposed to the cumbersome polling method of SNMP.

This facilitates zero-downtime upgrades and moving to a more centralized control and management plane, in-cloud or on-premises. It also facilitates more automation related to typical network administration tasks.

Everything is fully programmable, with intelligent root cause analytics and third-party customizations.

And scalability? A single customer can manage up to 100,000 network devices and 1 million client devices.

Several of these advantages are due to the efficient, hierarchical nature of the REST API.

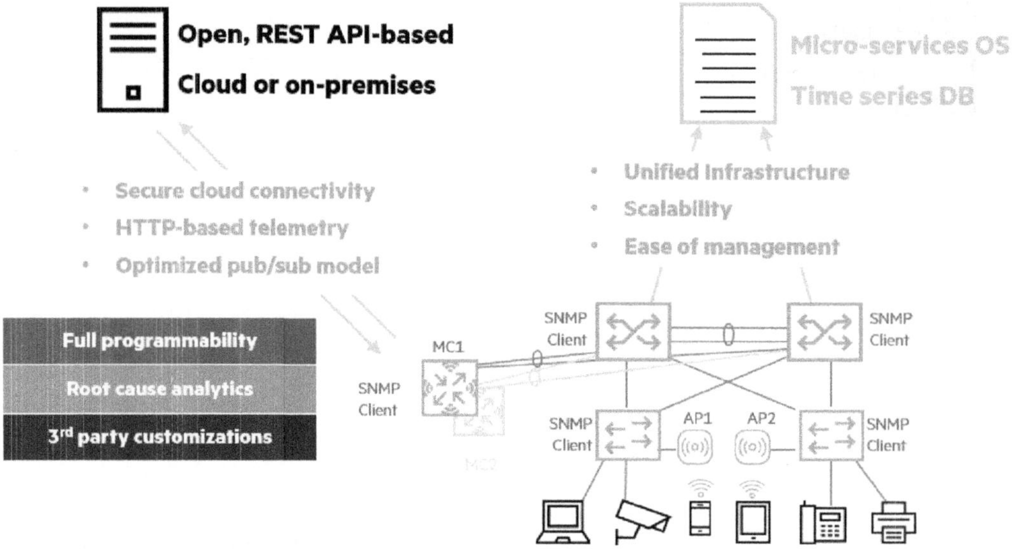

Figure 1-6: Modern management approach

The REST API and URIs

The REST API defines resources, each of which is a Uniform Resource Identifier (URI) that a client can access. For example, a switch that has the IP address 10.1.1.1 has the URL 10.1.1.1/rest/v1/system/bridge/vlans, which lists VLANs on the switch. A REST client can contact this URL to obtain information about this list or add to it.

As you see here, the URIs for resources follow a hierarchical structure. This means that a resource can also be a container for other resources. For example, the "system/vlans" resource contains multiple VLAN resources, each identified with an ID. The same concept holds true for the "system/interfaces" resource, which is a container for multiple interface IDs, and for several other resources on the switch.

The "interfaces" resource stores physical and lower-level information: admin status, UDLD (Unidirectional Link Detection), and more. The "ports" resource stores VLAN and IP/IPv6 protocol information. This is where you configure things like LACP, IP, OSPF, PIM, IGMP, and VLAN settings. AOS-CX is architected to optimize these programmatic functions, enabling the scripting, automation, and orchestration features we expect from modern systems.

For additional information, visit the Developer Hub: https://devhub.arubanetworks.com/

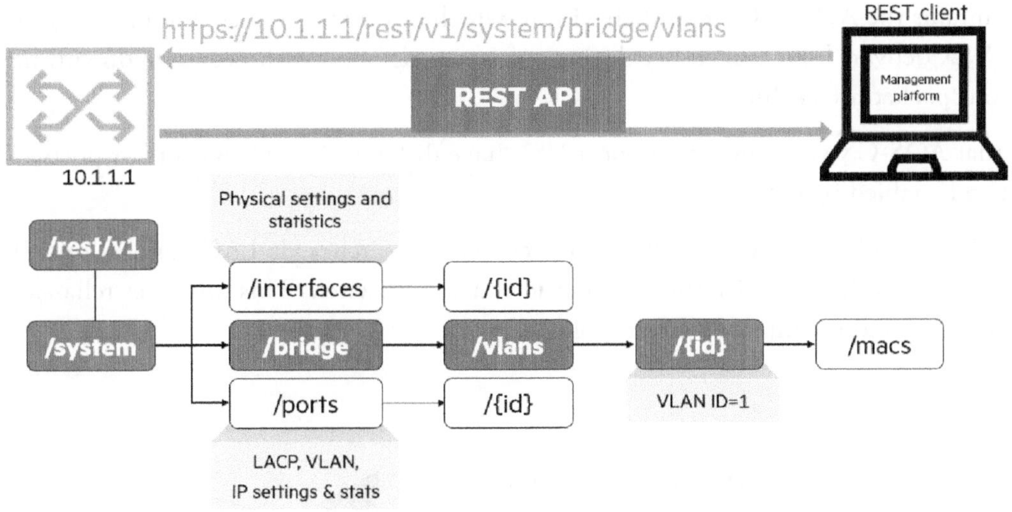

Figure 1-7: The REST API and URIs

AOS-CX switch virtualization: VSX connectivity versus VSF

VSX and VSF are similar each other, and even to earlier virtualization techniques like Virtual Router Redundancy Protocol (VRRP); multiple devices are perceived as a single device. This improves resiliency while reducing configuration and management overhead. But there are distinct differences between VSX and VSF. You use VSX at the campus network aggregation and core layers, and you use VSF primarily at the access layer. The differences in features and function are intentional because priorities differ between access and core.

Configuring VSF and backplane stacking is a plug-and-play affair: a switch, at default settings, automatically unites with another switch. This ease of deployment is especially important at the access layer, where you may have dozens or hundreds of access switches.

You must manually configure VSX on both switches. VSX maintains separate control and management planes, and much of the data plane. This mitigates the "shared fate" nature of common planes, for improved uptimes, giving you high availability and redundancy. This is especially important when aggregating all those access layer switches. And it remains easy to manage because you can synchronize the management planes, and optionally, other features like MAC and ARP addresses, and more.

The VSX fabric leverages both switches' control, management, and data planes, while appearing to other devices as a single switch. It supports an active-active path for Layer 2 traffic and Layer 3 unicast and multicast traffic. In other words, both switches actively switch and route traffic.

VSX maintains the AOS-CX default behavior in which ports are disabled and operate at Layer 3. Finally, VSX delivers high availability during software upgrades, with near zero downtime and continuous packet forwarding.

Notice that AOS-CX switches that support VSX have different behaviors on their ports, which are bridged and enabled by default.

Access switches and mobility controllers connect to core switches via LAG connections. If configured for LACP, they perceive the VSX fabric as a single switch. This mitigates reliance on STP to ensure stable, redundant, loop-free topologies.

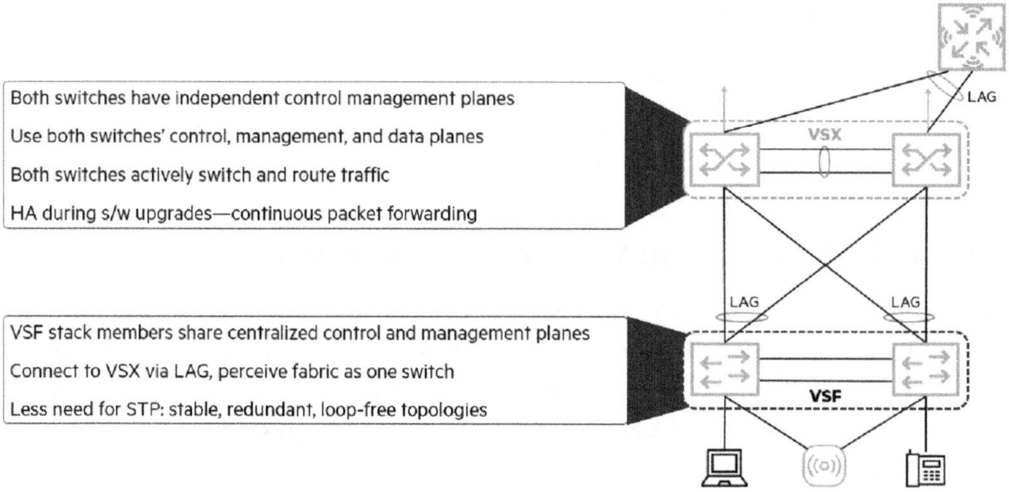

Figure 1-8: AOS-CX switch virtualization: VSX connectivity versus VSF

NAE and the time-series database

The AOS-CX includes the time-series database, which establishes a built-in network record for the state of various components on the switch. An NAE agent monitors the current state of one or more attributes in the configuration and state database and then writes the state to a Time-Series Database for an ongoing record. They can establish baselines of normal states and detect significant deviations from those. When they detect issues, they can trigger alerts and actions and collect further data. Customers can script, and they can create scripts to obtain information about NAE agents from the REST API.

Together, these components create a switch that delivers critical insights that can make the company more secure and competitive. Also, you can program the switch to act on those insights, permitting customers to transform their operations with automated workflows and business application integration.

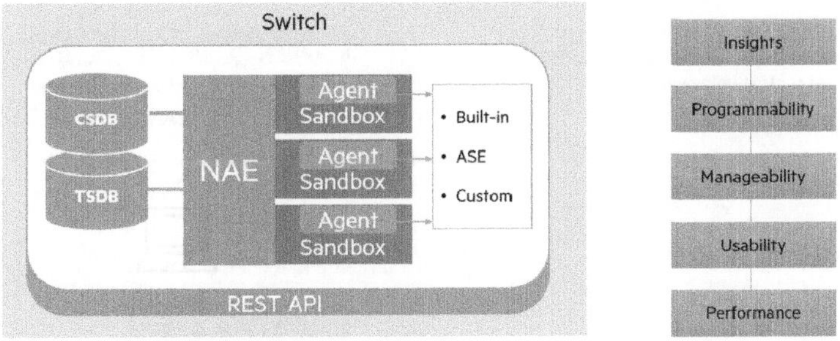

Figure 1-9: NAE and the time-series database

Dynamic Segmentation

In this section, we will introduce the concept of Dynamic Segmentation. We will explore the key features and provide examples along with visuals to enhance your understanding.

Dynamic Segmentation is an advanced networking feature prominently utilized in HPE switches, designed to enhance network security and simplify management. This technology addresses the challenges posed by the increasing complexity and diversity of devices in modern networks.

Colorless ports

One of the cornerstone concepts of Dynamic Segmentation is the use of colorless ports. Unlike traditional switch ports that require manual configuration for specific VLANs, colorless ports dynamically assign network policies based on the identity of the device connecting to the port. This capability allows for seamless and automatic network segmentation, eliminating the need for manual intervention and reducing the risk of configuration errors.

We are starting with colorless ports because it is essential to understand the following sections.

A colorless port is basically a port with default characteristics that are then dynamically changed based on device characteristics and authentication details. Before profiling, context assessment, and role assignment, the device type and context are largely unknown.

Manually configuring VLANs, QoS, ACLs, and more is tedious, time-consuming, and error prone—and you are now free from much of this!

Leave the ports at their default settings. Connected devices are automatically profiled, the context is assessed, and roles are assigned. Today, the finance person is connected to port 3 on the switch; tomorrow, they may connect to port 12. It does not matter: Dynamic Segmentation "colors" the port automatically.

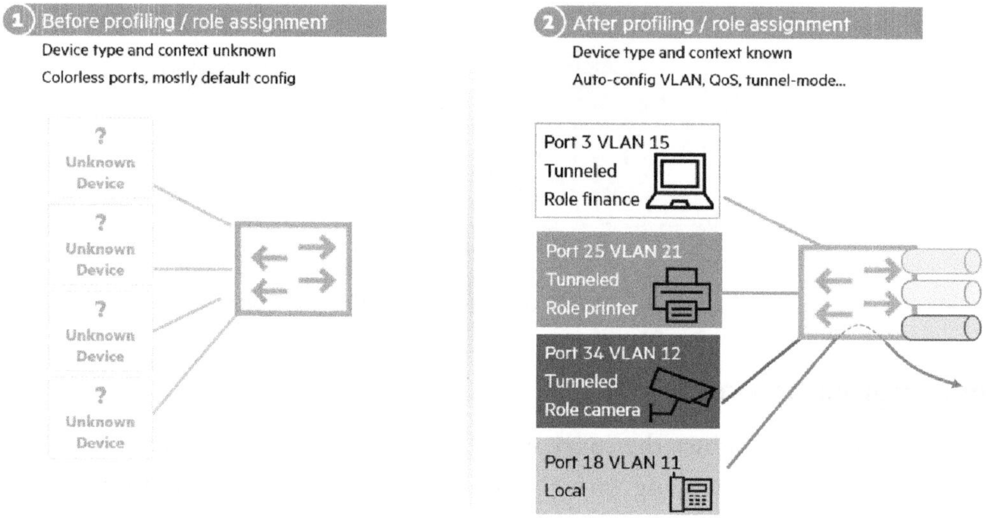

Figure 1-10: Colorless ports

Problem solved

Dynamic Segmentation tackles several critical issues:

- **Security:** By automatically assigning appropriate access policies, it minimizes the risk of unauthorized access and potential security breaches.
- **Management complexity:** It simplifies the process of network configuration and policy enforcement, making it easier to manage large-scale and diverse network environments.
- **Scalability:** It facilitates the addition of new devices without the need for reconfiguring the network, thereby supporting the growing number of IoT devices and mobile users.

All modern desktops, laptops, tablets, and smartphones support robust security mechanisms like 802.1X/EAP, but many devices lack these security mechanisms. This includes specialty devices and some devices that rely on Power over Ethernet (PoE), such as security cameras, payment card readers, Internet of Things (IoT), and medical devices. This lack of endpoint security can pose a serious risk to your infrastructure.

A switch with Dynamic Segmentation improves your security stance while easing administration. It authenticates these devices using ClearPass and tunnels the traffic to a mobility controller, which centralizes formerly disparate ACLs and firewall rules into single, intuitive policy system. You can continue to use these devices without compromising network security.

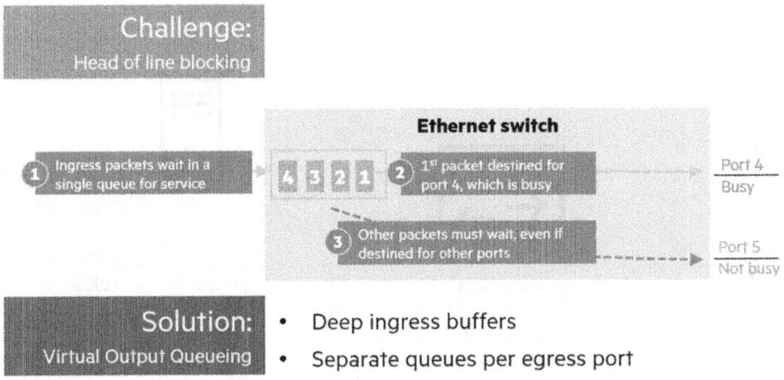

Figure 1-11: Problem solved

Operation overview

Segmentation operates by leveraging HPE Aruba Networking ClearPass Policy Manager Platform. When a device connects to a colorless port, it is authenticated and assigned a role based on its identity and context. This role determines the network policies, such as VLAN assignment, QoS settings, and access controls. The switch then dynamically configures the port according to these policies.

ClearPass profiles each device, deems appropriate endpoints as "acceptable," and sends accept messages to the controller. The messages include tunnel information, assigned VLAN, and secondary role. In the figure, a user was profiled into the Finance role, and so their port is configured for VLAN 15, and their traffic is securely tunneled to the controller. The controller applies centralized policy enforcement to limit finance traffic as appropriate. Printers and cameras are similarly treated, as shown in the following figure.

ClearPass can also indicate if device traffic should be locally forwarded by the switch, directly toward its ultimate destination, without tunneling to the controller. You sacrifice some of that centralized PEF (Policy Enforcement Firewall) enforcement, but you accommodate devices that are uber-sensitive to delay and jitter.

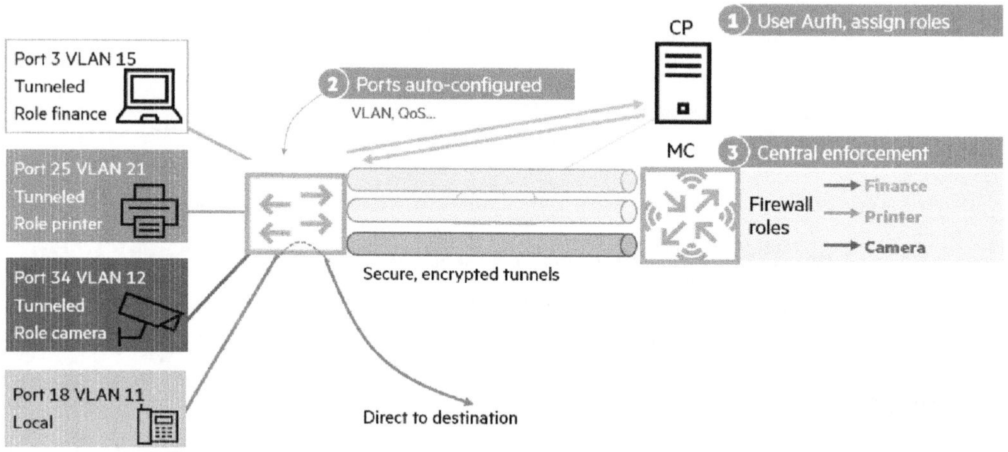

Figure 1-12: Operation overview

Tunneling options

Dynamic Segmentation provides the flexibility to apply policies not just at the edge but throughout the entire network. This is achieved through two primary tunneling options:

- **User-Based Tunneling (UBT):** also called Centralized Fabric or Centralized Overlay, UBT extends the role-based policies to HPE Aruba Networking Mobility Controllers, creating a secure tunnel from the switch to the controller. This ensures consistent policy enforcement regardless of the device's physical location.
- **Virtual Network-Based Tunneling (VNBT):** also called Distributed Fabric or Distributed Overlay, VNBT is another approach that focuses on segmenting network traffic based on virtual networks or VLANs, rather than individual user roles.

The following figure introduces you to tunneling options related to Dynamic Segmentation.

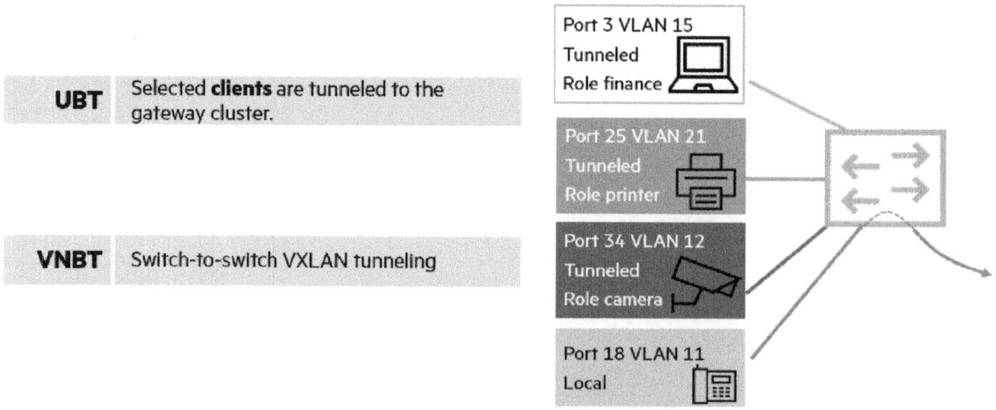

Figure 1-13: Tunneling options

UBT

User-Based Tunneling is designed to extend role-based access policies from the edge of the network to the core, ensuring consistent policy enforcement regardless of the device's physical location.

How it works:

- **Authentication and role assignment:** When a device connects to a switch, it is authenticated using ClearPassPolicy Manager.
- **Role determination:** Based on the device's identity and context, ClearPass assigns a specific role to the device.
- **Tunnel creation:** A secure tunnel is established between the switch and a mobility controller. This tunnel encapsulates the user traffic, applying the designated role-based policies.

Benefits:

- **Consistent policy enforcement:** Policies are uniformly enforced across the network, providing a seamless experience for users.
- **Enhanced security:** By tunneling traffic to a centralized controller, UBT ensures that sensitive data is handled securely.
- **Simplified management:** UBT reduces the complexity of configuring and managing policies at each switch.

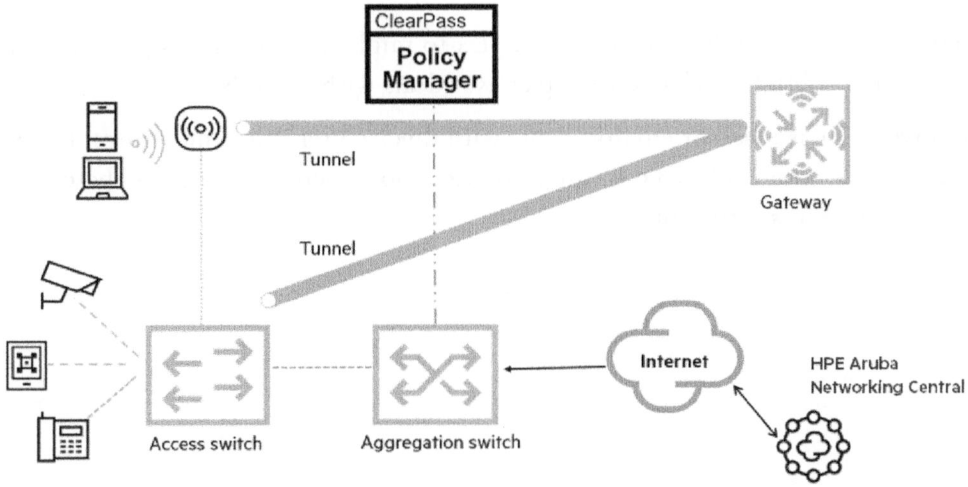

Figure 1-14: User-Based Tunneling

VNBT

Virtual Network-Based Tunneling is another approach that focuses on segmenting network traffic based on virtual networks or VLANs, rather than individual user roles.

How it works:

- **Virtual network assignment:** Devices are assigned to specific virtual networks or VLANs based on predefined criteria, such as device type, location, or function.
- **Tunnel creation:** Virtual networks are tunneled through the network infrastructure to a centralized point, such as a data center or a core switch.
- **Policy application:** Network policies are applied based on the virtual network, ensuring that all devices within the same VLAN receive the same level of access and security.

Benefits:

- **Network segmentation:** Provides robust network segmentation, isolating traffic based on virtual networks.
- **Scalability:** Easily scales to accommodate a large number of devices and VLANs.
- **Centralized control:** Offers centralized control over network policies and segmentation.

Differences between UBT and VNBT

- **Granularity:** UBT applies policies at the user level, offering more granular control, whereas VNBT applies policies at the VLAN level.
- **Complexity:** UBT tends to be more complex to implement due to its per-user policy application, while VNBT is relatively simpler as it deals with VLANs.
- **Use case:** UBT is ideal for environments with diverse user roles and stringent security requirements, whereas VNBT is suited for scenarios where segmentation based on device type or function is sufficient.

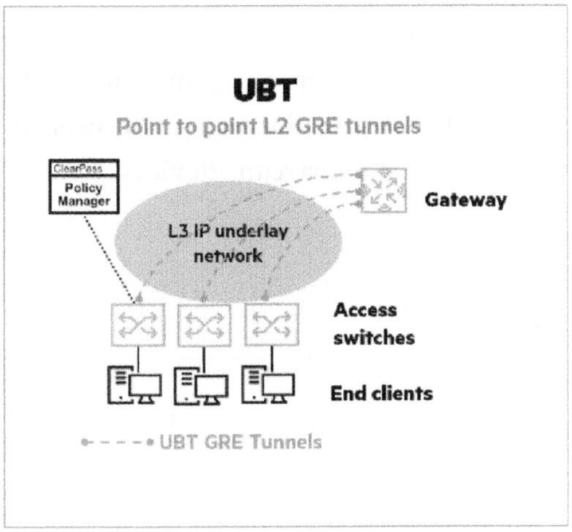

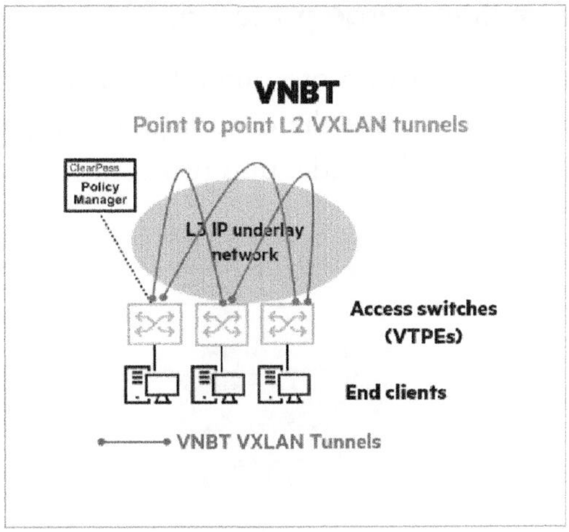

Figure 1-15: Differences between UBT and VNBT

Always-On PoE and quick PoE

Another valuable feature of the CX 6300 Series and CX 6400 Series switches relates to PoE. The CX 4100i Series, CX 6200 Series, CX 6300 Series, and CX 6400 Series support quick PoE and always-on PoE

Always-on PoE (persistent PoE)

Switches have supported PoE standards (IEEE 802.3af, 802.3at) for years, but now we want to maintain PoE power across planned and unplanned software upgrades—especially for planned upgrades. HPE calls this feature always-on PoE, which some may refer to as "Persistent PoE". This feature ensures that end devices do not power cycle on switch reboot. This is especially important for some IoT and Zigbee-based devices, which may behave oddly or take a long time to reboot after a switch upgrade.

HPE Aruba Networking quick PoE

Quick PoE (also known as "Fast PoE") can power devices the moment you connect switch power, without having to wait for the AOS-CX to completely boot. This is especially important for PoE-based lighting and other IoT devices.

The typical use case for this would be a device power outage. CX 6300 Series SKUs provide IEEE802.3bt or four-pair PoE. The fixed SKUs are two-pair PoEs, but the main board is hardware-designed to support a four-pair PoE controller and is four-pair (60W per port) ready.

PoE protection (surge protection)

One method of surge protection relies on a very quick-reacting electronic component called a Shockley Diode. When there is a negative surge, current flows from the switch's Power Supply Unit (PSU) to RJ line side, thus protecting the PSE chip and downstream devices.

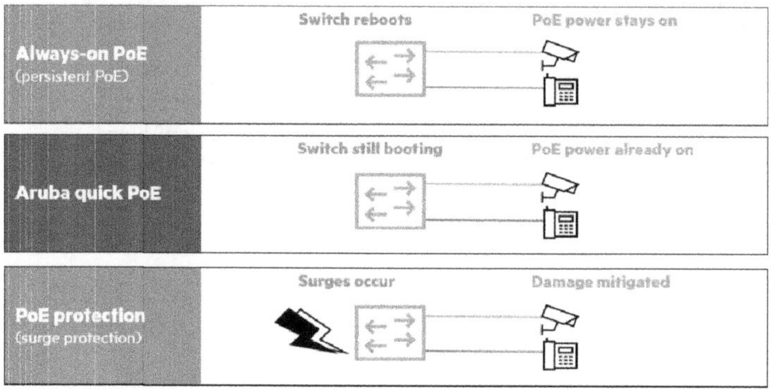

Figure 1-16: Always-on PoE and quick PoE

VOQ

Virtual Output Queuing (VOQ) is an advanced traffic management technique implemented in modern network switches, including the AOS-CX series. VOQ addresses one of the critical challenges in switch architecture: head-of-line (HOL) blocking, where a single congested output port; can impede the entire switch's performance. By maintaining separate queues for each output port within the input port's memory, VOQ ensures that packets destined for different egress ports are independently managed. This method significantly enhances the efficiency and throughput of the switch, allowing for more effective utilization of available bandwidth and reduced latency. The implementation of VOQ in AOS-CX switches leverages HPE's robust software architecture, delivering high performance, scalability, and reliability for demanding network environments.

Problem solved

Let's take a closer look at how traffic crosses the switch's backplane from one line card (LC) through the fabric to another line card.

The following figure shows that four packets have arrived at some interface, sitting in a queue, waiting for service. If the ingress buffer used a single queue, Head Of Line (HOL) blocking could delay traffic. This occurs when the first packet in the queue (at the "head of the line") is destined out a congested port, it delays all packets behind it, even those that are destined to non-congested ports.

HOL blocking is a performance-limiting phenomenon that can occur in network switches, routers, and other types of packet-switched networks. It happens when a queue of packets is held up by the first packet in the line, preventing subsequent packets from being processed even if they could be forwarded independently. This can lead to inefficiencies and reduced throughput in the network.

AOS-CX switches use an intra-switch queuing method called VOQ, it prevents this problem by providing deep ingress buffers with separate queues for each egress port. Physically, the ingress buffer consists of internal and external DRAM banks, with most traffic typically being buffered in the internal DRAM.

For example, consider a network switch with two output ports, A and B:

- If the queue at the switch holds packets for both ports but the first packet is destined for a congested port A, all packets behind it (even those for the free port B) will be delayed until the first packet is processed.
- With VOO, separate queues for port A and port B would allow packets destined for port B to be forwarded without waiting for the congested port A, thus avoiding HOL blocking.

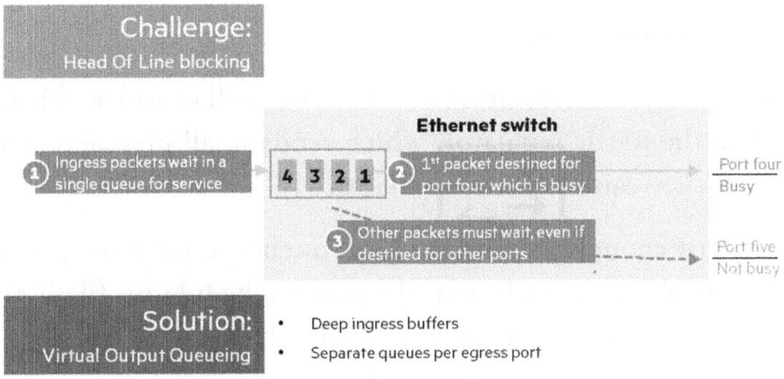

Figure 1-17: Example of problem solved by VOQ

VOQ and unicast traffic transmission across the fabric

This section covers congestion-aware global scheduling, where ingress and egress line cards communicate to manage traffic flow based on buffer availability, and dynamic load sharing crossbar fabric, which uses multiple paths and congestion-aware protocols to efficiently route and reassemble packets.

Congestion-aware global scheduling

On AOS-CX switches, traffic is scheduled for transmission across the fabric based on messages between the ingress and egress line cards. The egress line card sends credits to the ingress line card, indicating how many bits the egress buffer can receive based on current queue fullness. In this way, the scheduling is aware of global congestion.

If the ingress line card receives enough credits, it forwards the packet through the fabric to the egress card. If packets cannot be forwarded due to lack of room in the egress buffer, they are dropped at ingress. The ports have shallower egress queues because by the time traffic reaches that point, it should be rapidly forwarded. Packets are always dropped at the ingress buffer, never at the egress.

Dynamic load sharing crossbar fabric

The fabric uses a dynamic load share crossbar architecture to connect line cards. Larger packets are split into cells, which are dispatched across different paths based on both random assignment and a protocol that is aware of path congestion. Remember that each line card has 12 connections to each fabric card, so several paths are available. The egress card reassembles the cells into a packet.

> VOQ is cell-based on the CX 8400 Series and frame-based on the CX 6400 Series.

VOQs in the CX 6400 Series switch

In a VOQ architecture, the egress side selects which traffic will cross the fabric. In the CX 6400 Series, this block is called the traffic regulator, which sends small messages across the fabric to tell ingress VOQs how much they can send into the fabric.

The objective is to keep just enough packets in egress queues to keep the port fully busy given the port speed. Regarding how VOQs are selected, the port's schedule-profile is also a factor. For example, with Deficit Weighted Round Robin (DWRR), heavier-weighted queues get more selections.

Ingress

So, the ingress pipeline processes the frame with information like egress ports and queues, and drop eligibility (color). Then, the Ingress Admission Control puts the frame's descriptor record in the respective VOQ; the frame itself is now in the packet buffer.

Fabric

The process starts at the Control Loop Manager in the egress line card (destination node). Along with the Admission Control Process, it monitors all egress queues in the module, including congestion. Depending on the queue's available space, it sends "credit" messages to the ingress line cards (source node).

The Fabric Load Balancer (FLB) in the source node transmits the frames with most credits across the fabrics. Packets with the same credit level are selected using a source/destination node hashing algorithm. In multi-ASIC fabrics, packets do not have a pre-assigned dedicated path.

Egress

The Egress Admission Control places frames in their corresponding queue for transmission.

VOQ is supported on many switches, and is especially relevant on the CX 6400 Series and CX 8400 Series switches—modular switches.

Finally, it is important to underline that VOQs are supported on the CX 8400 Series and NOT on the CX 8325 Series.

Additionally, the CX 8360 Series, CX 6400 Series, and CX 6300 Series ASICs implement VOQ. In these cases, VOQ is relevant in the CX 6400 Series because it is the best way to manage traffic between modules, across the backplane/fabric.

Figure 1-18: CX 6400 Series

Learning check

Check your understanding about HPE AOS-CX switching portfolio by answering the following questions.

Question 1

Which AOS-CX platforms support Always-on PoE?

- O CX 6300 Series and CX 6400 Series
- O CX 9300 Series and CX 10000 Series

- CX 6300 Series, CX 6400 Series, CX 8320 Series, and CX 8325 Series
- All AOS-CX switches

Question 2

Which AOS-CX feature supports the time-series database to capture important statistics over a period and to take actions on the captured data?

- AFC
- Dynamic Segmentation
- NAE
- VSX

Question 3

Which AOS-CX feature includes dual management plane functions with zero-downtime during minor and major management upgrades?

- Blackplane stacking
- VSF
- VSX
- VSX and VSF

2 Virtual Switching eXtension

Welcome to the Virtual Switching eXtension (VSX) chapter.

This chapter is about the VSX feature and will focus on high availability technologies that help you simplify your network architecture, operations, and management while providing resiliency and scalability.

You will learn how to use HPE Aruba Networking Virtual Switching eXtension (VSX) and the benefits of using VSX to virtualize the aggregation/core of the campus network.

VSX provides advanced capabilities beyond what was previously available with the VSX-LAG feature.

You will learn how VSX meets customers' high availability needs to provide redundancy, resiliency, performance, and simplicity.

OBJECTIVES

After completing this chapter, you should be able to:

- ✓ Compare and contrast virtual switching technologies.
- ✓ Describe how VSX simplifies networks with resiliency and scalability.
- ✓ Explain how VSX meets high availability needs for customers.
- ✓ Implement best practices for VSX.

Virtual switching technologies

This section introduces the virtual switching technologies that HPE has designed for the HPE Aruba Networking CX Switch Operating System (AOS-CX).

Switch virtualization solution comparison

HPE Aruba Networking Virtual Switching eXtension is a virtualization technology running on AOS-CX. This solution lets the switches present themselves as one virtualized switch in critical areas. Configuration synchronization is one aspect of this VSX solution, where the primary switch configuration syncs to the secondary switch.

VSX virtualizes the control plane of two switches to function as one device at Layer 2 and as independent devices at Layer 3. From a datapath perspective, each device does an independent forwarding lookup to decide how to handle the traffic. Some of the forwarding databases, such as the MAC forwarding database and neighbor tables, synchronize between the two devices using a proprietary VSX control plane. Each switch independently builds some of the forwarding databases.

HPE Aruba Networking Virtual Switching Framework is an HPE Aruba Networking stacking technology that allows multiple switches of the same model to act as a single virtual device. VSF helps provide fast failover, scalability, manageability, and high availability.

A VSF fabric has one active management plane and one active control plane, which the management module in the conductor switch provides. The conductor proxies the control plane to the standby member. Both members of the fabric participate in the forwarding plane. Interface modules from both members combine as if over one large switch connected by the VSF link.

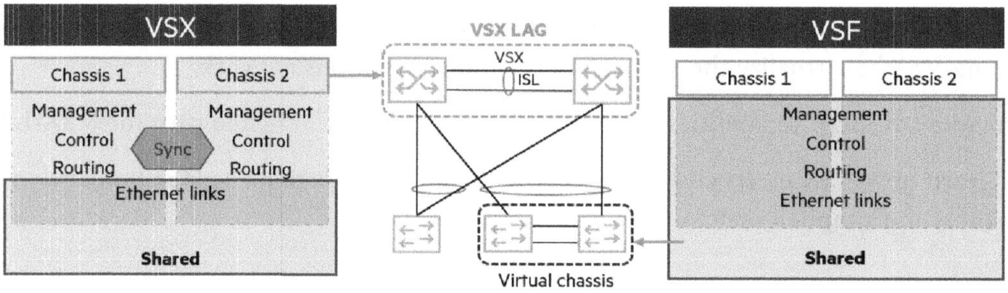

Figure 2-1: VSX vs. VSF

VSX and VSF are similar to each other, and even to earlier virtualization techniques like the Virtual Router Redundancy Protocol (VRRP)—multiple devices are perceived as a single device. This improves resiliency while reducing configuration and management overhead.

You use VSX at the campus network aggregation and core layers and VSF primarily at the access layer. The differences in features and function are intentional because priorities differ between access and core.

Configuring VSF and backplane stacking is a plug-and-play affair: a switch, at default settings, automatically unites with another switch. This ease of deployment is especially important at the access layer, since you may have dozens, or hundreds, of access switches.

You must manually configure VSX on both switches. VSX maintains a separate control plane and management planes, and much of the data plane. This mitigates the "shared fate" nature of common planes for improved uptimes; you get high availability and redundancy. This is especially important when aggregating all those access layer switches. And it remains easy to manage because you can synchronize the management planes and, optionally, other features such as MAC and ARP addresses, and more.

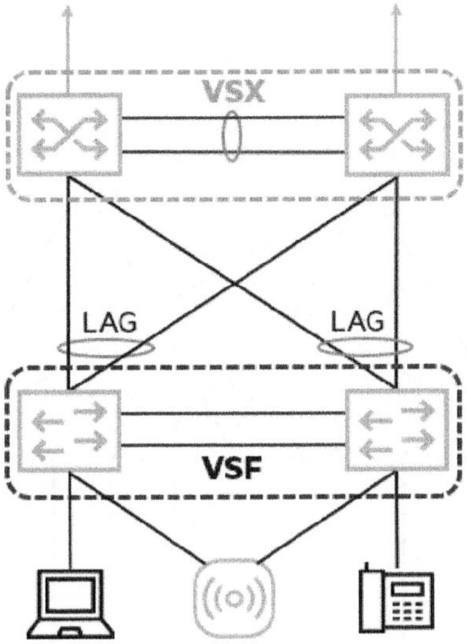

Figure 2-2: Solutions comparison

HPE Aruba Networking Virtual Switching Framework

Combine switches into a single virtual switch

HPE Aruba Networking Virtual Switching Framework (VSF) defines a virtual switch, which is composed of individual physical switches interconnected using Ethernet links. All member switches share a single control plane. Devices connected to the VSF stack perceive a single device. As you recently learned, this virtual switch behavior has significant benefits related to simplified management and improved connectivity.

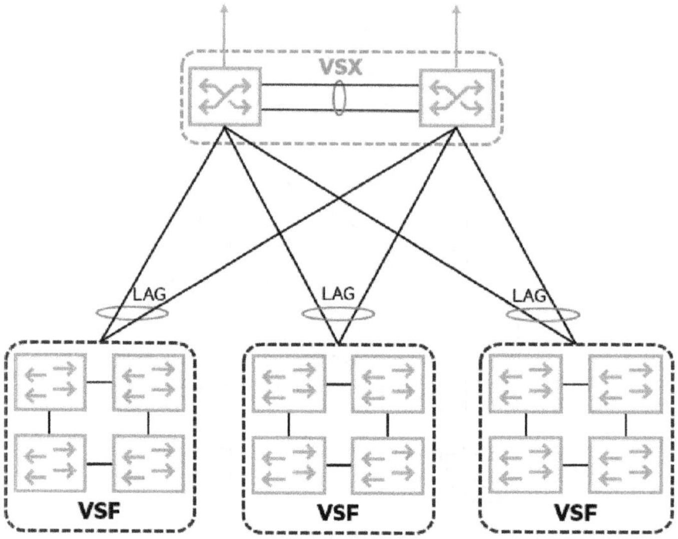

Figure 2-3: HPE Aruba Networking Virtual Switching Framework

In AOS-CX, VSF is supported on the CX 6200 Series and CX 6300 Series only; other higher-numbered models use a different technology, called VSX. You can configure a maximum of eight members using the CX 6200 Series of switches and 10 members in the CX 6300 Series family. This feature is enabled by default.

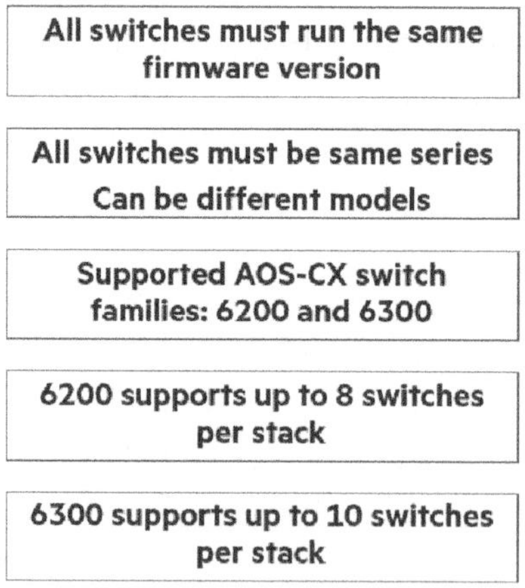

Figure 2-4: VSF features

AOS-CX switch virtualization: VSX connectivity

The VSX fabric leverages both switches' control, management, and data planes, while appearing to other devices as a single switch. They support an active-active path for Layer 2 traffic and for Layer 3 unicast and multicast traffic. In other words, both switches actively switch and route traffic.

VSX maintains the AOS-CX default behavior—ports are disabled and operate at Layer 3. Finally, VSX delivers high availability during software upgrades, with near zero downtime and continuous packet forwarding.

Notice that AOS-CX switches that support VSX have different behaviors on their ports, which are bridged and enabled by default.

Access switches and mobility controllers connect to core switches via LAG connections. If configured for LACP, they perceive the VSX fabric as a single switch. This mitigates reliance on STP to ensure stable, redundant, loop-free topologies.

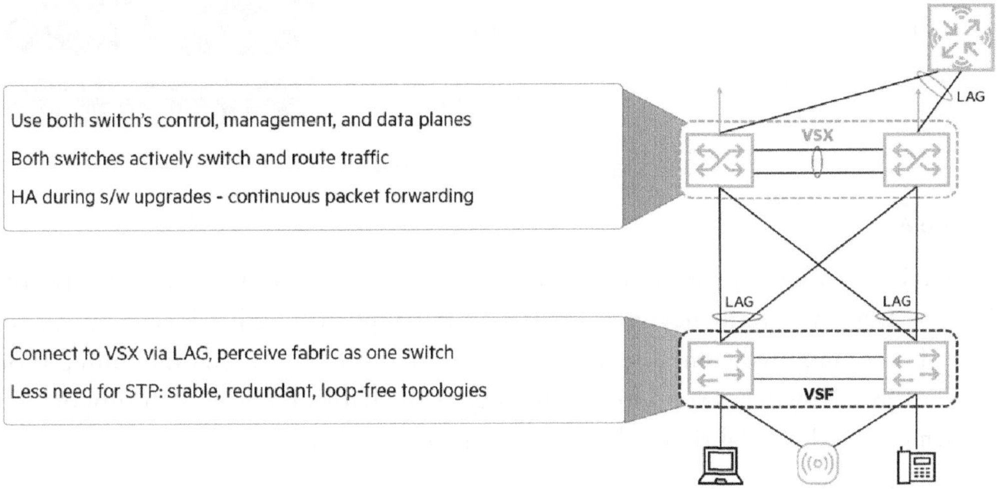

Figure 2-5: VSX connectivity

VSF versus VSX

The figure summarizes the differences between VSF and VSX, as a study and review aid.

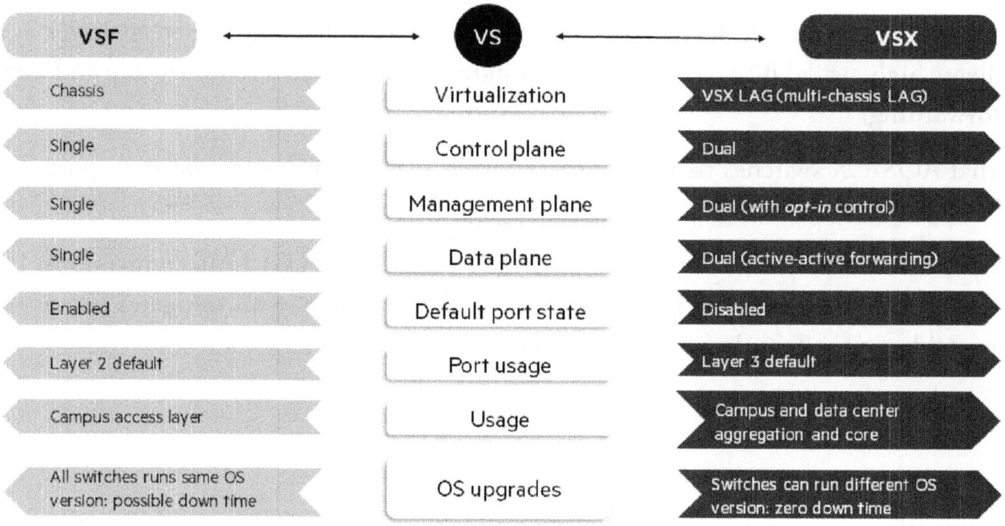

Figure 2-6: VSF versus VSX

VSX overview

AOS-CX switches, particularly when deployed with VSX, meet customers' high availability needs by providing redundancy, resiliency, performance, and simplicity. Redundancy consists of having at least two of every required component so that the system can continue functioning if one component fails. AOS-CX switches support redundancy at the hardware level with support for multiple management modules, fabric modules, power modules, and fans.

At the software level, VSX maintains two control planes, one in each switch, to ensure redundancy. Resiliency builds on redundancy by allowing the system to take advantage of redundant components. AOS-CX switches in a VSX fabric support link virtualization. This feature enables multiple redundant links, including ones on different members of the VSX fabric, to operate as one logical link, called a link aggregation (LAG). AOS-CX switches also support process resiliency; if a process encounters an error, the system can self-restart to the last known good state.

A good high availability solution should also provide good performance with fast failover in all scenarios. VSX minimizes the duration of traffic outages by providing low-latency link failovers. If one link in the LAG fails, traffic instantly fails over to other links. VSX supports live upgrades so the VSX fabric remains active as one member and then the next upgrades. In addition, the AOS-CX switches have a fast upgrade time, which minimizes the time that the network is at risk, with just one up member during the upgrade process.

Finally, a high availability solution must be simple to configure to lower the chances that operator error introduces unplanned downtime. It is easy to configure VSX on AOS-CX switches with simple CLI commands. VSX also supports configuration synchronization and consistency checks between the members of the fabric to lower the risk of error.

Table 2-1: What makes for a good high availability solution?

Requirement	Solution capabilities
Redundancy *"What if X breaks and Y fails?"*	**HW redundancy**—management modules, fabric, power, fans **SW redundancy**—dual vs. single control planes
Resiliency *"What if W sputters and Z hiccups?"*	**Link virtualization**—virtualize multiple links to one logical link aggregation **Process resiliency**—self-restart to last known good state
Performance *"How long is failover? In what scenarios?"*	**Low-latency link failovers**—minimize duration of traffic outage **Fast upgrade time**—minimize time-at-risk during upgrade
Simplicity *"How complex, with what chance for error?"*	**Easy to configure**—# entities to configure, # CLI commands **Low risk of error**—config sync & consistency checks

VSX benefits: Layer 2

While you can deploy VSX at the core if the topology requires, you may prefer to keep redundant core switches entirely independent. A dual control plane provides the best resiliency and software upgrades with near zero downtime. At the same time, VSX permits unified management on the features that administrators choose to synchronize.

Here is an example of an HPE-recommended topology in which you deploy VSX at the aggregation layer.

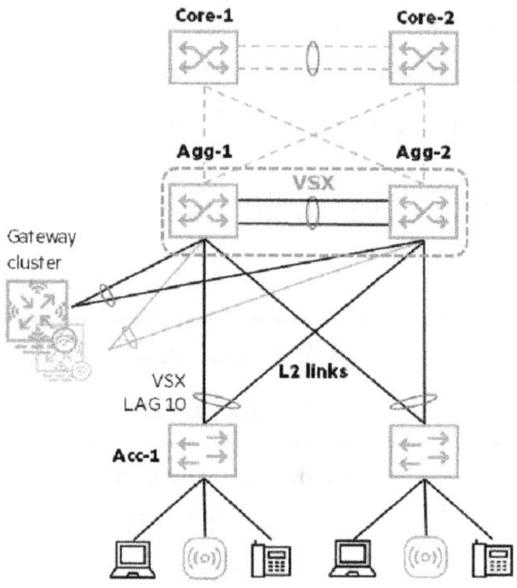

Figure 2-7: VSX benefits: Layer 2

The VSX pair uses link aggregations (LAGs) with links on both VSX switches to connect to access layer switches. This type of LAG is called a multi-chassis (MLAG), distributed, or VSX LAG. These VSX LAGs operate at Layer 2 because the access layer switches do not route, and the aggregation layer VSX pair acts as the default gateway for the access VLANs. The Layer 2 LAGs eliminate the need for Multiple Spanning Tree Protocol (MSTP) or Rapid Per-VLAN Spanning Tree (RPVST) on the redundant links. They provide a loop-free path that permits the use of all the links in the aggregation (active-active design). Configuring a LAG-based topology is simple, and if a link fails, failover occurs very quickly.

Control plane

- Dual control plane for better resiliency
- Unified management (synchronized configuration and easy troubleshooting)
- Independently software upgradable with near zero downtime
- In-chassis (8400/6400) and device-level redundancy

L2 distributed LAGs

- No spanning-tree
- Loop-free L2 multipathing (active-active)
- Simple configuration
- Rapid failover

Figure 2-8: VSX benefits: Layer 2 characteristics

VSX benefits: Layer 3

The VSX pair has Layer 3 links to the core. Each VSX switch can connect independently to each core switch on a routed-only port (ROP) or on a Layer 2 port associated with Layer 3 VLAN interfaces (also known as Switch Virtual Interfaces or SVIs) dedicated to that port. Alternatively, both VSX switches could connect to each core switch together on a VSX LAG. This LAG would operate at Layer 2 and associate with one or more VSIs. Whichever option you select, VSX can support active-active Layer 3 routing through Equal Cost Multipath (ECMP) and an efficient path for unicast traffic. With the final option, the combination of Layer 3 ECMP and the VSX LAG creates a highly fault-tolerant system.

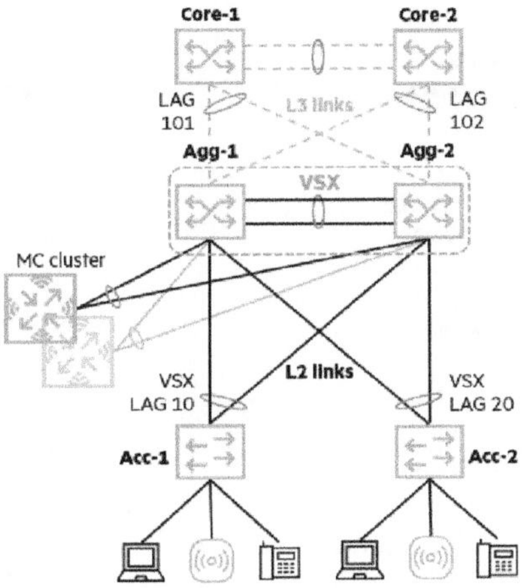

Figure 2-9: VSX benefits: Layer 3

Remember that the VSX pair acts as the default gateway for the access VLANs. To do so, the pair uses the active gateway feature. This feature allows each switch in the pair to act as an active default gateway for the VLAN using a shared virtual IP address (VIP) and virtual MAC address. It eliminates the need for the Virtual Router Redundancy Protocol (VRRP) or Hot Standby Router Protocol (HSRP). Simple to configure, the active gateway feature relies on VSX operations, so it does not add any protocol overhead. It also supports redundancy for DHCP relay functions.

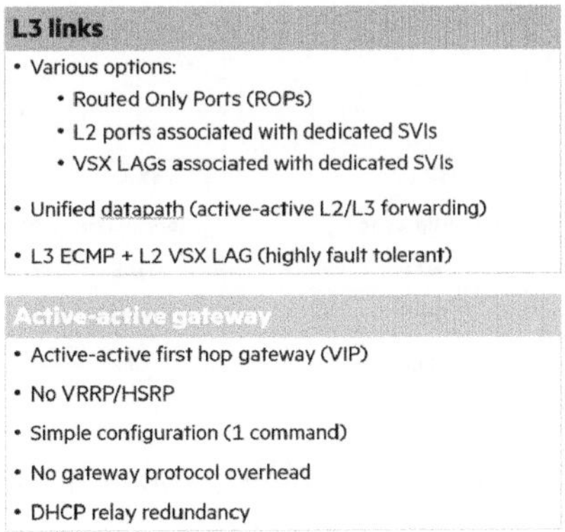

Figure 2-10: VSX benefits: Layer 3 characteristics

Data plane virtualization

Data plane virtualization supports VSX LAGs, so other LAG-connected devices perceive the VSX switch pair as a single device. In the following example, those are both downstream access layer switches and upstream core switches.

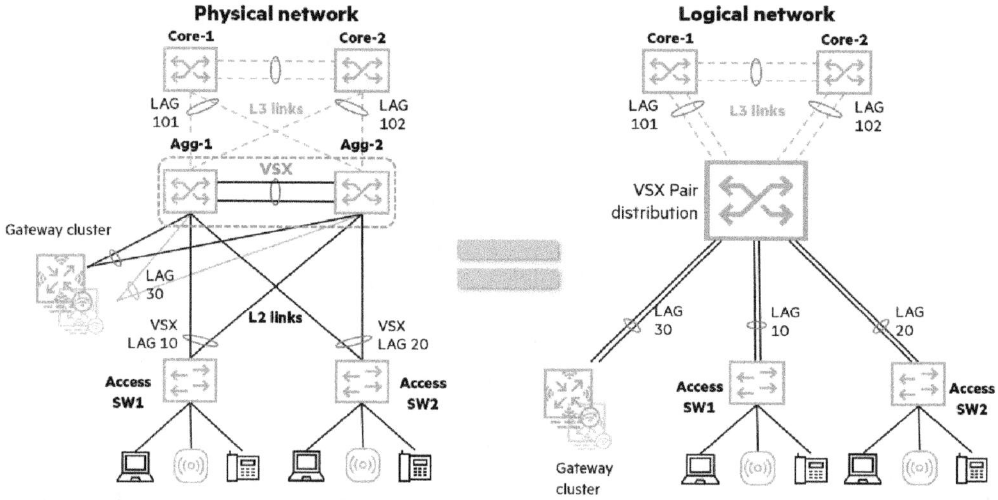

Figure 2-11: Example of data plane virtualization

Management plane synchronization

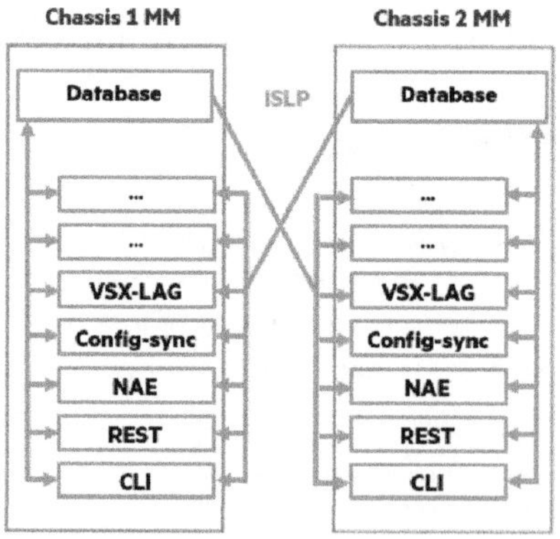

Figure 2-12: Management plane synchronization illustration

VSX's management plane synchronization provides a database-driving architecture, configuration and troubleshooting simplicity, hitless upgrade orchestration, and active-active analytics.

Database-driven architecture

The VSX dual management plane supports synchronization, made possible partially through AOS-CX's innovative database-centric design. VSX peers can synchronize portions of their database so that each peer can actively forward traffic while also knowing the state of its peer.

VSX simplifies management functions—the CLI, REST interface, and Web UI—exposing both control planes to admins and business applications. In addition to simplifying management, this design enables analytics across the redundant pair.

Config and troubleshooting simplicity

You can choose which specific components to synchronize, such as an interface or VLAN. Configuration for that component is then continuously synchronized for easier configuration and fewer misconfigurations. Show commands aggregate information from several configuration pieces across both switches and compare them, quickly picking out mismatched settings. You can add the `vsx-peer` option to many commands, to view information about both the local switch and the peer—you get a simple joint view of the VSX system.

Hitless upgrade orchestration

VSX synchronization also helps to orchestrate hitless live upgrades. When one member needs to undergo an upgrade, the other member knows, and the pair can proactively move traffic away from the member that needs to be upgraded toward the other member.

Active-active analytics

While each switch in the VSX pair maintains its own NAE agent, which stores sensor values locally, NAE is aware of VSX. The agent on each member can connect to the other member's database and cross-monitor it to detect discrepancies. For example, each NAE agent monitors the number of objects in the primary and secondary members' databases. If synchronization is working correctly, the number of objects should be identical. An alert can trigger if, for example, there are 20% more objects in the primary database than the secondary database for at least five minutes. Active-active analytics help to validate that the overall solution is healthy.

VSX components

VSX supports Multi-chassis Link Aggregation (MLAG): two or more links from two switches form a LAG that acts as a single logical interface. While the IEEE standard 802.3ad is limited to aggregating links on a single switch or device, VSX uses proprietary technology to overcome this limitation, to support link aggregation across multiple switches in the same VSX stack.

An Inter-Switch Link (ISL) connects the two switches, and a keepalive mechanism is used for resiliency.

Also, VSX provides node-level network redundancy for when one switch fails. You must configure the downstream device as a standard 802.3ad LAG interface; they perceive the VSX pair as a single device. Downstream devices can be anything that supports the 802.3ad Link Aggregation Control Protocol (LACP).

In VSX, one device acts as the primary and the other device as the secondary, but BOTH devices forward traffic, for an active-active gateway solution.

In a pair of VSX switches, one device is designated as the primary, while the other is the secondary. If you activate config-sync, the configuration on the primary switch will be synchronized automatically with the secondary switch. However, it is important to note that not all configuration elements will be synced; only those specifically enabled by VSX-sync will be synchronized.

And you will soon learn about the link-up delay feature, which enhances initial forwarding readiness.

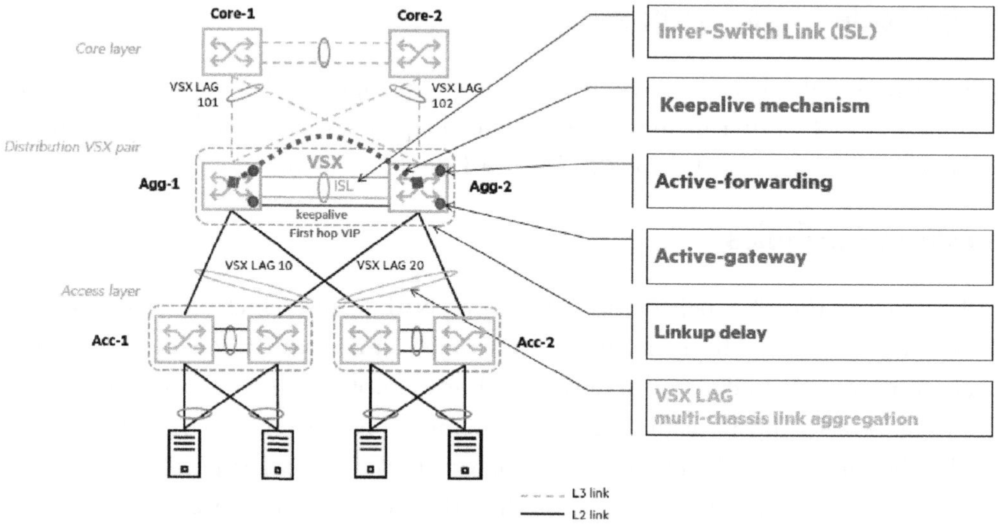

Figure 2-13: VSX components

VSX ISL

An Inter-Switch Link (ISL) is a Layer 2 interface between two VSX peer switches. You must configure each VSX switch with an ISL that directly connects to its peer VSX switch. The ISL is used at all three switch processing planes:

- Data plane: forward production traffic
- Control plane: exchange VSX protocol messages
- Management plane: synchronization

The ISL can consist of a single physical link, but HPE strongly recommends link aggregation. You can have up to eight physical links for resiliency, and all must be the same speed. You can use 10 Gbps but are strongly encouraged to use 40 or 100 Gbps to ensure adequate bandwidth for the datapath. The ISL uses front plane ports on the member switches and can use any media. If you select fiber media, the ISL can span long distances, depending on the fiber and transceiver type.

The ISL interface is a member of all VLANs by default. You can change ISL membership through the CLI, but you must ensure that any VLAN that is carried on a VSX LAG is also carried on the ISL. In the datapath, traffic is forwarded natively with no additional encapsulation, unlike with VSF. Also note that you can control the ISL with QoS and ACL policies, much as you do other interfaces on the switch. Of course, you must be careful that the ACL does not block required traffic. The ISL also preserves DSCP remarking. For example, you could configure a VSX LAG to re-mark traffic with a specific DSCP.

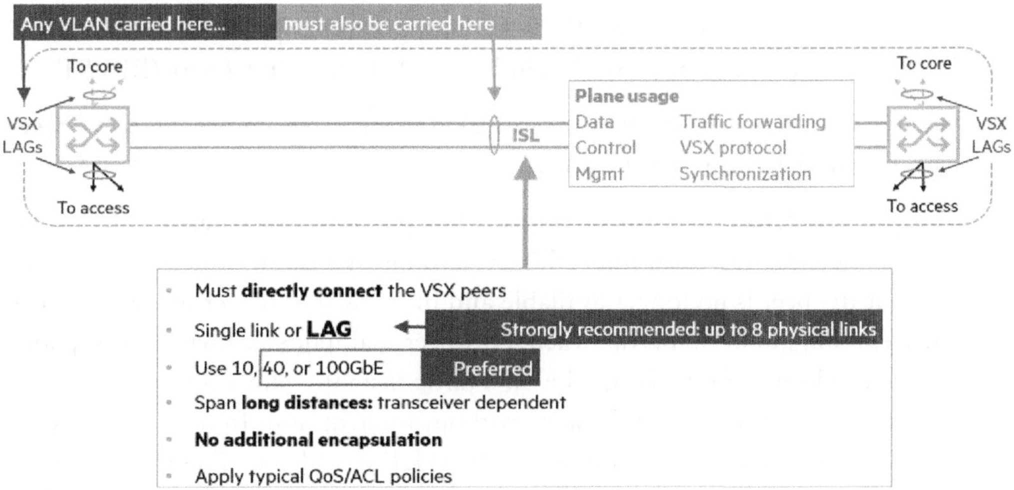

Figure 2-14: VSX ISL

VSX ISL resiliency

Recall that the VSX switches use LAG links to connect up to core switches and down to access switches.

If a local VSX switch's link fails, it transmits the traffic across the ISL so its peer can forward it. The packet keeps the DSCP value that it would have had if transmitted over the LAG locally. However, the ISL interfaces on the other peer must be configured to trust the DSCP settings on ingress frames.

This all works with the help of the Inter-Switch Link Protocol (ISLP).

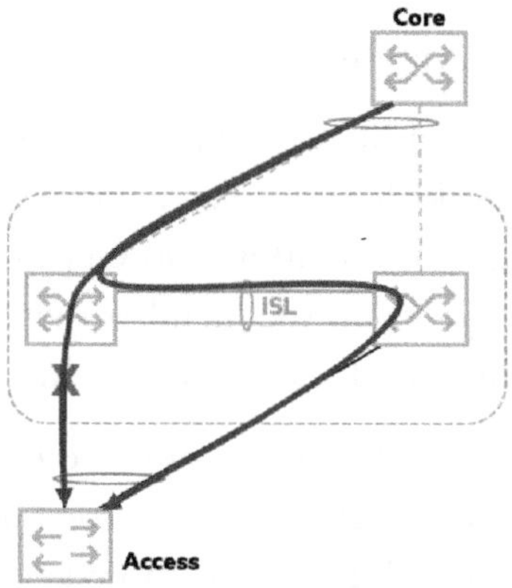

Figure 2-15: VSX ISL resiliency

VSX ISLP

The Inter-Switch Link Protocol (ISLP) runs over the ISL. When you first configure the VSX pair, ISLP synchronizes the switches. Then it periodically synchronizes LACP states, MSTP states, MAC tables, ARP tables, and config info. VSX also supports interoperability with Multiple Spanning Tree Protocol (MSTP), but not with Rapid Per-VLAN Spanning Tree+ (RPVST+).

ISLP runs at Layer 2. In other words, it is encapsulated in Ethernet frames. Since VSX switches are directly connected, there is no need for IP.

ISLP also helps peers to detect each others' status. They periodically exchange hello packets at one second intervals by default. The dead interval is 20 seconds by default; it determines when a switch decides that the peer is no longer available and initiates split detection. While the dead interval might seem long, understand that each VSX peer maintains its own control plane. If one peer goes down, all links in VSX LAGs go down. Traffic instantly fails over to the links on the other peer, which continues to actively forward without interruption. In most cases, peers determine that they must initiate split detection when the ISL link is down. The ISL link is considered down when all physical member links are down for the hold time, which is zero seconds by default.

ISLP uses version control and provides backward compatibility regarding VSX synchronization capabilities. Processes used for VSX include vsxd for VSXL AGs, vsx-synchd for synchronization, and vsx-mgmtd to synchronize REST, NAE, and configuration settings across the pair.

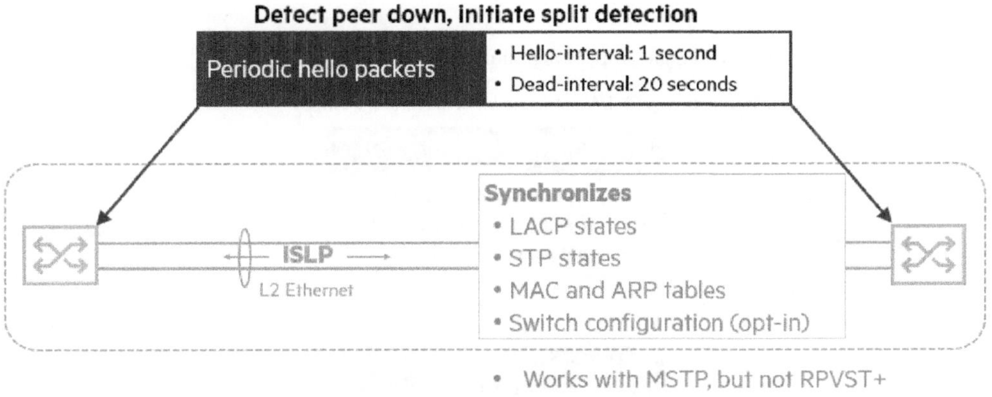

Figure 2-16: VSX ISLP

Keepalive

Protection from ISL failure

You should configure each VSX switch with a keepalive connection to the other switch. If the ISL goes down, each VSX peer uses the keepalive communications to identify whether the peer is still up, although unavailable over the ISL (that is, the VSX pair has a split brain). If the keepalive communications do indicate that the peer is still up, the primary VSX switch keeps its VSX LAG links up, and the secondary VSX switch forces its VSX LAG links down. The primary switch was selected when VSX was established, based on a user-configured role setting. If you configure the switches with the same role, the lowest MAC address acts as a tiebreaker.

Peers exchange keepalives over a routed network whose path can consist of a direct Layer 3 link or an indirect link through the upstream Layer 3 network. You can set the source for keepalive packets to a loopback interface for stability. In either case, make sure that keepalives do not pass through the ISL.

Keepalive packets are UDP-based. By default, they use port 7678, but the UDP port is configurable. Keepalive hello packets go out every one second by default, and you can configure the interval to a value from one to five seconds. The keepalive dead interval is three seconds, with a configurable range of 2 to 20 seconds. If a device does not receive a keepalive packet from its peer within the dead interval, it treats the peer device as out-of-service and does not implement the split-brain protection mechanisms.

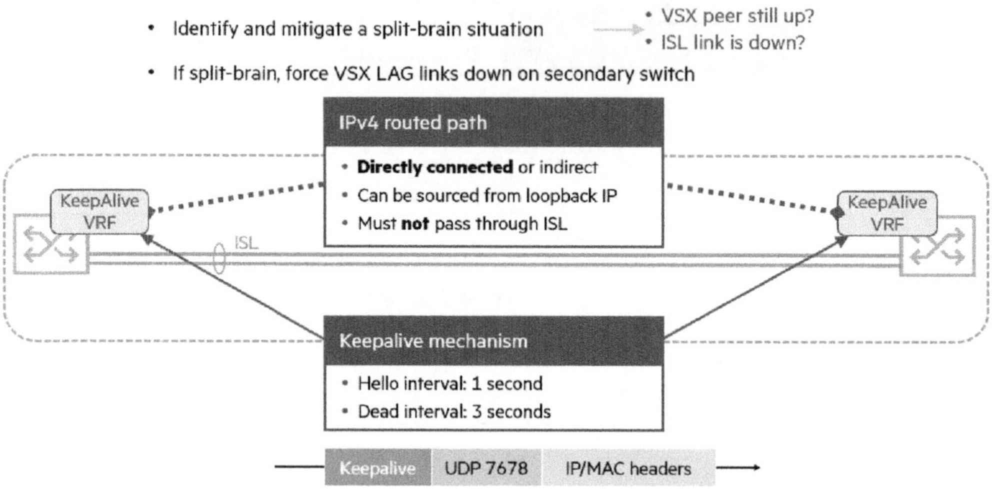

Figure 2-17: Keepalive: protection from ISL failure

Connection options

You have two options for establishing the keepalive links. The most crucial rule is that the keepalive communications must not pass through the ISL.

- Use a directly connected ROP on each peer or a directly connected route LAG. HPE recommends this type of configuration.
- Use the through-routed path option. Keepalives pass through the upstream Layer 3 network. HPE does not recommend passing the keepalives through the downstream network. The indirect option is cost-optimized because it requires fewer links.

You can choose an ROP, routed LAG, SVI, or loopback interface for the keepalive source. A loopback interface provides the most stability because the peers can always reach each other's loopback interfaces if even one Layer 3 path is available between each other. If you tie the keepalives to a particular physical interface or SVI, that interface might go down. Remember to advertise the source loopback interface in the dynamic routing protocol.

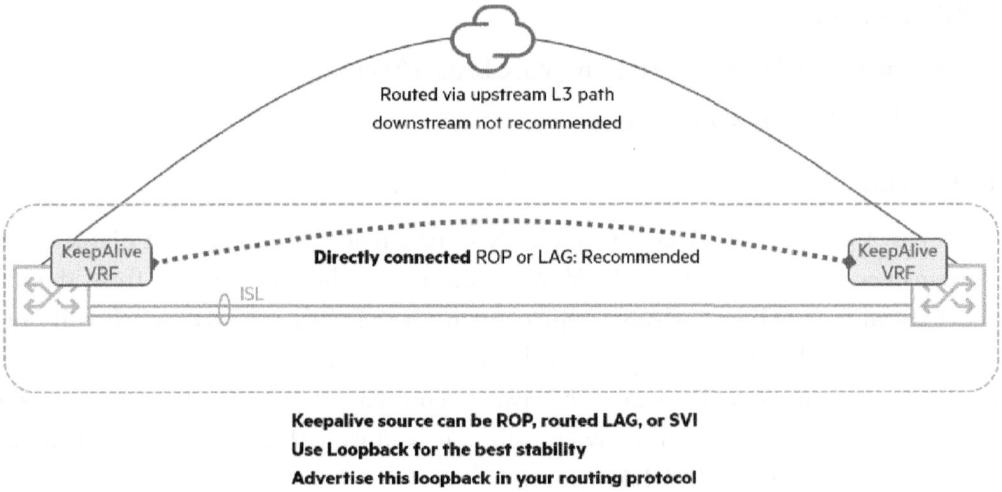

Figure 2-18: Keepalive: connection options

> For additional hardening, HPE recommends assigning a keepalive source interface to its own VRF.

VSX best practices

Let's take a look at the following best practices to keep in mind when configuring VSX.

- **MAC addressing**
- **Firmware/software version**
- **ISL LAG**
- **Keepalive link**
- **Cluster creation**
- **Split-recovery**
- **VLAN configuration**
- **Downstream VSX LAGs**
- **MSTP**
- **VSX LAG**
- **OSPF/BGP**
- **Multicast**

- **VRF transit VLANs**
- **Upstream VSX LAG configuration (VSX LAG 101/102)**
- **VSX linkup-delay-timer exclusion**

MAC addressing

One of the main VSX best practices is to set VSX system-mac. Do not leave it blank so the default system-mac is used; you want the VSX system-mac to be independent from the physical hardware MAC address. Thus, hardware replacement on the VSX primary switch will not affect your configuration, and so has no impact on the VSX secondary because the cluster ID remains unchanged. In other words, VSX primary hardware replacement is hitless for the VSX secondary. Otherwise, the VSX secondary would have to join a new cluster ID, ID from VSX primary, and would temporarily turn off its VSX LAG ports.

Use a locally administered unicast MAC address when assigning a system-mac or an active-gateway virtual MAC address. There are four ranges reserved for private use for unicast (with the second least significant bit of the first octet of the unicast address set to 1). x is any hexadecimal value.

- x2-xx-xx-xx-xx-xx
- x6-xx-xx-xx-xx-xx
- xA-xx-xx-xx-xx-xx
- xE-xx-xx-xx-xx-xx

Here is an example proposal to ensure that unique values are used in the administrative domain, where XX is the Unique Cluster ID in the function, and Y is the virtual MAC ID (0 to 15):

Function	System-mac	Active-gateway virtual MAC
Access/TOR Layer	02:00:00:00:XX:00	12:00:00:00:XX:0Y
Aggregation 02:01:00:00:XX:00 12:01:00:00:XX:0Y	02:01:00:00:XX:00	12:01:00:00:XX:0Y
Core/Spine	02:02:00:00:XX:00	12:02:00:00:XX:0Y

The scope of this VMAC is purely link-local. Consequently, the same virtual MAC address value can be used on any L3 VLAN interface (SVI).

If some servers or systems have dual-attachment to two different SVIs and you would like to see distinct MAC addresses for the next-hops over these separate interfaces, then 16 VMACs are available. For dual-stack IPv4 and IPv6, 16 VMACs can be used for IPv4, and the same VMACs can be used for IPv6. It is, however, a best practice to use only eight VMACs for IPv4 and eight different VMACs for IPV6.

Any other allocation rules can be chosen according to administrative rules in place by the network operational team. Multicast or broadcast MAC addresses must not be used for sssystem-mac.

Firmware/software version

Install the same version on both AOS-CX switches that will create the VSX cluster. It is better to avoid any version mismatch during the creation of the cluster, as a warning would appear.

ISL LAG

It is assumed that 2x40G, 2x50G, or 2x100G direct fibers / DACs are already interconnecting AGG-1 and AGG-2.

The best practice for ISL bandwidth is at least 2x40G (QSFP+), 2x50G (SFP56), or 2x100G (QSFP28). It is technically possible to use 2x10G or 2x25G; however, it is recommended to plan for any uplink failure and associated impact on the bandwidth requirement for the ISL. If the uplinks from AGG-1 fail, traffic from AGG-1 is redirected to AGG-2 over the ISL before reaching the upstream layer. This is fine if there is enough bandwidth remaining for the ISL protocol and control-plane communication. It is recommended to size the ISL bandwidth to be equal to, at least, the sum of the uplinks bandwidth of one VSX switch. The best practice rule is to size the ISL bandwidth according to the failure domain target.

The best practice for ISL physical ports is to select at least two ports of the same speed (2x40G, 2x50G, or 2x100G), and, in case of a chassis, to select these ports from different line cards.

The best practice for LAG numbering is to use the last available LAG ID (for example, 256 in AOS-CX 10.4) for the ISL, so that LAG ID=1 is used for connecting the Access Switch#1 on port 1/1/1, LAG 2 is used to connect the second access switch on port 1/1/2, and so on.

Other ISL LAG best practices

- VLAN trunking: Permit ALL VLANs, for simpler configuration. Specifying a restrictive list of VLAN IDs is entirely valid if the network admin wants more control.
- LACP timers: Keep the default long timer (30 s for lacp rate slow).
- Hashing algorithm: Keep the default l3-src-dst (alternative being l2-src-dst).

- MTU: Configure on all devices the appropriate size to support features such as Dynamic Segmentation, as well as other protocols/functions that require MTUs larger than 1500 bytes. Care should be taken to ensure that the IP path from access devices (switches or APs) can provide an MTU of at least 1564 bytes to the mobility controllers. Similarly, for data center server connectivity, largest MTU will ensure server jumbo frame traffic over ISL. Recommendation: Ethernet MTU = 9198 bytes.
- ACLs: Do not set any access-list on the ISL LAG to avoid designing complex and unnecessary ACL. The ISL is like a virtual data back-plane and security filtering is processed before or/and after crossing the ISL.
- QoS trust mode: Rely on the qos trust dscp that is globally configured on the Aggregation switches. If not configured globally (which is not the recommendation), qos trust dscp has to be set on the ISL LAG.

Keepalive link

The best practice for the keepalive connection is to use a direct L3 circuit, which can be a low speed port (1G transceiver is enough; 1GBASE-T works as well) between both VSX nodes. This circuit need not be directly connected, and the path can include active L2 and L3 equipment. Although this requires an additional dedicated port, it brings simplicity of configuration and operations. In case of a chassis switch (CX 6400 Series or CX 8400 Series), if possible, it is recommended to use a port from a different line card than the ones used for the ISL ports.

The best practice for keepalive routing is to use a dedicated VRF. This is entirely optional and the default VRF can also be used, typically for the single VRF model with UDP keepalive over the upstream L3 domain. Having a dedicated VRF for keepalive simplifies the operations and prevents any impact from a routing change on the default VRF.

The best practice for the keepalive subnet is to use a /31 subnet, as only two nodes will communicate together.

Cluster creation

Keep ISL timers (dead-interval, hello-interval, hold-time, peer-detect-interval) at default values—that is, no specific configuration.

The best practice for role (primary or secondary) is to have a meaningful relationship with the switch hostname/identification. Example: AGG-1 is VSX primary and AGG-2 is VSX secondary.

The best practice for vsx-sync includes vsx-global. Thanks to this vsx-sync FeatureGroup parameter, the VSX management-plane will synchronize the following VSX settings: inter-switch-link hello-interval, dead-interval, hold-time, peer-detect-interval, keepalive udp-port, hello-interval, keepalive dead-interval, system-mac, split-recovery, linkup-delay-timer.

VSX automatically tags the native VLAN configured on the LAG used for the ISL.

Split-recovery

The best practice for VSX split-recovery is to keep the default split-recovery enabled (no configuration change). This best practice might be revisited in case of VSX and VXLAN. (VXLAN is beyond the scope of this guide.)

VLAN configuration

The best practice for VLAN configuration is to configure the VLANs on the VSX primary with the vsx-sync attribute and let the VSX config-sync automatically synchronize the VLANs on the VSX secondary.

Downstream VSX LAGs

For simplicity, it is assumed that connected access switches are already configured with uplinks—link-aggregation and trunked VLANs. The best practice for a VSX LAG is to create the multi-chassis LAG interface on the VSX primary with all settings and then create the mirrored LAG interface on the VSX secondary. LAG interface settings (including description) will be synchronized automatically. Only `no shut` in the LAG interface context must be performed on the VSX secondary. Once the multi-chassis LAG interface is created, it can be assigned to the physical port.

The best practice for allowed VLANs is to exclude the native VLAN 1 from being propagated. This is a very robust method to avoid Layer 2 storm propagation due to a potential loop initiated on an access switch. In case of an access switch Zero Touch Provisioning use-case, this trunking exclusion is performed after the ZTP process.

The best practice for LAG numbering is to use LAG ID=1 for connecting Access Switch#1 on port 1/1/1, LAG 2 used to connect a second access switch on port 1/1/2, and so on.

The best practice for LACP timers on the VSX LAG is to keep the default long timer (30 s = lacp rate slow).

The best practice for an MTU is to configure the appropriate size to support features such as Dynamic Segmentation or server jumbo frame. Ensure that the IP path from the access devices (switches or APs) can provide an MTU of at least 1564 bytes to the mobility controllers and that the server jumbo packet of 9000 bytes can be encapsulated. Flexibility should be anticipated to perform VXLAN encapsulation from the access switch (9000+50 bytes) or VXLAN encapsulation from the aggregation layer MTU+50. So, the recommended Ethernet MTU is 9100 bytes for the downstream VSX LAG to the access layer and an MTU of 9000 bytes for endpoints or servers. The SVI IP MTU should match the MTU size on the aggregation layer, so the recommended IP MTU is 9100 bytes.

The best practice for a hashing algorithm on the VSX LAG is to keep the default l3-src-dst (alternative being l2-src-dst). This option has an effect only if at least two ports per VSX node are members of the same VSX LAG.

Usually, the VSX LAG includes only two links: one from the primary and one from the secondary. Consequently, hashing algorithm selection has no effect on the traffic path as it is forwarded to the local port of the VSX LAG on the receiving switch.

The `show lacp interfaces multi-chassis` command is very useful to get a complete status of the local LACP partnership as well as the VSX peer partnership details. Actor = local node, Partner = LACP neighbor (the access switch), Remote Actor = the VSX peer, Remote Partner = LACP neighbor of the VSX peer. Note that the port ID of the VSX secondary is equal to 1000+ID of the primary (in the example, 1001). ALFNCD LACP state-flags should appear on all entries.

The best practice for the LACP fallback feature is to enable it on the VSX LAGs for the following use-cases: PXE boot, access switch ZTP, server NIC driver migration from active/standby to LACP. When applied to the VSX primary, LACP fallback is automatically synced on the VSX secondary.

MSTP

The best practices on the aggregation layer are:

- Do not use loop-protect (MSTP used instead).
- Use the default common instance 0: MST0.
- Lower the spanning-tree priority to 4 to make VSX aggregation the STP root bridge (easier for support).
- Use root-guard on all downlinks to prevent any access switches from becoming the root bridge.
- Keep the default port-type admin-network.
- Let the VSX secondary be synchronized by the vsx-sync process.

The best practices on the access layer are:

- Use loop-protect for all endpoint access ports (not configured on uplinks). Set the re-enable timer to one hour.
- Keep the default common instance 0: MST0.
- Keep the default spanning-tree priority of 8.

- All endpoint access ports are admin-edge, should not receive any BPDU (BDPU guard), should not trigger any Topology Change Notification (tcn-guard).
- Use loop-protection on all endpoint access ports as an extra-protection mechanism (in case MSTP BPDUs are filtered by insertion of unmanaged switches, which creates a loop).
- Use loop-guard on all uplinks to prevent any floods due to failure of BPDU reception (fiber strand cut).

VSX LAG

- ACLs: If any ACL is used, the best practice is to have ACLs synchronized on the secondary through vsx-sync. Any ACL applied on a VSX LAG on the VSX primary will get applied on the VSX secondary as well.
- QoS: QoS Marking being performed on the access layer, the aggregation switch is configured in the global context with qos trust dscp. No further configuration is needed as this was already set in previously.
- SVI (VLAN L3 interface): The best practice for SVI active-gateway is to set the active-gateway Virtual IP and Virtual MAC on the VSX primary and get the value synchronized on the VSX secondary with the `vsx-sync` command.
    - The best practice for active-gateway VMAC is to use the same VMAC for all IPv4 SVIs. The scope of this VMAC is purely link-local. If some servers or systems have dual-attachment to two different SVIs, and the system administrator would like to see distinct MAC addresses for the next-hops over these separate interfaces, then 16 VMACs are available. For dual-stack IPv4 and IPv6, 16 VMACs can be used for IPv4, and the same VMACs can be used for IPv6. It is, however, a best practice to use only eight VMACs for IPv4 and eight different VMACs for IPV6.
- Mutlinetting: If it is used, set one VIP per secondary subnet and disable ip icmp redirect.
- IP MTU: Configure all SVI IP MTUs to match the size of the L2 MTU: IP MTU recommended value = 9100. This parameter must be identical and manually set on both VSX nodes.
- DHCP relay: Configure the ip helper-address on the VSX primary and let vsx-sync configuring the same on the VSX secondary.

OSPF/BGP

It is a best practice to create a dedicated transit VLAN between the VSX primary and the VSX secondary to exchange route information for subnets that are not attached to both VSX nodes (for example, loopback addresses of each VSX node). This dedicated transit VLAN provides better control and will not carry user data traffic in nominal situations, or very limited in case of east-west traffic between single-attached endpoints.

There are two strategies to inject endpoint subnets into the routing table—either through OSPF or through BGP:

- **OSPF:** Most campus deployments use OSPF to exchange route information for end-devices. This is simple and can scale very well with the appropriate usage of areas. This is the target of this current document.
- **BGP:** Many new data center deployments use BGP as a routing protocol due to the usage of EVPN-based VXLAN. Such a design can be used in the campus as well. Also, for more complex and granular routing engineering, BGP communities and route-map can offer a level of control that OSPF can not provide.

There are two options to inject end-user subnets into the OSPF database: using the `ospf` command on the SVI (VLAN L3 interface), or redistributing the connected into OSPF with route-map control.

The best practice is to use the `ospf` command on the SVI as offering a simpler configuration like for the area to which the subnets belong. This principle is selected as the OSPF best practice in the following described configuration. More details on OSPF best practices can be found in the *IP Routing Guide*.

The best practice for point-to-point interconnectivity subnet is to use the /31 subnet.

The best practice for OSPF configuration is to use the `vsx-sync ospf` synchronization option and have OSPF parameters automatically synced on the VSX secondary. As shown in the configuration step, very few elements must be configured on the secondary. To get such a benefit, the interface ID should be mirrored; that is, if interface 1/1/49 is used for uplink on the VSX primary, it is strongly recommended to use the same ID 1/1/49 on the VSX secondary, otherwise OSPF synchronization will not synchronize the proper interface.

The best practice for OSPF cost is to have VSX primary <-> VSX secondary cost lower than Core-1 <-> Core-2 cost, as the ISL bandwidth is frequently higher than the inter-core devices' bandwidth. In case of a single-attachment subnet on one of the VSX nodes and non-meshed topology, the traffic from the core would be sent to the VSX peer closest to the attached destination, avoiding consuming inter-core bandwidth.

The same concept applies for a south-to-north traffic pattern. In the following example, the OSPF cost for Transit VLAN over ISL is set to 50, and 1000 for core devices. The OSPF cost is synchronized from the VSX primary to the VSX secondary, so it is important to use the mirrored interface ID.

For BGP recommendations, read the *IP Routing Guide* for detailed information.

VSX can synchronize the full BGP configuration between the VSX primary and the VSX secondary. Most of the BGP configuration of the VSX secondary is the same as on the VSX primary, except for the configuration for iBGP peering between the VSX nodes inside the cluster, or for remote eBGP upstream peers with the neighbor IP address being the physical IP address of the L3 point-to-point circuit (ex: here 1/1/50 with 10.0.0.6/31 remote IP address). In such a case, the specific neighbor parameters are excluded from the VSX configuration synchronization with the following command on the VSX primary only: `neighbor <IP_ADDRESS> vsx-sync-exclude`.

Multicast

For multicast on the VSX cluster, the best practice is to configure PIM Dual-DR or active/active under the PIM router command. With the `active-active` command, the proxy-DR will also learn the multicast routes and will allow a fast recovery time if the actual DR fails.

Read the *Multicast Guide* for detailed information.

VRF transit VLANs

The best practice for upstream routing domain connectivity is to create one transit VLAN per VRF per upstream VSX LAG. Consequently, according to the proposed topology, for VRF1, VLAN 101 is created and configured on upstream VSX LAG connecting CORE-1 (VSX LAG 101); VLAN 102 is created on the upstream VSX LAG connecting CORE-2 (VSX LAG 102). Similarly, VLAN 201 and 202 are proposed for VRF2 and VLAN 301/302 for VRF3.

For simplicity and readability, only two VRFs are documented: VRF1 and VRF2. VRF3 is not included in the configuration to alleviate the document. Configuration for VRF3 or any additional VRFs is similar to VRF1 and VRF2.

These transit VLANs 101/102 and 201/202 will also serve for VSX intra-cluster routing (unlike the single VRF scenario where a dedicated transit VLAN (VLAN 2) was created for transit point-to-point routing between VSX nodes).

Upstream VSX LAG configuration (VSX LAG 101/102)

In this section, for simplicity, it is assumed that the core equipment is already configured with link-aggregation and trunked Transit VLANs.

The best practice for an upstream VSX LAG is to create the multi-chassis LAG interface on the VSX primary with all settings and then create the mirrored LAG interface on the VSX secondary. LAG interface settings (including description) will be synchronized automatically. Only `no shut` in the LAG interface has to be performed on the VSX secondary. Once a multi-chassis LAG interface is created, it is assigned to the physical port.

The best practice for allowed VLANs is to exclude the native VLAN 1 from being propagated and to allow only the Transit VLANs corresponding to the facing core device: for example, VSX LAG.101 permitting VLAN 101 and 201, VSX LAG 102 permitting VLAN 102 and 202.

The best practice example for LAG numbering is to use LAG ID=101 for connecting Core-1, LAG ID=102 for connecting Core-2. Any other numbering practice is possible as long as it does not introduce confusion with the downstream VSX LAGs.

The best practice for LACP timers on the VSX LAG is to keep the default long timer (30 s for lacp rate slow). The best practice for an MTU on these upstream VSX LAGs is to configure the maximum value (9198 bytes) like for the ISL.

The best practice for a hashing algorithm on the VSX LAG is to keep the default l3-src-dst (alternative being l2-src-dst), which would have an effect only if at least two ports per VSX node are members of the same VSX LAG.

VSX linkup-delay-timer exclusion

The best practice for VSX LAG exclusion for linkup-delay-timer is to exclude the upstream VSX LAGs that are used for Transit VLANs (in the previous topology, VSX LAG 101 and VSX LAG 102). By excluding these upstream VSX LAGs, the routing protocols can establish peering over upstream Transit VLANs as soon as the upstream ports (to L3 Core-1 and Core-2) are UP, without waiting for the linkup-delay timer to complete. The benefit is that routes from the L3 core are already learned when the ASIC is ready to forward traffic downstream.

For more information about an overview of VSX, best practices, and its configuration, see the *Virtual Switching Extension (VSX) Guide*. This has additional best practices based on using a multiple VRF routing model and connecting an access layer VSX cluster to an aggregation VSX cluster. You will also find useful information on maintaining and troubleshooting a VSX topology. And lastly, there are copious configuration examples included in the guide.

Configuration example for the primary switch

First, set up the ISL link—a normal Layer-2 LAG. When a native VLAN is defined, as shown here, the switch automatically executes the `vlan trunk allowed all` command to ensure that the default VLAN is allowed on the trunk. In this example, LAG 128 is being used as the ISL.

ISL interface

```
interface lag 128
no shutdown
no routing
vlan trunk native 1
vlan trunk allow x,y,x
lacp mode active
interface 1/4/28
no shutdown
lag 128
interface 1/4/32
no shutdown
lag 128
```

The same list of VLANs that are trunked over the VSX LAGs must be configured on the primary and secondary VSX switches in the global configuration. The list of VLANs can be synced to the secondary switch if the `vsx-sync` command is used in the VLAN context. Also, verify that the VLAN set is also permitted on the ISL on the primary and secondary VSX switches. To configure VLAN trunking on the ISL, enter the `vlan trunk allowed [<VLAN-LIST> | all]` command. If a native VLAN is defined, the switch automatically runs the `vlan trunk allowed all` command to ensure that the default VLAN is allowed on the trunk. To allow only specific VLANs on the trunk, enter the `vlan trunk allowed <VLAN-LIST>` command, for example: `vlan trunk allowed 2,3,4`.

Then, activate the LAG on the physical interfaces between the two switches. To follow best practices, you should minimally use two links in the LAG.

Next, define the Layer 3 interface that will be used for sending keepalives between the two switches. This can be a direct link between the switches. Each switch will need unique IP addressing.

Set up the keepalive interface

```
interface 1/1/5
vrf attach keepalive
ip address 192.168.100.1/24
```

Set up VSX

After defining the keepalive interface, you need to set up VSX. Enter the VSX context and define the role: primary or secondary. Next, specify the LAG that will be used for the ISL link. In the previous example, this is LAG 128.

That is all that is necessary to initially set up VSX. For detailed information on configuring, tuning, and troubleshooting VSX, see the *Virtual Switching Extension (VSX) Guide*.

```
vsx
role primary
system-mac 02:01:00:00:aa:bb
inter-switch-link lag 128
```

Creating a multi-chasis interface

To set up a multi-chassis LAG between this VSX cluster and another device, define a new LAG, but specify the multi-chassis parameter. This parameter tells the VSX cluster that there is at least one link on each switch connecting to a neighboring device, like a switch.

```
interface lag 1 multi-chassis
no shutdown
no routing
vlan trunk native 1
vlan trunk allowed 11
lacp mode active
interface 1/1/1
no shutdown
lag 1
```

VSX synchronization

Periodic synchronization

Every second, each VSX switch synchronizes its learned MAC addresses with its VSX peer using ISLP. Each VSX switch also buffers and sends its new neighbor ARP/ND entries to the VSX peer every time its buffer is full (64 entries) or when it hits a timeout of five seconds.

In a VSX scenario where all traffic from the core to the access layer flows through one VSX switch only, the other VSX switch still learns the necessary ARP/ND entries from its VSX peer. VSX has no functional impact on the VSX switches' normal datapath-based learning.

If a VSX split occurs and the switches then rejoin together, the switches perform a bulk sync; after that, periodic syncs resume.

Inter-VRF Route Leaking (IVRL) leaks neighbor entries from a source VRF to a destination VRF. This process creates two neighbor entries, one learned through the datapath in the source VRF and one that was induced in the destination VRF. The VSX switches do not sync the induced entries either during the initial ARP sync or the periodic syncs. Instead, each switch learns the ARP entries in the source VRF (either through the normal datapath process or through the ARP sync). The local IVRL process then induces the entries into the destination VRF on each VSX switch individually.

Keep one caveat about ARP synching in mind. If an admin clears ARP entries or enables and disables a VLAN interface on one VSX switch but not on the other, ARP entries will go out of sync between the switches. The admin must perform such operations on both VSX switches so that both devices will relearn entries and sync up.

So, if you enter one or more of the following commands on one VSX switch but not on the other VSX switch, the ARP entries on both switches will become unsynchronized.

- Clear ARP
- Interface VLAN
- Shutdown for a VLAN
- No shutdown for a VLAN

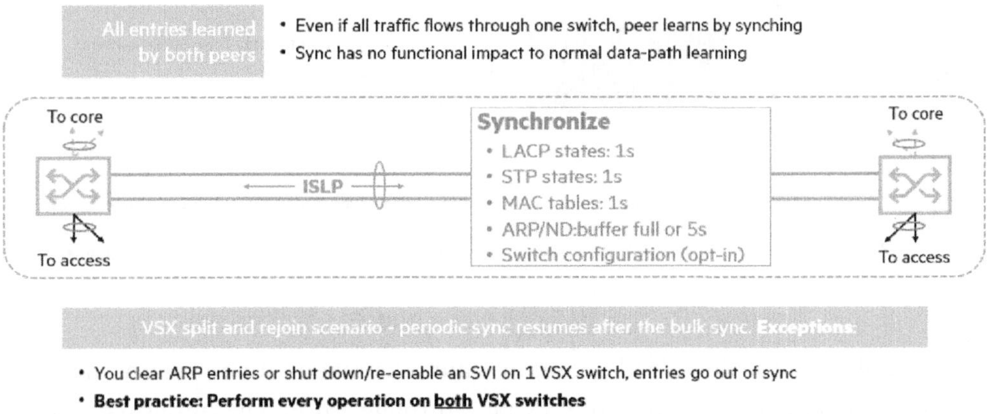

Figure 2-19: Periodic synchronization illustration

Initial sync and link-up delay

After a VSX switch reboots, it has no entries for ARP, MAC, and routes. If downstream VSX LAG ports activate before all this information is relearned, traffic drops. To avoid a traffic drop, VSX LAGs on the rebooted switch stay down until the restoration of LACP, MAC, ARP databases, and MSTP states, if MSTP is used.

The learning process for the VSX LAGs has two phases:

- **Initial sync phase:** The LACP states, MAC address table, ARP table, and potential MSTP states download from the forwarding switch to the freshly rebooted switch.
- **Link-up delay phase:** The downloaded entries are installed into the ASIC. Router adjacencies with core nodes and learned upstream routes are also established. You can configure the link-up delay phase with the `linkup-delay-timer <DELAY-TIMER>` command. The default value is 180 seconds. Set the link-up delay timer to the maximum value of 600 seconds for a network with many MAC addresses, a large ARP table, or a large routing table.

When both VSX switches reboot, the link-up delay timer is not used because both switches must relearn the LACP states, MAC address table, and ARP table. To get upstream router adjacencies established during the link-up delay, you must exclude the upstream LAGs (LAG101, say) from the scope of the link-up delay. Run the `linkup-delay-timer exclude lag-list <LAG-LIST>` for identifying the LAGs for exclusion. Until the link-up delay timer expires, all SVIs associated with VSX LAGs are kept in a pseudo-shut state.

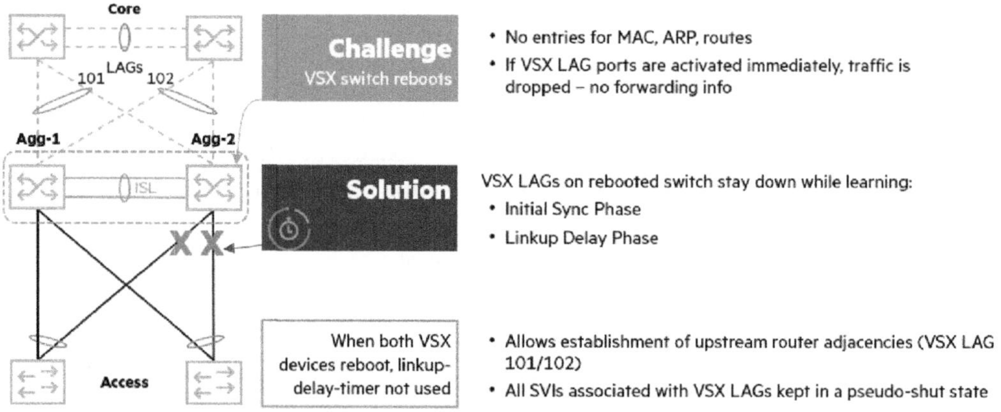

Figure 2-20: Summary of the two phases of link-up delay

The initial sync phase is the download phase, in which the rebooted node learns all the LACP, MAC, and ARP database entries from its VSX peer through ISLP. The initial sync timer is not configurable. It is the required time to download database information from the peer.

The link-up delay phase should last for as long as is required to install the downloaded entries to the ASIC, to establish router adjacencies with core devices, and to learn upstream routes. During this phase, links are forced down. The link-up delay timer's default value is 180 seconds. Depending on the network size, ARP table size, and routing table size, the timer may need to be set to a higher value—the maximum is 600 seconds.

Configuration synchronization

Dataplane sync and config sync are two different things. Dataplane sync was covered earlier. Let's dive into config-sync (vsx-sync).

You can use the vsx-sync parameter from the CLI or NetEdit to ensure that the configurations are "synced" between the VSX peers.

The two VSX switches should have nearly the same configuration, with a few exceptions. These exceptions include host IP addresses on Layer 3 interfaces, router IDs, hostnames, and a few other settings. Configuration synchronization makes it simpler to ensure identical configs. By default, VSX configuration synchronization is enabled along with VSX. However, enabling synchronization globally has no effect until you select individual configuration items on which to enable synchronization. The vsx-sync attribute enables synchronization on an item. You can disable configuration synchronization globally, in which case you can no longer apply the vsx-sync attribute to any configuration items.

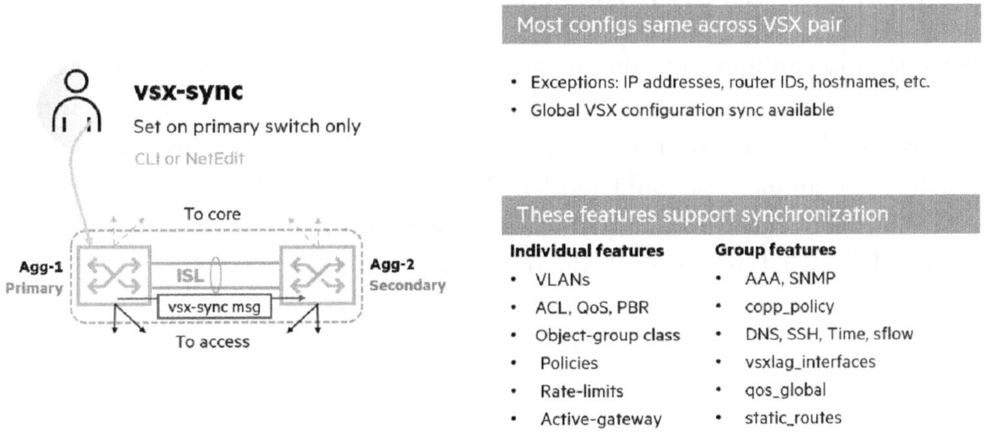

Figure 2-21: Individual and group features supported by vsx-sync

Configuration synchronization behavior

You must set the vsx-sync attribute on the desired items from the VSX primary switch. The primary switch then automatically pushes this config to the secondary switch. If the secondary switch is not available when you configure the primary, the configuration is pushed to the secondary switch after the switch comes up. If a certain config sync operation fails, the CLI status commands will indicate the failure. You must manually fix the inconsistency. There is no fail-safe mechanism that disables VSX if configuration synchronization fails.

You can configure vsx-sync on a secondary switch if config-sync is disabled or if the ISL is down.

The vsx-sync attribute uses this "opt-in" model due to customers' and partners' preference that they be able to control exactly which features are synched.

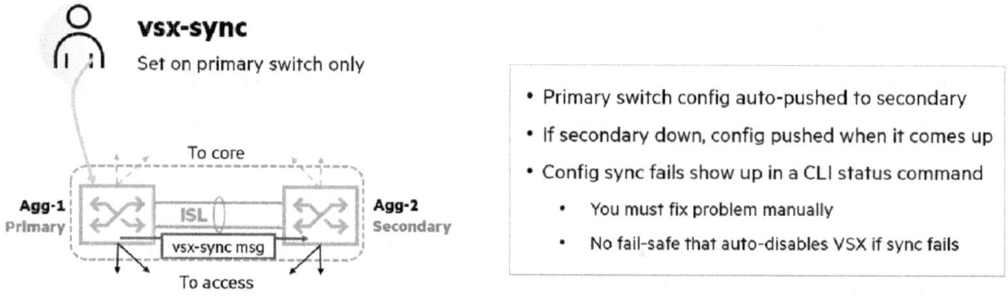

Figure 2-22: Configuration synchronization behavior

Configuration overlap and protection

It is perfectly valid to configure a setting manually on the secondary switch without that setting being synched from the primary switch. However, if vsx-sync is now enabled for this feature on the primary switch, the setting on the primary switch overrides the previous setting on the secondary switch. For example, you could add VLAN 10 on the secondary switch with description "employees." You then add VLAN 10 with no description on the primary switch and enable vsx-sync. The secondary switch now has VLAN 10, but with no description.

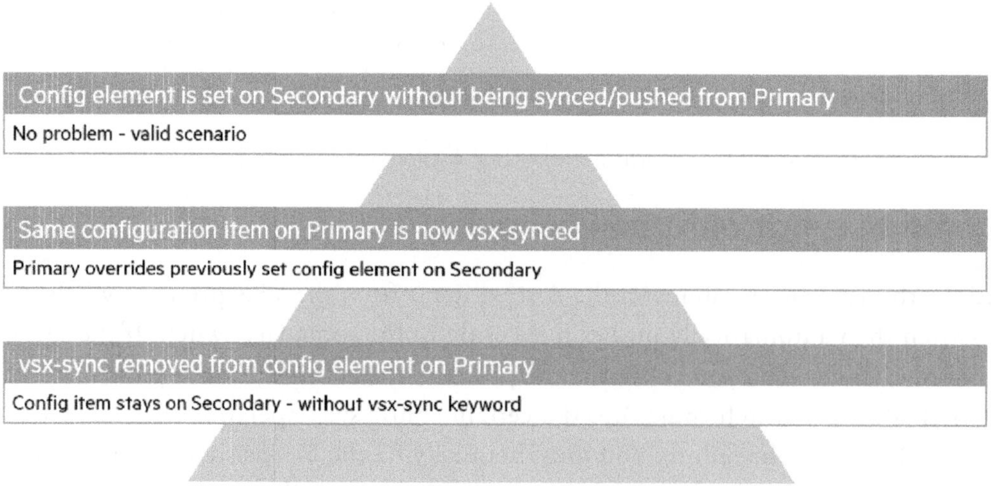

Figure 2-23: Configuration overlap and protection

Firmware upgrade

During a firmware upgrade, software versions will temporarily mismatch on the VSX peers. Therefore, the vsx-syncd process, which is still running, stops any synchronization attempts. It then performs an ongoing check to determine if it can sync. Once both VSX peers are running the same firmware, this ongoing check sees that it can restore the synchronization. The vsx-syncd process then continues the ongoing sync. The database connection between the peers works if the firmware versions are the same.

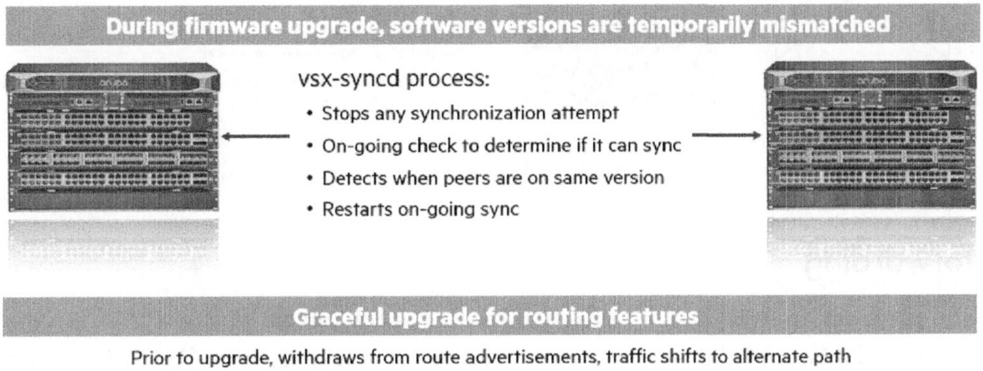

Figure 2-24: Firmware upgrade characteristics

Troubleshooting synchronization

While independent control planes and management planes for the VSX switches help to provide greater availability, they can complicate troubleshooting. You must collect protocol and datapath states from both VSX switches to analyze the issue holistically.

To simplify troubleshooting, the AOS-CX CLI offers a `vsx-peer` option for many show commands. When you enter a show command with this option, the output merges data from both the local switch and its VSX peer in a single output, making it easier to collect information and troubleshoot. The output also shows the distinct datapath tables on each switch, indicating how each switch will forward packets, which again helps pinpoint issues.

The `vsx-peer` option for show commands is not available during a firmware upgrade process.

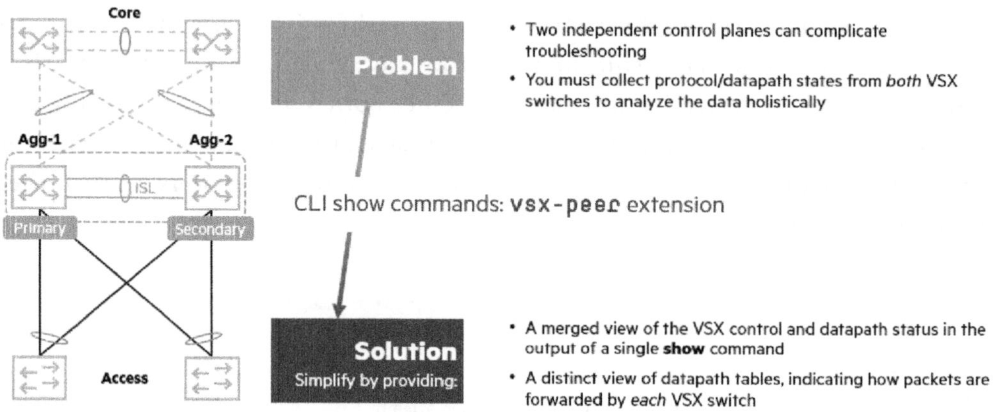

Figure 2-25: Troubleshooting synchronization

Traffic forwarding

This section provides an overview of traffic forwarding mechanisms within a VSX environment, focusing on VSX LAGs and the active gateway feature. You will see how VSX LAGs enable switches to connect to both physical peers in a VSX fabric, appearing as a single logical interface, and understand the synchronization and load-sharing processes involved. Additionally, we will discuss the active gateway protocol, which enhances network reliability by eliminating single points of failure for default gateway services. Finally, we will also discuss upstream unicast routing and connectivity options, including Route-Only Ports (ROPs) and various routing protocols like OSPF and BGP, providing a comprehensive guide for network design and high availability practices.

VSX LAG

VSX LAG allows you to connect a switch to both physical peers in a VSX fabric. In **Figure 2-26**, switch Acc-1 connects to both Agg-1 and Agg-2 of the VSX pair and perceives a standard LAG link—a single logical interface to a single logical switch. The same holds true for Acc-2 and the core switches.

The two VSX switches appear as a single peer ID to the upstream or downstream switches. Ports must have the same speed. The LAG can include up to four physical links per peer for eight links max per VSX LAG. As of the publication of this guide, VSX LAGs are Layer 2 only, and VSX LAGs must be LACP-based. A non-LACP or static VSX LAG is not supported.

The VSX peer switches synchronize their LAG states using ISLP. You can configure a LAG hash scheme, which dictates how the switches load-share traffic across the links in the LAG. (Note: The CX 8400 Series provides L4-based hashing, but the CX 8320 Series does not.) VSX LAGS are locally optimized, meaning if a switch with local links is in a LAG, then it constrains the LAG hash scheme to look at only local links, and it will only forward traffic over the ISL for transmission on the peer links in the LAG if local links of a VSX LAG are down.

The number of VSX LAGs supported is dependent on the switch model: 255 on the CX 8400 Series, 53 on the CX 8320 Series, and 55 on the CX 8325 Series.

> Originally, HPE used the term Multi-Chassis LAG (MC-LAG) to describe a LAG between an external device and a VSX switch pair. Today, HPE uses the term VSX LAG.

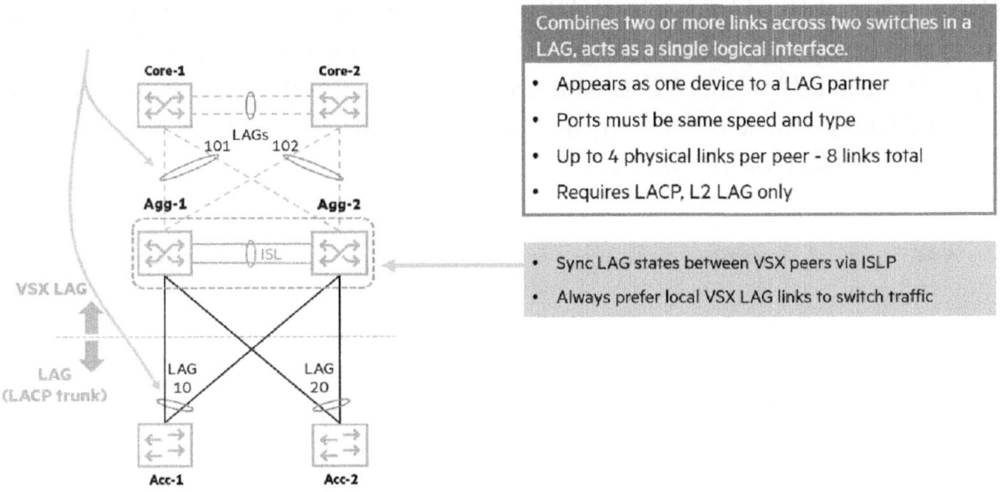

Figure 2-26: VSX LAG

Active gateway overview

Active gateway is a first-hop redundancy protocol that eliminates a single point of failure for default gateway services on an access network. The active gateway feature improves the reliability and performance of the host network by enabling a virtual router to act as the default gateway for that network.

To configure this feature, you simply configure a shared virtual IP address (VIP) and a shared virtual MAC address (VMAC) on an SVI on both VSX switches. The VIP/VMAC serves as the default gateway for each access VLAN. Which peer in the VSX pair receives a particular piece of traffic from an access VLAN depends on which link in the LAG the downstream access switch's LAG has selected. Whichever peer receives the traffic routes it across to the Layer 3 domain.

When you use active gateway on a VLAN, the VLAN has no need for VRRP; the features are mutually exclusive. As with VRRP, routed traffic from the VSX peer is sourced from the switch interface MAC, not the VMAC address. Every three minutes (non-configurable), each active gateway broadcasts a hello packet from the VMAC to avoid VMAC aging on the access switches. The hello from both peers transmits on the same VSX LAG, so no issues occur with the same MAC address appearing on different ports. The hello is a non-IP Ethernet packet using HP EtherType, 080009 Hewlett-Packard, and a destination reserved multicast address.

The VSX pair supports up to 16 different VMACs, which are not shareable between IPv4 and IPv6. For example, the pair could use eight VMACs for IPv4 simultaneously with eight VMACs for IPv6.

You should understand some key differences between VRRP and the active gateway feature. VRRP requires multiple configuration options, including the VRID on the VLAN, the roles on each VRID, the virtual IP address on each VRID, the advertisement timer configurations, and so on. Active gateway only requires a single line of configuration per VLAN interface: the configuration of the virtual IP/MAC. VRRP and active gateway also differ in terms of how they forward traffic. VRRP uses an active-standby dataplane with only the VRRP master routing traffic. With active gateway, both devices forward traffic.

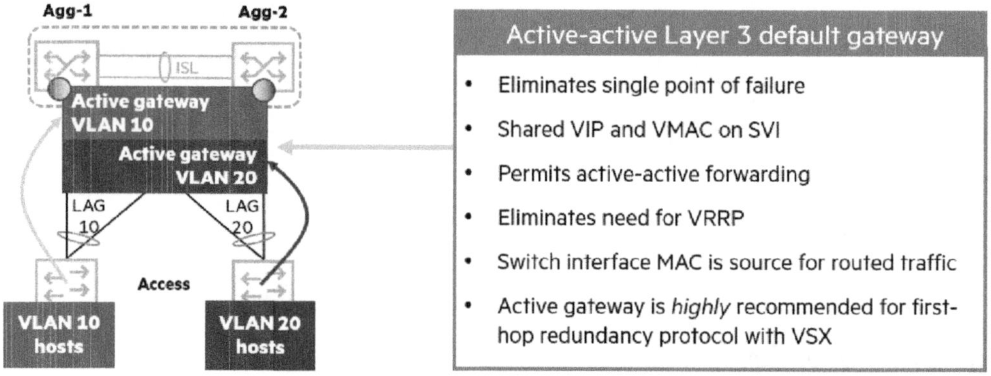

Figure 2-27: Active gateway overview

Active gateway: South-north unicast traffic

If you configure VSX LAGs with links on each peer, the normal operation unicast traffic sent between the access layer and the core (south-north) uses all access-to-aggregation links and does not flow over the ISL. From a Layer 2 perspective, VSX LAGs span two switches and operate in active-active mode. Traffic between the access layer and aggregation layer switches can be forwarded to any of the active links. There are no loops and no need for the spanning tree protocol or blocked ports.

From a datapath perspective, local link optimized usage ensures that when a VSX switch receives traffic, it always uses its local LAG links to forward the traffic to its destination. The VSX switch only uses the ISL link if the local LAG links are down. An active gateway ensures that the first VSX switch that receives traffic from the access layer can route the traffic across to the Layer 3 domain.

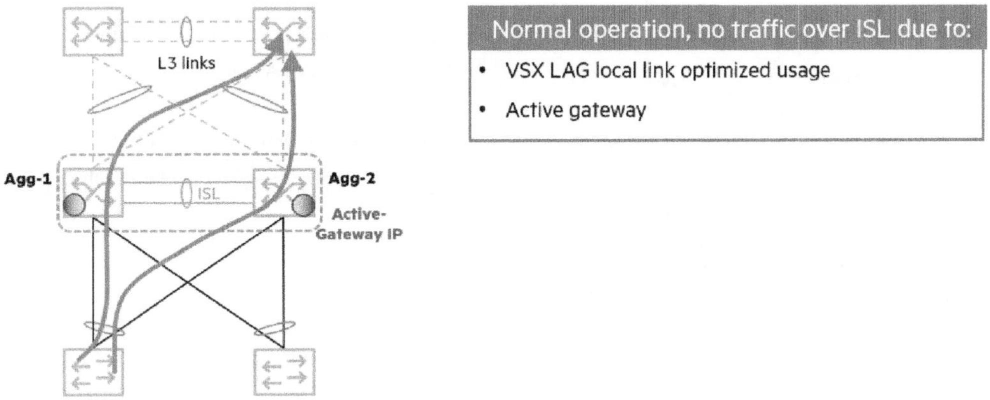

Figure 2-28: Active gateway: South-north unicast traffic

VSX LAG and upstream unicast routing

You will now examine how the VSX pair operates on the connections to upstream core switches, which are generally routed connections. Remember that each VSX peer runs an independent control plane with its own OSPF or BGP processes. Therefore, each VSX peer appears as a different router with its own router ID to the core switches. However, as far as the datapath is concerned, the VSX peers function as a single router and support active-active forwarding. This is possible because a core switch uses Equal Cost Multipath (ECMP) routing to learn routes to the subnets behind the VSX pair through both VSX peers and because the VSX pair supports a feature called "active forwarding," about which you will soon learn. Also remember that VSX LAGs can only operate at Layer 2.

Keeping these principles in mind, you can consider the options that you have for designing the upstream VSX LAGs and unicast routing. You could choose to connect with routed ports or Layer 2 ports associated with Layer 3 VLAN interfaces, or you can use Layer 2 VSX LAGs, which are associated with Layer 3 VLAN interfaces. For routing, you could choose static routing, OSPF, or BGP. You could use a single VRF or multiple VRFs, depending on the requirements of the scenario. Your choices will be informed by sizing limitations and by best practices for high availability.

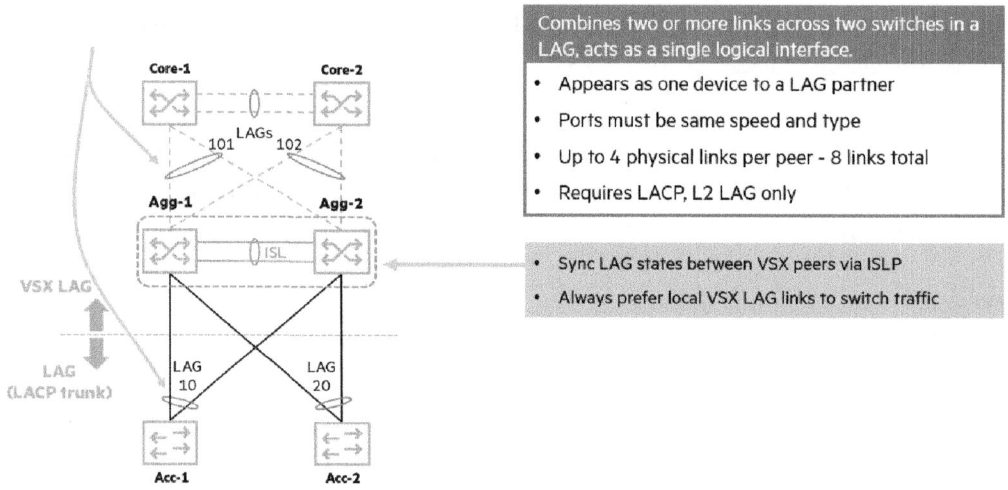

Figure 2-29: VSX LAG and upstream unicast routing

Upstream connectivity: ROP

There are several options for links between the aggregation VSX pair and the core: one involves a single VRF implementation, while the other options involve multiple VRFs.

The first option uses Route-Only Ports (ROP). It could also use routed LAGs, but those routed LAGs would only have links on the local switch; they cannot be multi-chassis. Each VSX switch has its own routed link or LAG to each core switch, and each of these four links is its own /30 or /31 subnet. In OSPF, this would be a point-to-point network. This simple VLAN-free configuration is best suited for a network that runs on a single VRF domain. If the network had multiple VRFs, each VRF would need its own ROP, which is less feasible.

The other options are beyond the scope of this guide.

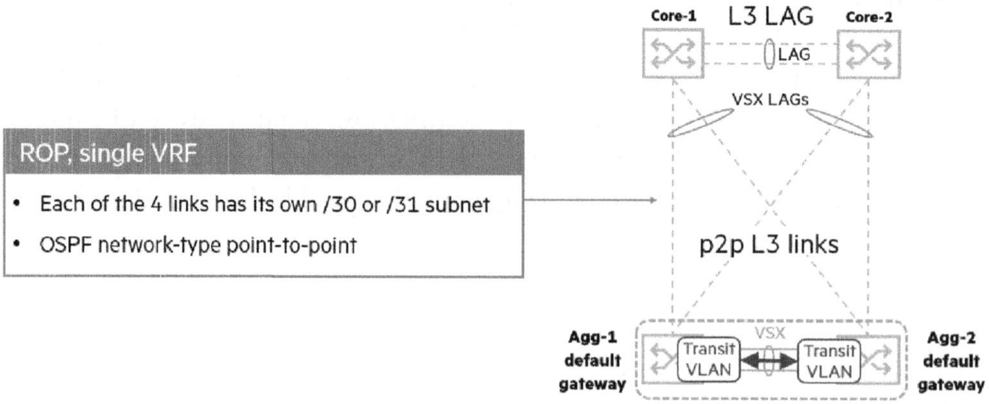

Figure 2-30: Upstream connectivity: ROP

VSX and upstream routing: Supported scenarios

The following chart summarizes the supported scenarios for the upstream connections between a VSX pair and the core. When you have a single VRF, ROPs provide the simplest option. However, Layer 2 ports associated with dedicated SVIs, as well as Layer 2 VSX LAGs associated with dedicated SVIs, work as well.

All these options work with static routing, OSPF, or BGP.

	Single VRF		
	L3 port (ROP)	L2 port (SVI)	VSX LAG (SVI)
Static	Yes (simplest)	yes	yes
OSPF	Yes (simplest)	yes	yes
BGP	Yes (simplest)	yes	yes

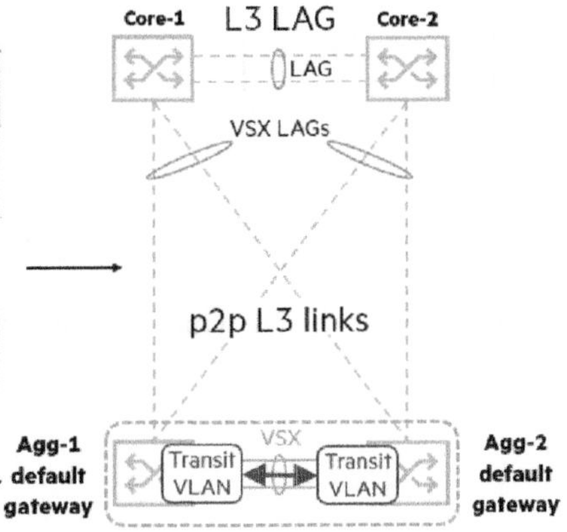

Figure 2-31: VSX and upstream routing: Supported scenarios

Split-brain scenario

The following diagram shows a situation in which the ISL between the two VSX switches is down. Both switches are still active, but the switches cannot exchange information, which causes the switches to become out of sync. This situation is called a split brain, and it occurs if the keepalive function is not enabled. The keepalive function prevents split brain issues by providing an out-of-band mechanism to detect if the link between the VSX switch is down or if the VSX switches are not working. The keepalive functionality brings down the link between Agg-2 and access switches. The traffic is forced to go from access switches to Agg-1 only.

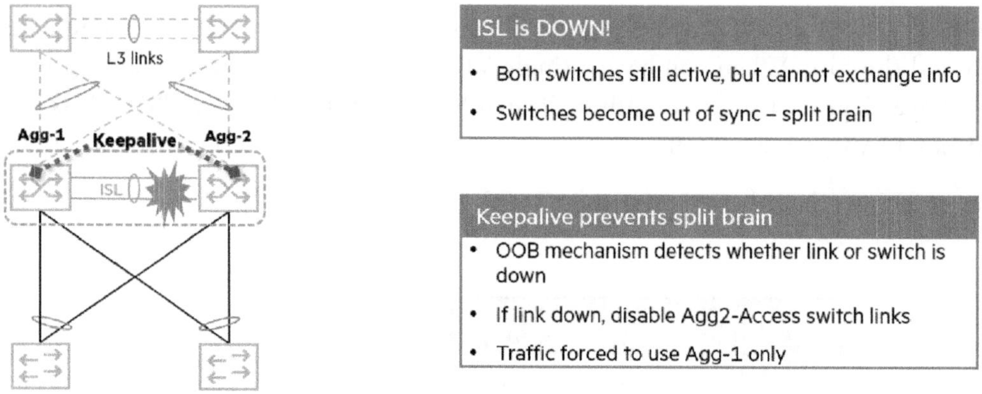

Figure 2-32: Split-brain scenario

The following figure shows what happens with various ISL and keepalive statuses on the VSX pair.

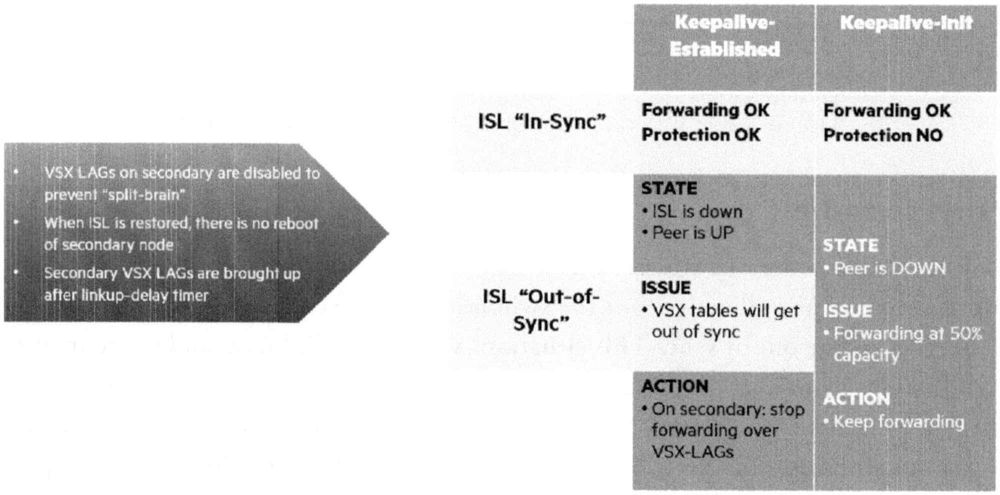

Figure 2-33: ISL and keepalive statuses on the VSX pair

ISL failure: SVI impact

An ISL cut has no effect on the up/down state of VLANs and associated SVIs that are not part of any VSX LAG but are part of at least one orphan port (beside the ISL LAG, which is not a VSX LAG).

During ISL cut (before initial sync)

If the secondary VSX node has a port that is a member of a VSX LAG, the associated SVI of the VLAN—transported by said VSX LAG—has the VSX secondary node shut down, whether or not there are orphan ports carrying the VLAN.

During link-up delay timer (after initial sync)

As long as the secondary VSX node has a port that is a member of a VSX LAG, the associated SVI of the VLAN—transported by said VSX LAG—is kept shut down on the VSX secondary node, whether or not there are orphan ports carrying that VLAN. The associated SVI of the VLAN transported by a VSX LAG restores ON/UP on the VSX secondary node if the following conditions are met:

- You have used the `link-up delay timer exclude lag` command to exclude the said VSX LAG, and
- No VSX LAG that is not part of the exclusion set allows the given VLAN.

In the below topology, after an ISL failure, the ports and SVI associated to a VSX LAG are automatically shut down; ports and SVIs not related to a VSX LAG are still operating. In this case, server 2 and VLAN 20 are still operating, but since there are no external uplinks and routes, server 2 is isolated. The server's port 3 remains UP, but VLAN 10 SVI is shut down due to ISL failure. As a best practice, always plan for redundant connections at the data center.

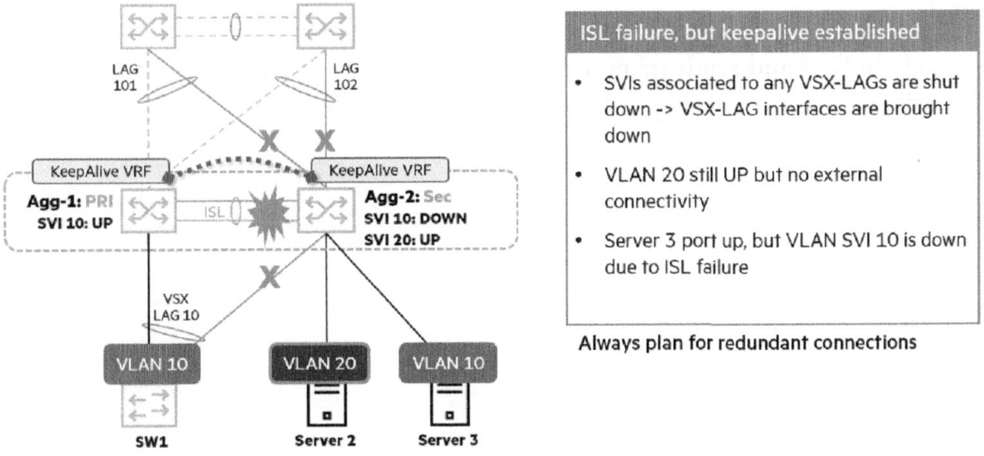

Figure 2-34: ISL failure: SVI impact

Learning check

Question 1

Which of the following are characteristics of HPE Aruba Networking Virtual Switching eXtension (VSX)? (Select two.)

- ☐ It allows for the centralized management of up to ten switches.
- ☐ The control plane for Layer 2 operation is unified/virtualized.
- ☐ The data plane is combined across VSX peers.
- ☐ Up to two switches can be part of a VSX system/group.

Question 2

Which VSX components prevent a split-brain situation in case of ISL failure? (Select two.)

- ☐ Simplify the network deployment.
- ☐ Minimize the need for routing.
- ☐ Smaller broadcast domains can improve performance.
- ☐ MAC address tables are smaller.
- ☐ Separate VLANs mitigate risk.

Question 3

Each VSX node buffers and sends its new neighbor ARP/ND entries to the VSX peer every time its buffer is full (64 entries) or when it hits a timeout of five minutes.

- ○ True
- ○ False

3 Layer 2 optimization

After examining the switching fundamentals and the HPE Aruba Networking Virtual Switching eXtension (VSX) feature of AOS-CX networks, you will now delve into the options available in AOS-CX switches to achieve Layer 2 optimization. These options build upon the fundamentals of switching such as link failure detection, VLAN, and network redundancy. Sometimes, failed links may remain undetected, leading to unidirectional flow. VLANs allow for segmenting the network. Private VLANs allow for a more specialized segmentation. The Spanning Tree Protocol (STP) allows you to maintain redundancy in the network without the risk of loops. But this technique is vulnerable to some attacks, which can impact the spanning tree calculation in an adverse manner. In this chapter, you will become familiarized with techniques to tackle these challenges and provide an optimized network architecture.

OBJECTIVES

After completing this chapter, you should be able to:

- ✓ Protect the network against failure due to unidirectional link behavior using the Unidirectional Link Detection (UDLD) protocol.
- ✓ Optimize Layer 2 network segmentation using private VLANs, an extension of the VLAN concept.
- ✓ Safeguard the STP-enabled networks from rogue devices using features such as root-guard and BPDU-guard.
- ✓ Explore Rapid Per-VLAN Spanning Tree+ (RPVST+) protocol features and its interoperability with Multiple Spanning Tree Protocol (MSTP).

UDLD

The Unidirectional Link Detection (UDLD) protocol is a special feature of AOS-CX switches that allows them to identify link failure from its unidirectional behavior. In this section, you will review how the protocol works and helps identify link failure. You will explore its available modes of operation and examine its configuration details.

UDLD overview

Let us review a simple scenario to understand the unexpected unidirectional behavior of a link in the network and how to resolve it using UDLD.

Unidirectional link behavior issue

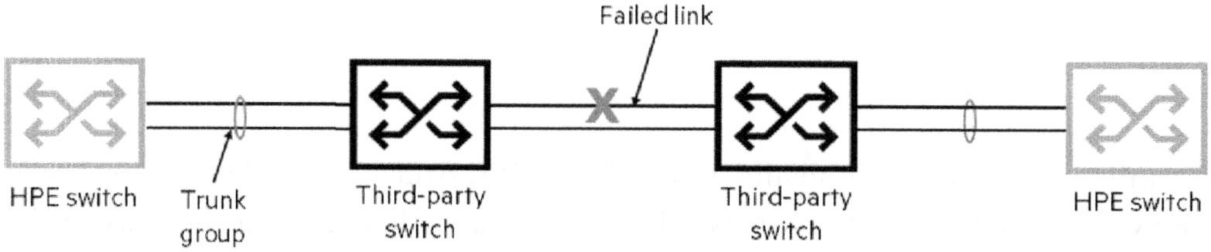

Figure 3-1: Unidirectional behavior due to link failure

Consider the scenario illustrated in **Figure 3-1** where a set of switches are connected using a trunk group—a group of ports configured to behave as one logical port. The distribution of traffic across the trunked links is based on the source address/destination address (SA/DA) pairs. When a link that is not directly connected to an HPE switch goes down, the failure goes undetected. As a result, the switches continue to send traffic to the affected ports. The link failure could be a result of a non-functioning port or link. The ports at one end of the link would continue to send traffic on the link, thus resulting in a unidirectional behavior of the link. This is where the Unidirectional Link Detection (UDLD) protocol benefits the network.

A similar risk exists in case of fiber-optic links that use two unidirectional fibers to transmit and receive packets. If a fiber breaks in one direction, a fiber port may assume the link is still good because the other direction is operating normally and, therefore, continue to send traffic on the connected ports.

Such type of unidirectional behavior may cause issues such as:

- **Formation of L2 loops:** A unidirectional flow may cause a blocking/alternate port in a spanning tree calculation to assume that there is no designated port at the other end of the link. This will cause the blocking port to transition to a forwarding state creating the risk of formation of loops in the network.
- **Traffic blackhole:** Another issue that may arise due to unidirectional links is loss of traffic. This may happen for the duration of keepalive timer set at the receiving end of the link.

Solution using UDLD protocol

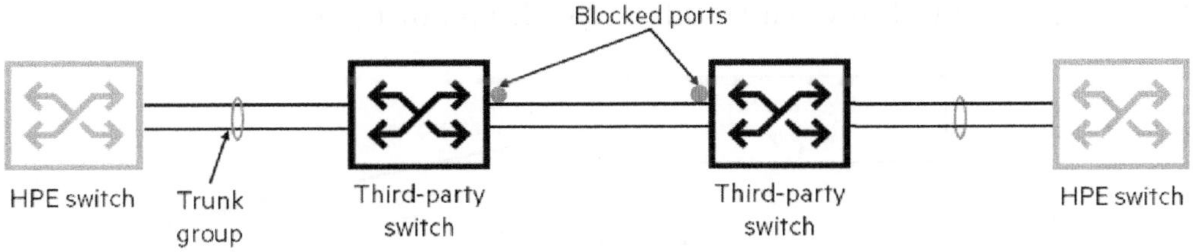

Figure 3-2: Solving the unidirectional link behavior issue by implementing UDLD

The solution to the problem of unidirectional link behavior is to enable the UDLD protocol on all the trunk ports on each of the switches. With UDLD enabled, when a link failure occurs, the switches detect the failed link and block the ports connected to the failed link. The traffic is redistributed across the remaining ports in the trunk group, thus avoiding unidirectional flow.

> For UDLD to work, both the connected devices must run the same UDLD protocol irrespective of the vendor.

UDLD also sees its implementation in monitoring fiber-optic links that use two unidirectional fibers to transmit and receive packets. UDLD-enabled ports will prevent traffic from being sent across a bad link by blocking the ports in the event that either the individual transmitter or receiver for that connection fails.

UDLD operation

The operation of the UDLD protocol is illustrated in **Figure 3-3**. The protocol operates as follows:

- To ensure successful UDLD operation, it is important to enable UDLD on both the connected devices, that is, on the ports on both ends of each link.
- Each pair of connected and UDLD-enabled ports exchanges health-check packets once every link-keepalive interval. The default value of this interval is seven seconds.
- If a port does not receive a packet from the port at the other end of the link within this interval, the port waits for four more intervals. By the end of the fifth interval, if there is no receipt of a packet, the port concludes that the link has failed and blocks both ports of the link.
- To ensure that the blocking of a port does not impact the operation of load balancing protocols such as the spanning tree protocol, the port-blocking event is recorded in the switch log. The port will remain in the blocked state until the link is unplugged, disabled, or

fixed.

- The port gets unblocked when UDLD is disabled on that port.

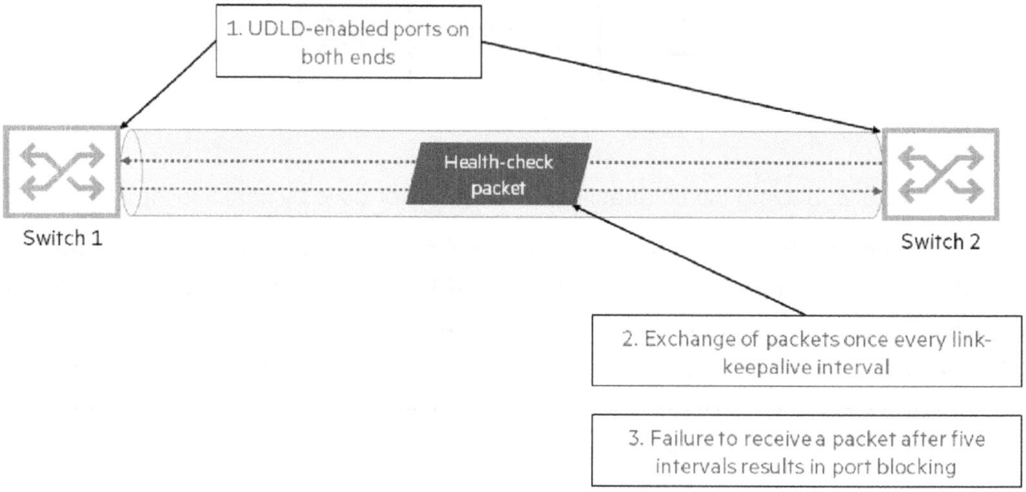

Figure 3-3: UDLD operation

UDLD modes

Port blocking and unblocking behavior depends on the UDLD mode set. The four types of available modes are listed below. Use the Aruba OS modes when interconnecting HPE switches, but use the RFC5171 modes when connecting with third-party switches.

Aruba OS verify-then-forward

In the verify-then-forward (VTF) mode, the ports are in a blocked state until the bidirectionality is confirmed. Once the bidirectionality is confirmed, the retries are performed after the specified interval and the port changes to blocked if no UDLD packets are received in this duration. Ports unblock if a UDLD packet is received, unless the (interval*retries) time is around six seconds.

Aruba OS forward-then-verify

The forward-then-verify (FTV) mode is similar to the VTF mode, except the ports start up as unblocked and proceed to check the bidirectionality.

RFC5171 aggressive

In the RFC5171 aggressive mode, the ports start up as unblocked. The ports wait for the specified retry limit to receive UDLD packets; if the packets are not received, this mode blocks the port. The aggressive mode implements a limited errDisabled recovery mechanism. In this mechanism, when the interface's state goes to errDisabled, a maximum of three attempts (five minutes apart) are triggered to try and bring up the interface in case the remote endpoint is still sending UDLD packets. After these three retries, the interface will remain blocked even if UDLD packets are received. The only way to unblock the interface when this occurs is to disable (and optionally re-enable) UDLD on the interface. The retry limit is reset once the interface becomes unblocked.

RFC5171 normal

Similar to the RFC5171 aggressive mode, the port starts up as unblocked in RFC5171 normal mode as well. The ports wait for the specified retry limit to receive UDLD packets. If the packets are not received, unlike the aggressive mode, the normal mode triggers an event.

UDLD configuration

To configure UDLD in a network, follow these simple steps:

1. Enable UDLD on the desired interface.
2. Set the values for mode, interval duration, and number of retries based on whether you want to use the default settings or apply a custom configuration.
3. Verify the values set for the parameters.

UDLD parameters

The following list specifies the key UDLD parameters along with the default values and commands to change them.

- **Packet transmission delay interval:**
  - Default value: 7000 ms
  - Command to change it: `udld interval`
- **Operating mode:**
  - Default value: forward-then-verify
  - Command to change it: `udld mode`
- **Retry count:**
  - Default value: 4
  - Command to change it: `udld retries`

Configuration example

Let us examine the configuration with the help of the scenario represented in **Figure 3-4**. This scenario describes how to use UDLD on a single physical interface as well as a LAG.

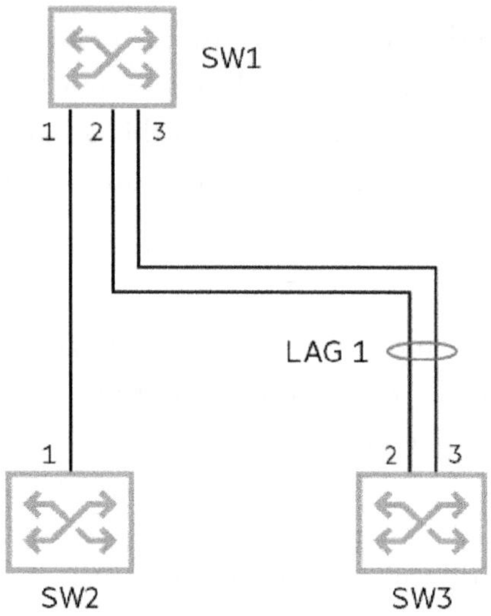

Figure 3-4: UDLD configuration for a single interface as well as a LAG

1. Configure UDLD between SW1 and SW2 with default settings.

   - Use the `udld` command to enable UDLD on interface 1/1/1 of both switches:

```
switch# config
switch(config-if)# interface 1/1/1
switch(config-if)# udld
switch(config-if)# exit
switch(config)#
```

2. Configure UDLD between SW1 and SW3 with custom settings.

   - On each of the four ports, set the mode, interval duration, and number of retries using the `udld mode`, `udld interval`, and `udld retries` commands respectively.

   - The configuration on SW1 looks like this:

```
switch(config)# interface 1/1/2
switch(config-if)# no shutdown
switch(config-if)# no routing
switch(config-if)# lag 1
switch(config-if)# udld interval 400
switch(config-if)# udld mode aruba-os verify-then-forward
switch(config-if)# udld retries 5
switch(config)# exit
switch(config)# interface 1/1/3
switch(config-if)# no shutdown
switch(config-if)# no routing
```

```
switch(config-if)# lag 1
switch(config-if)# udld interval 400
switch(config-if)# udld mode aruba-os verify-then-forward
```

- The configuration on SW3 looks like this:

```
switch# config
switch(config)# interface 1/1/2
switch(config-if)# no shutdown
switch(config-if)# no routing
switch(config-if)# lag 1
switch(config-if)# udld interval 400
switch(config-if)# udld mode aruba-os verify-then-forward
switch(config-if)# udld retries 5
switch(config)# exit
switch(config)# interface 1/1/3
switch(config-if)# no shutdown
switch(config-if)# no routing
switch(config-if)# lag 1
switch(config-if)# udld interval 400
switch(config-if)# udld mode aruba-os verify-then-forward
switch(config-if)# udld retries 5
```

- Verify the configuration using the `show udld` command. On SW1, the final configuration looks like this:

```
switch# show udld
Abbreviations:
 VTF - Verify-then-forward      FTV - Forward-then-verify
 NOR - RFC 5171 normal          AGG - RFC 5171 aggresive
---------------------------------------------------------------------
---
Interface  UDLD      UDLD      UDLD           UDLD    Mode  Interval
           Config    State     Substate       Link
---------------------------------------------------------------------
---
1/1/1      Disabled  Inactive  Undetermined   Unblock FTV   8000
1/1/2      Enabled   Active    Bidirectional  Unblock FTV   7000
1/1/3      Enabled   Active    Blocked        Block   FTV   7000
1/1/4      Enabled   Inactive  Uninitialized  Unblock NOR   7000
1/1/5      Enabled   Active    ErrDisabled    Block   AGG   7000
1/1/6      Disabled  Active    Detection      Unblock NOR   7000
------------------------------------------------------------------
Retries  Tx       Rx       Rx           Rx           Transitions
         Pkts     Pkts     Pkts disc.   Pkts drop.
```

```
-------------------------------------------------------------
4       4        54       123      123       1
7       1234567  1548421  23214    1878981   3
4       3        77871    2157     81878     1
5       50       0        0        0         0
3       150      25       0        2         1
3       6        54       123      23        1
```

Private VLAN

We know by now that VLANs segment the network and keep them isolated. In this section, you will learn the specific use case for implementing private VLANs (PVLANs). PVLAN is an extension of the VLAN concept. You will review the PVLAN mechanism and also examine the process of configuring PVLANs and how to extend them in a network.

VLAN recap

Recall that if a switch receives a broadcast on any port, it floods it out all other ports, except the ingress port. This is known as a broadcast domain, which is simply a group of devices directly connected to different ports of a physical switch. All ports are part of a single LAN. Now consider that you want to restrict the devices that would receive the broadcast based on the transmitting device. This is achieved by configuring a virtual LAN (VLAN) and assigning the desired ports to that VLAN. This can be viewed as dividing a physical switch into multiple virtual switches.

Consider the sample switch SW1 and its connections represented in **Figure 3-5**. Ports 1 to 6 of the physical switch SW1 are assigned to VLAN 10 and ports 11 and 12 to VLAN 20. The rest of the ports are part of the default VLAN—VLAN 1. A message from the device connected to port 1 is only forwarded to the devices connected to ports 2 to 6. Similarly, a message from the device connected to port 11 is only forwarded to port 12 and vice versa. As a result, you have created two separate broadcast domains on a single physical switch. Configuring VLANs benefits the network by allowing to use the ports of a single switch to create multiple broadcast domains, thus removing the need to use multiple switches. Furthermore, the VLANs can be extended over multiple switches by creating trunk links between them.

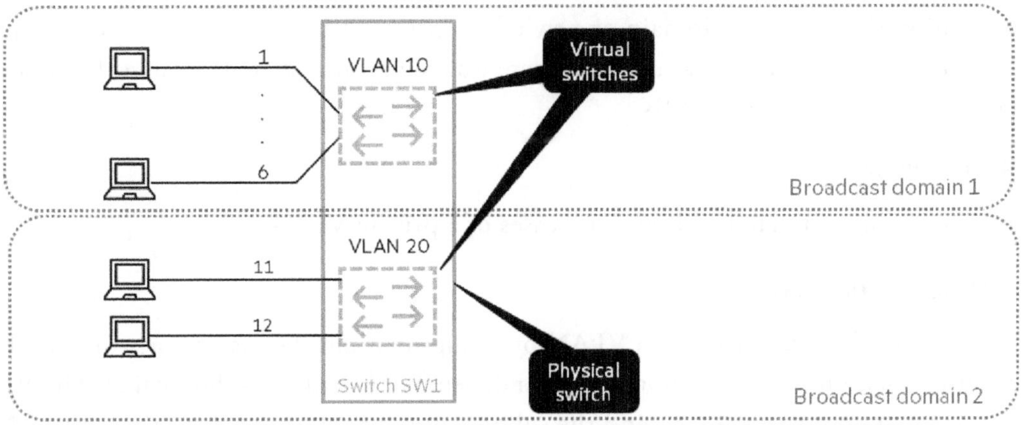

Figure 3-5: VLAN using virtual switches inside a physical switch

VLAN versus private VLAN

Figure 3-6 compares the features of a VLAN and a private VLAN.

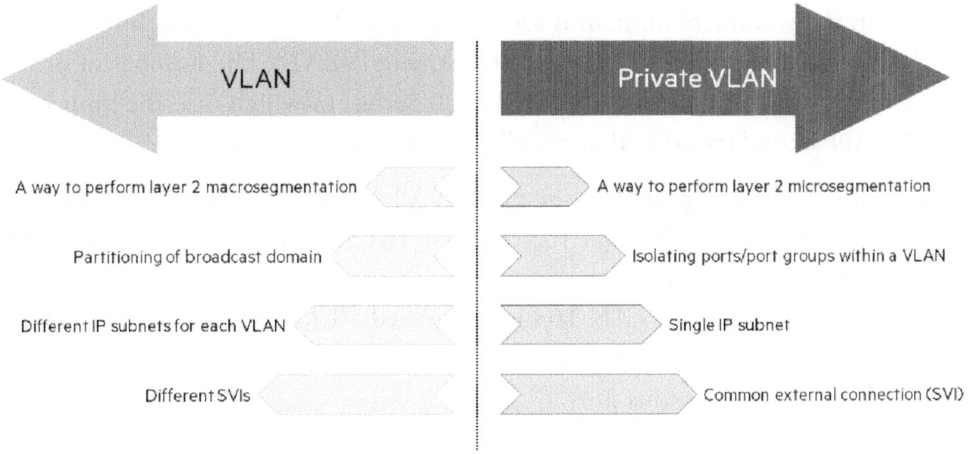

Figure 3-6: Comparison of a VLAN and a private VLAN

There could be scenarios in which segmenting a LAN using VLANs may not be an effective solution. VLANs segment the network into different IP subnets, each managed by a separate interface, known as the Switch Virtual Interface (SVI). In order to further optimize Layer 2 segmentation, you can create subgroups inside a single VLAN called private VLANs. These subgroups share the same IP subnet, thus reducing the management overhead. At the same time, the traffic can be partitioned into independent broadcast domains that do not interfere with one another. A VLAN can be considered to perform a Layer 2 macrosegmentation, whereas a PVLAN offers one way to perform Layer 2 microsegmentation.

VLANs partition the broadcast domain of a switch. A private VLAN isolates the ports or groups of ports within the same VLAN. As a result, these isolated groups share the same IP subnet and management interface, unlike VLANs.

PVLAN overview

Let us review the operational details and use cases of a private VLAN.

Operational overview

The private VLAN feature partitions a VLAN by grouping multiple sets of ports that need Layer 2 traffic isolation from one another into independent broadcast subdomains. The VLAN to be partitioned is the primary VLAN, and the subdomains carved out of this primary VLAN are secondary VLANs. These secondary VLANs are also regular VLANs, with a subgroup of ports from the original VLAN and identified by a unique VLAN ID.

Depending on the level of isolation provided, secondary VLANs can be further classified into isolated and community VLANs. You can associate up to 24 secondary VLANs with a primary VLAN. Traffic in the secondary domain is seen inside the PVLAN domain, and when the traffic leaves the PVLAN domain, it will only go on the primary VLAN. The number of primary VLANs is restricted to 32 instances (except in the CX 10000 Series, in which case the limit is 512). The number of PVLAN secondary ports that are allowed per line card is 24.

In the scenario illustrated in **Figure 3-7**, the primary VLAN 100 is subdivided into three secondary VLANs—VLANs 101, 102, and 103. VLANs 101 and 102 are community VLANs. Devices in a community VLAN can communicate with each other through L2 switching and form a single broadcast domain. VLAN 103 is an isolated VLAN. Devices in an isolated VLAN cannot communicate with each other through L2 switching; traffic in an isolated VLAN only travels to and from the promiscuous port.

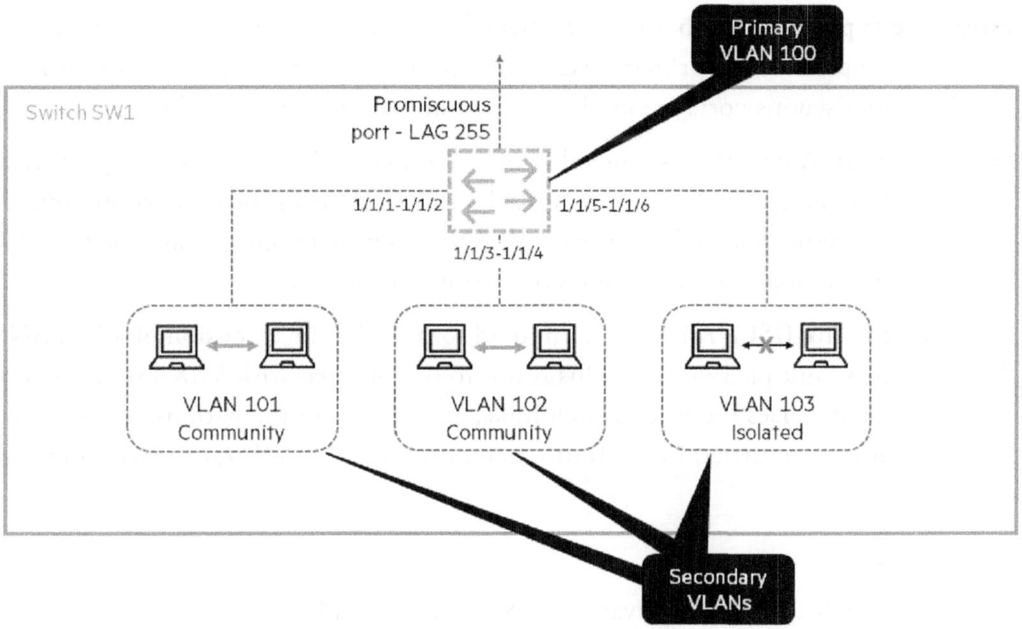

Figure 3-7: Private VLAN operational overview

A promiscuous port is a port of the primary VLAN that can communicate with all the ports of that primary VLAN, including the ports of the secondary VLANs associated with that primary VLAN. This port is used to communicate outside the PVLAN domain, for example, to connect to a routing device or a gateway. Depending on whether the PVLAN needs to be extended across switches, the outgoing connection can be through a promiscuous port or an Inter-Switch Link (ISL).

Components of a private VLAN

The components of a private VLAN are described as follows.

- **Primary VLAN:** Root of the PVLAN domain. There can be multiple secondary VLANs associated with a primary VLAN, which uses the primary VLANs to communicate with hosts outside the PVLAN domain.
- **Secondary VLAN:** Term for both isolated and community VLANs.
- **Isolated VLAN:** A secured VLAN where hosts cannot communicate with each other through L2 switching.
- **Community VLAN:** A VLAN where hosts can communicate with each other through L2 switching. There can be multiple community VLANs associated with a primary VLAN.

- **Promiscuous port:** A port that is a member of the primary VLAN. These ports can send packets to all ports of the primary VLAN and ports of associated isolated and community VLANs. Promiscuous ports are used to communicate outside the PVLAN domain.
- **Secondary port:** A port that is a member of a secondary VLAN. Secondary ports that belong to community VLANs can send packets to all ports of the same community VLAN and to promiscuous ports of its primary VLAN. A port belonging to an isolated VLAN can only send traffic to promiscuous ports of its primary VLAN.
- **Inter-Switch Link (ISL):** A generic term used to describe the extension of a PVLAN domain across multiple switches. This is not to be confused with VSX-ISL. As secondary VLAN traffic traverses the inter-switch links, all switches in the path use the secondary VLAN information carried in the traffic to ensure that PVLAN traffic forwarding rules are preserved.

PVLAN use cases

Some scenarios of applications of a private VLAN are described below.

Providing privacy in hotels

A very common use case is in the hospitality domain. Consider a hotel-like site consisting of multiple rooms, each equipped with a device with a wired connection. Similar management and security policies need to be applied to each room's connection. Creating separate VLANs for each room would mean having to configure each VLAN's (room's) interface separately. An isolated PVLAN would best suit this scenario. All rooms will be part of a single isolated PVLAN; they cannot communicate with each other, but the network is routed so as to successfully access the internet.

Isolating hypervisors in data centers

Private VLANs are also implemented in data centers to secure the applications running on different Virtual Machines (VMs). In a data center, there could be multiple VMs hosting different applications. A breach in one of the applications may leak into some other application. PVLANs provide a way to isolate the VMs and as a result, the applications running on those VMs are protected to some extent. Nowadays, many hypervisors come equipped with the PVLAN feature.

Securing headless devices

In a campus architecture, a typical implementation of a PVLAN is to secure the headless devices including cameras and IoT devices. These devices are common in campus environments such as a university or an office. If any of these devices were to get compromised and gain access to undesired applications or servers, that might impose a serious security risk. PVLANs allow to apply one level of security by configuring such devices to be part of an isolated VLAN so that they can only communicate with one backend device.

Shopping centers

An example of the application of a community VLAN is a shopping complex consisting of multiple shops, each shop equipped with a set of wired devices. The shops need isolation from other shops but require connectivity among their devices. Community PVLANs would best suit this scenario as they provide just the required amount of isolation.

PVLAN configuration process

PVLAN configuration is a straightforward process as illustrated in **Figure 3-8**.

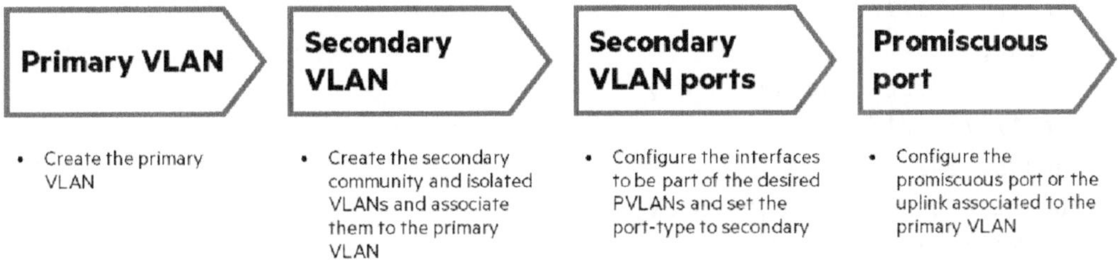

Figure 3-8: PVLAN configuration process

1. Start by configuring the primary VLAN and explicitly declaring that it is the primary VLAN.
2. Then configure the secondary VLAN. You must specify if it is isolated or community and also specify the primary VLAN it is associated with.
3. Next, configure the secondary VLAN ports. Associated physical ports must be members of the secondary VLAN, and the port-type must be secondary.
4. The promiscuous port can be a LAG or a physical interface. It must support the primary VLAN (not the secondary), and this is typically the uplink to a core switch or another Layer 3 device such as a firewall.

The commands for each of these steps are described as follows.

Primary VLAN

```
vlan 100
  name PRIV-PRIM
  private-vlan primary
```

Secondary VLAN

```
vlan 101
  name PRIV-COM101
  private-vlan community primary-vlan 100
vlan 102
  name PRIV-COM102
  private-vlan community primary-vlan 100
vlan 103
  name PRIV-ISOL
  private-vlan isolated primary-vlan 100
```

Secondary VLAN ports

```
interface 1/1/1-1/1/2
  vlan access 101
  private-vlan port-type secondary
interface 1/1/3-1/1/4
  vlan access 102
  private-vlan port-type secondary
interface 1/1/5-1/1/6
  vlan access 103
  private-vlan port-type secondary
```

Promiscuous port

```
interface lag 255
  vlan trunk allowed 100
  private-vlan port-type promiscuous
```

Extending a PVLAN

An important aspect of VLANs is to create an isolated broadcast domain that can also span across multiple switches. Private VLANs can also benefit from this feature of VLANs. There could be different scenarios where we would need to extend the PVLAN architecture across layers or switches, described as follows.

Extending the primary VLAN

Consider a scenario with a two-tier campus architecture consisting of an access switch and a core switch (illustrated in **Figure 3-9**). The access switch has the primary VLAN 100 configured on it, which in turn has the secondary VLANs 101 and 102 associated to it. For the devices to successfully transmit data to the core switch, a trunk link tagged with VLAN 100 needs to be created; it could be a single port or a LAG. This port/LAG should be configured as promiscuous.

Remember that for a link to act as an uplink and carry the PVLAN traffic outside of the primary VLAN, the port-type has to be set to promiscuous. In such a case, when the traffic from any secondary VLAN—say PVLAN 102—reaches the promiscuous port, the switch retags the traffic with the VLAN ID (VID) of the primary VLAN, that is, 100. Such traffic, when it reaches the core switch, is directly taken to the VLAN interface and then routed to the rest of the world. Similarly, when the traffic from the core switch tagged with VID 100 reaches the access switch, it is retagged based on the destination address with the ID of the respective PVLAN.

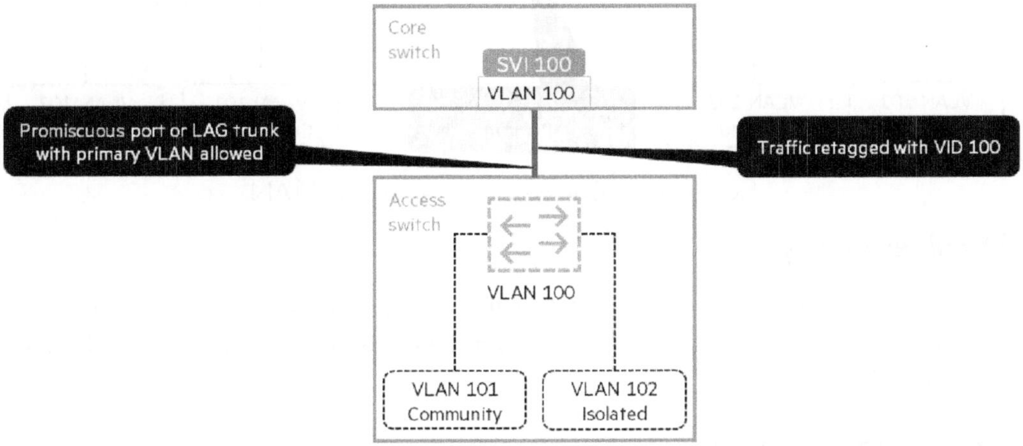

Figure 3-9: Extending the primary VLAN

Extending the secondary VLAN

In another scenario (illustrated in **Figure 3-9**) where we want to extend the secondary VLAN across the switches in the same layer—for example, hotel rooms on different floors connected to different switches but sharing the same isolated VLAN—the key is to not define the port/link as promiscuous. The link connecting the two switches will be a normal trunk link with VLAN 100 (primary VLAN) allowed. When the traffic from any secondary VLAN from switch 1 reaches the trunk link, it is forwarded as is without changing the VID tag, thus extending the secondary VLAN to the other switch. It is important to have the same set of VLANs configured on switch 2 for this to work seamlessly.

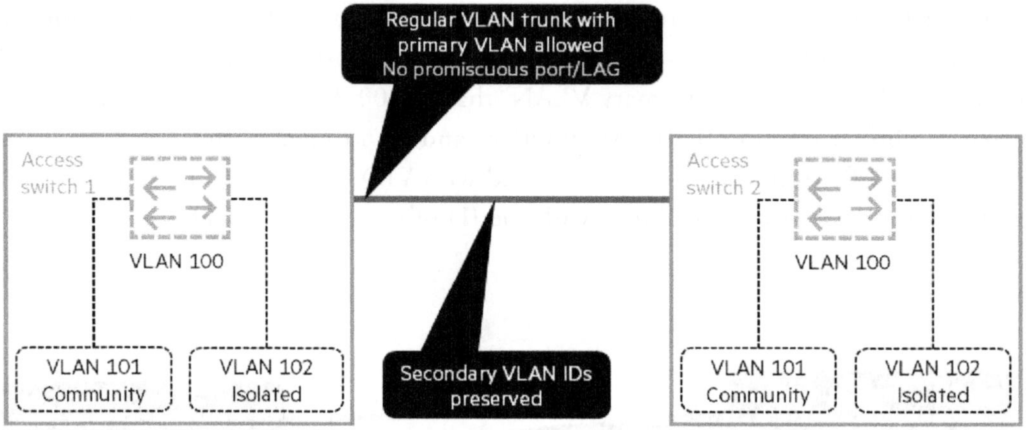

Figure 3-10: Extending the secondary VLAN

PVLAN troubleshooting

To troubleshoot a private VLAN, you can perform a variety of tasks from validating the running configuration to verifying the logs. The sample commands to execute different troubleshooting tasks are given as follows.

Troubleshooting commands

To validate the PVLAN configurations, use the `show running-config private-vlan` command as follows:

```
show running-config private-vlan
  vlan 100
     private-vlan type primary
  vlan 101
     private-vlan type community primary-vlan 100
  vlan 102
     private-vlan type community primary-vlan 100
  vlan 103
     private-vlan type isolated primary-vlan 100
interface 1/1/1
    vlan access 101
    private-vlan port-type secondary
interface 1/1/3
    vlan access 102
    private-vlan port-type secondary
interface 1/1/5
    vlan access 103
    private-vlan port-type secondary
interface lag255
```

```
    vlan trunk native 1
    vlan trunk allowed 100
    private-vlan port-type promiscuous
```

To ensure that the secondary VLANs are associated with the correct primary VLAN, use the `show private-vlan association` command as follows:

```
show private-vlan association
----------------------------------------------
Primary   Isolated          Community
-----------------------------------------------
100       103               101-102
```

To confirm the PVLAN configuration applied to physical ports, use the `show private-vlan port-type` command.

```
show private-vlan port-type
----------------------
Port      Port-type
----------------------
1/1/1     secondary
1/1/3     secondary
1/1/5     secondary
lag255    promiscuous
```

Event and debug logs

Use the `show event -r -d vland` command to view PVLAN-related logs.

```
show event -r -d vland
2023-11-16T03:46:58.763713+00:00 6200-BLDG02-F1 vland[621]: Event|2115|LOG_
INFO|||
VLAN 103 is down due to pvlan misconfig reason pvlan_no_primary_
```

You can enable a PVLAN-related debug with the `debug pvnet all` command.

```
debug pvnet all
show debug buffer
2023-11-16:04:11:01.901900|vland|LOG_ERR|MSTR|1|VLAN|VLAN_TRACE| pvlan_handle_
sec_vlan_misconfig: No primary association found for sec vlan 103
```

Basics of spanning tree protocol

In this section, you will review some fundamentals of the Spanning Tree Protocol (STP), including its operation and the possible network improvements achieved through its implementation.

STP overview

Let us first examine the need for a specialized algorithm like STP followed by the benefit achieved through STP's implementation.

Shortcomings of redundant links

Redundant links can provide load balancing and fault tolerance in networks. However, connecting Layer 2 switches with redundant links creates Layer 2 loops. If not properly handled, these loops cause serious problems that can effectively disable a network. These loops can cause broadcast storms and multiple frame copies, which can in turn cause instability of the MAC address table. Let us discuss these issues one by one.

Broadcast storm

You have learned that switches flood broadcasts out all interfaces in the same VLAN, except the ingress port. This can cause problems on a redundant, looped network like the one in **Figure 3-11**.

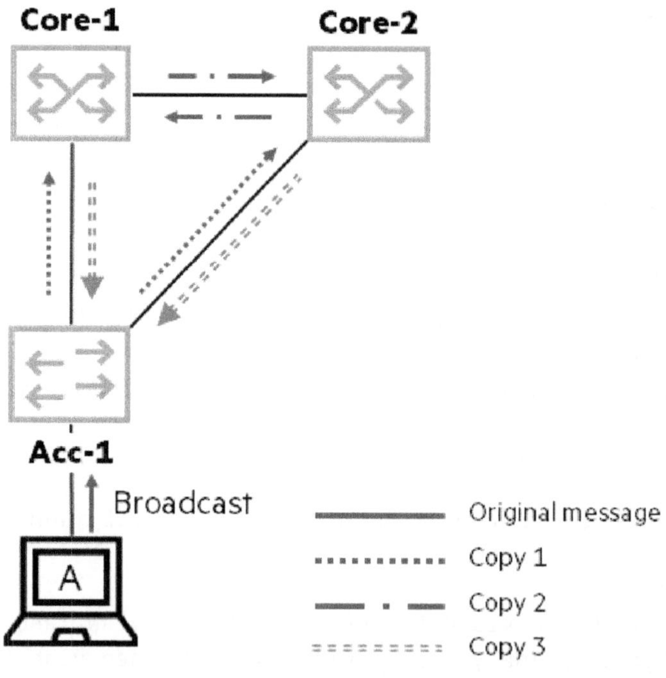

Figure 3-11: Broadcast storms caused by L2 loops created by redundant links

Suppose that host A sends a broadcast frame. Switch Acc-1 receives this frame and floods it out all interfaces (Copy 1). Core-1 and Core-2 receive this and flood it out their connections to each other (Copy 2). Core-1 and Core-2 receive the second copy and flood them out their other ports (Copy 3). Acc-1 receives the third copy, and the cycle continues. These frames circle around the network continuously. As a natural part of network and endpoint operation, nearly all devices send broadcast frames, often many times per minute. Every broadcast from every device circulates around the network forever. Soon, all available bandwidth and CPU cycles are used up by processing the broadcasts. No resources are available to process normal data communication frames. The network is effectively unusable.

Multiple frame copies

You may recall that switches not only flood broadcasts out all ports in the VLAN (except the ingress port), but they do the same thing with unicasts to an unknown destination. If a switch receives a unicast packet to some destination MAC address, and that address has not yet been learned in the MAC address table, then the switch floods the frame out all ports in the VLAN (except the ingress port). These can also circulate and waste bandwidth and CPU cycles. They also cause other issues. Consider the scenario illustrated in **Figure 3-12**.

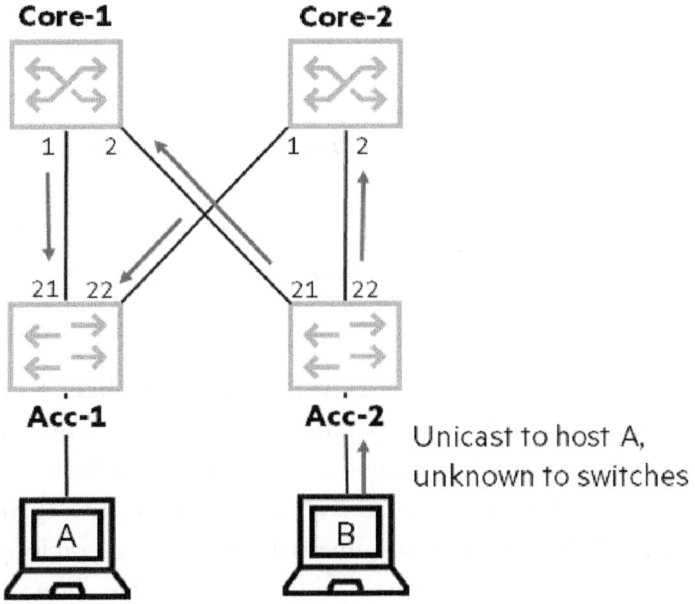

Figure 3-12: Multiple frame copies created by the L2 loops caused by redundant links

Suppose that host B sends a frame to host A. Acc-2 has not yet learned host A's MAC address, and so it floods this frame out ports 21 and 22. Core-1 and Core-2 both receive a frame with the source MAC of host B, inbound on port 2. Both switches send a copy to Acc-1. Acc-1 now believes that it can reach host B via ports 21 and 22. Acc-1 is thus confused because host B can only exist in one place, but it appears to exist in two places. The multiple frame copies generated by Acc-2 create MAC database instability on Acc-1.

Loop-free topology using STP

It can be concluded that there is a serious downside to using redundant links. Although redundant links are required for reliability, this redundancy can bring your network to its knees. This is where the spanning tree protocol helps save the network.

The STP algorithm runs on all switches, and as a result, the redundant links are disabled to avoid loops. However, if a failure occurs, STP converges on a new topology of active links, which are used to forward frames. Thus, a loop-free topology is always maintained.

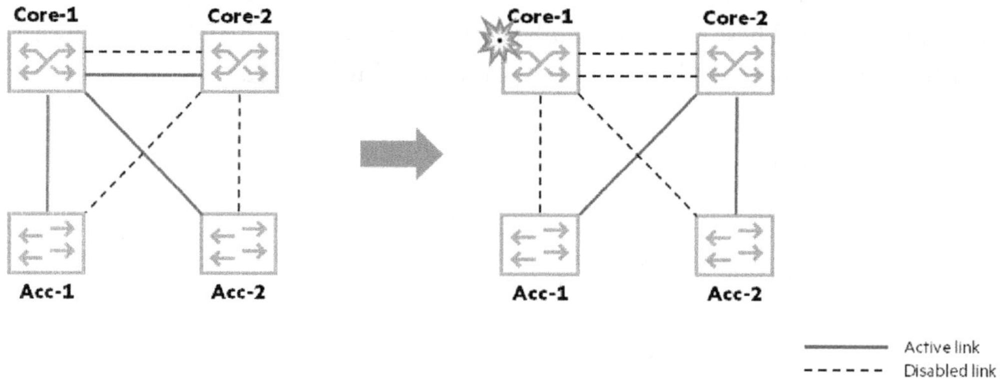

Figure 3-13: Loop-free topology achieved using STP

STP operation

To describe the operational flow, consider the revised version of the basic STP algorithm—the Rapid Spanning Tree Protocol (RSTP). The operational flow of this algorithm is illustrated in **Figure 3-14**, which shows a highly redundant network for resiliency and fault tolerance. These redundant links could cause loops and their associated issues. However, once a spanning tree protocol is engaged, certain redundant links are automatically disabled. There are no loops, so there are no problems with broadcast storms or MAC address table instability. However, these disabled links remain available to provide redundancy as needed.

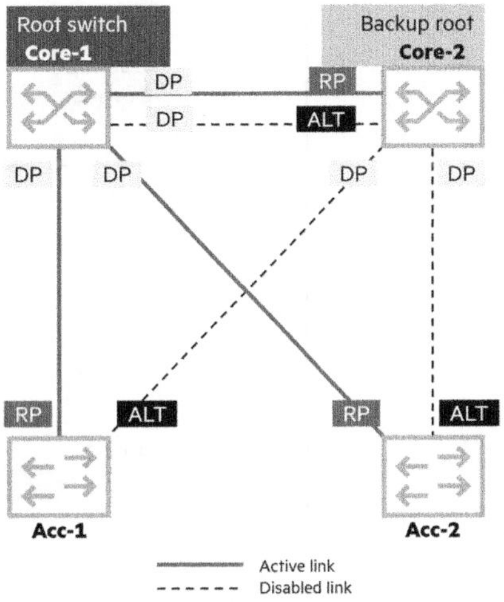

Figure 3-14: Operation of rapid spanning tree protocol

To accomplish this goal, one switch in the spanning tree domain is elected as a root switch and another as a backup switch based on the bridge IDs (BIDs) of the switches. Based on the path costs and port proximity, each port is either assigned as a root port (RP) or a designated port (DP). These ports help decide the closest path to the root switch. All the remaining ports are assigned as alternate ports (ALT) and are in a disabled state. All switches that participate in the algorithm exchange control messages called Bridge Protocol Data Units (BPDUs). In RSTP, all switches originate BPDUs with their current information every two seconds—the default hello-time period. Thus, if a port stops receiving BPDUs for three consecutive hello timers, the switch quickly knows that it has lost connectivity to its neighbor. It ages out the protocol information and begins to converge. Because each switch originates BPDUs, this becomes a true keep-alive mechanism. Failure detection will take no longer than six seconds.

MSTP overview

Multiple Spanning Tree Protocol (MSTP) enhances RSTP by building an additional, new topology or instance of the spanning tree protocol. Different instances can carry different VLAN traffic allowing load balancing and maintaining optimal paths. The example in **Figure 3-15** illustrates this. It shows two topologies created on the same physical network. In the topology on the left, VLANs 1 to 20 share a single iteration of RSTP. Core-1 is the root, and the tree grows from there. VLANs 1 to 20 share these forwarding links with the unused links available for redundancy. In the topology on the right, VLANs 21 to 40 share a different iteration of RSTP. Core-2 is the root, and the topology emanates from that switch. The active links are shared by VLANs 21 to 40, and the other links are available for backup. The dormant links for VLANS 1 to 20 are the active links for VLANs 21 to 40, and vice versa.

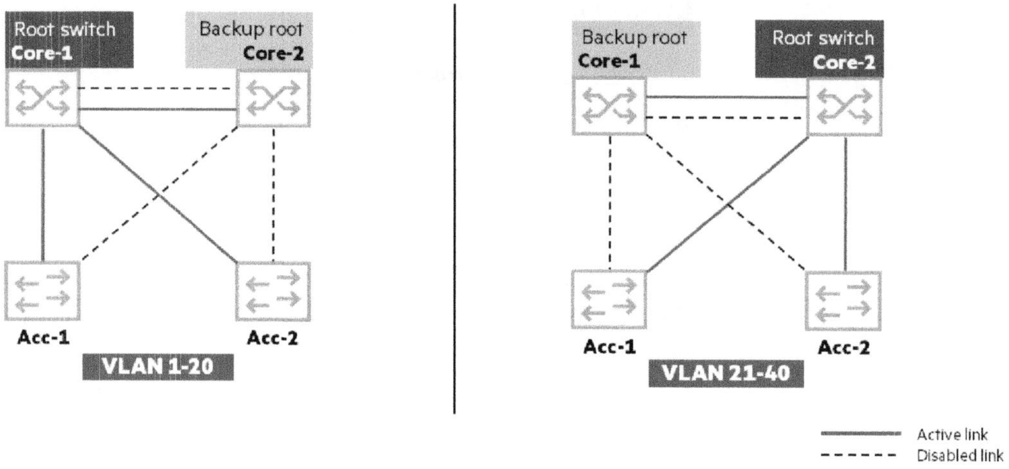

Figure 3-15: MSTP overview

Additional features of spanning tree protocol

STP relies on BPDUs, the control messages sent by the participating switches. This has its own disadvantages such as slower device onboarding as well as risk of network failure due to an undetected rogue device joining the network. In this section, you will examine some features available for STP implementations such as root guard, BPDU guard, and loop protection that will help prevent these issues.

BPDUs and edge ports

BPDUs are an important part of the keepalive mechanism of the network. At the same time, BPDUs from unexpected port connections may alter the topology, affecting the performance in a negative way.

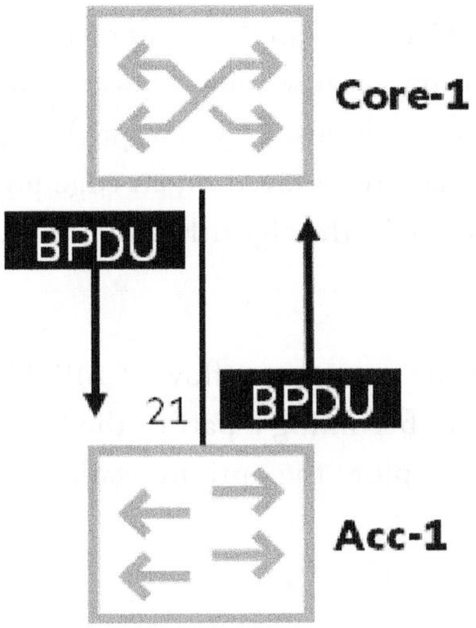

Figure 3-16: BPDU overview

All switches that participate in the spanning tree algorithm exchange control messages called Bridge Protocol Data Units (BPDUs). These messages carry information required for the calculation of the spanning tree such as bridge IDs and path costs. Upon initialization of a network, each device periodically generates and sends configuration BPDUs with itself as the root bridge. After network convergence, only the root bridge generates and sends configuration BPDUs. The other devices only forward the BPDUs.

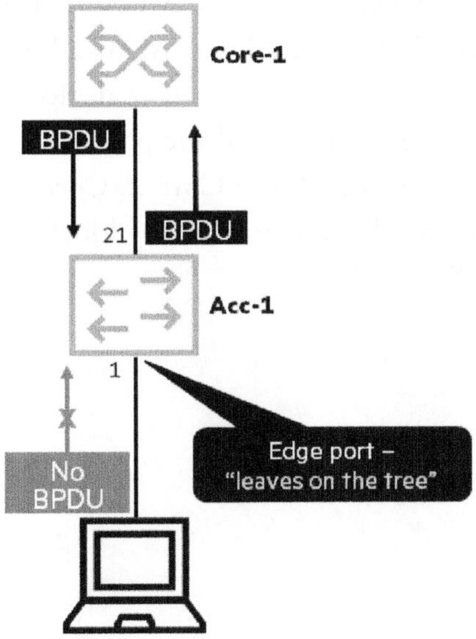

Figure 3-17: Edge ports overview

Edge ports connect to endpoint devices, and so they do not receive BPDUs. They need not participate in the spanning tree algorithm because they cannot cause a loop—they are "leaves on the spanning tree." Since they cannot cause loops, they can quickly transition to a forwarding state, with no intermediate steps. If BPDUs are received on an edge port, then the port will act as a normal spanning tree port and participate in the algorithm to prevent Layer 2 loops.

Safeguarding STP

With this keepalive mechanism of the continuous flow of BPDUs, there is an associated risk of unwanted ports sending BPDUs and becoming a part of the STP calculation as a result. Let us analyze this risk in more detail and explore the options available in AOS-CX switches to overcome it.

BPDU vulnerability

By design, STP allows any network device to join the spanning tree. While this feature ensures that all loops are eliminated, it leaves the network vulnerable to manipulation from unauthorized devices. Because BPDUs have no authentication, it is easy for rogue devices to send legitimate BPDUs and join the spanning tree. Whether the rogue device is a malicious attacker or simply an unauthorized device such as an unmanaged switch added by a user, it can negatively affect the network in a significant number of ways. For example, it can affect path selection, inactivate correct links, activate incorrect links, and even become the tree's root.

Figure 3-18 illustrates one such scenario. The rogue switch that has joined the network at Acc-2 has a priority of zero. It also happens to have a lower MAC address than the desired root switch (Core-1). As a result, the BPDU received from the rogue device is considered to be superior to the other BPDUs by Acc-2. Acc-2 changes the type of the port connected to this device to root port and advertises the superior BPDU to Core-1 and Core-2. Core-1 and Core-2 reconverge the topology based on the rogue device being the new root. They block the high-speed link between them, creating a Denial of Service (DoS) attack.

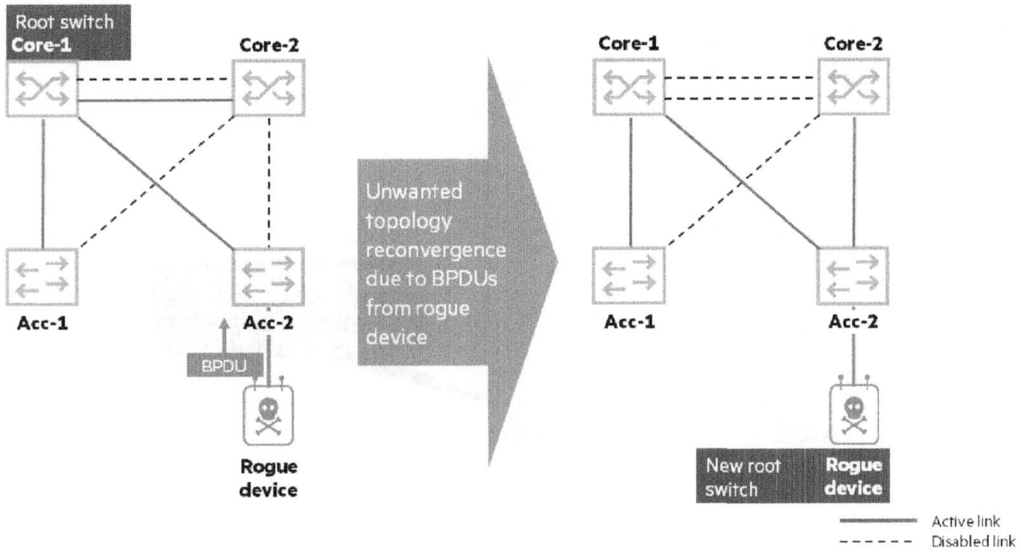

Figure 3-18: Rogue device undesirably becoming a root switch

Next, you will explore some features available in AOS-CX switches to protect the network against such attacks.

Solution 1: BPDU protection

AOS-CX switches offer BPDU protection to prevent the above situation from occurring in a network that uses MSTP or RSTP. This feature operates at the port level and is best implemented on edge ports, which should not be receiving BPDUs.

This feature is illustrated in **Figure 3-19**. The BPDU protection feature monitors a port for incoming BPDUs. If the port receives a BPDU, the switch shuts down the port, protecting the network from the rogue device. You can configure the amount of time the port remains disabled using a number of seconds that ranges from 0 to 65535. Configuring a value of zero (0) specifies an infinite timeout period. In other words, the port remains disabled until manually re-enabled (such as with the `interface <ID> enable` command). If STP BPDU packets are received on a protected port, the BPDU guard disables the port, and an alert is sent.

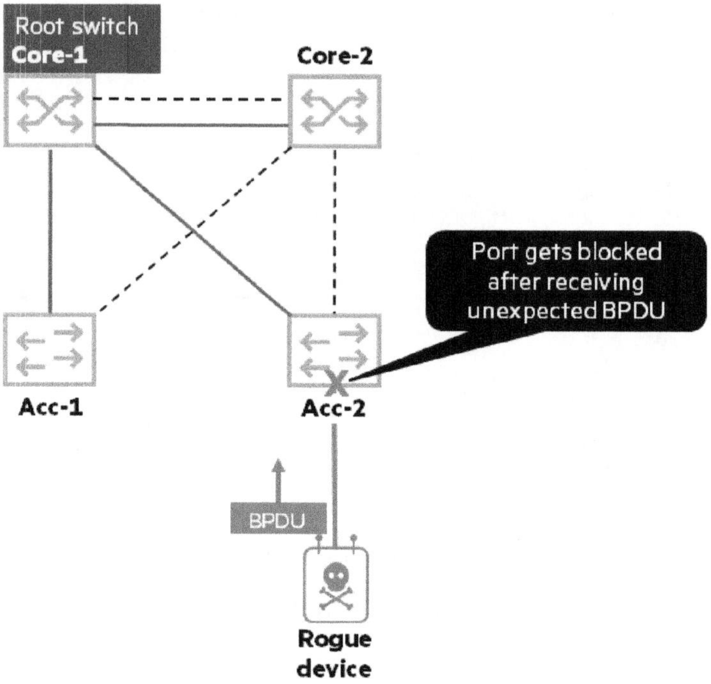

Figure 3-19: BPDU protection feature overview

A BPDU can be configured on a port using the `spanning-tree bpdu-guard` command. A protected port blocked due to receipt of unwanted BPDUs shows the state as `Disabled Bpdu-Error`.

Solution 2: Root protection

Another method of protecting the network from a rogue device becoming the root is using root protection (also known as root guard). Just like BPDU protection, root protection is also applied on a port-by-port basis. But instead of applying the protection on every edge port, root protection is enabled on ports that should never receive superior BPDUs. For example, in the topology shown in **Figure 3-20**, the switches in the core layer are the potential root switches. These switches should never receive BPDUs containing superior information (that is, superior BIDs or path costs) from the access layer switches. Applying root protection on ports 1/1/1 and 1/1/2 of both the core switches will protect the network from converging to any root switch-based topology other than these two switches.

Consider the same scenario as in the previous case. The rogue device joined at Acc-2 sends superior BPDUs, which propagate to the core layer through Acc-2 on ports 1/1/2 of both Core-1 and Core-2. Since these ports have root guard enabled on them, they discard these superior BPDUs.

At the same time, the ports transition to the Root-Inc (inconsistent) state, implying that they are temporarily blocked by the root guard. The root guard will change the port state from Root-Inc to its previous state when the port stops receiving superior BPDUs.

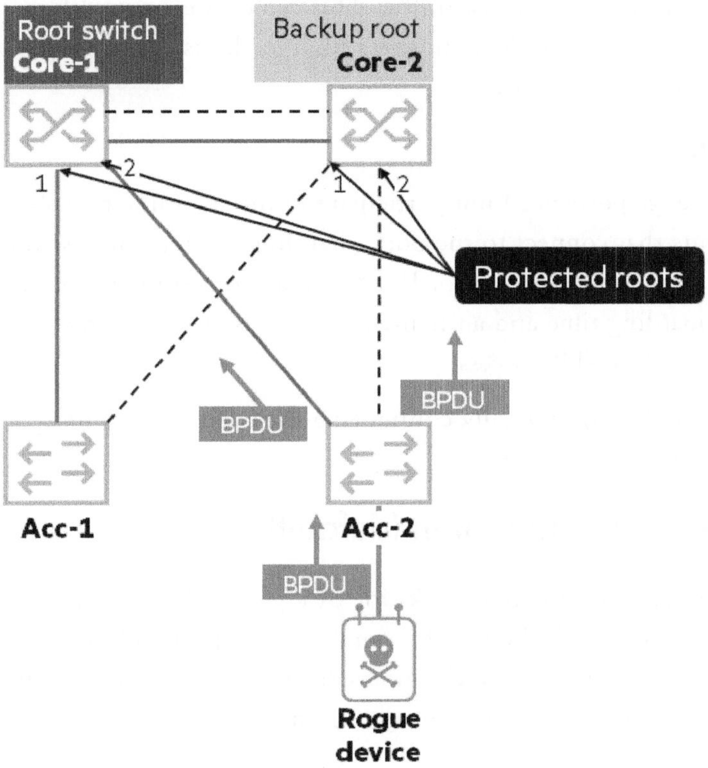

Figure 3-20: Root protection feature overview

Root protection can be configured on a port using the `spanning-tree root-guard` command.

Additional BPDU protection feature: BPDU filter

Another option available with the AOS-CX switches is enabling BPDU filter for an interface. This feature allows to control the spanning tree participation on a per-port basis. It can be used to exclude specific ports from becoming part of spanning tree operations. A port with the BPDU filter enabled will ignore incoming BPDU packets, does not transmit BPDU, and stays locked in the spanning tree forwarding state. All other ports maintain their role.

You can enable this feature using the following command:

```
switch(config)# interface 1/1/1
switch(config-if)# spanning-tree bpdu-filter
```

This feature is useful not only for protecting a network from DoS attacks caused due to BPDU spoofing, but also to prevent unnecessary topology changes. However, there is a risk associated with the use of this feature. Because the activated ports permanently remain in the forwarding state, they permit all broadcast traffic. This can create a network storm if there are any loops using these ports. If you suddenly notice a high load, disconnect the link and disable the BPDU filter using the no command.

STP on edge ports

As discussed earlier, edge ports need not participate in the spanning tree algorithm. This is because the endpoints that connect to the ports that do participate in a spanning tree (STP) may experience DHCP assignment timeouts or IP address assignment delays. Moreover, it may lead to extended client onboarding time and authentication issues. These problems occur because the port participates in the full STP process.

Such issues can be avoided by manually configuring the edge ports to not participate in STP using commands as illustrated below.

Using the admin-network option (default)

In the case of admin-network, the port looks for BPDUs for the first three seconds after the link is up. If no BPDUs are received, the port becomes an edge port and immediately forwards frames. If BPDUs are detected, the port becomes a non-edge port and participates in normal STP operation. The detailed commands to configure an edge port using the `admin-network` command are as follows:

```
switch(config)# interface 1/1/1
switch(config-if)# spanning-tree port-type admin-network
```

Using the admin-edge option

The admin-edge option does not wait for the receipt of any BPDUs. The port immediately transitions to the forwarding state on startup. The detailed commands to configure an edge port using the `admin-edge` command are as follows:

```
switch(config)# interface 1/1/1
switch(config-if)# spanning-tree port-type admin-edge
```

Loop protection

The access layer switches remain vulnerable to a variety of loop-causing incidents. As you may well know, employees who use spare Ethernet jacks to plug in personal devices such as unmanaged switches can inadvertently cause loops that, in turn, can bring your network down in an instant. Often, the spanning tree protocol applied on these switch ports can block loops created on the edge ports. For example, in the scenario where a user connects two Ethernet jacks, the switch can detect its own BPDU and, as a result, block one of the ports. However, STP cannot protect the network from loops created by unmanaged devices that drop BPDU packets. In addition, STP might fail to detect loops on ports that enforce client authentication. In such cases, the authentication mechanisms have the ability to lock down a port to accept traffic from the authenticated device MAC address. Ports on which the device is already authenticated may drop BPDUs, and any loop created may go undetected. In such cases, loop protection can help ensure that your edge network is loop-free.

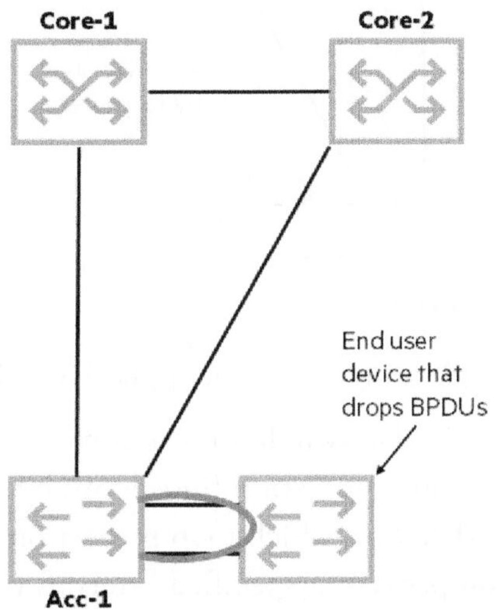

Figure 3-21: Edge ports being vulnerable to loop creation

The solution to this problem is applying loop protection on edge ports. This feature is illustrated in **Figure 3-22**.

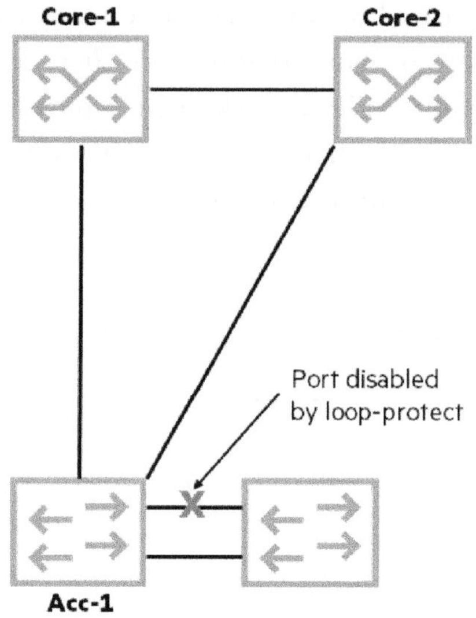

Figure 3-22: Loop protection applied on edge ports

The loop protection feature protects edge switches from downstream loops by transmitting loop protocol packets out of the ports. Loop protection finds loops by sending loop protection packets on each port, LAG, VLAN, or VXLAN on which loop protection is enabled. The enabled ports automatically send loop protection packets at specified intervals (five seconds by default). If a loop protection packet is received by the same switch that sent it, it indicates a loop exists and one of the following actions is taken.

The actions supported on AOS-CX switches are listed as follows.

- **Do not disable:** No ports are disabled. On every transmit interval, the loop will be detected and the detection will be reported via an SNMP trap and an event log message. The value to be set for the action parameter is **do-not-disable**.
- **Transmission disabled:** The port that transmitted the loop detection packet is disabled. This is the default action. The action is set to **tx-disable**.
- **Transmission and reception disabled:** The ports that transmitted and received the loop detection packet are disabled. The action is set to **tx-rx-disable**.

Loop protection is configured on a port as follows:

```
switch# config
switch(config)# interface 1/1/1
switch(config-if)# loop-protect
switch(config-if)# loop-protect action {do-not-disable | tx-disable | tx-rx-
disable}
```

RPVST+

In this section, you will examine a version of the spanning tree algorithm called RPVST+, which allows AOS-CX switches to work seamlessly with other devices in heterogenous networks. This protocol was introduced as an enhancement to Rapid Spanning Tree Protocol (RSTP) to improve the link utilization issue and require less configuration overhead.

RPVST+ overview

You have observed how an MSTP algorithm provides enhancements to RSTP by allowing to create multiple instances of the tree with different configurations. These instances can carry different sets of VLANs. Another algorithm called Rapid Per-VLAN Spanning Tree+ (RPVST+) is one more modified implementation of RSTP. It enables the creation of a separate spanning tree for each VLAN on a switch and ensures that only one active, loop-free path exists between any two nodes on a given VLAN.

Spanning tree protocols are used to prevent loops from occurring when multiple paths exist between the devices on a network. They are also used to provide redundancy, enabling data to use an alternative path when one link to a device fails. This is also achieved in the basic implementation of RSTP. However, as discussed in the case of MSTP, RSTP creates a tree at the network level, which is common for all the VLANs. This may result in some of the links being blocked throughout the operation, causing inefficient resource utilization. RPVST+, similar to MSTP, creates different instances of spanning trees that can have different root switches and thus different topologies. This way, we can achieve load sharing of traffic across redundant links. On the other hand, RPVST+ creates a separate spanning tree for each VLAN, whereas MSTP maps one or more VLANs to one instance. This is illustrated in **Figure 3-23**. Due to the per-VLAN feature of RPVST+, topology changes affecting a particular VLAN are propagated only inside that VLAN. RPVST+ is also VLAN-aware. The spanning tree only extends on links that actually support the VLAN, preventing misconfiguration and accidentally blocked links.

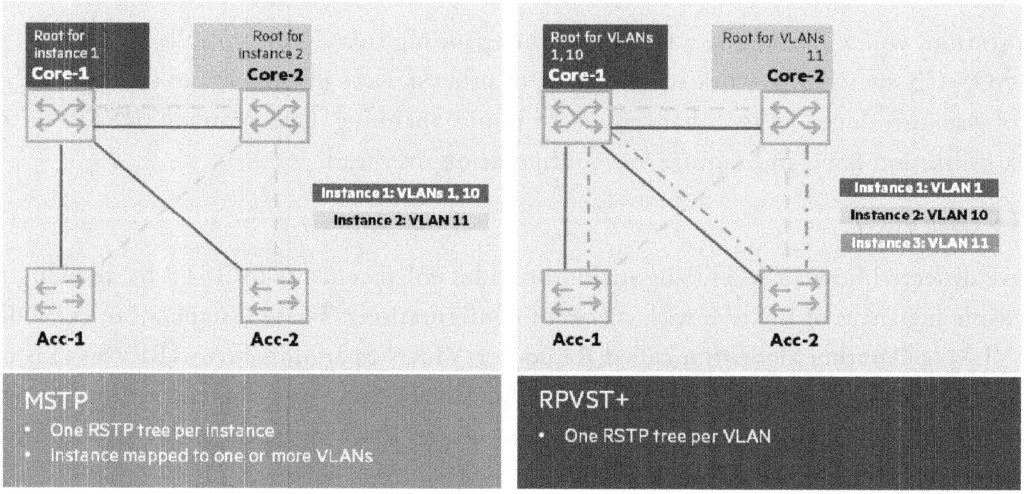

Figure 3-23: RPVST+ overview

Need for RPVST+

This version of the spanning tree protocol is required to configure heterogenous networks for the following reasons:

- Not all Cisco switches and IOS versions support MSTP. Many companies with Cisco networks use the default mode on Cisco switches, which is RPVST+.
- Many vendors' devices also support RPVST+.
- To administer a heterogeneous network, network administrators who are very familiar with RPVST+ may not want to transition to MSTP, even if the latter is an option.

In order to provide the ability to implement and manage such heterogenous networks, RPVST+ is supported on AOS-CX switches. The predecessor of RPVST+, the PVST+ protocol, uses a set of Cisco proprietary technologies such as UplinkFast and BackboneFast. RSTP has similar mechanisms built into it to speed up convergence. Each RPVST+ spanning tree uses an RSTP just like MSTP. Therefore, AOS-CX switches support RPVST+ without the need to support the other Cisco technologies.

RPVST+ interoperability with MSTP

The following considerations need to be undertaken while implementing RPVST+ in a network:

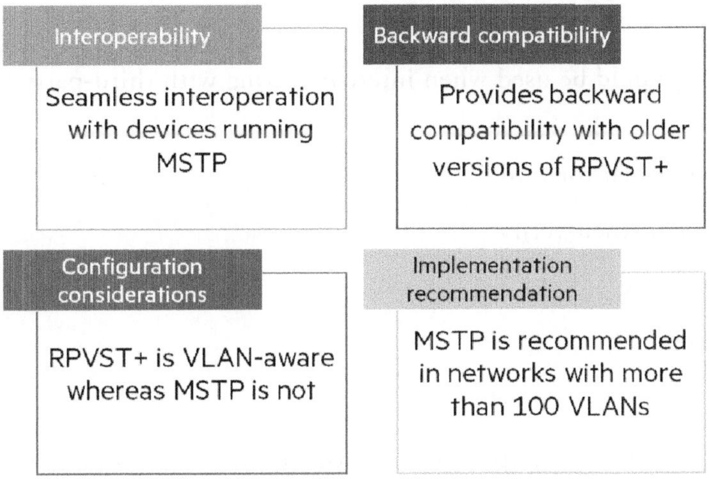

Figure 3-24: RPVST+ interoperability features and considerations

- RPVST+ interoperates with devices that run legacy IEEE 802.1D STP and MSTP-IEEE 802.1s.
- RPVST+ was preceded by PVST+, a similar Cisco standard that was based on STP. Most modern networks use RPVST+, and this is the version of the protocol supported on AOS-CX switches when you enable PVST+ mode. However, if the AOS-CX switch connects to a switch that uses only the basic Per-VLAN Spanning Tree+ (PVST+) protocol, the AOS-CX switch interoperates in much the same way that an RSTP switch interoperates with an STP switch.
- As described earlier, RPVST+ is VLAN-aware, which means this protocol only creates a spanning tree on links that actually support the VLAN. This gives the ability to simplify configuration—you only need to propagate VLANs in a network where they need to go and not on all switch-to-switch links as you do in the case of MSTP.
- For each VLAN in a network, RPVST+ creates a new spanning tree, even if many VLANs may use the same path. With a high number of VLANs, the switches need to process more BPDUs and maintain more spanning trees, increasing the processing load on the CPU. With MSTP, on the other hand, a single untagged BPDU carries information for every instance. This enables load sharing across the network while keeping the switch's CPU load at a moderate level. Therefore, in networks with 100 or more VLANs, it is best to migrate to MSTP from RPVST+.

Learning check

Check your understanding of this material by answering the following questions.

Question 1

Which UDLD modes should be used when interconnecting with third-party switches? (Select two.)

- ☐ Aruba-OS verify-then-forward
- ☐ Aruba-OS forward-then-verify
- ☐ RFC5171 aggressive
- ☐ RFC5171 normal

Question 2

Promiscuous port is a port or a group of ports on the primary VLAN on which a secondary VLAN is created.

- ○ True
- ○ False

Question 3

A network topology consisting of Core and Access layers needs to be protected from rogue devices. There is a risk of unwanted devices joining the network at edge ports and becoming the root switch by unexpectedly participating in STP calculations. Which command should be used for configuring the edge ports?

- ○ `spanning-tree root-guard`
- ○ `spanning-tree bpdu-guard`
- ○ `spanning-tree port-type admin-edge`
- ○ `spanning-tree port-type admin-network`

Question 4

AOS-CX switches need to be interconnected with some third-party switches implementing PVST+ spanning tree protocol. Which statement is true regarding the implementation of such an architecture?

- ○ AOS-CX switches cannot interoperate with devices implementing PVST+ protocol.
- ○ The resulting heterogenous architecture will have one topology for all VLANs.
- ○ AOS-CX switches can use RPVST+ protocol to interoperate with the third-party switches.

4 Advanced OSPF

After learning how to manage and optimize the L2 traffic, you will learn how to implement Open Shortest Path First (OSPF) routing in multiple area designs. You will examine several advanced scenarios and learn ways to configure OSPF to meet their needs. In the *AOS-CX Switching Fundamentals* guide, you learned how to implement OSPF in a small, single-area design. This chapter will quickly review what you learned, but then expand upon this for a large campus design. Design features that you will explore include implementing multiple areas, LSA types, and area types for optimizing the LSA traffic, authentication, and redundancy. You will also learn how to redistribute routes from an external network deployed with other routing protocols into OSPF, as well as some additional OSPF features.

OBJECTIVES

After completing this chapter, you should be able to:

- ✓ Describe the use of multi-area OSPF for scalable and stable enterprise networks.
- ✓ Redistribute external networks' routes into OSPF networks.
- ✓ Explain failover in OSPF networks.
- ✓ Implement virtual links to connect to the backbone through a non-backbone area.
- ✓ Implement OSFP authentication.

OSPF overview

Many companies have multiple routers and routing switches that send traffic across the campus. Dynamic routing protocols enable routers to discover and update routes automatically.

As an Interior Gateway Protocol (IGP), OSPF enables route info exchange among routers in the same autonomous system (AS)—a system under a single organization's control—like your company.

OSPF is a link state protocol, where routers advertise information about their connected Layer 3 interfaces, or links, along with the cost (or metric) associated with that link. These Link State Advertisements (LSAs) propagate across participating routers, which use the advertisements to build a network topology, a map of the network: all known links, the routers connected to them, and associated costs. Using the topology, routers calculate their own lowest cost route to each destination subnet.

Link state protocols like OSPF tend to choose better routing paths and to converge faster than distance vector protocols like Routing Information Protocol (RIP). OSPF uses Dijkstra's algorithm, also called the Shortest Path First (SPF) algorithm, to find the lowest-cost, loop-free paths.

For scalability, OSPF supports a two-layer hierarchical design.

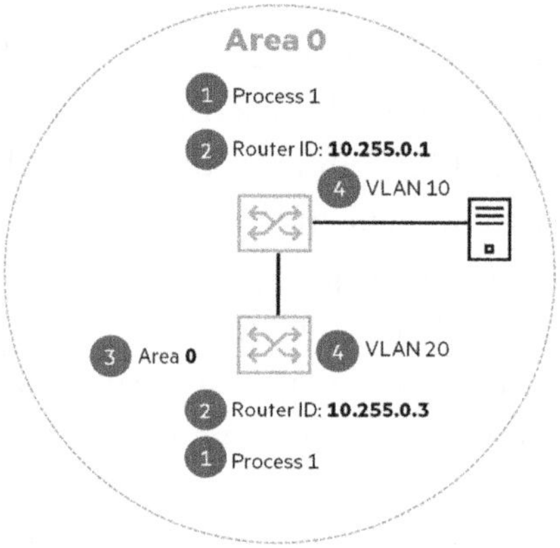

Figure 4-1: Single OSPF

Basic steps for setting up OSPF

1. Create an OSPF routing process

An arbitrary number with local significance that need not match between routers is used to identify an OSPF process in a router.

2. Optionally define a router ID

Unlike the process ID, this needs to be different for every router in your OSPF network. For more stability, use a loopback interface to capture its IP address as a router ID.

3. Create an area

Your initial configuration will use one area, Area 0, and you will then learn about multiple area solutions for scalable OSPF networks.

4. Enable OSPF on Layer 3 interfaces

For AOS-CX switches, these are either routed or VLAN interfaces. Enabling OSPF on an interface has two effects:

a. The routing switch sends OSPF messages on the interface and can form neighbor relationships with other routers on the same interface. Neighbors that fully exchange their topology information are called "fully adjacent."

b. The router advertises this interface in OSPF. For example, if the interface is associated with IP address 10.255.0.1/24, the router advertises that it has a link to 10.255.0.0/24.

5. VLANs that need to be advertised in OSPF include:

a. Those that connect routers (like VLAN 255 on Access-1 and Core-1).

b. Those that support endpoints (like VLAN 10 on Core-1).

c. If you are only enabling OSPF on the VLAN because you want to advertise its subnet, and you do not want the router to form adjacencies with other routers on the VLAN, configure the VLAN as a passive OSPF interface.

d. When you enable OSPF on a VLAN, you also specify the VLAN's OSPF area. The area ID must match on all OSPF routers connected on the same VLAN or subnet.

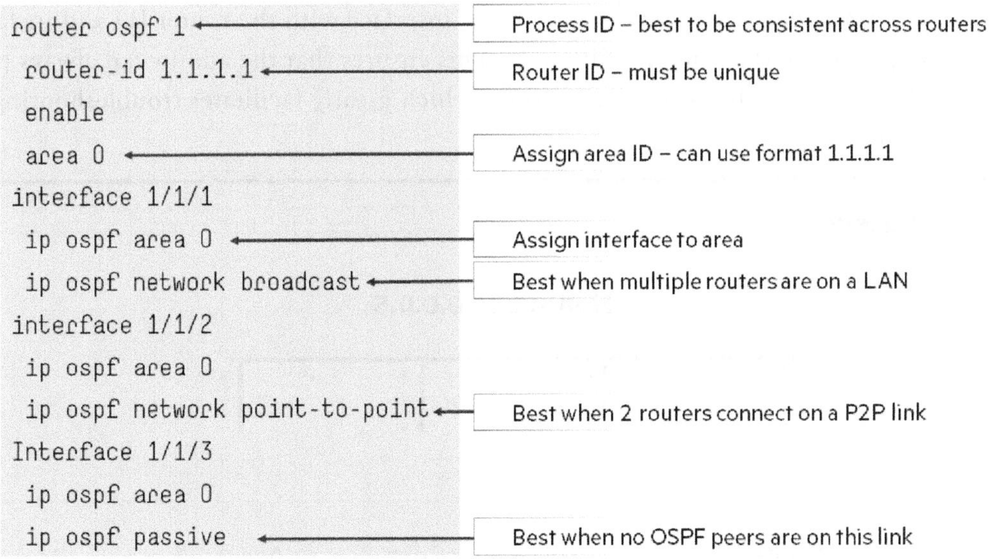

Figure 4-2: Single area OSPF configuration

The previous example shows a basic OSPF configuration. After you complete these steps, your routers automatically create a topology of your network and learn routes to every subnet within it. If a link fails, the routers inform each other of the change and learn new routes.

By default, OSPF is enabled when you create the routing process the first time. If you disable it with the `disable` command, use the corresponding `enable` command to re-enable it. This allows you to temporarily turn off OSPF without losing your OSPF configuration.

Router ID

The router ID is a 32-bit number that uniquely identifies the router within the OSPF system. This ID has the format of an IP address, and it can be an IP address on one of the routing device's interfaces. However, you can also set an ID manually, and this ID does not necessarily need to match an IP address on the router. You could also define a logical loopback interface on each router, and set the router ID to match that loopback IP address. This ensures that the associated IP address is always reachable, even if some of the router's links fail.

Selection criteria for the router ID are as follows:

1. Manually defined router ID
2. Highest IP address on a loopback interface of the router
3. Highest IP address of an active interface

It is a best practice to set up a loopback interface with the router ID and include the loopback interface in an OSPF area. This ensures that the admin can always ping the address, even if links' statuses change, which greatly facilitates troubleshooting OSPF problems.

Neighbor adjacency

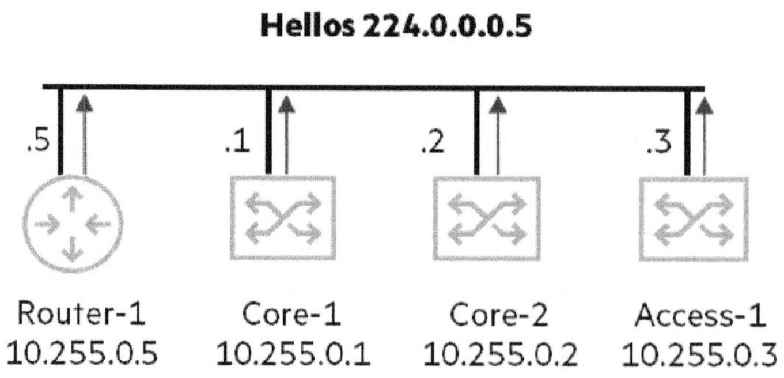

Figure 4-3: Neighbor adjacency

Match requirements

- Same area ID
- Same Network ID
- Same hello and dead interval timers
- Same stub area flag
- Same authentication password
- Same Interface MTU size

OSPF network types

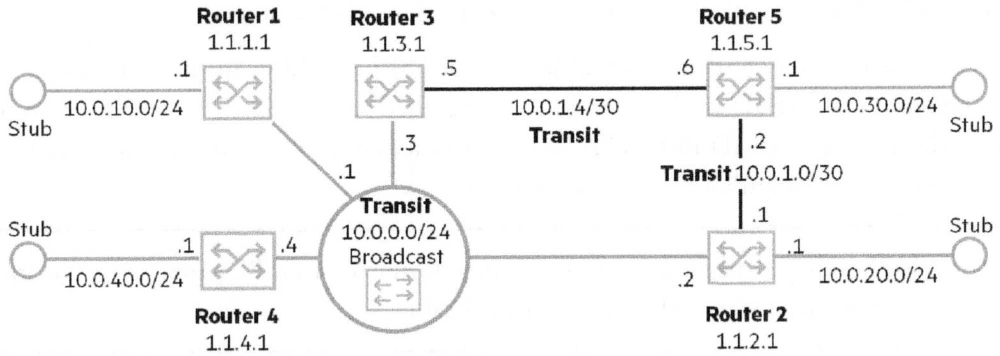

Figure 4-4: Different link types

1. Broadcast network

- Multiple routers
- Any size subnet
- DR/BDR elected
- Default for physical and VLAN

2. Point-to-point network

- Multiple routers
- Any size subnet
- DR/BDR elected
- Default for physical and VLAN

Now you will review how an OSPF router discovers its neighbors. The router sends hellos on each OSPF interface to multicast address 224.0.0.5—the "all OSPF routers" address. Therefore, all other OSPF routers in the same subnet will hear the hello.

An OSPF interface is a Layer 3 interface with an IP address on which OSPF is enabled. The hello includes information about the router's OSPF settings on the interface, and a list of neighbors discovered on the interface. The neighbor list is empty until the router receives hellos from neighbors.

A receiving router checks the settings in the hello to make sure that they match its own settings on the interface. **Figure 4-4** shows the settings that must match for the router to accept the hello. Routers continue to transmit hellos periodically, so the neighbor will receive the hello with the mismatched settings and remove the neighbor relationship. If these do not match, the OSPF routers will fail to form an adjacency and will be stuck in an INIT state.

But if a peer's settings match, the router adds it as an OSPF neighbor in the INIT state. It also adds the neighbor's router ID in the neighbor list included in its hellos. When an OSPF router receives a hello with its own ID in this list, it knows that it has bidirectional connectivity with the neighbor. It moves the neighbor to the 2WAY state.

OSPF point-to-point networks converge faster than OSPF broadcast networks and therefore are more preferred when applicable.

DR, BDR, and DROTHER

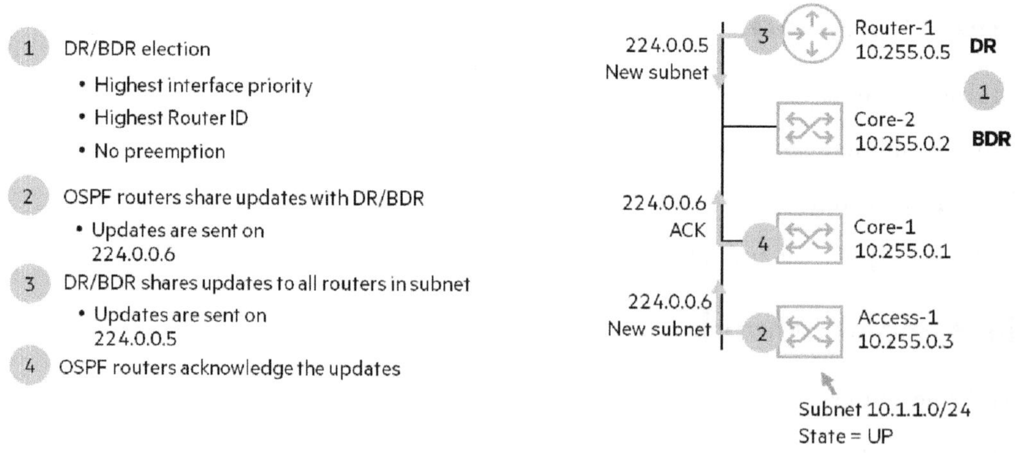

Figure 4-5: OSPF topology with DR and BDR

DR

- Achieves adjacency with all other routers on the subnet
- Advertises Type 2 (Network) LSAs
- Receives LSUs from other routers, sends its own LSUs
- Advertises LSUs on 224.0.0.5 (all routers)

BDR

- Achieves adjacency with all other routers on the network
- Receives LSUs from other routers, sends LSUs to the DR
- Takes over DR role if the DR fails

DROTHER

- Achieves adjacency with DR and BDR
- Has 2WAY relationship with other routers
- Sends LSUs to 224.0.0.6 (on which the DR and BDR listen)

An OSPF router will not form adjacencies to just any router. Instead, a client/server design is implemented in OSPF on each broadcast segment. For each multi-access broadcast segment, such as Ethernet, there is a Designated Router (DR) and a Backup Designated Router (BDR), as well as other OSPF routers, called DROTHERs. As an example, if you have 10 VLANs/subnets in your switched area, you will have 10 DRs and 10 BDRs. Remember, point-to-point links only have two routers, so there is no advantage or need for DR/BDRs.

The OSPF router with the highest priority becomes the DR for the segment. If there is a tie, the router with the highest router ID (not IP address on the segment) becomes the DR. By default, all routers have a priority of 1 (priorities can range from 0 to 255—it's an 8-bit value). If the DR fails, the BDR is promoted to DR and another router is elected as the BDR. If a router does not become a DR or BDR, it is referred to as a designated router-other, or commonly DROTHER, for short. There is no preemption of the DR/BDR role. If a DR/BDR exists and a router with a higher router ID joins the segment, it will not preempt the existing DR/BDR. Election only occurs if there is no DR or BDR on a segment.

This feature was specifically designed based on the hardware (or lack thereof) capabilities of routers in the 1980s. Today, this is a moot point, and you should not be concerned with which of your routers play which role.

When an OSPF router comes up, it forms adjacencies with the DR and the BDR on each multi-access segment to which it is connected; if it is connected to three segments, it will form three sets of adjacencies. Any exchange of routing information is between these DR/BDR routers and the other OSPF neighbors on a segment (and vice versa). An OSPF router talks to a DR using the IP multicast address of 224.0.0.6. The DR and the BDR talk to all OSPF routers using the 224.0.0.5 multicast IP address.

OSPF routers use Link State Advertisements (LSAs) to communicate with each other. One type of LSA is a hello, which is used to form neighbor relationships and as a keep-alive function. Hellos are generated every 10 seconds. When sharing link information (directly connected routes), links are sent to the DR (224.0.0.6), and the DR disseminates this to everyone else (224.0.0.5) on the segment. On point-to-point links, since no DR/BDR is used, all OSPF packets are addressed to 224.0.0.5.

Adjacency

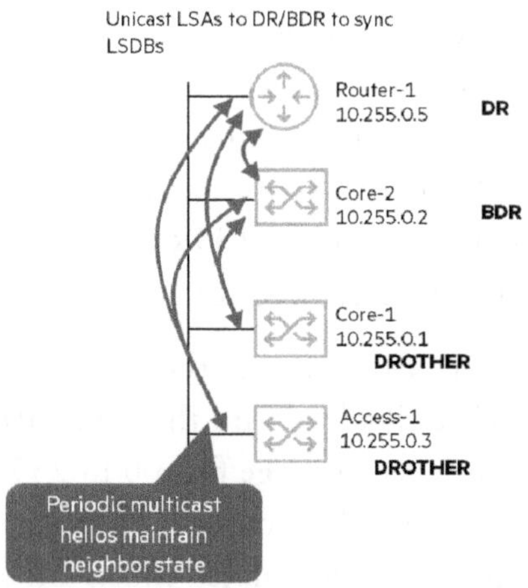

Figure 4-6: Showing OSPF topology with full adjacency

The DR and BDR form a full adjacency with all other routers. Or, you could say that all DROTHER routers from a full adjacency with the DR and BDR. DROTHER routes only reach the 2WAY state between each other.

To achieve full adjacency, OSPF neighbors send unicasts to describe their databases to each other and exchange LSAs until their Link State Databases (LSDBs) match. In this way, information about the complete topology spreads from neighbor to neighbor throughout the network.

OSPF routers must establish full adjacency with:

- The single neighbor on a point-to-point interface
- The DR and BDR of a broadcast or Non-Broadcast Multi-Access (NBMA) network

Periodic hellos (every 10 seconds, by default) maintain the 2WAY or FULL state. If the OSPF router misses four hellos from a neighbor, it moves the neighbor to DOWN. You will learn about how OSPF routers detect that neighbors are down in more detail later.

SPF algorithm

Within an OSPF area, the Type 1 and Type 2 LSAs give each router enough information to create a topology (sometimes called a directed graph) a bit like the one in the following figure. That is, they know which routers connect to other routers by the links that they share, and they know which routers offer gateways to which stub networks, which are networks advertised by only that router.

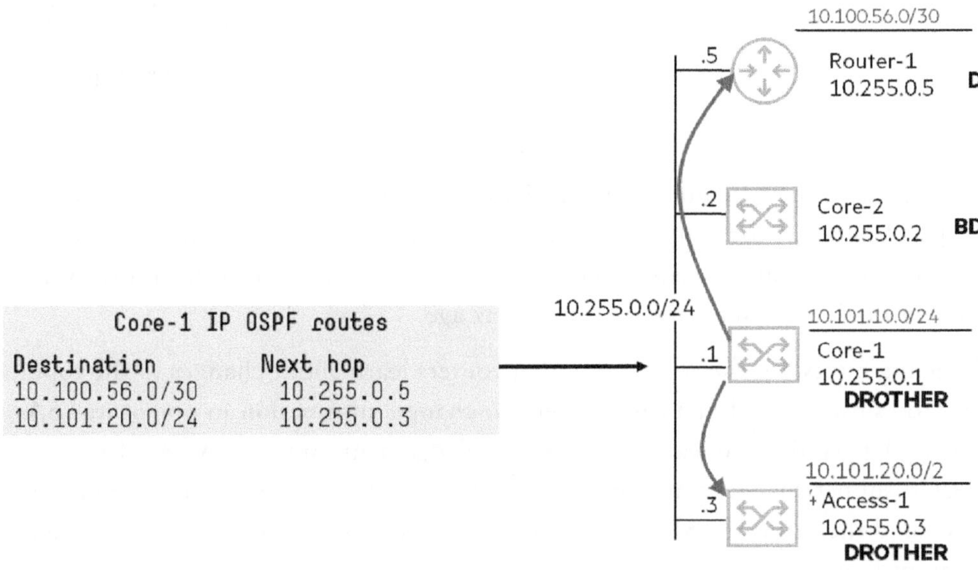

Figure 4-7: SFP algorithm

Each router then calculates the best route to each network—each advertised link—based on this topology. Specifically, they run the Shortest Path First (SPF), or Dijkstra's algorithm, which determines the lowest cost path between the router and the network. In **Figure 4-7**, Core-1 has learned that it has a route to 10.100.56.0/30 through next hop 10.255.0.5 This is the actual Router-1 IP address on the transit network, not the router ID, although they are the same in this instance.

Link State Update (LSU)

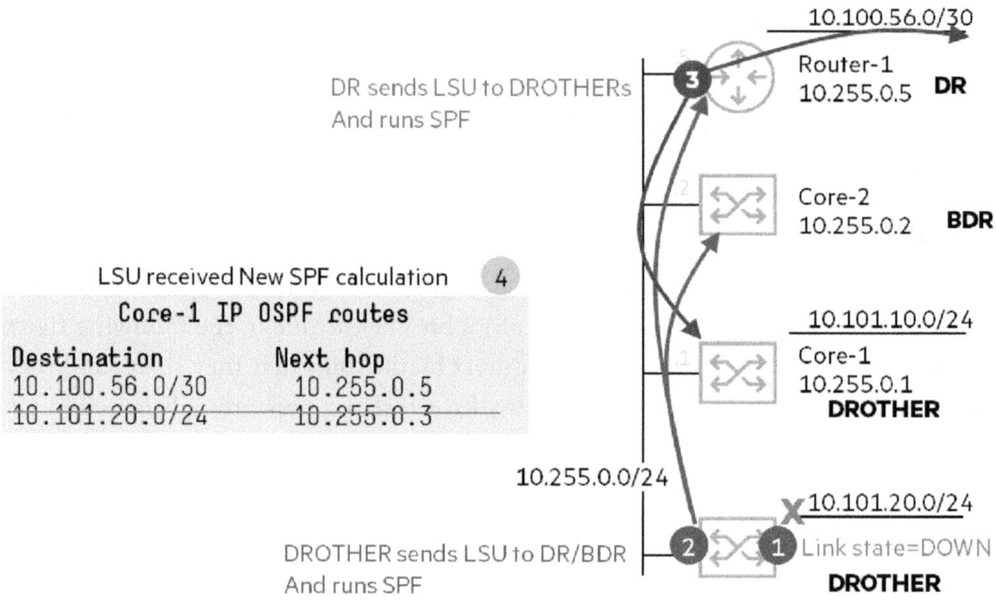

Figure 4-8: LSU

LSAs have an age, and they age out of the database when the LSA reaches the max age (standard 1800 seconds). Routers periodically readvertise the LSAs that they have generated to reset the age and maintain them. If a router needs to withdraw one of its LSAs, it can flush it from the system by readvertising the LSA with an age set to the max age.

But an LSA aging out is not the typical way that routers learn about changes to the topology; this would be far too slow. Instead, if a router needs to change information in one of its LSAs, it sends a Link State Update (LSU) and waits for an acknowledgement. Suppose Access-1's 10.101.20.0/24 subnet changes state from UP to DOWN. It sends this information in a Type 1 (router) LSU, and then runs the SPF algorithm to calculate its new best routes and then adjust its route table accordingly.

The DR receives this update and floods the LSU out all other interfaces, and to DROTHERs on the 10.255.0.0/24 subnet. DROTHERs like Core-1 receive this update, flood it out any other links they may be connected to, and run SPF to calculate their new best paths. In the previous figure, Core-1 has removed the 10.101.20.0/24 subnet from its route table. In a more complex and redundant topology—such as the ones you will explore later—the router might find a new next hop for the route.

Multi-area OSPF

Multi-area OSPF will be deployed with a two-tiered hierarchy (Area 0 backbone, and Area 1 and Area 2 are non-backbone areas), allowing for more efficient routing. The backbone area (Area 0) connects all other areas, providing a central point for inter-area routing. As the network grows, new areas can be added without affecting the entire network. This modular design supports scalability without requiring changes to the entire OSPF domain.

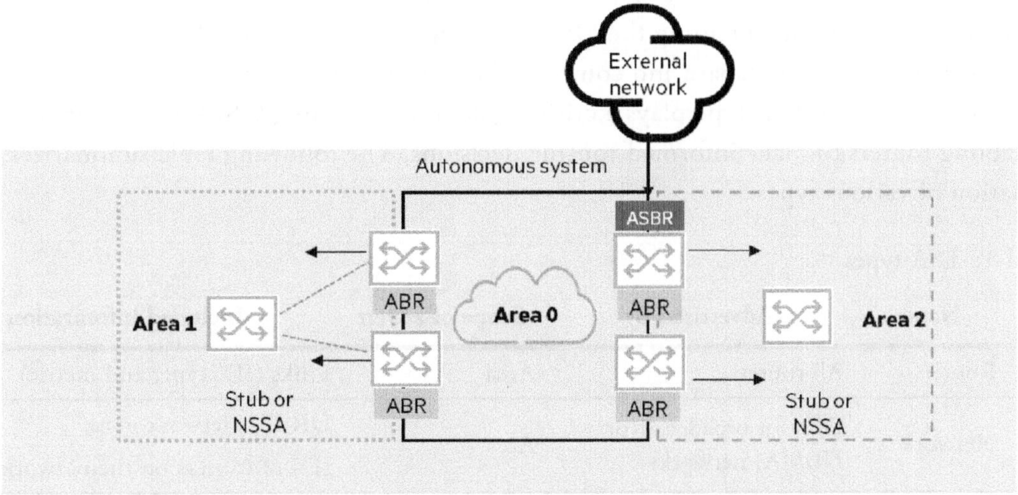

Figure 4-9: Multi-area OSPF with ABR and ASBR routers

Typically, you will have a backbone area (Area 0) connecting to other areas using an Area Boundary Router (ABR). An ABR connects OSPF areas, summarizes routes between areas, and maintains separate LSDBs, whereas an Autonomous System Border Router (ASBR) connects OSPF to external routing domains, redistributes external routes into OSPF, and advertises OSPF routes to external domains. In a multi-area OSPF network, these routers work together to ensure efficient routing within OSPF areas and connectivity to external networks.

You also need to define different types of areas, each serving specific functions within the OSPF routing domain. These area types are used to create a hierarchical structure within OSPF networks, allowing for more efficient routing and better network management. The backbone area is the core of the OSPF network. Every OSPF network must have an Area 0; it serves as the transit area for all other areas. Standard areas are all areas other than Area 0. These areas are connected to the backbone area.

Stub areas reduce the size of routing tables by limiting the amount of routing information received from the backbone area (Area 0). Totally stubby area and NSSA are extensions of a stub area that further reduce the size of routing tables by limiting external routing information to be imported into the OSPF network.

The choice of OSPF area types depends on network requirements, such as reducing routing table sizes, controlling external routes, and simplifying routing in certain areas. Each area type offers a different level of control and optimization for OSPF routing.

Link State Advertisements in multi-area OSPF

Link State Advertisements (LSAs) are used to exchange routing information between OSPF routers. There are 11 different types of LSAs available in OSPF. Each LSA type serves a specific purpose in building and maintaining the OSPF topology database. These LSAs work together to build the OSPF topology database and compute the shortest path tree for efficient routing within an OSPF domain. Each LSA type plays a critical role in maintaining OSPF's link-state database and enabling routers to make informed routing decisions. The following table summarizes key information of various types.

Table 4-1: LSA types

Type	Name	Advertised by	Scope of advert	Included infomration
1	Router	All routers	Area	Links (ID, type, and metric)
2	Network	DRs for broadcast (or NBMA) networks	Area	DR IP, Network mask IDs of routers on the network
3	Summary	ABRs	Area	Network in another area (network IP, mask, and metric)
4	ASBR	ABRs	Area	ASBR router ID and metric
5	AS External	ASBRs	All normal areas in the AS	External network, mask, metric, metric type
7	NSSA External	ASBRs	Area	External network, mask, metric, metric type
9	Opaque Link Local	All capable OSPF routers	Link	Extensions (graceful restart)
10	Opaque area-local	All capable OSPF routers	Area	Extensions (traffic engineering)
11	Opaque AS	Capable ASBRs	Non-stub areas	Extensions

Type 1 LSAs are router advertisements

This LSA carries the information of all directly connected routes of the advertising router, and it always stays within the area.

Type 2 LSAs are network advertisements

The DR generates this LSA on an NBMA segment and always stays within the area DR advertising its connected subnets.

Type 3 LSAs are route summary advertisements

These packets are generated by an Area Border Router (ABR) to summarize its directly connected area and advertise inter-area router information to other areas the ABR is connected to, with the use of a summary prefix. Type 3 LSA packets are flooded to multiple areas throughout the network and help with OSPF's scalability with the use of summary prefixes.

Type 4 LSAs are ASBR summary advertisements

These packets are generated by the ABR, and it advertises the presence of an Autonomous System Border Router (ASBR) to other areas. This LSA will not be used by the ASBR itself within its local area.

Type 5 LSAs are ASBR external route advertisements

These packets are generated by the ASBR to advertise external redistributed routes into the OSPF's AS.

Type 7 LSAs are Not-So-Stubby Area (NSSA) external route advertisements

These packets are generated by the ASBR to advertise the external distributed routes into NSSA areas. LSA Type 7 packets act as a mask for LSA Type 5 packets to allow them to move through NSSA areas and reach the ABR that is able to translate LSA Type 7 packets back to LSA Type 5 packets

Types 9, 10, and 11 are opaque LSAs

These are extended or opaque LSAs, which enable features such as OSPF graceful restart and traffic engineering. Routers only advertise opaque LSAs to neighbors that indicate that they are opaque capable during adjacency establishment. LSA Type 9 in OSPFv2 (IPv4) is defined as a link scope opaque LSA for carrying OSPF information. For OSPFv3, it is redefined to handle a communication prefix for the stub area. LSA Type 10 packets are used to flood OSPF information through other area routers. Even if these routers do not process this information in order to extend OSPF functionality, this LSA is used for traffic engineering to advertise MPLS and other protocols. LSA Type 11 packets serve the same purpose as LSA Type 10 packets but are not flooded into stub areas.

The following figure illustrates the OSPF routers that create each form of LSA. You see Type 1 and Type 2 LSAs inside an area. ABRs inject Type 3 summary LSAs into an area. ASBRs send Type 5 external LSAs, and sometimes Type 7 LSAs, while Type 4 LSAs are information about those ASBRs.

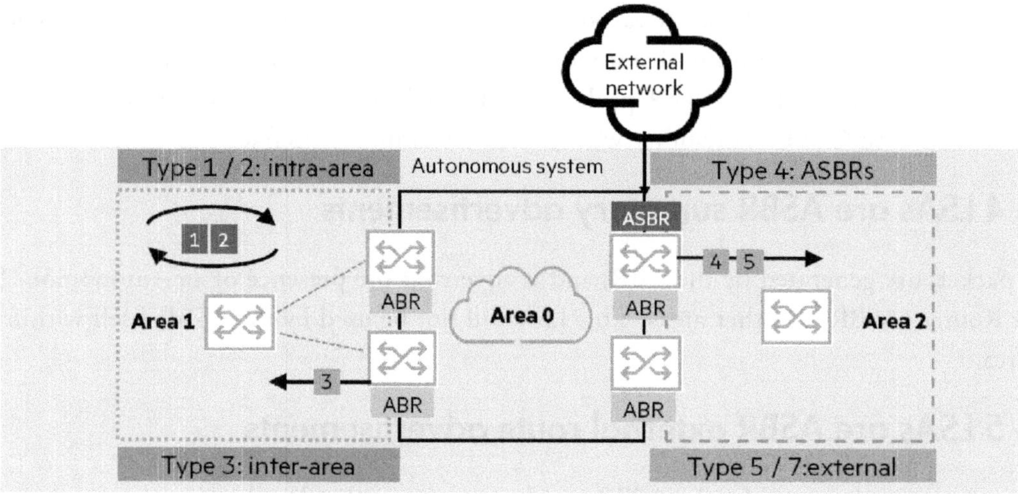

Figure 4-10: Routers that generate different LSA types

High scalability with multi-area OSPF

OSPF areas segment the domain into more stable and manageable pieces. Each OSPF routing device maintains the LSDB and runs the shortest path algorithm only for its own areas. The following figure shows how ABRs control the flooding of Type 1 and Type 2 LSAs into other areas.

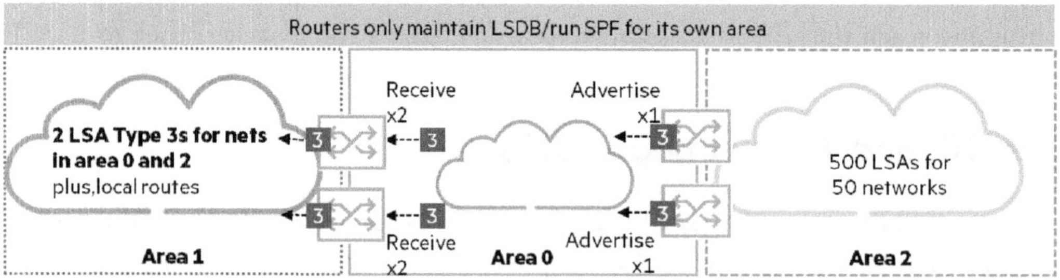

Figure 4-11: Multi-area OSPF

OSPF ABRs forward routing information between areas. Type 1 and Type 2 LSAs do not flow across area boundaries. ABRs summarize this huge number of Type 1 and Type 2 LSAs into a few Type 3 summary LSAs.

This decreases LSDB size and improves stability. Routers do not have to rerun the SPF algorithm as often, and when they do, it runs on a smaller LSDB. Understand that ABRs automatically summarize LSAs, the advertisements about routes. They do not automatically summarize the routing tables themselves. If you had 100,000 routes in your routing table with a single area design, you will still have 100,000 entries after using areas.

However, now that you have ABRs, you can configure route summarization, which can drastically reduce your routing table sizes. Routers in Area 1 may only have the a dozen or hundred routes in that area, plus a default route (0.0.0.0) to reach all other subnets via the ABRs. These benefits still apply even for powerful routers that could support very large areas.

This is especially true as relates to network stability. With ABRs and route summarization, you have effectively segmented your network into smaller failure domains. A flapping route in Area 1 has absolutely no effect on Area 0 or Area 2. This can be a big boost in network stability and ease of troubleshooting.

Area Border Router (ABR)

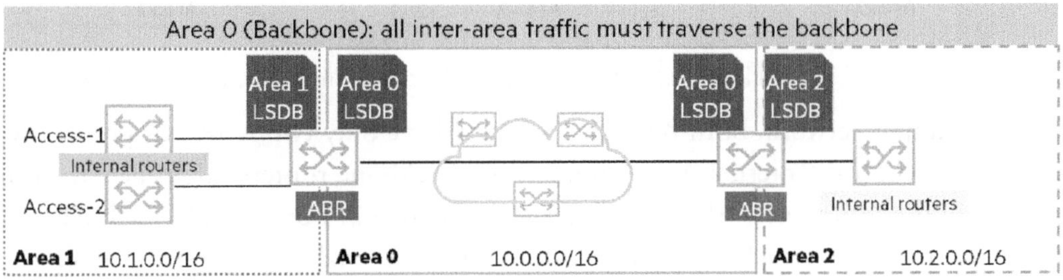

Figure 4-12: Multi-area OSPF topology with ABRs

As show in **Figure 4-12**, ABRs receive Type 1 and Type 2 LSAs from one area, summarize them into a Type 3 summary LSA, and inject them into another area. Area 0 has a special role in multi-area networks—the backbone. All areas must connect directly to the Area 0 backbone, and all inter-area traffic must traverse the backbone. In the figure, Area 1 reaches Area 2 via the backbone, and vice versa, always through an ABR. Each ABR must have at least one interface in Area 0, and at least one interface in some other area. So, ABRs have interfaces in at least two areas, one of which is Area 0. OSPF design requires all areas to connect to this backbone.

Each ABR maintains one LSDB per connected area. Think about it: All those Type 1 and Type 2 LSAs are synchronized among all routers in Area 1, so all Area 1 routers have an identical LSDB. This is true for all areas. These Type 1 and Type 2 routes are summarized into Type 3 LSAs and injected into the other areas. So, Area 2 does not have any topology information for Area 1 (connectivity detail), but Area 2 knows about all Area 1 networks. It simply knows that it can reach these subnets via any connected ABR.

In the previous example, Core-1 is an ABR. It synchronizes Type 1 and Type 2 LSAs with other OSPF routing devices in Area 1. The figure shows just two other routing devices, Access-1 and Access-2, for simplicity. In the real world, the area could have dozens or hundreds of routers. Access-1 and Access-2 reside entirely within Area 1 and are called internal routers. That connected ABR, Access-1, and Access-2 have identical Area 1 LSDBs.

LSA aggregation between areas (summarization)

Simply dividing an AS into multiple areas begins to decrease the size of OSPF routers' LSDBs and to reduce the complexity of SPF calculations. However, to achieve the true benefits of multiple areas, you should further reduce the number of LSAs advertised between areas.

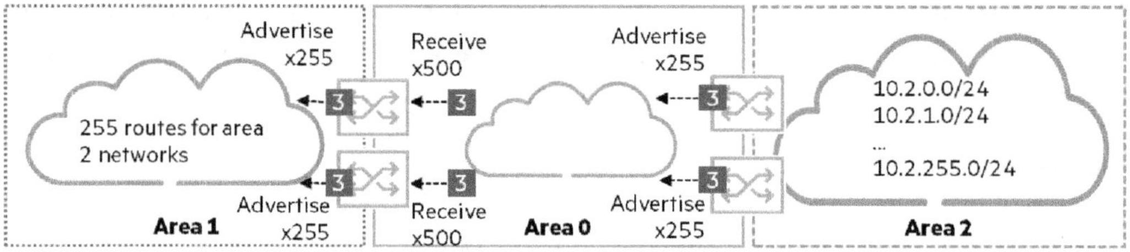

Figure 4-13: Multi-area OSPF topology with route summarization on ABRs

In the above example network, both Area 1 and Area 2 have only two ABR exit points toward the backbone and other areas. Internal Area 1 routers and backbone routers might benefit from some level of detail in the information that they receive about Area 2. For example, each ABR's Type 3 LSAs reflect the metrics of reaching the network in question, enabling routers in other areas to select the best ABR for reaching each network in Area 2. However, with only two entry points into Area 2, routers (and, in particular, internal Area 1 routers) are probably receiving too much granularity about each subnet. This information clutters the routers' LSDB and routing tables.

In addition, small changes in the Area 2 topology could flood LSUs through the backbone and Area 1, causing routers to rerun the SPF algorithm. For example, network 10.2.40.0/24 in Area 2 becomes unavailable. The two Area 2 ABRs withdraw their Type 3 LSA for that specific network in Area 0, and then the two Area 1 ABRs do the same in Area 1. However, the Area 1 routers have no other path to this network. It might be best to simply not inform routers beyond Area 2 of the small change. If they continue to forward traffic for this network, the Area 2 ABRs simply drop it.

You should aggregate these Type 3 LSAs to hide the unnecessary information and to smooth over small changes, thereby gaining the true benefits of a multiple area OSPF configuration.

This problem can be resolved by using route summarization on the ABR; the solution is explained in the following example.

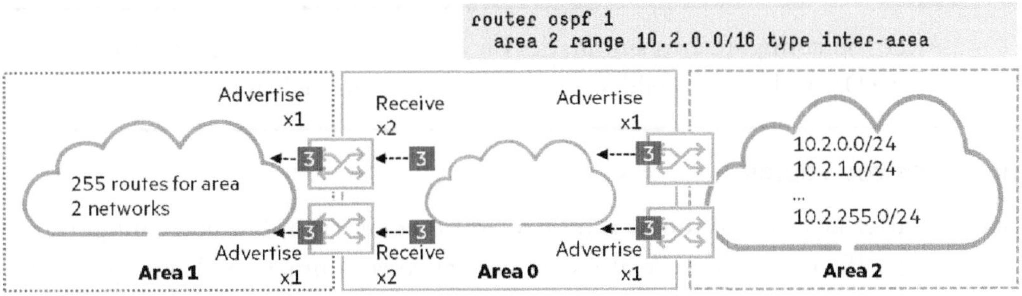

Figure 4-14: The `route summarization` command

You can configure an ABR to aggregate its summary routes. Rather than generate a Type 3 LSA for every network in an area, it generates a few Type 3 LSAs, each of which includes many networks in its range. In other words, the Type 3 LSA advertises a network with a shorter prefix length, such as a /20 or a /16 subnet that includes many /24 subnets. You configure area summaries on the ABR that connects to the area to be aggregated. In the figure, you configure area summaries for Area 2 on the two ABRs connected to Area 2. These ABRs then advertise the summaries to other backbone routers and ABRs in Area 0. The other ABRs can then copy them into their own areas. In this example, the aggregated summary 10.2.0.0/16 shows up as a route on ABR A and on internal routing switches in Area 1.

Similarly, you could configure the ABRs connected to Area 1 to summarize Area 1 routes and inject them into Area 0. All ABRs connect to Area 0, so you can configure them all to summarize Area 0 routes and inject them into another area. You see the drastic reduction in routes: Before summarization, Area 1 had 255 entries for Area 2. Now it has two entries.

Figure 4-14 shows the syntax to configure this aggregation. When you enter the `area <ID> range` command, the ABR automatically suppresses the generation of individual Type 3 LSAs for subnets within the specified network. However, it continues to advertise individual Type 3 LSAs for networks that do not fall into the range. Thus, it is best to assign contiguous IP addresses within an area to ensure easy route aggregation.

The ABRs for Area 1 now receive one Type 3 LSA with the aggregated summary from each of the two ABRs that connect to Area 2. Each Area 1 ABR selects the best LSA for the summary range and creates an inter-area route to 10.2.0.0/16 through the best next hop.

Each Area 1 ABR then generates an Area 1 Type 3 LSA to advertise this inter-area route to routers in Area 1. The internal Area 1 routers now have two LSAs for all Area 2 networks, and they learn routes to only the one aggregated network (up to two routes if the ABRs advertise equal metrics and the internal routers support Equal Cost Multipath [ECMP]).

After you configure area summaries on an area's ABR, you can see the summaries in the LSDB and routing tables for routers not connected to that area.

Filtering inter-area routes

Note that the ABRs advertise all routes from one area to another, unless you configure an area summary for a specific range. You can also configure an area summary with the special no-advertise option, which prevents networks from being advertised outside of their local area.

Suppose your company has a strict security policy for one of the sites in OSPF Area 2. The policies allow only local access to high-security subnets 10.2.128.0/24 through 10.2.255.0/24 (10.2.128/17). Devices in other areas should still be able to reach 10.2.0.0/24 through 10.2.127.0/24 (10.2.0.0/17). Of course, a firewall can filter out prohibited traffic, but in this case, you want to prevent any routing of traffic from other areas to the restricted networks.

As shown in the following example, you need to configure two area summaries on the Area 2 ABRs, one for 10.2.0.0/17 and one for 10.2.128.0/17. ABRs advertise the first but not the second, as shown in the figure. If you wanted to re-enable advertising the filtered subnets later, you could simply re-enter the command, but without the no-advertise option. Simply removing the non-advertised summary would enable the advertisement of individual routes without aggregation.

Note that the filtering occurs at the Area 2 boundary; you cannot apply the filtering to some areas and not others. For example, you cannot use this configuration to block inter-area routes to the Area 2 subnets in Area 1 but allow them in Area 0.

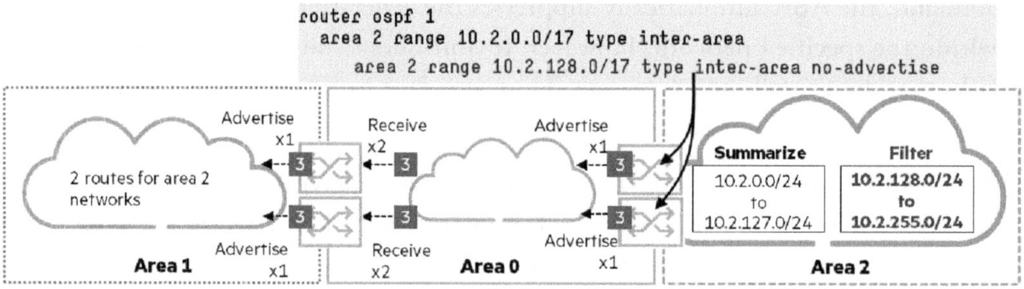

Figure 4-15: Inter-area route filtering

OSPF interface cost

OSPF path selection is based on link cost, which is based on the reference bandwidth, which defaults to 100,000. This is shown in **Table 4-2**, with the output of the `show ip ospf` command.

AOS-CX devices divide this reference bandwidth by interface link speed in Mbps to determine the cost. So, the default reference of 100,000 Mbps / 1000 Mbps (1 Gbps) = 100. This is proven in the output of the `show ip ospf interface` command.

Table 4-2: Default reference bandwidth

Bandwidth	Cost
1 Tbps	1
100 Gbps	1
20 Gbps	5
10 Gbps	10
4 Gbps	25
2 Gbps	50
1 Gbps	100

One problem with this default is shown in this table. The default reference of 100,000 Mbps / 100,000 Mbps (100 Gbps) = 1. OSPF cannot use fractions, and so all interfaces faster than 100 Gbps will also have a cost of 1. Thus, OSPF will perceive a 100 Gbps path and a 1 Tbps path as being equal cost. If you have no links faster than 100 Gbps, this is not an issue.

If you DO have links faster than 100 Gbps, then you should change the reference bandwidth to a higher value. In the previous table, you have changed the reference bandwidth to 1,000,000. Thus, 1 Tbps links have a cost of 1, while 100 Gbps links have a cost of 10, and so on, as shown in the table. OSPF can now distinguish between these fast links.

Another issue is that by default, AOS-CX VLAN interfaces have a link speed of 1000 Mbps (1 Gbps), and so have a cost of 100. Because every VLAN interface has the same cost, route selection becomes a matter of hop count, rather than true lowest cost path, if you are using Layer 3 VLAN interfaces. If you modify the reference bandwidth to 1,000,000 as shown, then all VLAN interfaces would have a cost of 1000.

Changing the reference bandwidth does not mitigate the issue of VLAN interfaces having the same cost. If this is an issue, you can change the individual cost values on a per-interface basis:

```
interface vlan41
ip ospf cost <1-65535>
```

You can increase the reference bandwidth by using the following command syntax:

```
router ospf 1
reference-bandwidth 1000000
```

Table 4-3: Changed reference bandwidths

Bandwidth	Cost
1 Tbps	1
100 Gbps	10
20 Gbps	50
10 Gbps	100
4 Gbps	250
2 Gbps	500
1 Gbps	1000

As you can see in **Table 4-2**, you can set each individual interface to an OSPF cost value of your choice—any value between 1 and 65,535 inclusive. Preferably, you want to set the interface costs globally for the OSPF process by changing the bandwidth reference with the following configuration:

```
switch(config)# router ospf <PROCESS_ID>
switch(config-ospf-1)# reference-bandwidth <MBPS>
```

If the OSPFv2 interface cost is not explicitly set, then the cost of all the OSPFv2 interfaces is recalculated based on the reference bandwidth and link speed of the interface. For VLAN interfaces, the link speed value is taken as 1 Gbps if the OSPFv2 interface cost is not explicitly set. The bandwidth range is 1-4,000,000 Mbps, where the default is 100,000 Mbps.

You can use this command to set the cost on a Layer 3 interface, which will override the reference calculation:

```
Switch(config)# interface <INTERFACE_ID>
Switch(config-if)# ip ospf cost <COST>
```

When this command is applied, the router no longer considers the cost formula; it simply uses the manual value that was entered.

Route redistribution using ASBR

As a link state protocol, OSPF routers advertise network interfaces that run OSPF. But sometimes an OSPF router has other routes in its IP routing table, which must be advertised to OSPF neighbors.

Perhaps one or more routers connect to an ISP with static default or BGP routes.

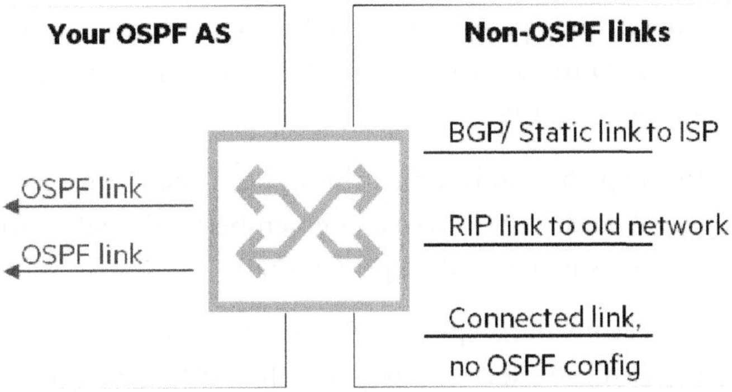

Figure 4-16: OSPF router connected with external network

You must inject these routes into your OSPF domain. Or, suppose you are migrating your network from RIP to OSPF. During the upgrade, some segments run RIP. You must inject these routes into the already-migrated OSPF AS, and the old RIP segments must know about the OSPF networks.

For security and efficiency reasons, you might choose not to enable OSPF on a stub network. So you redistribute connected routes into OSPF to be advertised to the rest of your AS. Use route redistribution to connect OSPF to any foreign routing protocol.

Autonomous System Border Router

The router that redistributes external routes is an Autonomous System Border Router (ASBR) because it communicates routing information between the OSPF AS and other routing processes. A redistributed route becomes an external OSPF route, sometimes referred to as an AS external route. All routes calculated from Type 1, Type 2, and Type 3 LSAs, including both intra-area and inter-area routes, are internal OSPF routes.

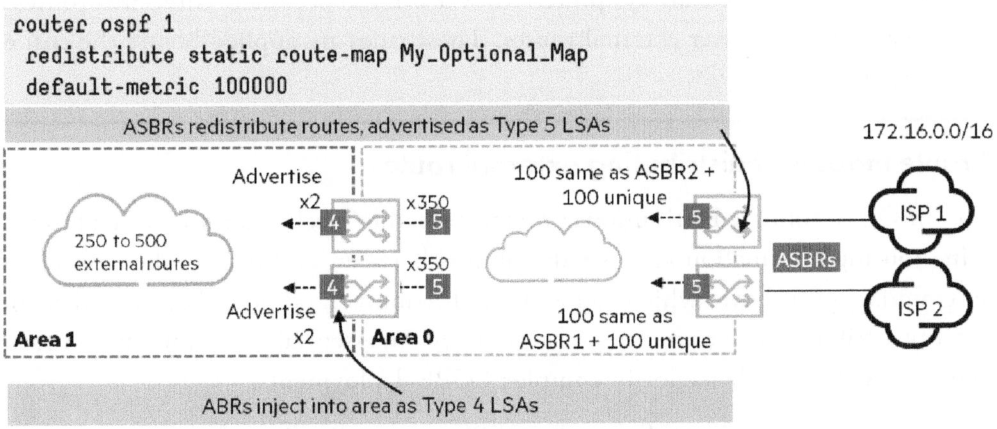

Figure 4-17: Multiarea OSPF topology with ASBR

ASBRs advertise external routes in Type 5 LSAs. The LSA includes the network address for the redistributed route, the router ID of the advertising ASBR, and the metric, which you assign when you redistribute the route into OSPF.

OSPF routers propagate the Type 5 LSA in LSUs throughout the entire AS, making Type 5 LSAs different from Type 1, Type 2, and Type 3 LSAs. (Remember: although Type 3 LSAs advertise inter-area routes, each area has its own set of Type 3 LSAs.)

To calculate an external route from the Type 5 LSA, a router refers to the LSA's advertising ASBR. The router has calculated the shortest path to each ASBR. The forwarding interface and next hop for that path become the forwarding interface and next hop for all external routes advertised by that ASBR.

In other words, routers do not add the external networks to the directed graph for the shortest path algorithm. Instead, they add them to a list of networks for which the advertising ASBR is the gateway. Therefore, changes to the particular external networks advertised by an ASBR do not affect the area topology.

Figure 4-17 shows a configuration syntax for ASBRs to redistribute static routes into OSPF. You could also specify connected (directly connected links not running OSPF) or BGP. The route map syntax is optional.

By default, all routes of the specified type are redistributed unless you employ route maps. When you redistribute routes into OSPF, they are assigned a seed metric to them: a beginning cost value. If you specify a default metric with the `default-metric` command, it would be applied to redistributed routes; if you want to have different metrics for different external routes, then use a route map to accomplish this task.

Intra-area routes are always preferred over inter-area routes. Inter-area routes are always preferred over external routes. These rules are applied before the cost evaluation.

Need of route maps in redistributing external routes

Route maps, in OSPF route redistribution, provide a flexible mechanism for controlling and manipulating routing information as it is redistributed between OSPF and other routing domains. They offer granular control over which routes are redistributed and how they are redistributed, and can modify attributes such as metrics or tags. They are essential for maintaining routing integrity and optimizing network behavior in complex OSPF deployments.

Route maps offer the following benefits to the OSPF route redistribution:

- Selective Redistribution:

  Route maps allow you to selectively redistribute routes into OSPF based on specific criteria such as prefix, prefix length, or route tags. For example, you can configure a route map to redistribute only routes from a specific routing protocol (like BGP or EIGRP) into OSPF, while filtering out others.

- Filtering Routes:

  Route maps can filter routes based on various criteria such as prefix, prefix length, or route tags. This filtering capability allows you to prevent certain routes from being redistributed into OSPF or to control the types of routes that are redistributed.

- Modifying Route Attributes:

  Route maps enable you to modify attributes of redistributed routes before they are injected into OSPF. You can adjust metrics, set administrative distances, or apply route tags to redistributed routes, influencing how OSPF routers evaluate and use these routes.

- Policy Enforcement:

  Route maps provide a means to enforce routing policies during route redistribution. For example, you can use a route map to prioritize certain routes over others or to implement route filtering policies based on organizational requirements.

By default, when you redistribute a certain type of route into OSPF, the AOS-CX switch redistributes all routes of that type. However, you can filter the routes that are redistributed at the ASBR by applying a route map to the `redistribute` command, as shown in the following figure.

```
ip prefix-list PL_Net10 seq 10 permit 172.16.0.0/16
route-map RM_Net10 permit seq 10
 match ip address prefix-list PL_Net10
router ospf 1
  router-id 1.1.4.1
  area 0
  redistribute static route-map RM_Net10
```

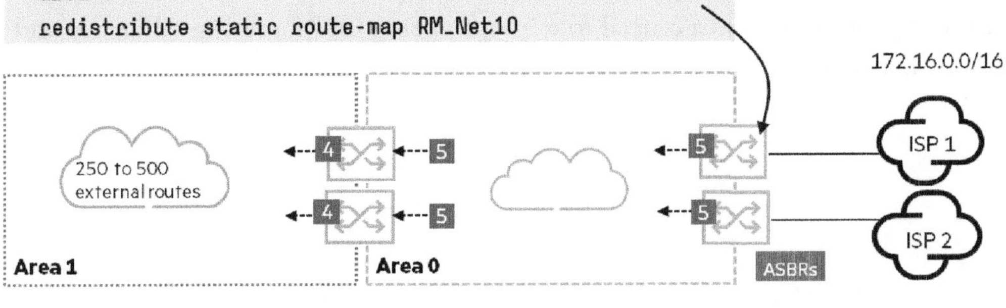

Figure 4-18: Use of route maps

The route map refers to an IP prefix list that specifies the networks to be redistributed into OSPF. In this example, the ASBR has several static routes, but you only want to redistribute 10.10.0.0/16 into OSPF. Optionally, you can use various set parameters with match criteria to modify how routes are advertised. You will learn more about this later in this chapter.

Verifying external routes

```
ip prefix-list PL_Net10 seq 10 permit 172.16.0.0/16
route-map RM_Net10 permit seq 10
 match ip address prefix-list PL_Net10
router ospf 1
  router-id 1.1.4.1
  area 0
  redistribute static route-map RM_Net10

ASBR# show ip ospf lsdb external
OSPF Router,ID (1.1.4.1) (Process ID 1 VRF default)
=============================================
AS External Link State Advertisements
-------------------------------------
LSID          ADV Router    Age  Seq#        Checksum
----------------------------------------------------
172.16.0.01.1.4.1      305  0x80000002 0x0000adef
```

```
ASBR# show ip route
..ipv4 routes for forwarding
'[x/y]'denotes [distance/metric]
10.8.0.0/16, vrf default
via  10.1.100.1,  [1/0], static
10.9.0.0/16, vrf default
via  10.1.100.1,  [1/0], static
172.16.0.0/16, vrf default
via  10.1.100.1,  [1/0], static
```

Figure 4-19: Verifying external routes

Figure 4-19 shows the ASBR routing table. Notice that the last static route matches the prefix list, as highlighted in the figure. The figure also shows the LSDB entry for the redistributed route. Note that on the ASBR, the local static route, because its administrative distance is 1, is placed in its local routing table. However, other adjacent OSPF routers will receive the LSBD redistributed route and place that in their routing table if it happens to have the best distance and metric.

Route summarization and summary ASBR advertisements

Route summarization on the ASBR in OSPF provides benefits such as reduced routing table size, controlled advertisement scope, improved convergence, optimized SPF calculations, and mitigation of routing loops. It plays a crucial role in optimizing OSPF routing efficiency and scalability in large networks.

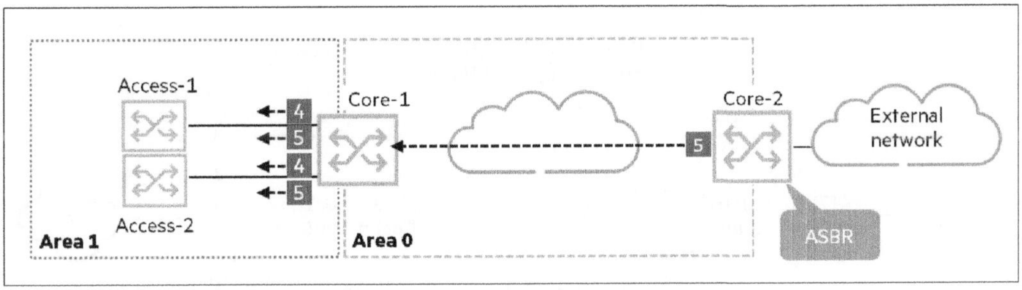

Figure 4-20: Summary ASBR advertisement

When a router receives a Type 5 LSA from an ASBR in the same area, it can use the area's topology to calculate the shortest path to that ASBR: the ASBR route. The next hop in this route becomes the next hop for all external networks for which that ASBR is the gateway. Routers in different areas from the ASBR also receive the Type 5 LSA, but they need help determining how to reach the ASBR: Type 4 LSAs are used to denote the ASBRs themselves. The Type 5 LSA indicates the advertising ASBR by router ID. You have learned how to make router IDs routable. However, because OSPF does not require routable router IDs, it must provide another mechanism for informing routers in other areas how to reach the ASBR.

Also, even if OSPF did require routable router IDs, an OSPF router in another area might not want to use its typical inter-area routes to reach the ASBR. In areas with multiple ABRs, the ABR with the lowest cost for the broader area range might not provide the best path to the ASBR within the area.

OSPF defines the Type 4 summary LSA so that the routers in other areas can learn the best ABR through which to reach particular ASBRs. For each area, an ABR generates one Type 4 LSA for each ASBR that it can reach in another area. As the metric for this LSA, the ABR specifies the cost for its shortest path to the ASBR. For example, in **Figure 4-20**, Core-1 would specify its cost for reaching Core-2. Internal routers that receive the Type 4 LSAs use them to select the best path to each ASBR in other areas. In this example, the choice is simple because Area 1 has only one ABR.

Figure 4-20 shows an external OSPF route learned by an internal Area 1 router from the ASBR in Area 0. You can see both the Type 4 LSAs (about the ASBR) and the Type 5 LSAs (about routes advertised by the ASBR).

OSPF default metric for the external routes

Consider how OSPF routers choose the best external route in networks with redundant ASBRs, as well as redundant paths to the ASBRs. Multiple ASBRs might send Type 5 LSAs for the same network. The Type 5 LSA includes a metric and metric type, helping other routers to choose the best ASBR for reaching that network.

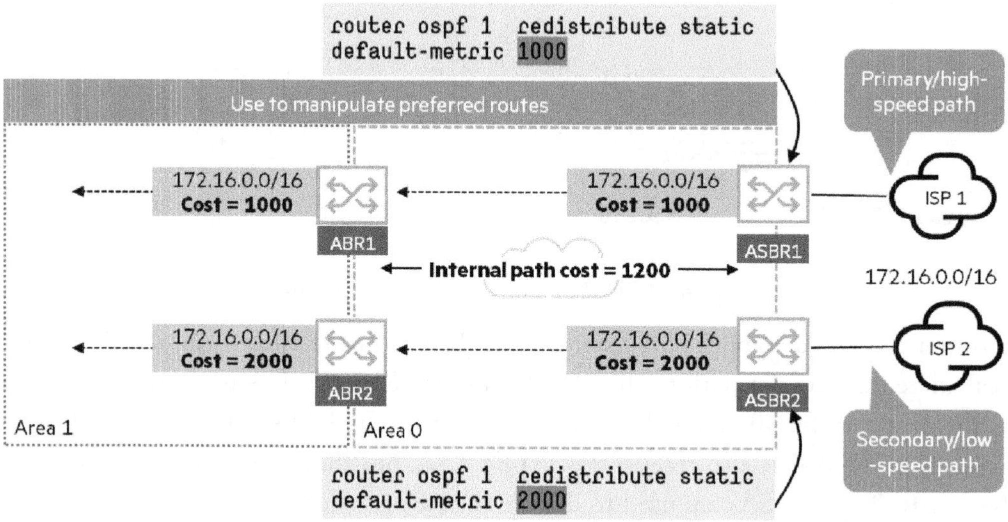

Figure 4-21: External routes with default metric

The ASBR that advertises the Type 5 LSA chooses the metric type and sets the default metric, often called the seed cost or seed metric. In **Figure 4-21**, ASBR1 is configured with a seed metric of 1000, while ASBR2 has a seed metric of 2000. Thus, all routers will prefer to use ASBR1 to reach external networks, only using ASBR2 if the path via ASBR1 fails. Perhaps ISP 1 is your primary, high-speed link path to 172.16.0.0/15 subnets, while ISP 2 is a lower-cost, slower link, only to be used in emergency/outage situations. Or perhaps you know that ISP 1 is simply a better, faster path, with lower latency and a better Service Level Agreement (SLA). You could also use this technique for load sharing, where ISP 1 is the primary path for 172.16.0.0/16 subnets, while ISP 2 is the primary path for 172.17.0.0/16 subnets.

Did you notice something strange about the example? ASBR1 and ASBR2 advertise their default or seed metric of 1000 and 2000 respectively. This advertisement traverses Area 0 to reach ABR1 and ABR2, with a cost of 1200. However, ABR1 and ABR2 did not add this cost before they advertised routes toward Area 1 internal routers! They still advertise the same metric advertised by the ASBRs, without considering their cost to reach the ASBRs. Why is that? The answer relates to how ASBRs inject external routes into OSPF as either External Type 1 (E1) or External Type 2 (E2) routes.

E1 and E2 types of external routes

E1 type external routes:

E1 routes in OSPF are external routes redistributed into OSPF with the cost calculated as the sum of the OSPF internal cost to reach the ASBR and the cost advertised with the external route. They offer granular control over routing decisions based on the internal OSPF topology and are preferred when you want to consider internal OSPF costs in routing decisions.

E2 type external routes:

E2 type external routes in OSPF are redistributed external routes with a fixed metric throughout the OSPF domain. They provide a consistent routing metric for external destinations within the OSPF domain, making them useful for scenarios where a fixed-cost path is preferred or where the internal OSPF topology does not significantly influence route selection.

External Type 2 is the default on AOS-CX and most other devices. However, if multiple LSAs tie for the lowest value, the metric for reaching the ASBR is a tie breaker. If a tie occurs again, the router uses ECMP. So, every external route advertised in a Type 5 LSA has two important properties: the metric type, which is an "external" Type 1 or Type 2, and the metric.

By default, OSPF prefers routes based on the following order:

1. Internal OSPF routes
2. External Type 1 routes
3. External Type 2 routes

Let us look at how a typical, modern network might be deployed.

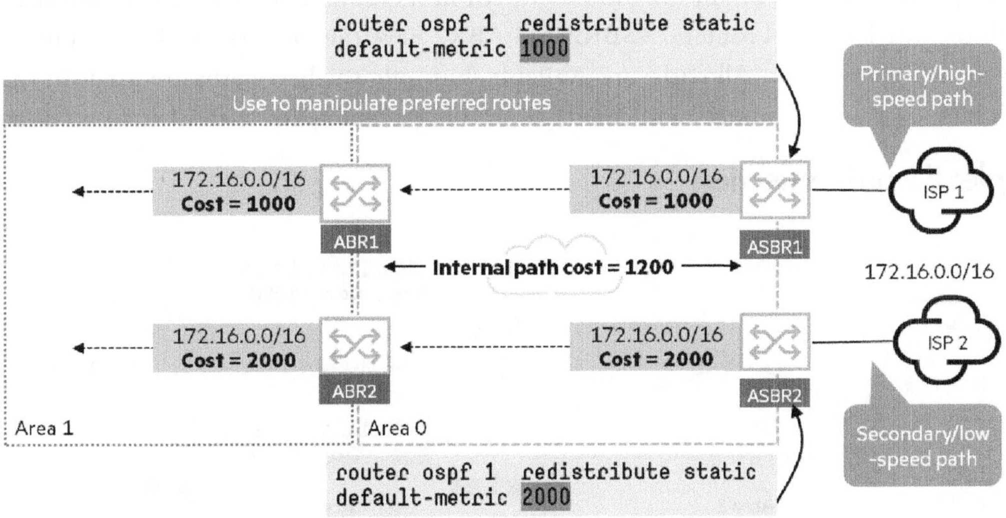

Figure 4-22: E1 and E2 type routes

Typical deployment of E1 type routes

Figure 4-23 shows a common deployment scenario, where both ASBRs inject external routes as type E1, with the same default metric.

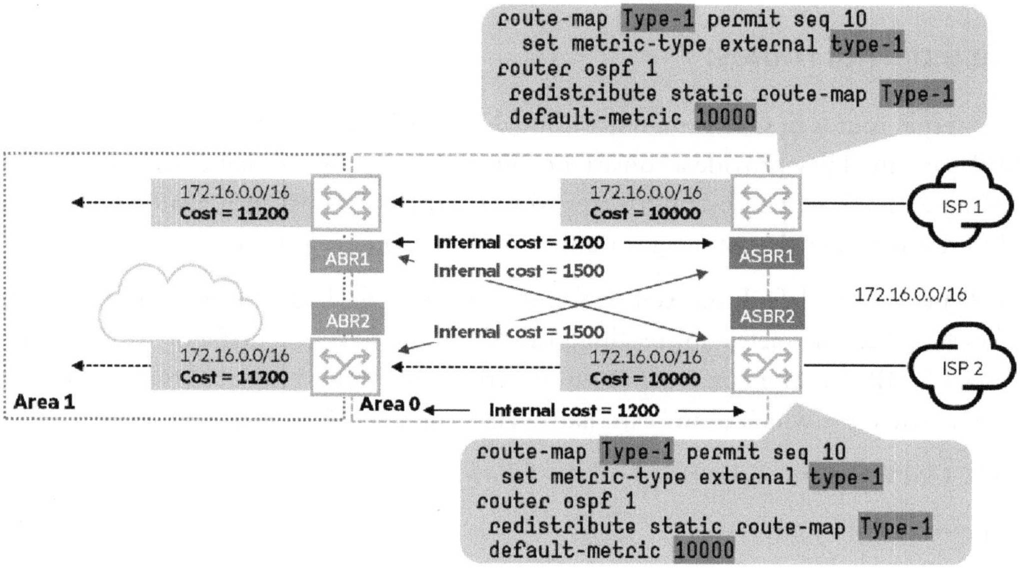

Figure 4-23: Deployment of E1 type routes

ABR1 has an internal cost of 1200 to reach ASBR1, and 1500 to reach ASBR2. Thus, total cost = 11,200 to reach 172.16.0.0/16 via ASBR1, and 11,500 to reach those subnets via ASBR2. It maintains both paths in its LSDB, but only the best path in its route table. ABR1 then advertises this best path to Area 1 internal routers. ASBR2 and ABR2 have the same story. This is a nice solution for many deployments. All paths are available, but only the best paths are used. In case of failure anywhere in the path, routers will converge on the alternate path with little to no downtime.

Example: E1 path selection

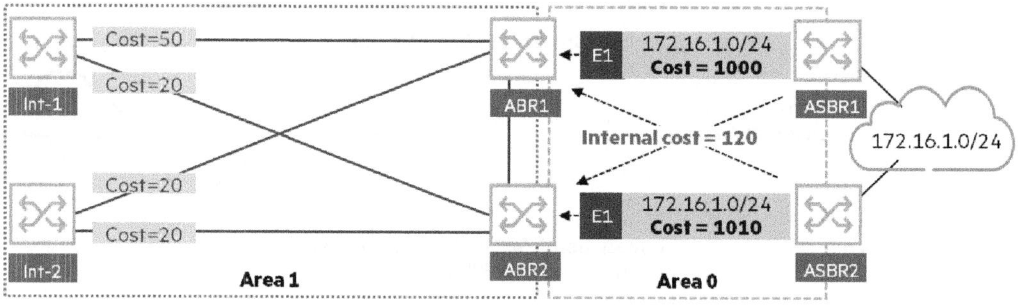

Figure 4-24: E1 type external routes

Figure 4-24 shows a scenario, which you can use to solidify your understanding of how an OSPF router chooses between Type 1 external routes to the same network.

Recall these facts:

- The router uses the Type 5 LSA to calculate the external route, setting the correct forwarding interface and next hop for the path to the advertising ASBR.
- The router sets the external route's metric equal to the metric advertised in the Type 5 LSA plus the metric for the forwarding path to the advertising ASBR.
- When the OSPF routing table already includes an external (Type 2) OSPF route to this network, the router selects the route with the lower metric.
- If both routes have the same metric, the router installs multiple routes in the OSPF routing table. Otherwise, it uses a mechanism such as router ID to break the tie.

Now study the figure. All routers support ECMP. Which ASBR does internal router Int-1 select to reach the external 172.16.1.0/24?

Solution

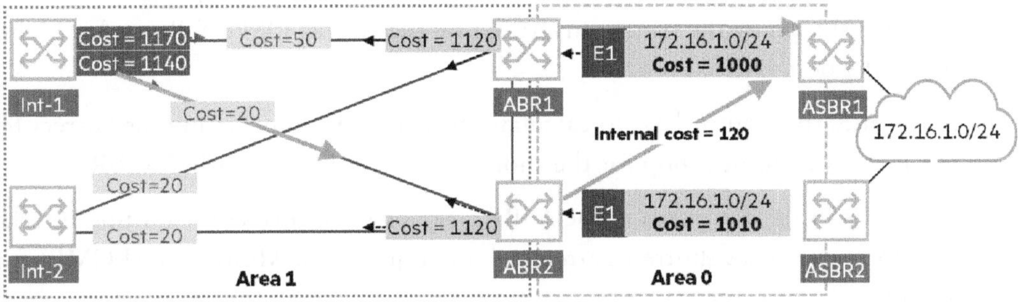

Figure 4-25: Best path selection with the E1 routes

Which ASBR does internal router Int-1 select to reach the external 172.16.1.0/24?

Figure 4-25 shows the answer to this question. To figure it out, you must know that ABR1 learns of 172.16.1.0/24 from ASBR1, with a cost of 1000; and from ASBR2 with cost of 1010. The cost to reach both ASBRs is 120, so actual cost for ABR1 is 1120 via ASBR1 and 1130 via ASBR2. For ASBR1, the path via ASBR1 is best. The same holds true for ABR2—it learns of 172.16.1.0/24 from both ASBRs, with ASBR1 being the lowest-cost path.

This is advertised into Area 1. So, Int-1 learns of 172.16.1.0/24 via both ABR1 and ABR2, both with a best path of 1120. However, the cost to reach ABR1 is 50, for a total cost of 1170, while the cost to reach ABR2 is 20, for a total cost of 1140. This is Int-1's best path.

Example: E2 type path selection

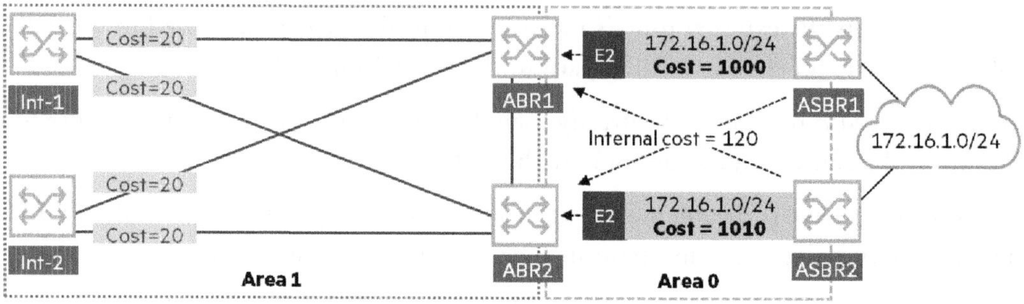

Figure 4-26: E2 type external routes

Figure 4-26 shows a scenario where ASBRs advertise Type E2 routes. Keep the following in mind.

Type 5 LSAs use type E1 and type E2 metrics. When routers examine a type E2 metric, which is the default, they consider only the metric advertised by the ASBR. OSPF follows this high-level process to choose the best external route:

- The router uses the Type 5 LSA to calculate the external route, setting the correct forwarding interface and next hop for the shortest path to the advertising ASBR.

  If the router is in the same area as the ASBR, it knows the correct forwarding interface and next hop from the area's shortest path topology. Equal-Cost Multi-Path (ECMP) routing might apply.

  - If the router is in a different area from the ASBR, it calculates a metric for the ASBR route by incrementing the metric in the Type 4 LSA with the metric for reaching the advertising ABR. If multiple ABRs advertise a Type 4 LSA to the same ASBR, the router chooses the ASBR route with the lower total cost.
  - A router might have multiple equal cost paths to an ASBR. Multiple ABRs might offer the router an equal-cost path to the ASBR, and the router has a different next hop for each ABR. Or the router might have multiple equal-cost paths to one ABR.
  - The best path might not match the IP route to the ASBR's IP address in the routing table. For example, the ABRs in the figure above advertise the same cost to the ASBRs. However, in Figure 6-26, router A and router B might advertise the same cost for the Area 0 summary. However, ABR-B advertises a better Type 4 LSA for ASBR-E. Therefore, the internal Area 1 routers will use their best path to router B to calculate the path to the ASBR, rather than load-balance over the link to both ABRs.

- The router sets the external route's metric equal to the metric advertised in the Type 5 LSA. The advertising ASBR added this metric when it redistributed the route into OSPF.
- When the OSPF routing table already includes an external (Type 2) OSPF route to this network, the router selects the route with the lower metric.
- If both routes have the same metric, the router examines the metric for the forwarding path to the ASBR. It selects the route associated with the ASBR to which it has a lower cost path.
- If forwarding paths to the ASBRs have equal costs and the router supports ECMP, the router installs multiple routes in the OSPF routing table. (Otherwise, it uses another mechanism such as the router ID to break the tie.)

When all ASBRs use Type 2 metrics, an OSPF router will always select the external route through the ASBR that advertises the lowest metric for it, even if the router is closer to another ASBR that advertises the same route. However, if multiple ASBRs advertise the route with the same metric, other routers do take the cost of reaching the ASBR into account. The reasoning behind this behavior is that the cost of reaching the ASBR through the OSPF AS may be negligible compared to the cost of reaching the external network. In addition, the ASBR might not be assigning metrics to the external routes that correspond to the metrics used in the OSPF AS. In other words, a metric of 10 for the external route might not mean the same thing that a 10 metric on an OSPF link means, in terms of how strongly the company prefers using a particular path.

Figure 4-26 examines the selection process when OSPF is choosing between identical external OSPF routes. Usually, a network advertised in a Type 5 LSA should not be associated with an internal OSPF route, but a misconfiguration could cause this situation to occur. In that case, the OSPF always prefers internal (intra-area and then inter-area) routes.

Which ASBR does router Int-1 select to reach the external route, 172.16.1.0/24?

What is the next hop in Int-1's route table to reach 172.16.1.0/24? (This router supports ECMP.)

Solution

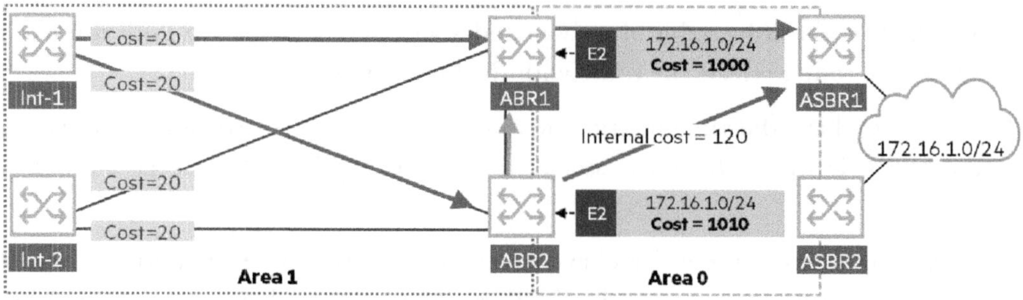

Figure 4-27: Best path selection with the E2 routes

- Which ASBR does router Int-1 select to reach the external route, 172.16.1.0/24?

  The figure shows the paths that Int-1 would take in this scenario – all equal cost paths.

- What is the next hop in Int-1's route table to reach 172.16.1.0/24? (This router supports ECMP.)

  Both ABR1 and ABR2 would therefore appear in the router Int-1's route table as next hops.

How to choose E1 and E2 type routes?

- if you want to leverage all paths to destinations, you need to use E2 with the same default metric.
- If you want to use the closet exit point, you need to use E1 with the appropriate seed metric.
- If you have a primary exit with a secondary backup, you need to choose E2, the primary with a lower default metric than the secondary.
- If you need to ensure traffic via a firewall is seen in both directions, you need to choose E2 with the appropriate seed metric to prefer one primary path.

OSPF area types

Each OSPF area type has specific characteristics and serves different purposes within an OSPF network. The choice of OSPF area types depends on network requirements, such as reducing routing table sizes, controlling external routes, and simplifying routing in certain areas. Each area type offers a different level of control and optimization for OSPF routing. it supports the following area types:

Backbone Area (Area 0)

The backbone area is the core of the OSPF network. Every OSPF network must have an Area 0; it serves as the transit area for all other areas. All other areas must connect to Area 0 (either directly or through other areas using a virtual link). Area 0 must be contiguous; it cannot be split into multiple parts, and it is responsible for interconnecting all other areas in the OSPF network.

Standard Area (Non-Backbone Areas)

Standard areas are all areas other than Area 0. These areas are connected to the backbone area which contains routers and networks within a specific functional group, location, or department. This reduces the size of OSPF routing tables by dividing the network into smaller units, and it also summarizes routes before advertising them to other areas. Each non-backbone area must have at least one ABR connecting it to Area 0.

Stub Area

Stub areas reduce the size of routing tables by limiting the amount of routing information received from the backbone area (Area 0). Stub areas do not allow external LSAs (Type 5 LSAs) from outside the area. An ASBR cannot exist in a stub area. Routes to external destinations are forwarded to a default route (0.0.0.0) within the stub area.

Totally Stubby Area

This is an extension of a stub area that further reduces the size of routing tables. It is similar to a stub area, but also blocks Type 3 LSAs (Summary LSAs) from entering the area. Only a default route (0.0.0.0) is advertised to routers within the totally stubby area.

Not-So-Stubby Area (NSSA)

NSSAs allow for limited external routing information to be imported into the OSPF network. It is similar to stub areas, but allows for the import of external routes in the form of Type 7 LSAs. An ASBR within the NSSA translates Type 7 LSAs into Type 5 LSAs for distribution within OSPF. Routers within the NSSA receive external routes as Type 5 LSAs from the ASBR.

Totally NSSA

This is similar to an NSSA, but further restricts external routes. It blocks Type 3 LSAs (summary LSAs) from entering the area, like a totally stubby area. Only a default route (0.0.0.0) is advertised to routers within the totally NSSA.

What is the need for eliminating excessive external routes?

Look at the following scenario, where 200 routes from ISP 1 and 150 routes from ISP 2 are injected into Area 0 as Type 5 LSAs. The ABRs then advertise these into Area 1, increasing route table size by 350 routes, along with associated memory and CPU usage. The internal routers in some OSPF areas rarely require all external routes.

Look at each non-backbone area: Does it really need all external routes?

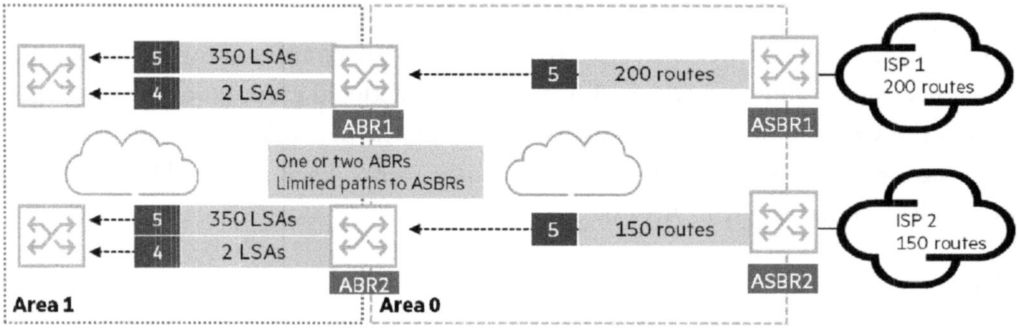

Figure 4-28: Multi-area OSPF with ASBRs

Remember the primary benefit of individual Type 4 and Type 5 LSAs. Type 5 LSAs help routers choose the best ASBR for reaching an external network. If that ASBR is in another area, Type 4 LSAs further help the routers choose the best ABR to reach that ASBR, and perhaps to select the best ASBR. You should examine each non-backbone area and determine whether routers in the area truly need to choose between paths for particular external networks.

First, consider whether the area includes any ASBRs. If not, then all external traffic must traverse the ABRs. If there's only one ABR, there's only one path for external traffic. A default route through the ABR provides the same benefit as 350 individual external routes. If the area has multiple ABRs, internal routers might benefit from choices for external paths. But ABRs often offer roughly equivalent paths to external networks, so internal routers can simply receive a default route from both ABRs.

Stub areas eliminate external routes

To eliminate external routes and advertisements from an area, define the area as a stub area. ABRs for stub areas do not forward Type 4 or Type 5 LSAs into those areas, and internal routers in those areas do not generate or accept them. You must define the stub setting on every OSPF router in the area—ABRs and internal routers. Use the syntax shown in the following example. Routers with mismatched area types will not become neighbors, and the routing is broken.

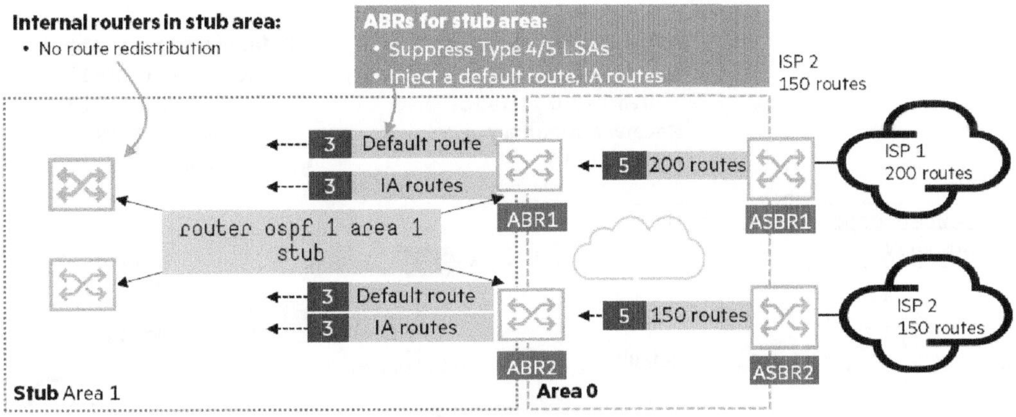

Figure 4-29: Multi-area OSPF topology with a stub area

When you configure an ABR area as a stub, it automatically generates a default route and injects it into the stub area as a Type 3 LSA. As usual, the ABR assigns a metric to the default route, and you can manipulate metrics to set up one ABR as a preferred exit for external traffic.

Notice that Type 3 LSAs for Inter-Area (IA) routes are injected into the area, just like a normal area. Internal stub area routers receive summary LSAs for other areas through ABRs; the summary LSAs include individual or aggregated routes as configured on the ABRs. Therefore, internal routers in the stub area can continue to select the best path for inter-area traffic, based on the cost of the summary routes advertised by the ABRs. In larger networks, you might also want to prevent these IA routes from being injected into the area.

Totally stubby area, eliminates IA and external routes

Figure 4-30 shows a scenario where Area 1 is a stub. All traffic destined to it terminates there. The area has only one ABR exit for all non-local traffic, including traffic destined to the backbone, to other OSPF areas, and to the ISP. In such an environment, the internal Area 1 routers do not gain a benefit from multiple inter-area routes. You can configure the OSPF system to suppress all IA routes in this area in addition to all external routes.

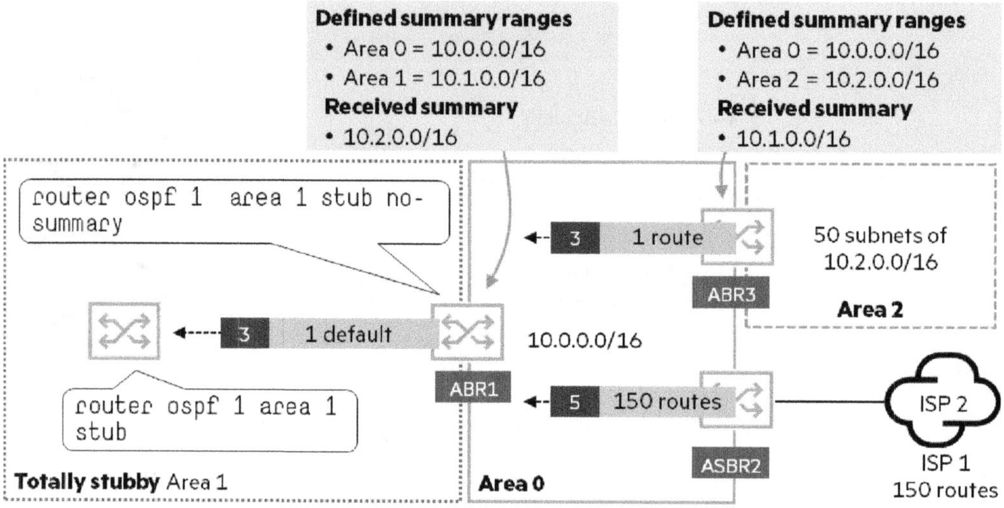

Figure 4-30: Multi-area OSPF with a totally stubby area

You define a totally stubby area by configuring the ABR or ABRs connected to that area to suppress summaries. Use the OSPF configuration shown in the figure on all ABRs that connect to the area. Configure internal, non-ABR routers in the area with the regular Area 1 stub command. The no-summary option transforms a stub area to a totally stubby area; it suppresses all external route advertisements as normal for a stub area. It also prevents the ABR from generating non-aggregated IA summary routes for this area. For example, the ABR in the figure would not generate Area 1 summary LSAs for known routes in Area 0 and Area 2. Instead, the internal routers in Area 1 use the default route to reach the other areas.

The number of routes eliminated may seem fairly trivial in this example, but in more complex networks, totally stubby areas can greatly reduce the size of LSDBs and route tables.

Not so stubby area advertises external routes into stub area

You may need to create a stub or totally stub area to reduce table sizes and improve stability. Remember, these area types do not allow Type 5 LSAs. This means that you cannot have routers in a stub area connected to external networks doing route redistribution, and thus creating Type 5 LSAs.

Sometimes you need this functionality. You want the area to disallow all Type 5 areas like a stub, and perhaps all Type 3s, like a totally stubby area. However, you need this area to be connected to external subnets that use a foreign routing protocol (static, BGP, etc.). And remember, any redistributed route is defined as an external route. Therefore, you would encounter a similar issue if you decided to redistribute connected routes into OSPF rather than configure the associated Layer 3 interfaces as passive OSPF interfaces.

You need the area to act stubby in one sense, and not so stubby in another sense.

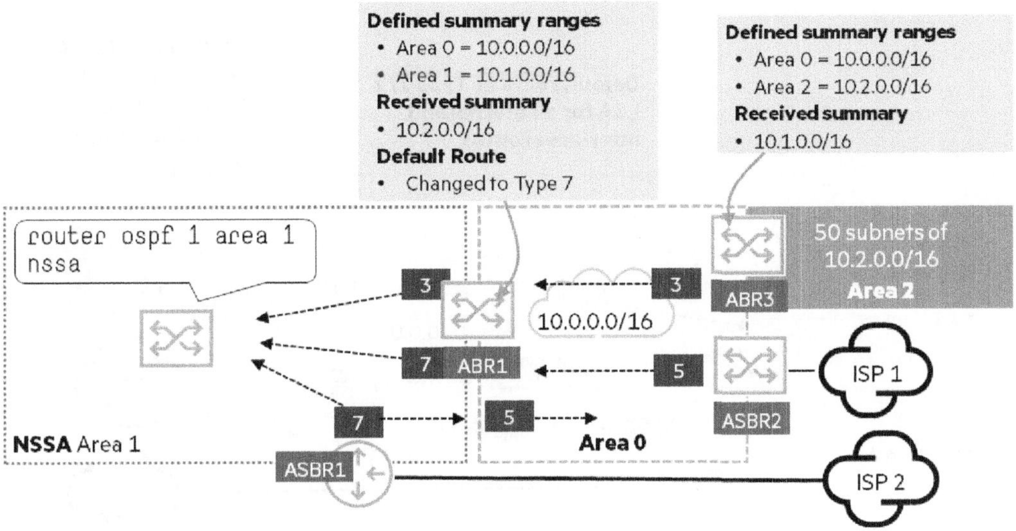

Figure 4-31: Multi-area OSPF with an NSSA area

A not-so-stubby area (NSSA) furnishes the solution for this problem. You must define this area type on every OSPF router in the area, as shown in **Figure 4-31**.

An NSSA acts like a stub area. However, the area permits external routes that are advertised in OSPF by a local ASBR. These external routes are carried in Type 7 LSAs. Of course, you must configure the local ASBR (ASBR1 in the figure) to redistribute routes, advertise a default route, or both, as you learned in the section on configuring ASBRs and external routes. Like other ASBRs, the NSSA ASBR assigns a metric type to the external routes that it injects into OSPF, which you can adjust to influence route selection. For NSSA external routes, these are designated as N1 and N2, instead of the normal E1 and E2 metric types.

The ABR translates the Type 7 LSAs (N1 and N2) into Area 0 as Type 5 LSAs (E1 and E2) that it appears to originate. The ABR still filters Type 5 LSAs from other areas and does not advertise them in this area. It also continues to advertise a default route in the NSSA, but as a Type 7 LSA rather than a Type 3. If the NSSA's ASBR advertises a default route as well, you must adjust metrics such that routers choose the intended routes.

Totally not-so-stubby areas (NSSAs)

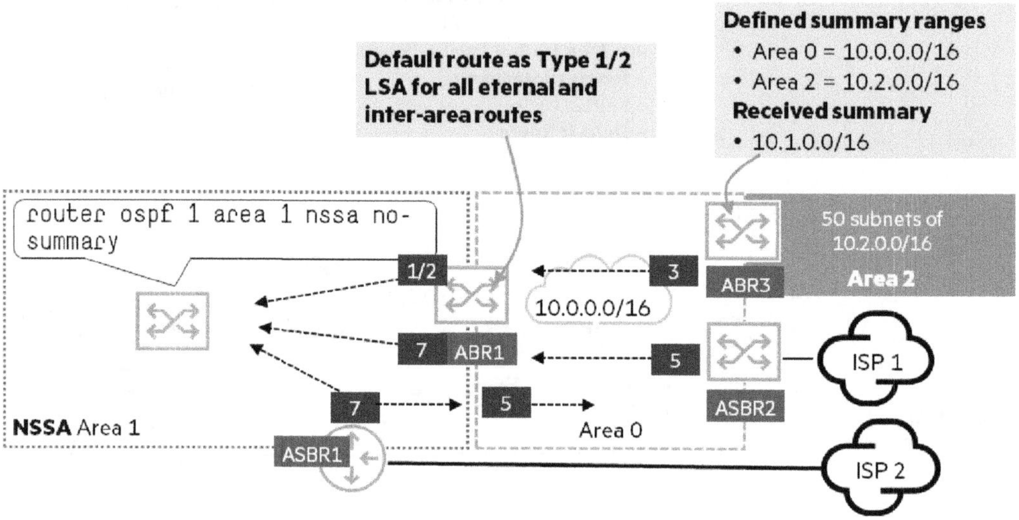

Figure 4-32: Multi-area OSPF with a totally NSSA area

You can also set up an NSSA area as a totally not-so-stubby area, similar to a totally stubby area. You only need to add the no-summary option to the area configuration, as shown in **Figure 4-32**.

As a totally NSSA, the ABR prevents all external or inter-area routes from entering the area. Instead, the ABR injects a default route into the area for all external and inter-area routes.

Factors that affect OSPF route selection

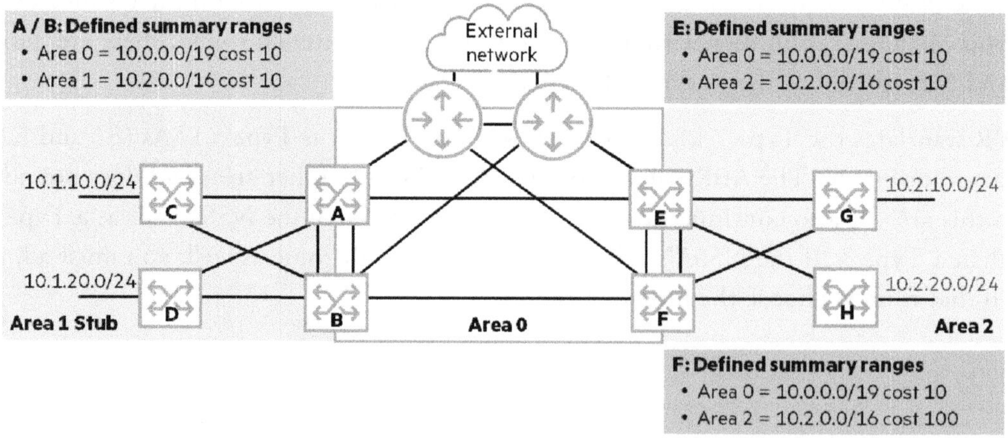

Figure 4-33: Multi-area OSPF topology

What factors affect Router C's next hop for 10.1.20.0/24?

Consider Router C, an internal router in Area 1. Area 1 is a stub area with two ABRs, Router-A and Router B. For intra-area destinations like 10.1.20.0/24, link costs alone determine the best path and selected next hop. In other words, if the cost of the C-to-B link + the B-to-D link is less than the cost of the C-to-A + the A-to-D link, C will use B as the next hop. If the costs are the same, C will learn two routes and use a hash to assign conversations to each (ECMP).

What factors affect Router C's next hop for 10.2.20.0/24?

Link costs also play a large factor in determining how C forwards inter-area traffic such as that destined to 10.2.20.0/24. Remember that C uses the 10.2.0.0/19 route for all traffic destined to Area 2 (10.2.0.0 to 10.2.31.255). You must consider the link costs from C to its ABRs and then from those ABRs to the Area 2 ABRs. Then you add the cost of the 10.2.0.0/19 summary advertised by the Area 2 ABR. In other words, the cost configured for the summary range on the Area 2 ABR influences the path for the inter-area traffic destined to that area, but a router's cost for reaching that ABR is also taken into account.

What factors affect Router C's next hop for external destinations?

Because this is a stub area, Router C uses a default route to reach external destinations. It chooses the default route based on the metric advertised by the ABR and the cost for reaching the ABR. The next hop is the next hop in the best path to the advertising ABR. So, the next hop for the default route is affected by link costs within Area 1 and the cost set in the Area 1 stub command on the ABRs.

What factors affect Router-A's next hop for external destinations?

More factors affect Router-A's choice of next hop for routes to external destinations. The metric type (1 or 2) advertised by the ASBRs plays a role, as does the advertised metric itself. Link costs within Area 0 can also affect the choice. If ASBRs use type 1 LSAs, these costs are incremented with the advertised metric when Router-A chooses the best of two external LSAs for the same destination. Even if ASBRs use type 2 LSAs, the Area 0 costs will still affect Router-A's best path to the ASBR that advertises the best LSA.

If you change Area 1 to Totally Stubby, what factors affect how Router C forwards inter-area traffic?

As a final review, pretend that you have changed Area 0 to a totally stub network. In this case, Router C uses its default route for both inter-area and external OSPF routes. Therefore, the ways to influence how Router C forwards inter-area traffic are to manipulate the link costs within Area 1 and the cost in the Area 1 stub command on the ABRs.

OSPF redundancy

In a topology with redundant links, OSPF automatically handles reconvergence if a link fails. You should consider how your design and configuration can affect reconvergence time. We will now analyze the factors that affect whether connectivity is disrupted and for how long.

What happens when a router or link fails? The reconvergence time depends on the following factors:

- How far do LSUs need to propagate?
- How many routers must rerun the SPF algorithm?
- How quickly do neighbors detect lost connectivity?
- How quickly can routers generate, transmit, and receive LSUs?

In the case of a dedicated VLAN on a point-to-point link, detecting an L2 issue and bringing down the VLAN interface takes less than a second. However, in the case of an Ethernet link with multiple network routers or intervening L1 or L2 devices, the dead interval (duration) for removing the neighbour relationship takes between 45 seconds and 1 minute.

LSUs and SPF algorithm for convergence

For the first two questions, consider how what you have learned about multi-area designs helps to stabilize the system and enhance convergence.

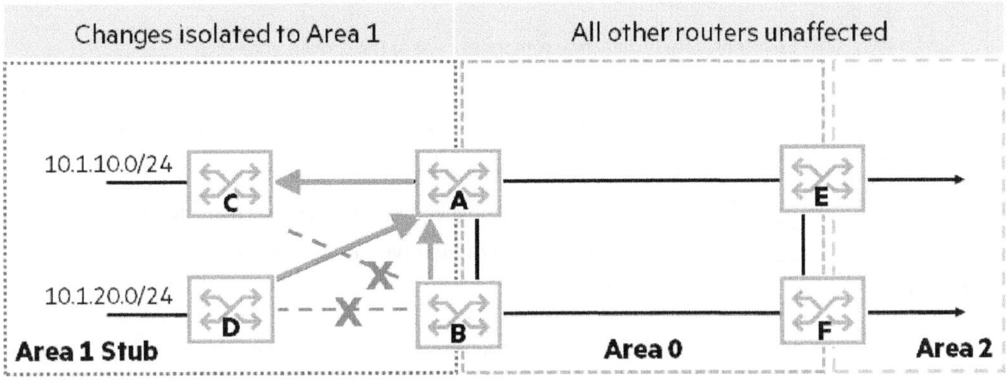

Figure 4-34: Link/router failure is isolated to Area 1

Area borders act as boundaries for changes, and often LSUs generated by a change within the area will not prompt LSUs for other areas. In addition, only changes within the OSPF area cause a router to rerun the full SPF algorithm. Updates to summary, summary-ASBR, and external LSAs require just a partial rerun. The router simply adds a new network to the list of networks reachable behind the ABR, removes a network from the list, or adjusts the cost for reaching the network. The router does not need to recalculate and run the shortest path algorithm because it cares only about the segment of the inter-area path between itself and the proper ABR for reaching the inter-area network. Those paths have not changed.

In **Figure 4-34**, the link between Routers C and B is down, as is the link between Routers D and B. Routers in Area 1 learn this information and converge on alternate paths. Routers D and B route all traffic to 10.1.10.0/24 through Router-A. However, Routers A and B still have a route to this network, so they do not withdraw its Type 3 LSA for the subnet. They might need to update the cost, but usually it will not even need to do that because it is aggregating the Area 1 summaries anyway. As far as routers in Area 0 and Area 2 are concerned, nothing has changed. As you see, changes and any attendant disruptions are isolated to Area 1.

Directly connected neighbors

Now consider how quickly routers can generate, transmit, and receive LSUs. The default timers often provide sufficiently speedy LSU generation and propagation. When very fast convergence is required, you can adjust OSPF timers like the LSA generation interval. However, in most cases, you should leave the intervals at their defaults; tuning them is beyond the scope of this guide.

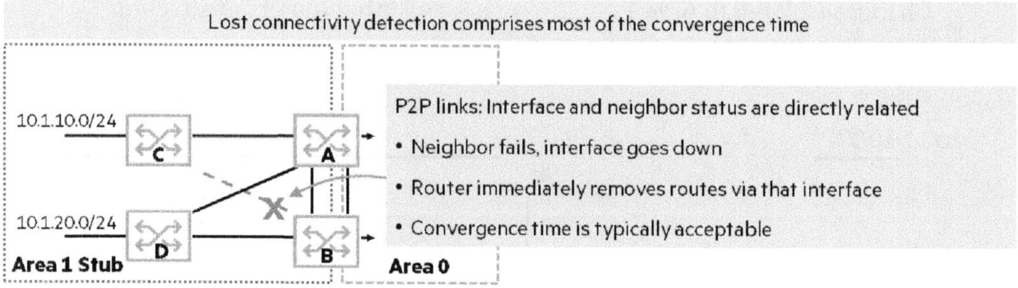

Figure 4-35: Reconvergence over P2P link

In a well-designed OSPF AS with proper area boundaries, the time to detect failed neighbor relationships often determines how seamlessly OSPF routes fail over.

Suppose an OSPF router loses its next-hop connection to one of its routes. Either a link failed, or the neighbor as a whole is down. To remove the invalid route and use an alternate, the router must first detect the primary path failure. This often takes up the bulk of the time required for routes to fail over.

When two neighbors have a point-to-point connection on a direct link, the status of the interface correlates to the status of the neighbor. If the neighbor fails, the interface with the direct connection goes down. The router can immediately remove any routes through that interface. Similarly, when a VLAN interface is dedicated to a single physical link, the failure of that link brings the VLAN interface down. The routing switch can then remove any routes through that interface. In either case, the time to detect the link layer problem affects the time for detecting the failed neighbor relationship. This time varies depending on the media and is acceptable for many companies' needs. However, sometimes the status of an OSPF interface and the status of an OSPF neighbor do not directly correlate.

Hello and dead timers

Hello and dead timers facilitate the dynamic and robust operation of OSPF by ensuring efficient neighbor discovery, neighbor relationship maintenance, and rapid detection of network changes.

Hello timers: Used for neighbor discovery and maintenance of neighbor adjacencies by exchanging hello packets at regular intervals.

Dead timers: Used to detect network failures by determining when a neighboring router is unreachable and triggering OSPF convergence.

Sometimes interface and neighbor status do not directly correlate, relying on hello/dead timers. The following example explains it.

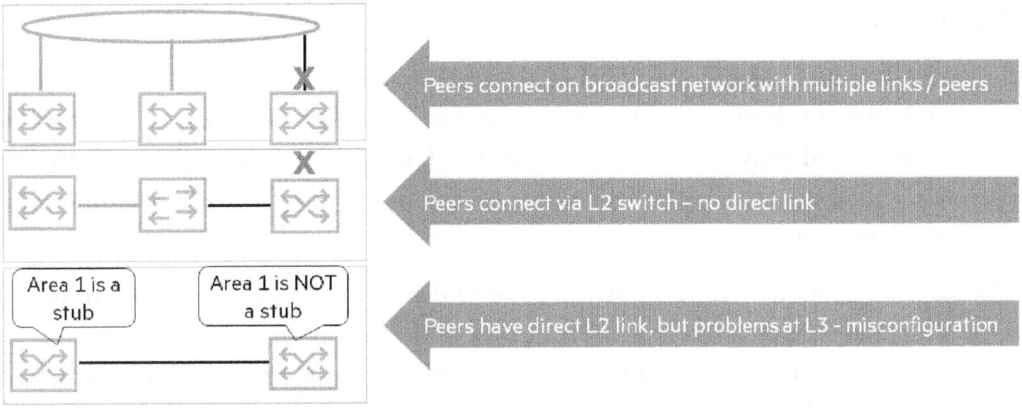

Figure 4-36: Different cases where routers rely on hello/dead timers to determine peer status

Examples relevant to AOS-CX switches include:

- The neighbors or peers connect on a broadcast network such as a VLAN but the VLAN has multiple links, and possibly multiple neighbors, associated with it. In **Figure 4-36**, a router is down (red X), but the other router's links remain up (green).
- The neighbors are connected via a non-routing Layer 2 switch, and so they do not have a direct Layer 2 connection. In the figure, a router is down (red X), but the other router's connection to the switch remains up (green).
- The neighbors have a direct Layer 2 connection, but problems occur at Layer 3. For example, a misconfiguration has prevented one of the neighbors from implementing OSPF on the network. In the figure, one router in the area is configured as a stub, but the other is not. You might go back to the first section of this chapter and review the criteria in OSPF hello packets that must match for neighbors to form an adjacency.

In the cases found in **Figure 4-36**, OSPF routers must rely on timers to determine the neighbor status. If the dead time expires without a hello from the neighbor, the router considers the neighbor down and removes routes using that neighbor as a next hop from the routing table. The OSPF dead timer is always four times the hello timer. Because the default hello timer is 10 seconds, the default dead timer is 40 seconds, which is far too long for contemporary networks. You can adjust the hello and dead timers, but the lowest standard hello timer is one second, making the lowest dead timer four seconds, still too long for many contemporary networks.

Take care when setting the dead interval timer. The dead timer, by default, is four times the hello timer. Typically, it is not recommended to set the hello and dead interval timers to one second each, since this could cause inadvertent missed hellos.

BFD with OSPF

Bidirectional Forwarding Detection (BFD) is a protocol used in conjunction with OSPF to provide fast and efficient detection of link and neighbor failures. BFD operates independently of the routing protocol and provides a low-overhead, fast failure detection mechanism. In OSPF, BFD is primarily used to enhance the reliability and responsiveness of OSPF neighbor adjacencies and link failure detection.

BFD enhances the reliability and responsiveness of OSPF-based networks, leading to improved network stability and reduced convergence times. In OSPF, it can be configured to monitor OSPF neighbor adjacencies and detect link failures. BFD can be used in conjunction with OSPF's built-in timers to provide an additional layer of fault detection and redundancy.

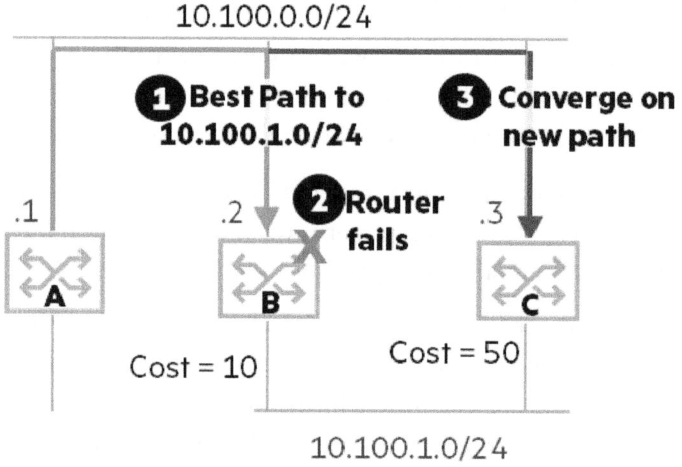

Figure 4-37: OSPF reconvergence with BFD enabled

BFD tests the connectivity between two IP addresses in a BFD session. BFD reports to OSPF when connectivity is lost. Routers use that information to take appropriate actions, depending on the functions tied to BFD. When you enable BFD on an OSPF Layer 3 interface, BFD assumes responsibility for testing neighbor connectivity. If the BFD session fails, the router removes the BFD neighbor relationship, and so can remove all LSAs through that neighbor and update routing paths. The interface still sends out OSPF hellos, but BFD will generally detect any problems first.

In **Figure 4-37**, three routers connect on subnet 10.100.0.0/24. Routers B and C also connect to network 10.100.1.0/24. Router-A uses Router B as the next hop for 10.100.1.0/24 because Router B advertised the lowest cost.

If BFD is not in use and Router B fails or its link on 10.100.0.0/24 fails, Router-A continues sending traffic to 10.100.1.0/24 through 10.100.0.2 until its neighbor relationship times out. This could be 40 seconds if the dead timer is at its default value, or as fast as four seconds if you lower the dead timer to the recommended safe minimum. Because 10.100.0.2 is no longer up, this traffic does not reach its destination.

With BFD, Router-A quickly detects that 10.100.0.2 is no longer available—typically, this is less than one second. BFD removes the neighbor relationship, causing Router-A to lose adjacency with Router B. Router-A can then immediately start using Router C as the next hop to 10.100.1.0/24.

If an OSPF neighbor does not support BFD, the switch continues to try to establish a BFD session with that neighbor. It lists the session as down, but because the session was never up, the down state does not cause the switch to tear down the neighbor relationship.

BFD in echo mode

When you configure BFD, echo mode is enabled by default. In echo mode, routers periodically send BFD echo packets. The peer router returns the received BFD echo packets back without processing them. If the sender does not receive BFD echo packets from the peer within the specified interval, the session is considered down. Because peers respond to echo packets without processing them, detection is FAST.

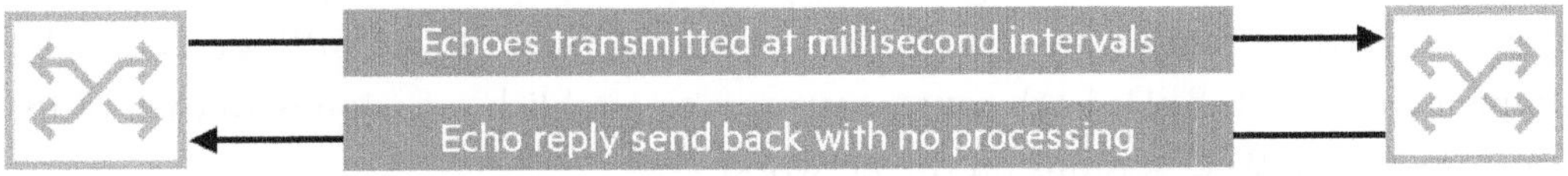

Figure 4-38: BFD echo mode operation

BFD configuration

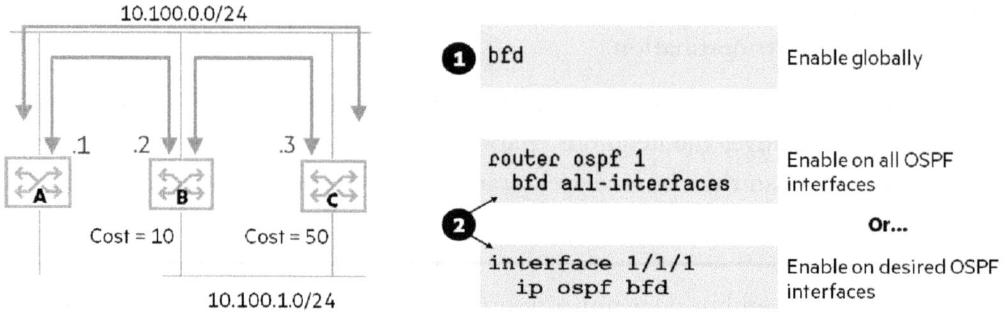

Figure 4-39: BFD configuration

- First, enable BFD globally from global config mode, as shown in **Figure 4-39**. When you enable BFD, it automatically enables in echo mode. There is a `bfd echo disable` command, which disables echo mode, but this is not recommended. Use the `no` form of the command to re-enable.
- Next, you can either enable OSPF BFD on all interfaces in the OSPF context, or you can enter the desired Layer 3 interface context and enable BFD on that interface specifically, as shown in the previous figure.

Because AOS-CX switches use active mode, the interface automatically attempts to establish a BFD session with each OSPF neighbor on that interface. It transmits a unicast BFD control packet on UDP port 3784 to each neighbor. As you see in the previous figure, a BFD session is point-to-point. In this example, the 10.100.0.0/24 network has three OSPF routers, each of which implements OSPF BFD. Each router attempts to establish a session with each of its two neighbors for six total BFD sessions on the network.

To establish a BFD session, the neighbors simply determine that both support BFD and negotiate timers.

BFD control messages and timers

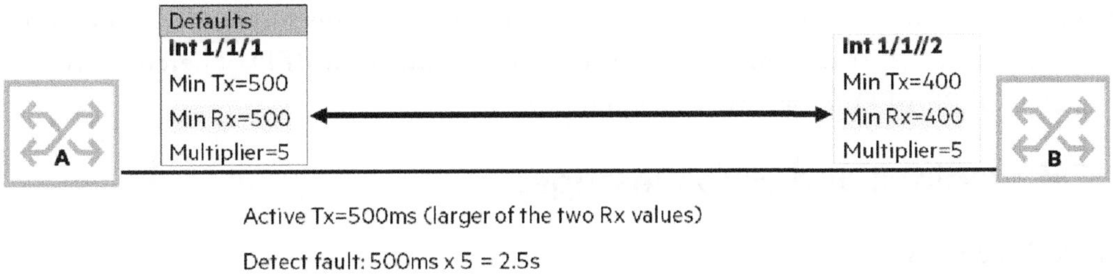

Figure 4-40: BFD timer settings

Each neighbor interface has three settings that control convergence time.

- Minimum Transmit (Tx) timer: Minimum time between sent echo (control) packets. Default = 500 milliseconds (ms)
- Minimum Receive (Rx) timer: Minimum time between expected echo packets received from peer. Default = 500 ms
- Detection Multiplier: Number of consecutive missed packets received before declaring the session dead and informing OSPF to converge on an alternate path

For most deployments, the default values should not be changed. If you must change these settings, use this syntax:

```
(config-if)# bfd min-transmit-interval=500
(config-if)# bfd-min-echo-receive-interval=500
(config-if)# bfd detect-multiplier=5
```

In **Figure 4-40**, Router-A has been left at the default values, while Router B's timers have been lowered to 400 ms. Interestingly, this has no effect on convergence times. The two routers share echo/control packets and compare values. The actual "Active Tx" timer used is the larger of the local peer setting versus the remote peer settings. Thus, the actual transmit interval is negotiated to 500 ms. Of course, this scenario is only to help you understand how timers work. You should be consistent with your configurations and typically leave timers at default values.

The multiplier is 5, and so 5 consecutive packets missed, at 500 ms apart = 2.5 seconds before the BFD session is torn down, which causes the OSPF neighbor relationship to be declared down, even if the dead timer has not yet expired. You can lower the multiplier to 4 or 3 if you have a reliable link that loses little traffic. If you set the Min Tx and Min Rx timers to the minimum value (one second), do not set the detect multiplier lower than 3. Understand that lowering these values can be dangerous—especially if you have links that occasionally become congested. This could cause routers to converge when links are down, and reduce overall system stability.

The active Tx timer and detect time on the two neighbors in a BFD session do not have to match. However, they often do. When the switch uses the control message timeout alone to determine when a session fails, it operates in asynchronous mode. In this mode, BFD does not provide sub-second detection of issues, which is often what you want from BFD. Therefore, you should usually enable OSPF BFD to operate in echo mode.

BFD: ICMP redirect and echo messages

- **ICMP redirects**

  It is also a best practice to disable ICMP redirects on the switches that use BFD in echo mode. Before you enable OSPF BFD, you should disable ICMP redirects on the AOS-CX switch. In some cases, the echo could have a source and destination on the same subnet, which would usually trigger the switch to send an ICMP redirect. The extra processing can cause issues on the switch. Disabling ICMP redirects prevents these issues.

  ```
  Switch(config)# no ip icmp redirect
  ```

- **Echo messages**

  If you want, you can adjust the BFD timers, but HPE recommends the default values (see the *Command-Line Interface Guide* for more details on how to do this).

  You should set up BFD in the same way on all the OSPF routers connected on the interface. In echo mode, control messages still maintain the BFD session, but echo messages, which are transmitted at sub-second intervals, allow BFD to respond to issues much more quickly. The interface implementing BFD sends an echo message to the other side of the session. The other side of the session simply echoes the message back without processing it. If the interface does not receive its echo message back, BFD takes down the session and the OSPF neighbor relationship. (The switches still exchange control messages in echo mode, but these become less important for determining when the session has failed.)

- **Echo source IP address**

  By default, AOS-CX switches support echoing echo messages back to their BFD neighbors at a 500 ms interval. However, they will not send echo messages themselves until you set a global BFD echo source for the packets, which is not set by default on the CX 8400 Series. Until you set the address, the BFD session will have echo mode disabled. You should choose a unique "dummy" IP address that is not used on this device.

On the CX 8400 Series switches, you should next set a global echo mode IP address, which the switch includes as the source IP address for the echo packets in all sessions. Remember that the destination is the switch's own IP address, so the source is just a "dummy" address. It should be an IP address on a subnet you're not using on the switch or the network. You cannot set this address on CX 8320 Series or lower switches.

```
Switch(config)# echo-src-ip-address <IPV4-ADDR>
```

This sets the source IPv4 address for BFD echo packets. This address is used in all echo sessions. The source IP address must not be on the same network segment as any switch interface; otherwise, a large number of ICMP redirect packets may be sent by the remote device, causing network congestion.

The switch uses this as the source for the echo packets. The destination is the switch's own IP address, which allows the neighbor to send a message back without processing it. Setting the source IP address to a different subnet ensures that the neighbor does not have to process the packet and send ICMP redirects.

Using OSPF with VSX

Using OSPF with VSX in AOS-CX switches provides a highly available, scalable, and resilient network infrastructure with simplified management and fast convergence times. It enables dynamic routing capabilities, load balancing, and adaptability to changing network conditions, ensuring optimal performance and reliability for enterprise networks.

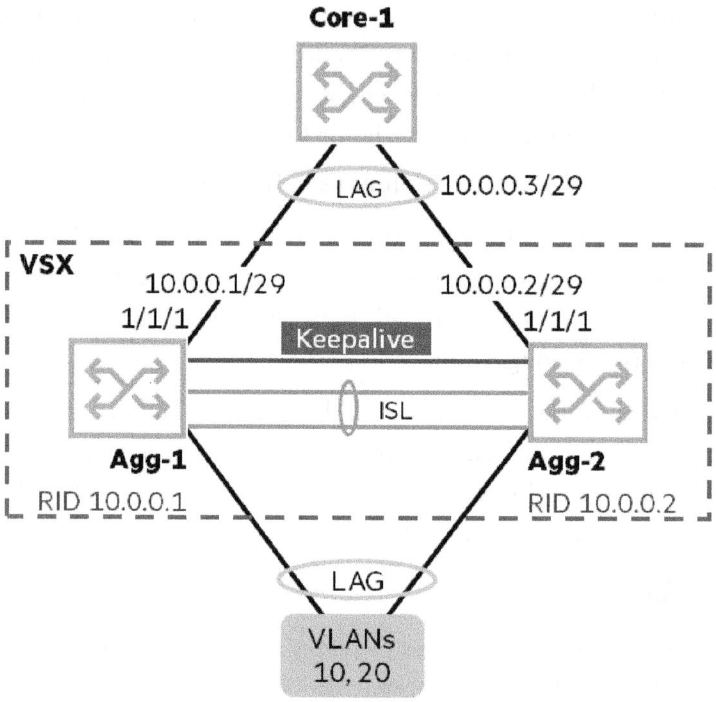

Figure 4-41: VSX enabled in the aggregation layer

Each VSX member runs OSPF separately. In **Figure 4-41**, aggregation layer switches Agg-1 and Agg-2 are a VSX pair. They have a VSX LAG to Core-1, and each switch has an IP address on the Layer 3 VLAN interface assigned to the LAG. If you were using multiple VRFs, the LAG would carry multiple VLANs; in this example, however, the LAG has just one VLAN.

Core-1 learns ECMP routes to VLAN 10 and 20 subnets through both Agg-1 and Agg-2. Because the ECMP next hop selection algorithm will not necessarily match the link aggregation selection algorithm, the core switch might send traffic through next hop 10.0.0.1 but on the link to Agg-2. The active forwarding, which is set per VLAN, enables Agg-2 in this situation to route the traffic itself, even though the traffic was destined to 10.0.0.1. You should typically enable active forwarding on every OSPF interface that is a transit network.

The VSX pair is probably the default gateway for subnets such as those associated with VLANs 10 and 20. You should typically set up the active gateway feature on those VLANs. Recall that this feature enables the VSX switches to act as active-active default gateways for a subnet. You might run OSPF on these subnets, but in passive mode—VSX peers need not become OSPF neighbors on all subnets. Alternatively, you could redistribute connected routes into OSPF.

Additional OSPF features

Additional features in OSPF provide network administrators with powerful tools to optimize OSPF routing operation, enhance network security, improve scalability, and ensure reliable routing performance in diverse network environments. By leveraging features like graceful restart, passive interface, virtual link, and authentication, OSPF can effectively address the evolving requirements of modern enterprise networks.

Graceful restart

OSPF graceful restart enables OSPF routers to maintain routing adjacencies and forwarding tables during router reloads or restarts. Graceful restart minimizes disruption to network traffic and ensures uninterrupted routing operation during planned maintenance activities.

OSPF disruptions with redundant solutions

Challenge with the redundant solution in AOS-CX switches

- Fabric conductor maintains route/protocol info, proxies to standby.
- If conductor fails, standby has all info to continue routing traffic.
- But OSPF requires a new adjacency to be established.
- Unnecessary outage during reconvergence.

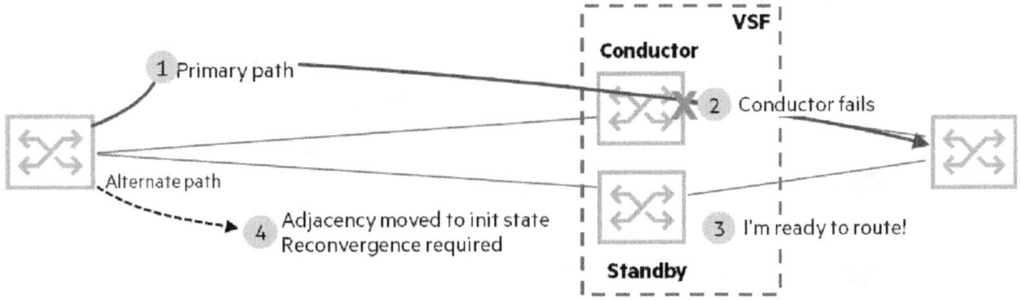

Figure 4-42: OSPF implementation with VSX

Consider a scenario where some OSPF routers must also implement other protocols for high availability:

- HPE Aruba Networking Virtual Switching Framework (VSF) (preferred when possible on the 6300s)
- Virtual Router Redundancy Protocol (VRRP)
- VSX

However, even though default gateway functions are maintained, failure of the VSF master or VRRP master could still disrupt routing and connectivity. OSPF graceful restart ensures nonstop routing as the standby member takes over as the master. This is illustrated in **Figure 4-42**.

The VSF fabric master maintains routing tables and protocol information in its control plane. In a chassis switch, the master's primary management module maintains this information. The master proxies this information to other members' control planes, but if the master fails, the routing processes must restart on the new master.

The VSF members have routes proxied to the Forwarding Information Bases (FIBs) in their control planes (or the control planes of their LPUs, in the case of chassis-based switches). They can continue to route traffic with these stale routes until the new master builds the routes in its own control plane—and, when the VSF fabric is using static routes or Routing Information Protocol (RIP) routes, they do seamlessly.

However, standard OSPF does not allow OSPF neighbors to entirely resynchronize the LSDBs without tearing down the old adjacency relationship. Therefore, when the new master tries to re-establish adjacency, the VSF fabric's neighbors take down the neighbor relationship and send out LSUs as if the fabric were no longer available. Routes reconverge away from the VSF fabric—even though the fabric is perfectly capable of routing the traffic—causing unnecessary disruption to the traffic flow and possible interruption to service.

Solution for this challenge:

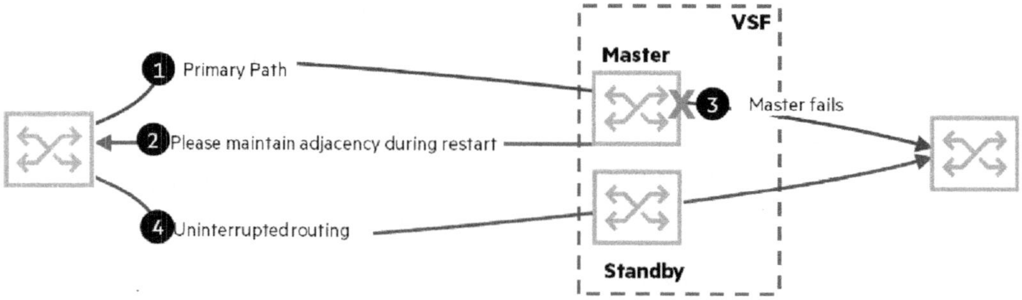

Figure 4-43: Implementation of graceful restart

OSPF graceful restart enables the VSF fabric to ask its neighbors to maintain the neighbor and/or adjacency relationship undisturbed while the new master restarts the OSPF process. Thus, traffic continues toward its destination undisturbed.

An OSPF router with redundant management modules should also implement OSPF graceful restart. The failover of the primary management module to the standby module is analogous to the failover of the VSF master to the standby member. Although the OSPF router or routing switch can use its stale routes to forward traffic, it must also send OSPF graceful restart requests to inform its neighbors not to drop its LSAs while the new management module rebuilds the adjacencies.

OSPF graceful restart also helps to maintain routing system stability in any circumstance in which an OSPF router must restart its OSPF processes, including when a network administrator restarts the process for maintenance purposes (such as applying a new router ID).

Graceful restart process

When an OSPF router restarts the process—for example, because a main management module fails and a redundant one takes over—it must re-establish neighbor relationships. This triggers neighbor routers to clear the router and its LSAs out of the topology. Routing is disrupted across the network until the router finishes restarting the process. AOS-CX switches support OSPF graceful restart, which lets an OSPF router restart the process without causing disruptions for traffic.

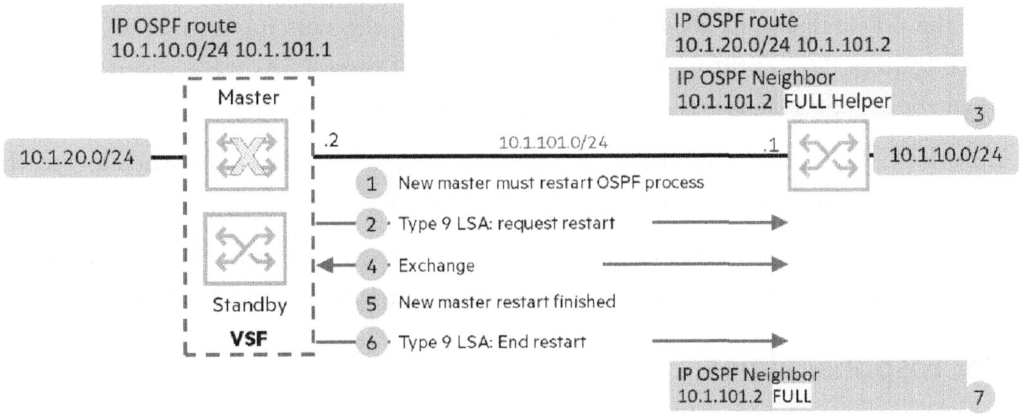

Figure 4-44: Graceful restart process

The IETF standard graceful restart process:

- The GR Restarter is a router that uses graceful restart. It determines the need to restart its OSPF process.
- The GR Restarter sends a Type 9 opaque LSA (grace LSA) to neighbors, asking them to move to graceful helper mode. The LSA includes:

- A grace period timer, which specifies how long the neighbors stay in graceful helper mode
- A graceful restart reason:
    - 0 = Unknown
    - 1 = Software restart
    - 2 = Software upgrade
    - 3 = Failover from an active to a standby management module, or, for VSF, failover from master to standby
- The GR Restarter's IP address

■ Neighbors must be configured as IETF standard GR Helpers. They place the GR Restarter in graceful restart mode. In this mode, GR Helpers act as if they have full adjacency with the GR Restarter, regardless of what is occurring. Thus, Helpers continue routing traffic to the Restarter and maintain LSAs from it, trusting that the Restarter can truly continue to route traffic with its stale routes.

During this time, the GR Restarter is establishing neighbor relationships and adjacencies with the GR Helpers. As always during this process, the switch obtains LSAs advertised by OSPF routing devices throughout the area. Some of these LSAs were advertised by itself (or the previous VSF master); the Restarter labels those LSAs as stale.

■ After the GR Restarter establishes an adjacency with all neighbors, it sends a grace LSA, asking Helpers to leave helper mode; that is, grace period time is set to 0.

■ The GR Helpers leave helper mode. However, because they have achieved adjacency with the GR Restarter, they continue to route traffic to it without interruption.

Restarter responsibilities

1. **Recalculate routes**
    - Calculates routes from received OSPF neighbor LSAs, replaces stale routes
    - Remaining stale routes are deleted, thus routing with most current info
2. **Change stale routes to active**
    - Generates its own LSAs and compares them to its own stale LSAs received from Helpers

- Removes the stale designation from those LSAs, no further action—peers already know the LSAs
- Unless its connections have changed, Restarter LSAs match stale LSAs

3. **Updates neighbors**
    - Advertises new LSAs in an LSU, as normal
    - If stale LSAs remain, deletes/withdraws them in an LSU message

GR Restarter has continued routing traffic with stale routes. However, it must ensure the truth of what it asked the GR Helpers to trust: it knows the correct routes, and the topology that it advertised before the restart remains accurate. During and after the process of establishing adjacency, the GR Restarter recalculates routes, changes stale routes to active routes, and determines whether any updates are necessary:

The GR Restarter calculates routes from the LSAs that it received from OSPF neighbors. The new routes replace the stale routes. If any stale routes remain, they are deleted as no longer valid. The GR Restarter now knows that it is routing traffic with the most up-to-date information.

The GR Restarter generates its own LSAs and compares them to its own stale LSAs, which it received from the Helpers. Unless its connections have changed, the GR Restarter's LSAs will match its stale LSAs. The GR Restarter removes the stale designation from those LSAs and takes no further action (its neighbors already know the LSAs).

If the GR Restarter has generated any new LSAs, it advertises them to neighbors as normal in an LSU message. Similarly, if any stale LSAs remain (no matching current LSA), the GR Restarter deletes those LSAs and withdraws them in an LSU message.

The GR Restarter and the GR Helpers now know the proper LSAs, network topology, and routes. Throughout the process, all of the OSPF routing switches continued routing traffic with stale routes, only removing the routes if they were demonstrated to be truly invalid. Because the stale routes typically prove to be valid, the routing switches typically transition seamlessly from routing with stale routes to routing with active routes.

The OSPF graceful restart standard dictates that GR Helpers must leave helper mode if the topology changes in a way that would affect the GR Restarter. That is, the GR Helper generates or receives an LSA that it would normally send to the GR Restarter. This provision prevents routing loops.

Configuring graceful restart

OSPF graceful restart deployment is rather simple. Configure VSF fabrics, and any routing devices that must support non-stop routing during management module failover, as GR Restarters. As a best practice, you could also enable the restarter function on other routing devices that support it, protecting the network in case an administrator accidentally restarts the OSPF process in a live environment.

All neighbors of the GR Restarters must act as GR Helpers. As a best practice, you should generally ensure that all OSPF routing devices in the AS support this function.

Configuration

Graceful restart itself is always enabled and cannot be disabled. The graceful restart helper is enabled by default, but can be disabled. However, you can adjust the graceful restart timer (defaults to 120 seconds) and enable or disable the helper function.

After this timer expires, the switch removes any stale routes—that is, routes not recalculated from the reconstructed LSDB. The restart interval dictates the amount of time that the switch promises its neighbors that it can continue to route traffic while it achieves adjacency and reconstructs the LSDB. Therefore, the rapid switchover time must exceed the restart interval. (For example, exceed the restart interval by 3 seconds.)

```
router ospf 1
graceful restart helper
graceful-restart restart-interval 120
```

Max LSA feature

The max LSA feature is another feature than can smooth OSPF convergence. This feature tells the OSPF router to send out a Type 1 Router LSA with the maximum metric either permanently (until you remove the command) or for a period after the switch starts up. The max metric signals to other OSPF routers that the router is not available.

You would typically use this feature on an ASBR that redistributes routes from another protocol. You want to ensure that other OSPF routers do not send traffic through the ASBR until BGP is fully operational. BGP can take longer to load the full table than OSPF. Simply set the max LSA time period to a period a bit longer than you think it will take the ASBR to load all routing protocols.

Another use case relates to a secondary VSX member that requires time to boot up. The feature allows the secondary VSX member to learn upstream routes while the VSX uplink delay is keeping the VSX LAGs down. So when the VSX LAGs come up, the routing table is populated. Later, the secondary member will reduce the LSA metric, so it can also attract traffic from backbone for the local users.

Configuring max LSA

```
router ospf 1
max-metric router-lsa
max-metric router-lsa on-startup
max-metric router-lsa on-startup 600
```

Implementing OSPF and VRRP

As you have learned, you should use VSF to provide redundancy for the default router role whenever possible. VSF provides redundancy while allowing the VSF fabric to act as a single entity in the routing protocol, making implementing VSF with OSPF almost as easy as implementing OSPF without redundancy. However, sometimes customer requirements might force you to implement OSPF and VRRP.

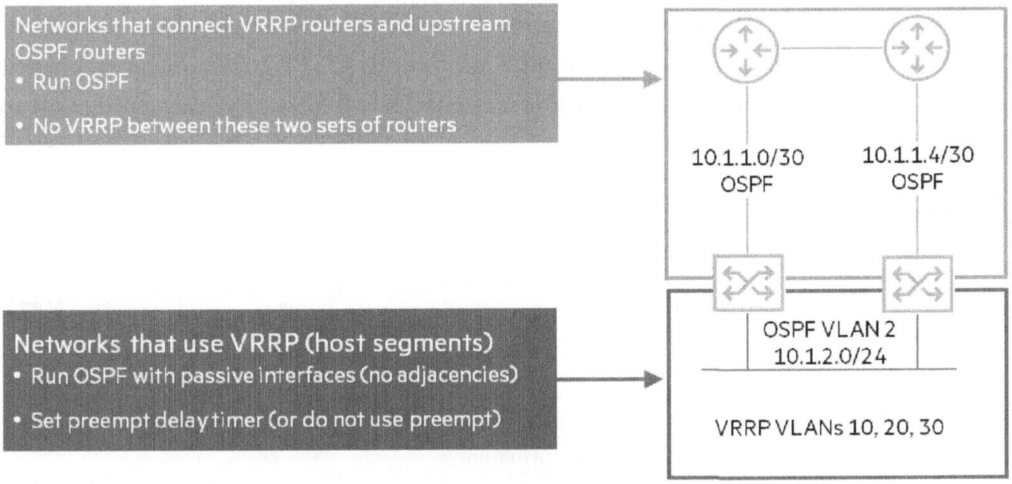

Figure 4-45: Implementation of OSPF with VRRP

You must design the solution carefully to ensure that the OSPF and VRRP functions will work together. When a router implements VRRP and OSPF on the same interface, it uses its actual IP address to run OSPF. However, the protocol interoperations can cause complexities. To minimize these complexities, you should avoid running OSPF and VRRP on the same interfaces.

First, connect VRRP routers to upstream OSPF routers on their own networks and do not run VRRP on these networks. This design helps an upstream router to immediately determine when it loses connectivity to a VRRP router.

Second, do not establish OSPF adjacencies on the interfaces that run VRRP (where the hosts reside). Because VRRP provides redundancy for default gateway services, VRRP runs on directly connected networks, which are generally stub networks. You can redistribute the directly connected networks as external networks. (Make sure to set up the area as a normal area or an NSSA.) HPE recommends that you configure the interfaces as passive OSPF interfaces.

Third, choose the VRRP design in which none of the VRRP routers own the virtual IP address. Otherwise, when the VRRP owner fails and comes back up, it immediately takes over routing on the interfaces for which it owns the VRRP address. However, it has not yet established adjacencies with upstream OSPF routers, disrupting connectivity on those VLANs. By configuring a virtual IP address that none of the VRRP routers own, you can prevent this problem. Either disable preempt mode or set a preempt mode delay time that is longer than the time required to establish adjacency with all neighbors and calculate the OSPF routes (generally, two or three minutes).

Virtual link

A virtual link in OSPF is a logical connection between two non-backbone areas through Area 0 (the backbone area) when a direct physical connection between the non-backbone areas is not available. Virtual links allow for the creation of a contiguous OSPF backbone area, ensuring network connectivity and proper OSPF operation in scenarios where a physical connection to Area 0 is not feasible.

How it works

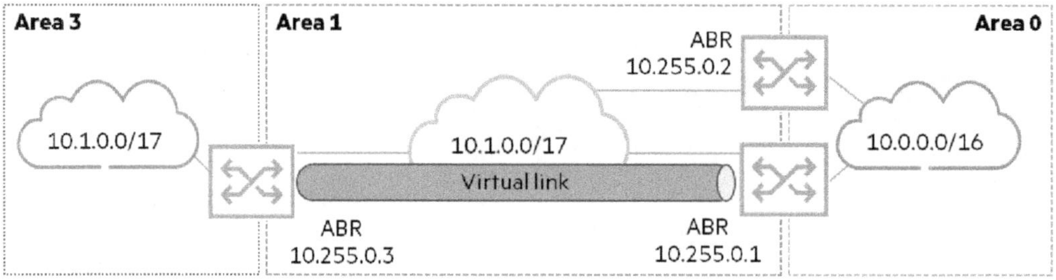

Figure 4-46: How virtual links work

- Establishes adjacency between two ABRs in the same area
- Like a point-to-point OSPF network
- Should be used a temporary solution

You establish the virtual link across a transit area (Area 1, in this example). ABR3 connects Area 3 to Area 1, and ABR1 connects Area 0 to Area 1. You form a virtual link between these two ABRs in Area 1. The virtual link acts like a shared network interface, allowing the two OSPF routing devices to exchange hellos and establish adjacency. Logically, the virtual link acts as a point-to-point OSPF network on which both ABRs have an interface.

You should not generally rely on virtual links as part of your design because they introduce complexity. They also negate some of the benefits of multi-area designs by leaking backbone routing information into non-backbone areas. Instead, use virtual links as a patch when you have no other way to create the necessary link to the backbone at the current time.

How to implement

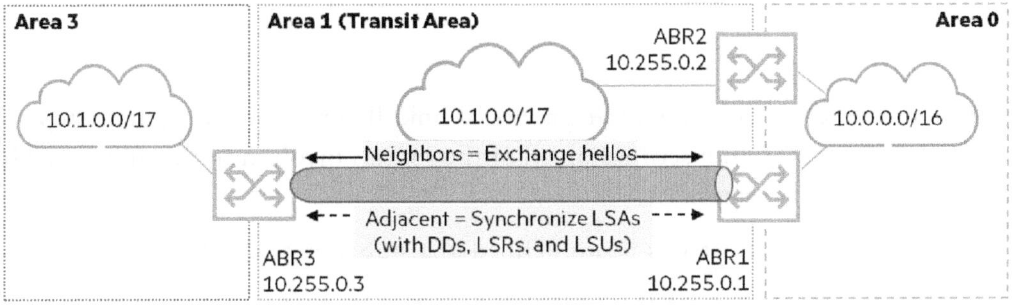

Figure 4-47: Virtual link implementation

Well-designed OSPF systems should ensure that you can always connect new areas directly to the backbone. However, sometimes the company expands in unexpected ways, and you cannot meet this requirement.

Figure 4-47 shows a scenario where the company must expand to new sites (10.1.128.0/17). Routers in Area 1 are available for connecting to the new routers. However, Area 1 has grown so large that you would prefer to establish the new sites in Area 3. However, the rules of OSPF prevent you from creating the new area—all areas must connect to Area 0, and all inter-area traffic must be via Area 0.

So, you create a virtual link between two ABRs in Area 1.

Similar issues can arise when you need to add a redundant link between an area and the backbone Area 0. As you have learned, you can create link aggregations between internal area routers and the ABR, which might be one router or, for even better redundancy, two physical routing switches in a VSF fabric. You can also connect internal area routers to two ABRs in Area 0. You should choose one of these solutions whenever possible. In this case, however, a second connection between a particular site and an ABR at the backbone is not available. You must establish the link through a transit site.

Configuring virtual link

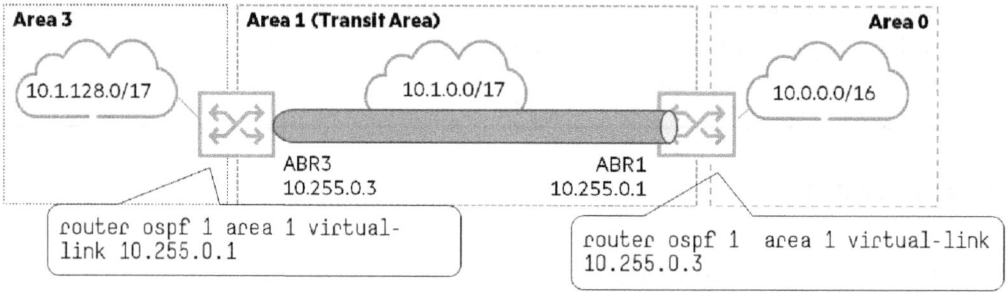

Figure 4-48: Configuring a virtual link

Figure 4-48 indicates the process for adding a virtual link. Before completing these steps, you should configure the basic OSPF settings on all of the routers involved in the solution. This section provides more detailed steps.

1. Make sure that the transit area is not defined as a stub area or an NSSA.
2. Configure the virtual link on the ABR that does not connect to Area 0.
   a. As mentioned previously, this ABR should have its basic OSPF settings defined:
      i. Router ID
      ii. The area without direct connection to the backbone (area 3 in the example)
      iii. The transit area (Area 1 in the example)
      iv. OSPF enabled on the appropriate interfaces in each area
      v. Any other settings required for the OSPF solution (such as aggregated area summaries)
   b. In the transit area (Area 1 in the example), configure a virtual link. Specify the router ID of the device at the other end of the link as shown in the figure.
3. Configure the virtual link on the ABR that connects to Area 0.

a. Again, this ABR should have its basic OSPF settings defined:
    i. Router ID
    ii. Area 0
    iii. The transit area (Area 1 in the previous example)
    iv. OSPF enabled on the appropriate networks in each area
    v. Any other settings required for the OSPF solution

b. In the transit area, configure a virtual link, specifying the router ID of the device at the other end of the link, as shown in the previous figure.

4. You must check consistency for the typical settings required for two OSPF routers on a link to achieve adjacency. These settings include the OSPF timers. (The virtual link does not have a network associated with it, so you do not have to check consistency for its type.)

Note that the OSPF routing devices have connectivity within the transit area. As always, if you want to be able to ping the router ID, use an IP address on a loopback interface that runs OSPF.

There are other options you can configure for virtual links. See the *IP Routing Guide* for more details.

Authentication

You implement authentication for OSPF processes on router interfaces. OSPF includes authentication data in all OSPF packets sent, and checks this data in all received OSPF packets. Only OSPF routers with matching authentication type and keys can establish neighbor relationships.

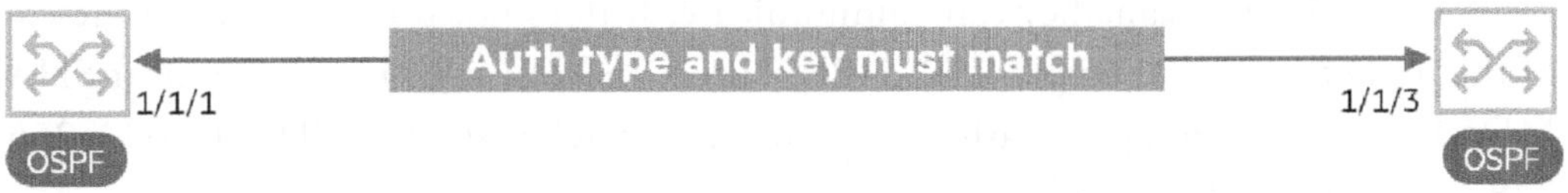

Figure 4-49: Authentication between two OSPF routers

Figure 4-49 shows the three authentication types, is described as follows:

Null

The default authentication type, in which OSPF interfaces do not enforce authentication. The OSPF packets have an authentication field, but the field does not include data. An OSPF interface with null authentication does not examine the authentication field; it simply processes all valid OSPF packets.

Simple

The OSPF interface places the authentication key in every OSPF packet header. Other OSPF routers examine this field and accept or reject packets accordingly. The actual password or key is in plain text, so anyone who can capture OSPF packets can learn the key. You should use this form of authentication only if you primarily aim to eliminate misconfigurations and routers inadvertently connecting on the wrong networks. You should not use it to secure your OSPF system from unauthorized routers and hackers.

Cryptographic

- The OSPF interface hashes the OSPF packet data and the key. It then places the hash at the end of the packet. The authentication field includes the key ID, a length for the hash, and a cryptographic sequence number; however, it does not include the key itself, protecting it from eavesdroppers. A router receiving the OSPF packet hashes the packet with its secret key, identified by the key ID. If the hash matches, the router accepts the packet. The cryptographic sequence number provides a degree of protection against replay attacks.

  An OSPF interface only accepts packets with a sequence number equal to or higher than the number for the most recent accepted OSPF packet from this neighbor. The sequence number might be a counter but is usually based on the system clock. In either case, the number always increases. If a hacker obtains a legitimate OSPF packet and attempts to retransmit it later, the packet's cryptographic sequence number is no longer valid, and the other routers reject it. A hacker could replay a packet in the window before the legitimate neighbor sends a packet with a higher sequence, but this window is quite small.

  Cryptographic authentication provides authenticity, ensuring that only OSPF interfaces configured with the same key can communicate. It does not provide privacy: the contents of the packet are not encrypted. All HPE routers and routing switches support SHA and MD5 for the cryptographic hashing algorithms. Switches support SHA-1, SHA-256, SHA-348, and SHA-512 key hashing.

Example:

- Plaintext:

  ```
  interface 1/1/1
  ip ospf authentication simple-text
  ```

- SHA:

  ```
  interface 1/1/1
  ip ospf authentication hmac-sha-256
  ```

- MD5:

```
interface 1/1/1
ip ospf authentication message-digest
```

- **Null**—The default authentication type, in which OSPF interfaces do not enforce authentication. The OSPF packets have an authentication field, but the field does not include data. An OSPF interface with null authentication does not examine the authentication field; it simply processes all valid OSPF packets.
- **Simple**—The OSPF interface places the authentication key in every OSPF packet header. Other OSPF routers examine this field and accept or reject packets accordingly. The actual password or key is in plain text, so anyone who can capture OSPF packets can learn the key. You should use this form of authentication only if you primarily aim to eliminate misconfigurations and routers inadvertently connecting on the wrong networks. You should not use it to secure your OSPF system from unauthorized routers and hackers.
- **Cryptographic**—The OSPF interface hashes the OSPF packet data and the key. It then places the hash at the end of the packet. The authentication field includes the key ID, a length for the hash, and a cryptographic sequence number; however, it does not include the key itself, protecting it from eavesdroppers. A router receiving the OSPF packet hashes the packet with its secret key, identified by the key ID. If the hash matches, the router accepts the packet. The cryptographic sequence number provides a degree of protection against replay attacks.

  An OSPF interface only accepts packets with a sequence number equal to or higher than the number for the most recent accepted OSPF packet from this neighbor. The sequence number might be a counter but is usually based on the system clock. In either case, the number always increases. If a hacker obtains a legitimate OSPF packet and attempts to retransmit it later, the packet's cryptographic sequence number is no longer valid, and the other routers reject it. A hacker could replay a packet in the window before the legitimate neighbor sends a packet with a higher sequence, but this window is quite small.

  Cryptographic authentication provides authenticity, ensuring that only OSPF interfaces configured with the same key can communicate. It does not provide privacy: the contents of the packet are not encrypted. All HPE routers and routing switches support SHA and MD5 for the cryptographic hashing algorithms. Switches support SHA-1, SHA-256, SHA-348, and SHA-512 key hashing.

Example:

- Plaintext:

```
interface 1/1/1
```

```
ip ospf authentication simple-text
```

- SHA:

```
interface 1/1/1
ip ospf authentication hmac-sha-256
```

- MD5:

```
interface 1/1/1
ip ospf authentication message-digest
```

Learning check

Check your understanding of Advanced OSPF by answering the following questions.

Question 1

What is the primary advantage of configuring multi-area OSPF in a network?

- O It simplifies the routing table by summarizing routes within each area.
- O It allows for the use of different routing protocols within each area.
- O It increases the overall bandwidth available for network traffic.
- O It eliminates the need for route redistribution between different OSPF instances.

Question 2

Two OSPF routers are on the same network. One defines the dead interval as one second and another uses the default. What will happen?

- O The router that defines the network as point-to-point becomes DR.
- O The router that defines the network as broadcast becomes DR.
- O The routers cannot successfully become neighbors.
- O The two routers cannot become fully adjacent neighbors and cannot install each other's LSAs in the LSDB.

Question 3

Which type of routes should you expect internal routers in a stub area to learn?

- ○ Intra-area routes only.
- ○ Intra-area routes and a default route.
- ○ Intra-area routes, a default route, and inter-area routes.
- ○ Intra-area routes, a default route, inter-area routes, and external routes.

Question 4

In an OSPF multi-area configuration, which statement is true regarding the role and responsibilities of an Area Border Router (ABR)?

- ○ An ABR is responsible for summarizing routes between OSPF areas.
- ○ An ABR can exist in only one OSPF area at a time.
- ○ An ABR does not participate in the backbone area (area 0).
- ○ An ABR does not exchange routing information between areas.

Question 5

What purpose does configuring OSPF BFD on an AOS-CX switch Layer 3 interface fulfill?

- ○ The switch can restart OSPF without affecting connectivity during the restart.
- ○ The switch detects when it has lost connectivity with an OSPF neighbor more quickly.
- ○ The switch avoids connecting to a rogue or unauthorized OSPF device.
- ○ The switch detects OSPF configuration errors and automatically mitigates their effects.

5 Border Gateway Protocol

After learning how to configure and manage the advanced OSPF, you will learn how to implement Border Gateway Protocol (BGP) routing. This chapter introduces you to BGP, a powerful routing protocol designed for ISPs and very large enterprise networks.

You will learn to describe and deploy BGP in customer networks, along with strategies for making BGP meet the specific requirements—BGP basics and use cases, set up neighbors, and control advertisements. You will learn how to influence route selection with route maps and understand how route reflectors can improve BGP scalability. Of course, some lab activities tie it all together.

OBJECTIVES

After completing this chapter, you should be able to:

- ✓ Understand the BGP protocol and it's route metrics
- ✓ Establish and monitor BGP sessions
- ✓ Advertise an IP block to multiple ISP routers
- ✓ Configure a BGP router to advertise a default route in OSPF
- ✓ Filter routes to prevent transit networks

BGP overview

In order for autonomous systems to share routing information and route traffic between networks, BGP is a crucial protocol for interdomain routing on the internet. It offers the policy-based routing capabilities, scalability, and flexibility required to manage the intricate internet routing infrastructure.

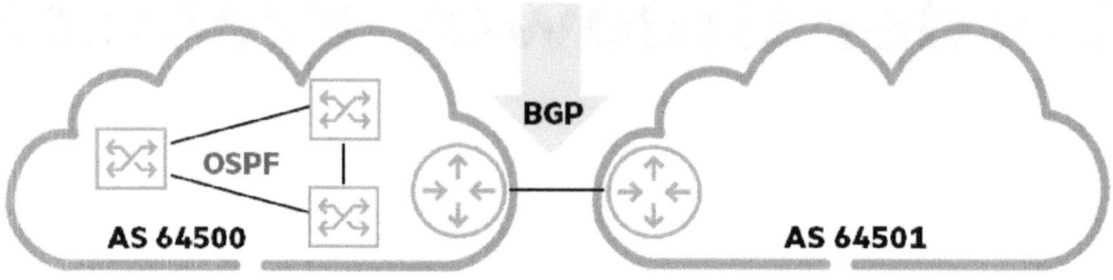

Figure 5-1: BGP

Some of its features are as follows:

Path vector protocol

BGP is a path vector protocol that uses a vector of autonomous system (AS) numbers to represent the path to reach a destination network. It takes into account multiple factors, such as policy decisions and route attributes, to determine the best path for routing traffic between ASes.

Interdomain routing

BGP is primarily used for interdomain routing, facilitating the exchange of routing information and traffic between different autonomous systems on the internet. It enables ISPs and organizations to advertise their network prefixes (IP address ranges) to other ASes and receive routing updates from them.

Policy-Based Routing

BGP allows network administrators to implement policy-based routing decisions based on factors such as route preferences, traffic engineering, and business requirements. Policies can be defined to influence route selection, control traffic flow, and enforce security measures.

Reliable and scalable

BGP is designed to be highly reliable and scalable, capable of handling the vast and complex internet routing infrastructure. It uses TCP as its transport protocol, providing reliability and error detection mechanisms for exchanging routing information.

Internet backbone protocol

BGP is the protocol that forms the backbone of the internet routing infrastructure, enabling connectivity between thousands of networks and ASes worldwide. It allows ISPs to establish peering relationships and exchange routing information with each other to ensure global connectivity.

Primary use cases of the BGP protocol

- Multi-homed ISP connections:
  - Multiple connections to multiple (or single) ISP
  - Granular path manipulation and control needed
  - You must share routing information with ISPs

Figure 5-2 illustrates BGP's primary use case: a company must communicate with multiple ISP routers. Your company might need to advertise different IP blocks over different connections. This allows the ISPs to advertise your company's networks to the rest of the world, such that some paths may be best reached via ISP 1, while others are best reached via ISP 2. You only need to do so when the company specifically needs to communicate routing information with the ISP.

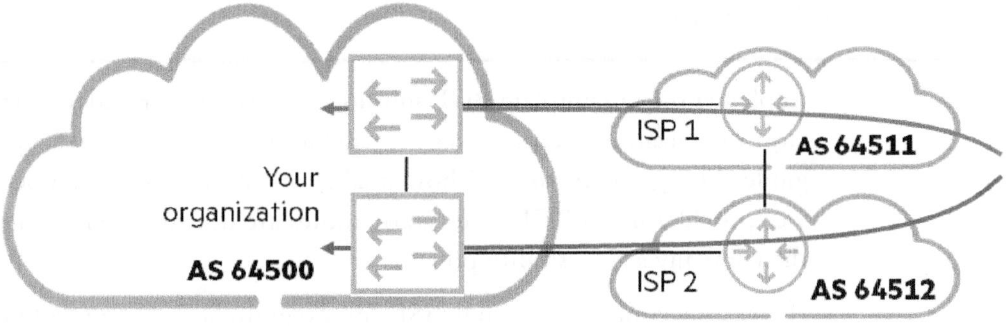

Figure 5-2: BGP with multi-homed ISPs

The figure's dual-ISP example shows a different router connected to each ISP. This improves resiliency over a single connection but can be slightly more complex because it requires two BGP routers, which communicate with each other as well as the ISP routers.

- BGP for ISPs and large company networks:
  - Within their own domains
  - With other ISPs
  - With customers and partners

Figure 5-3 illustrates another BGP use case: when a large company connects to a single provider and you need more granular path manipulation and control than is available with simple static routes. In other words, you only need to do so when the company specifically needs to communicate routing information with the ISP.

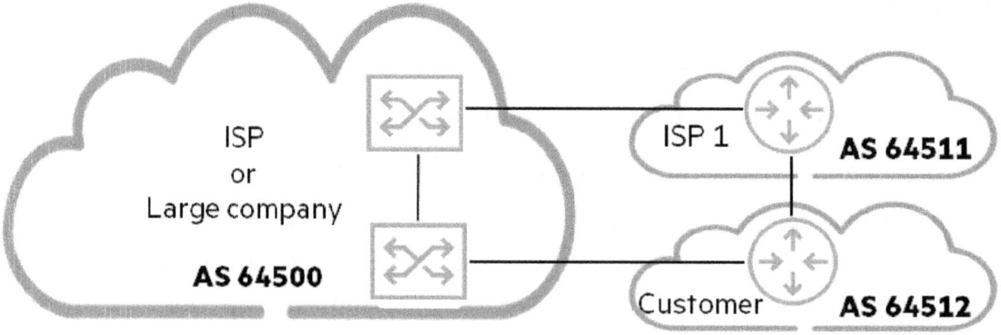

Figure 5-3: BGP with a single ISP or a large company network

ISPs and some bigger companies can have very large, complex networks—too large for OSPF. These organizations may use BGP to share routes within their domains, with other ISPs, and with customers and partners.

Implementing a complete BGP solution for a service provider or ISP, or for a large organization that involves confederations, is beyond the scope of this guide. This guide focuses on how smaller organizations can connect to a service provider using BGP when static routes are not sufficient or do not meet the company's accessibility policies.

If you only connect to a single ISP and no granular path control or manipulation is needed, BGP is not required. You can simply use static routes, reaching all internet sites via your ISP routers. The ISP handles announcing the block of public IP addresses assigned to the company.

BGP: a path vector routing protocol

Like Routing Information Protocol (RIP), BGP is a kind of distance-vector routing protocol. However, with RIP, the "hop count" is based on the number of routers to traverse or hop over. With BGP, it is based on the number of ASes to traverse. As shown in the following figure, using ISP 1, your company is three ASes away from any subnet inside AS 64514 (including the target AS). However, your company must only traverse two ASes to those subnets via ISP 2. All other things being equal, the shortest path is considered the best path. You will soon learn to use other criteria to control BGP path decisions—you will soon be the BGP puppet master!

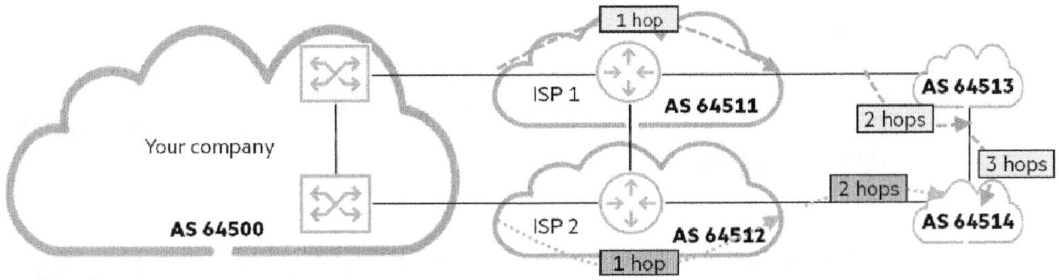

Figure 5-4: Path vector routing protocol

So, if you don't care about pathing into or out of your AS, you can simply use static routes. Some paths may be sub-optimal, but in many scenarios, it will not affect the user experience, nor your performance or security requirements. However, if you need this level of control, you need BGP.

You can also use VSF to combine two physical AOS-CX switches that support BGP, such as CX 6300 Series switches. You can then connect to the two ISPs on different physical chassis, but still retain the simplicity of a single (logical) BGP router for both connections.

BGP uses autonomous system (AS)

The AS forms the fundamental building block of BGP, and each organization, such as a company or ISP, generally has its own AS, identified by a unique number. Originally, BGP supported only 16-bit AS numbers, ranging between 0 and 65535. The following table shows how AS numbers are used.

Table 5-1: Distribution of the AS numbers

Purpose	AS numbers
Reserved	0 and 65535
Public IANA-assigned	1 – 64495
Documentation/examples	64496 – 64511
Private	64512 – 65534

Because 16-bit numbers allow for only 65,535 unique numbers, Request for Comments (RFC) 4893 establishes 32-bit AS numbers, increasing the range of AS numbers to 4 billion.

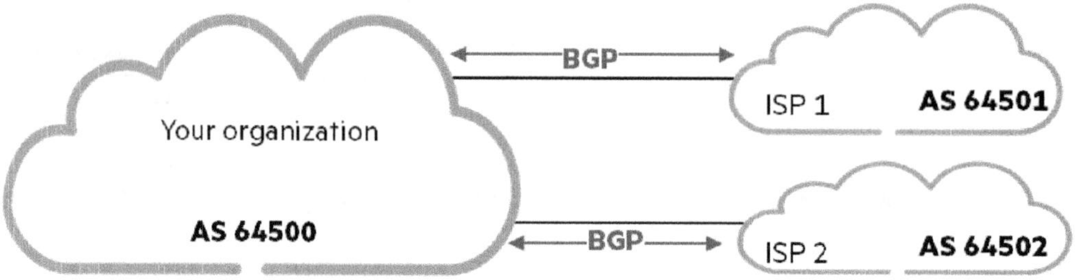

Figure 5-5: BGP deployment with AS numbers

AOS-CX switches support 32-bit numbers in one of two formats:

- 1-4294967295: The Autonomous System Number (ASN) for the domain
- 0.1-65535.65535: The ASN for the domain in dotted decimal format

When a company determines that it needs to run BGP in a multi-homed scenario, it must obtain a valid public AS from the proper regional authority.

iBGP and eBGP

BGP operates in one of two forms, depending on the autonomous systems to which the communicating neighbors belong:

- Neighbors in different autonomous systems communicate with external BGP (eBGP).
- Neighbors in the same AS communicate with internal BGP (iBGP).

The same router might use eBGP with one neighbor and iBGP with another, as shown in the following figure.

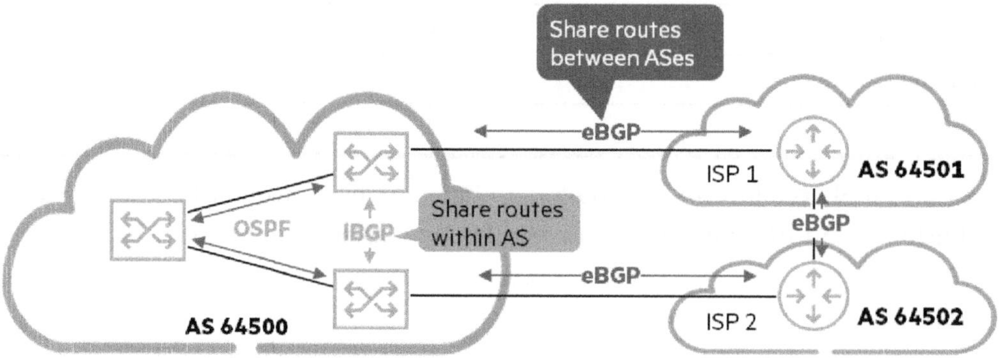

Figure 5-6: iBGP and eBGP deployment

Understand that iBGP is not the only routing protocol running within the AS. The AS also runs an Interior Gateway Protocol (IGP), such as OSPF or RIP, which is localized to the AS and handles routing the local traffic. Typically, just a few routers in the AS—the routers that connect to other autonomous systems and perhaps some core routers—run BGP in addition to the IGP.

BGP neighbor connections

A router that shares a common BGP session with another router is referred to as a BGP neighbor in the context of computer networking. A BGP interface that is configured on both routers, along with matching BGP settings, including an ASN, are required for the two routers to establish this connection. The two routers can exchange routing information and communicate with one another as needed once they have established a BGP neighbor relationship.

It's important to note that just because two routers can establish a BGP neighbor relationship, it does not necessarily mean that they will exchange routing information; this depends on whether or not their respective networks have been configured to do so. In summary, a BGP neighbor is simply another router with which one has established a successful BGP session.

BGP sessions

When you enable OSPF on an interface, it automatically discovers neighbors. With BGP, you must manually define BGP neighbors to establish connections. BGP neighbors establish a TCP session on port 179. They then open a BGP connection within the session.

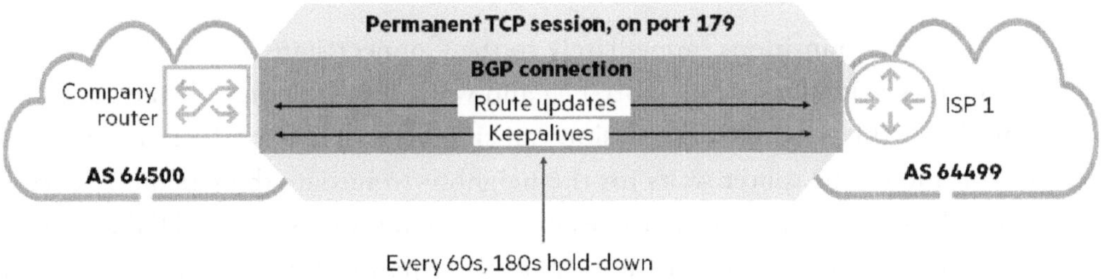

Figure 5-7: BGP connection

As shown in **Figure 5-7**, this connection carries all messages and routing updates between the neighbors, and the TCP session provides a degree of reliability for the communications. The BGP connection also features keepalives, which help each router determine its neighbor's reachability. AOS-CX switches send keepalives every 60 seconds by default and have a hold-down timer of 180 seconds, three times the keepalive interval. With this BGP connection in place, routers can exchange their BGP routing tables.

BGP connection states

Understanding the BGP connection states is essential to BGP neighbor adjacencies and ensuring the stability and reliability of BGP routing in networks. These connection states help in understanding the progress and stability of BGP neighbor adjacencies, and monitoring the transition between these states helps in diagnosing connectivity issues and ensuring smooth BGP operation.

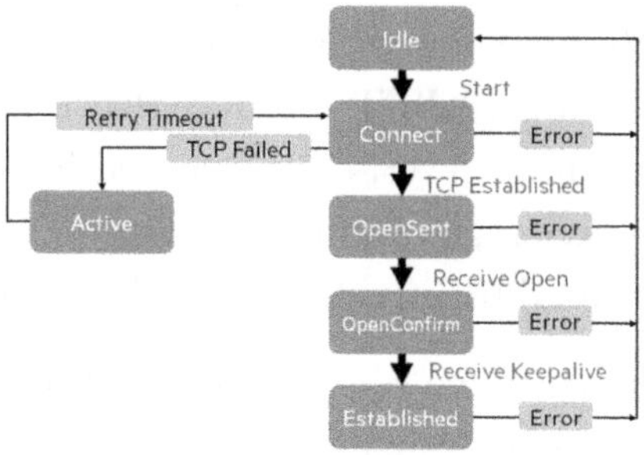

Figure 5-8: BGP neighbor states

To establish the BGP connection, the neighbors first open the TCP session on port 179. They then send BGP open messages with a list of parameters for the session and, assuming the parameters match, finally send the first keepalive of the session.

You can trace the states through which a BGP connection might pass:

The BGP connection begins in the Idle state and returns to this state when the session establishment times out without success or an error occurs.

The connection typically transitions immediately to the Connect state, in which the router attempts to contact its neighbor at TCP port 179 and waits for a response. If the router is set to Passive TCP mode for this neighbor, the session transitions to Active mode, rather than to Connect mode. In this mode, the router waits for the neighbor to initiate the session. The connection also moves to the Active state if the router does not receive a response to its TCP session initiation. In this state, the router waits for the neighbor's response and also, after a timer expires, might try to initiate the connection again. The router might wait in the Active state for a significant period of time if the TCP session fails to establish. If you see a router in Active mode after several minutes, the neighbor might be unreachable.

After the BGP neighbors have established the TCP session, the router sends an open message, either immediately or after a configurable delay. The connection also transitions to the OpenSent state or directly to the OpenConfirm state, depending on whether the router has already received the neighbor's open message.

In the OpenSent state, the router waits for the neighbor's open message. Once it receives that message, it verifies that necessary parameters match (for example, the neighbor's AS matches the remote AS configured on your router). Assuming that the parameters match, the router sends a keepalive and transitions to the OpenConfirm state. If the parameters do not match, the router sends an error notification and returns to the Idle state. Therefore, if you see a router stuck in the Idle state, you might want to check that you configured the local AS and neighbor AS correctly.

In the OpenConfirm state, the router waits for the neighbor's keepalive message and also sends its own keepalive. It also listens for error notifications such as one indicating that the open message parameters do not match.

After receiving the neighbor's keepalive, the connection transitions to the Established state. The connection remains in this state during normal operation. The neighbors continue to send keepalives. If a BGP router fails to receive a number of keepalives, it ends the connection. Note that the keepalive period is quite long (five minutes, by default).

New eBGP neighbors exchange their full BGP routing tables. After the eBGP neighbors have exchanged their BGP routing tables, they only exchange updates. For example, a BGP router sends an update if configuration changes affect routing policies or if a topology change has occurred.

Routes remain in the BGP routing table unless an update occurs. A router can also request a route refresh from a neighbor, which might be necessary so that the router can apply new policies to routes that it has already received.

Idle: The BGP connection begins in the Idle state and returns to this state when the session establishment times out without success or an error occurs.

Connect: The connection typically transitions immediately to the Connect state, in which the router attempts to contact its neighbor at TCP port 179 and waits for a response.

Active: BGP is actively trying to establish a TCP connection, but the connection is not yet established. Router enters into this state after attempting to establish a TCP connection in the Connect state but it fails.

OpenSent: BGP has sent an OPEN message to its peer and is waiting for an OPEN message in response. Router enters into this state when the BGP router sends an OPEN message to its peer.

OpenConfirm: BGP has received an OPEN message from its peer and is waiting for a KEEPALIVE message. Router enters into this state when the BGP router receives an OPEN message from its peer.

Established: BGP peers have exchanged OPEN messages and confirmed the session. The BGP session is established, and routing information exchange can begin. Router enters into this state after successful completion of the OpenConfirm state.

Neighbor authentication

- Uses MD5 authentication
- Hashes IP payload
- Ensures that the peer knows the correct password
- Provides data integrity for all BGP messages

The ISP might require your router to authenticate. If so, you simply add a password when you define the neighbor. The BGP neighbors use this password to hash the TCP payload of packets that carry the BGP messages, appending the hash to the packet.

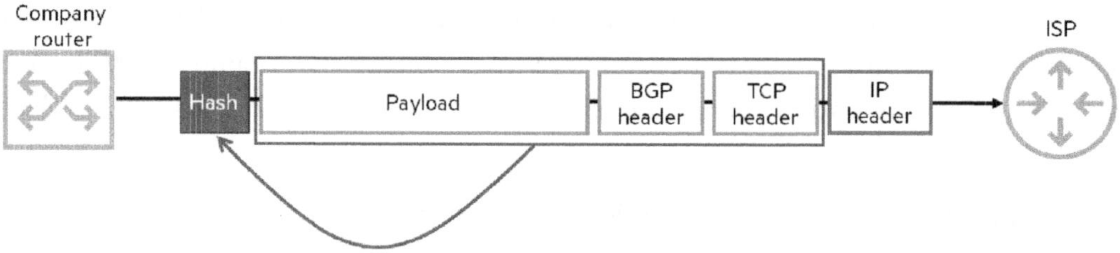

Figure 5-9: Neighbor authentication

If the neighbors' passwords do not match, the TCP session for the BGP connection fails to establish.

Because the neighbors add the MD5 hash to each packet sent on the TCP session, MD5 authentication provides data integrity for all BGP messages and for the underlying TCP session. In other words, each neighbor knows that the proper neighbor sent the message and that no one has altered the message. However, MD5 does not provide protection against eavesdropping.

iBGP neighbors

Unlike other protocols, iBGP neighbors need not be directly connected. The iBGP peers seen in **Figure 5-10** are directly connected, but there could be a large cloud of routers running OSPF between them. If you peer iBGP routers using a physical interface, and that physical interface or link fails, the peer is down, and routing is broken.

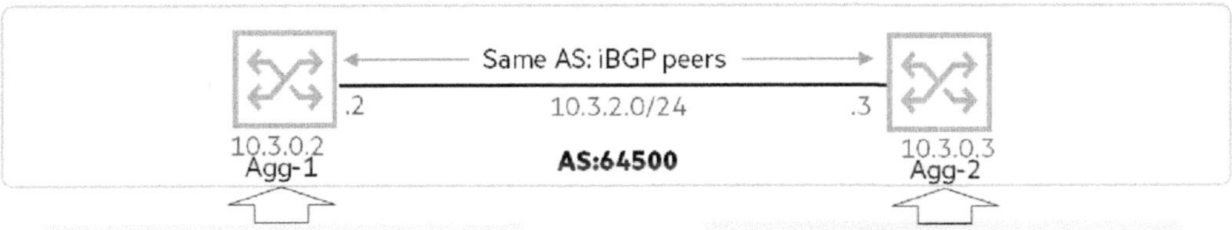


```
interface loopback0
  ip address 10.3.0.2
255.255.255.255
router bgp 64500
 bgp router-id 10.3.0.2
 neighbor 10.3.0.3 remote-as 64500
 neighbor 10.3.0.3 update-source
loop 0
 address-family ipv4 unicast
   neighbor 10.3.0.3 activate
```

```
interface loopback0
  ip address 10.3.0.3
255.255.255.255
router bgp 64500
 bgp router-id 10.3.0.3
 neighbor 10.3.0.2 remote-as 64500
 neighbor 10.3.0.2 update-source
loop 0
address-family ipv4 unicast
   neighbor 10.3.0.2 activate
```

Figure 5-10: iBGP peer configuration

Instead, you configure iBGP routers to peer to each other's loopback address, and they rely on a highly redundant OSPF system to route packets between these loopbacks. If one path fails, the peer session can be maintained via another path.

The configuration shown in **Figure 5-10** is very similar to an eBGP multihop configuration. Please take a moment to review. Especially important is to use the `neighbor update-source` command and activate the neighbor.

iBGP peers do not advertise routes learned from other iBGP peers. This may seem odd if you are used to OSPF, or any other IGP. OSPF is based on the fundamental idea that OSPF Router-A shares what it learns from OSPF Router B with all other routers—not so with iBGP. This leads to the iBGP full-mesh requirement; every iBGP router must peer with every other iBGP router in your AS.

Most organizations only have a few BGP routers, so this is not a problem. Two routers – two peers, three routers – six peers (two per router), four routers – 12 peers (three per router).

But large organizations may have dozens or even hundreds of iBGP routers. Imagine how large and unwieldy the configuration shown in **Figure 5-10** would be if you had to configure 900 peers on each router! Later in this chapter, you will learn about route reflectors and confederations to mitigate this issue.

Understanding the peer details

Figure 5-11 shows output for Agg-2's `show bgp ipv4 unicast neighbor` command, edited to focus our attention. The output is split into two sections, one for each BGP neighbor or peer.

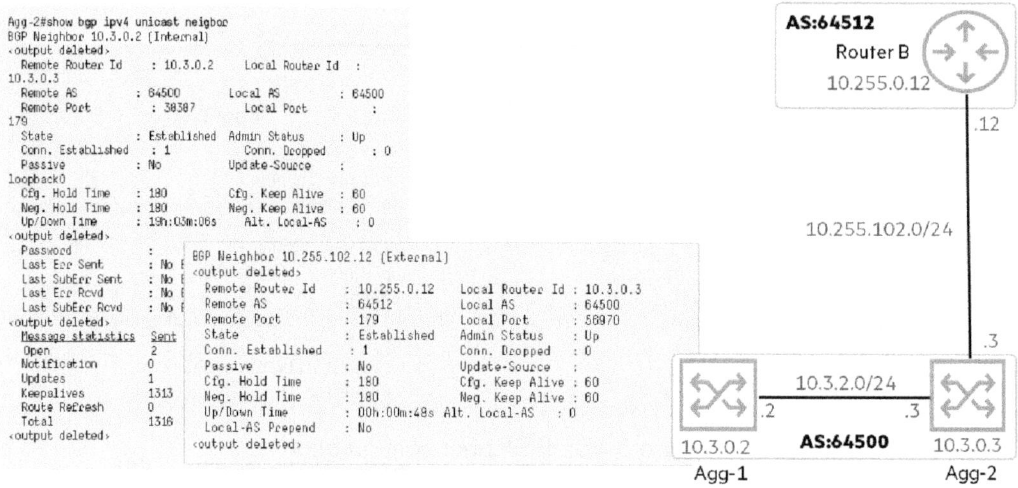

Figure 5-11: Neighbor details of Agg-2 switch

Neighbor 10.3.0.2 is an internal (iBGP) peer. You see the local and remote router IDs and ASes, and you see that, indeed, BGP uses TCP port 179.

The state/status is Established/Up. You can validate that the update source is loopback 0, as well as the hold and keepalive timers of 180 and 60 seconds, respectively. Pay attention to the Up/Down Time: consistently short up times mean that the peers are consistently losing connectivity, meaning you may have a network stability issue. You can see that there is no password configured. Finally, you see the messages you recently learned about: Open, Notification, updates, keepalives, and route refresh.

The next section is for neighbor 10.255.102.12, Agg-2's eBGP peer, which holds the same information as the first section.

Interpreting the peer summary

The detail you get with the `show bgp ipv4 unicast neighbor` command is often quite useful. However, you can see a nice summary of this information, often all you really need, with the command `show bgp ipv4 unicast summary`.

The `summary` command proves that you have successful BGP peering, but the `show bpg ipv4 unicast paths` command shows that no routes have been learned.

Focus on the sections in **Figure 5-12**'s CLI output labeled "1," "2," and "3" while studying the information below.

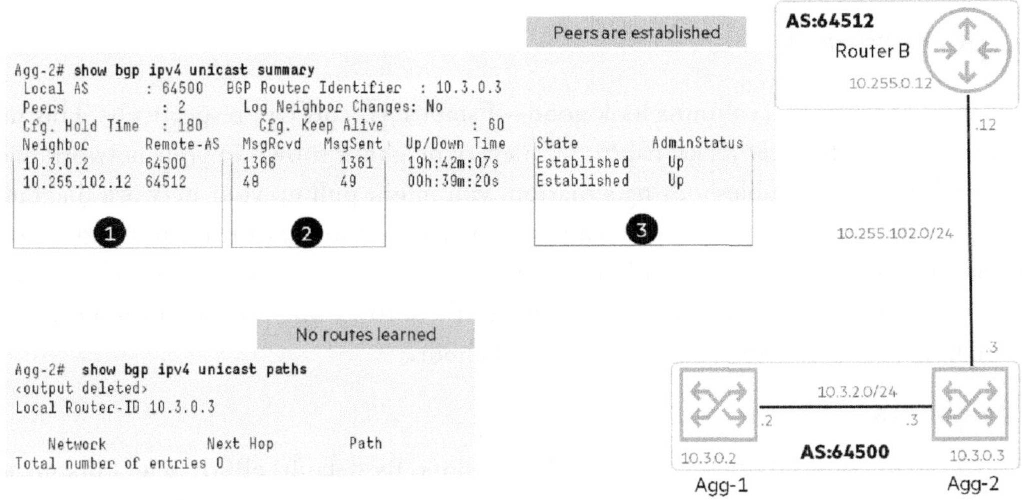

Figure 5-12: Output of neighbor summary

CLI output section 1

The neighbor column shows who this router is peering with. It is very effective for both learning and troubleshooting to compare the output of `show` commands to a network diagram. Play a game: Look at the diagram, and try to predict what you will see in the output. The diagram shows that router Agg-2 has two peers. One is an eBGP peer connected on a single physical link, so the peer is formed with the physical address 10.255.102.12. The other peer is an iBGP neighbor, and so should peer with the loopback address 10.3.0.2.

You know how to confirm whether a peer is iBGP or eBGP, yes? The first line of the output shows that Agg-2's Local AS is 64500. Neighbor 10.3.0.2 has the same AS—iBGP. Neighbor 10.255.102.12 is in a different AS: 64512. Thus, you know this is an eBGP neighbor.

CLI output section 2

The Messages Received and Sent columns are greater than zero, so you know that the peers are successfully communicating with each other. You will likely see these number slowly increment over time.

The Up/Down Time column can help you with forensics. Neighbor 10.3.0.2 has been up for over 19 hours and 42 minutes, while the other peer has only been up for about 39 minutes. You might ask, "What happened 39 minutes ago?" Perhaps this router was just brought up for the first time, so this is normal. Or maybe this peer was established months ago. You might explore log files and other resources to determine why this connection was dropped.

CLI output section 3

The State and AdminStatus columns look good—Established and Up, respectively. This means you have fully successful peer relationships with every neighbor shown in your network diagram. During a high-pressure troubleshooting situation, you might pull up your network diagram and see that Agg-2 should have three connected neighbors, so you look at the output and validate that you see the rows with Established and Up. If you only see two rows, you know that one router is missing (or your diagram is out of date). Or you see three rows, one for each neighbor, but one peer is stuck in one of the other states you learned about.

eBGP neighbors

Figure 5-13 shows connectivity for two eBGP neighbors. By default, eBGP neighbors must establish the session on a direct Layer 3 connection. The eBGP protocol enforces this criterion by checking that the neighbor IP address is on a network connected to the local BGP router and by using messages with a TTL of 1.

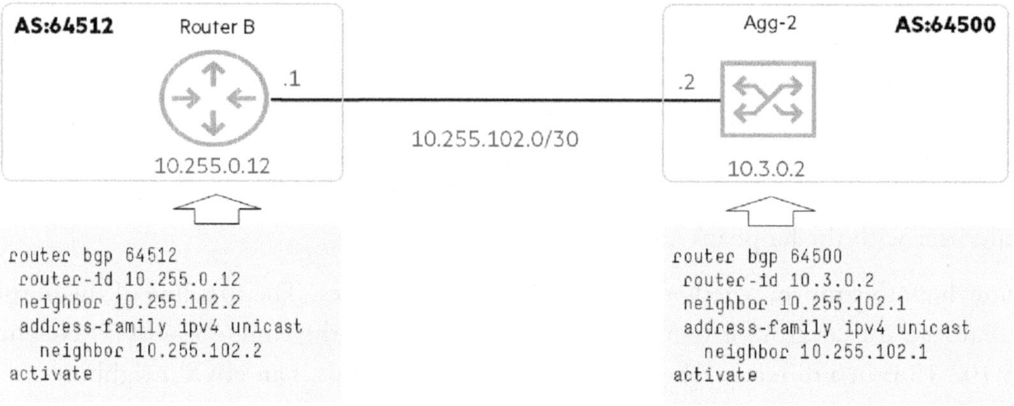

Figure 5-13: eBGP neighbor configurations

The configuration is shown in this figure. You configure each router with their respective AS. Since the AS numbers are different, the peers automatically know that this is an eBGP connection. Then you specify a router ID, just like with OSPF, often a loopback address. Now you specify the physical IP address of the directly connected neighbor. Finally, you must activate the neighbor. This begins the TCP port 179 handshake and establishes a BGP session.

eBGP multihop peers

Figure 5-14 shows an exception to the directly connected eBGP rule. If you have multiple physical links between eBGP peers, they will benefit by peering via their loopback addresses to maximize resiliency.

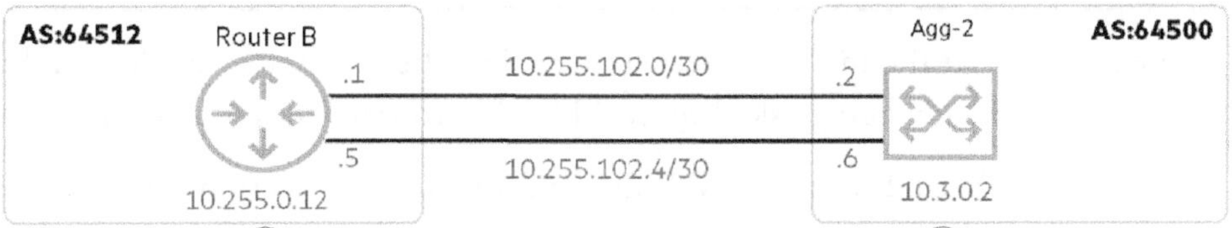


```
router bgp 64512
 router-id 10.255.0.12
 neighbor 10.3.0.3 remote-as 64500
 neighbor 10.3.0.3 ebgp-multihop 2
 neighbor 10.3.0.3 update-source
loop 0
 address-family ipv4 unicast
   neighbor 10.3.0.2 activate
ip route 10.3.0.3/32 10.255.102.2
ip route 10.3.0.3/32 10.255.102.6
```

```
router bgp 64500
 router-id 10.3.0.3
 neighbor 10.255.102.12 remote-as 64512
 neighbor 10.255.0.12 ebgp-multihop 2
 neighbor 10.255.0.12 update-source
loop 0
 address-family ipv4 unicast
   neighbor 10.255.0.12 activate
ip route 10.255.0.12/32 10.255.102.1
ip route 10.255.0.12/32 10.255.102.5
```

Figure 5-14: eBGP multihop peers

As shown in this figure, you configure eBGP peers to use their loopback addresses, just like you typically do with iBGP peers. The difference is that with eBGP, you must configure the `neighbor ebgp multihop` command highlighted in the figure. The number 2 at the end of the command is a hop count: it sets the TTL field to 2, so the peer can only be two router hops away.

The static routes are configured so BGP can actually reach the loopback addresses. Router B has two static routes to Agg-2's loopback address via the directly connected links shown in the figure. Similarly, Agg-2 has two static routes to Router B's loopback interface.

As configured, both links have the same administrative distance. The router will load-share across these links. If desired, you could adjust administrative distance or some other BGP metric, so that one link is the primary link. The other link is only used if the primary link fails.

Also understand that by default, each router sends BGP packets with its physical interface IP as the source address. Router B would send a packet to Agg-2 with source address 10.255.102.1 or .5. Agg-2 would receive this and think, "Who is this? I'm configured with neighbor 10.255.0.12." Agg-2 would discard this packet, and peering fails. Thus the `neighbor x update-source loopback 0` commands: You instruct Router B to send all BGP packets to Agg-2 with source address 10.255.0.12, which matches Agg-2's configured neighbor. Packets are accepted, and peering succeeds.

Another use case for eBGP multihop is when you must reach the ISP router that runs BGP through another router or routers. To set up this solution, you must set the eBGP multihop to the correct number (2 if one intermediate router intervenes, 3 if two routers intervene, and so on). Your router also requires a route to the ISP router's IP address, of course. However, it is not recommended to have intervening routers between eBGP neighbors, and you should avoid this setup whenever possible.

Understand that the multihop feature described here is fine to deploy, and quite common. You are only being asked to avoid having routers between eBGP peers.

As of AOS-CX 10.2, routing switches support BGP connections over GRE tunnels.

BGP route advertisements

You learned how to configure BGP peers, and you told the routers who to talk to. Now you must ensure they know what to say (advertise). What routes will BGP neighbors advertise? How do you get them to advertise routes as desired?

You know that iBGP routers do not advertise iBGP-learned routes, thus the full-mesh iBGP requirement. But they do advertise eBGP learned routes. So, how do you get a router to originate a route advertisement that will propagate through the system?

If you want BGP routers to advertise any other routes, such as routes to local networks, you must add those routes to the BGP routing table explicitly. For example, your BGP router needs to advertise the company's IP block to the ISP. You must add a route to that block to the BGP routing table.

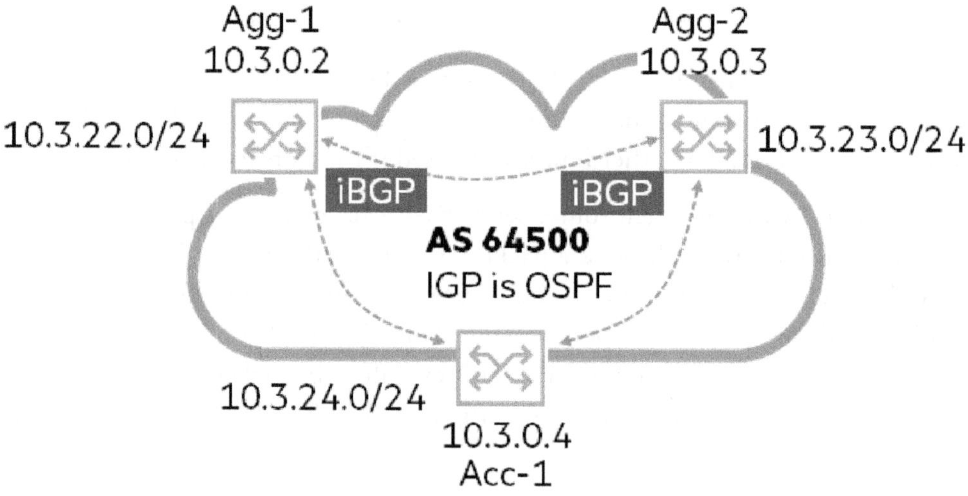

Figure 5-15: BGP advertising routes

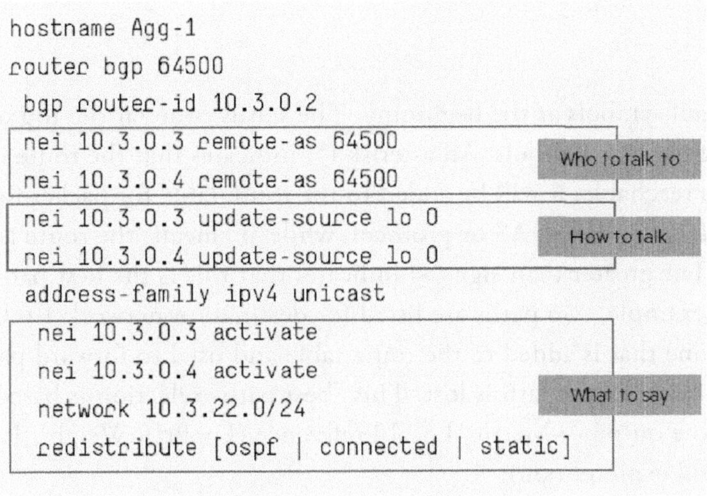

Figure 5-16: How to advertise the routes

Figure 5-16 shows the BGP configuration for Agg-1, with new syntax—network 10.3.22.0/24. You are telling Agg-1 that it should advertise this route. This makes sense, since this network is directly connected to Agg-1. Similarly, Agg-2 has network 10.3.23.0/24, and Access-1 has network 10.3.24.0/24.

Another way to get routers to originate a network is to redistribute it from another routing protocol: OSPF, directly connected, or static routes.

Let's look at some details about how BGP advertises these routes from the following figure.

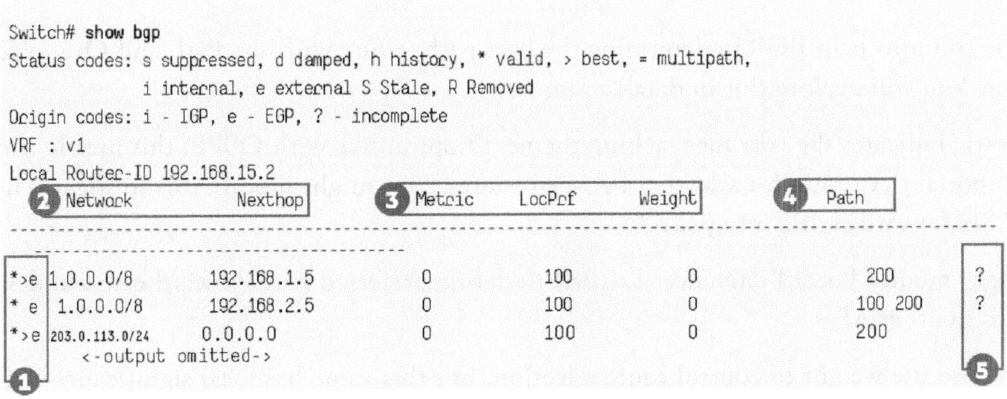

Figure 5-17: BGP routing table

Now your BGP routers are establishing a connection, advertising routes, receiving routes, and building a BGP routing table. You should be good at interpreting this table.

Focus on the number-labeled section of the CLI output in the figure while reading the information below.

#1

Each route has several symbols at the beginning. The status codes at the top of the output summarize the meaning of these symbols. An asterisk (*) indicates that the route is valid. This means that the next hop is reachable; it will be added to the route table for packet forwarding. The "e" means the route is from a foreign AS or protocol, while "i" means the route is a native BGP route from the local AS. The greater than sign (>) indicates that this is the best path to reach a destination. In the example, two paths are listed for destination network 1.0.0.0/8. The path via 192.168.1.5 is the one that is added to the route table and used to forward packets. The other path is not used unless the best path is lost. This "best path" selection is based on the other columns shown in the output—Metric, Local Preference (LocPrf), Weight, Path, and Origin Code (which you will explore soon).

#2

These fields indicate the destination network address and the next hop. You can use these fields to validate the expected networks exist, and that the next hop is as expected. You compare your network documentation to the output of this command. If you expect 100 networks and only 99 are listed, you determine the missing route and trace it back to its source. If the route exists but the next hop is incorrect, there would not be an asterisk (*) in the first column. You may have forgotten to add the `next-hop-self` parameter to one of this router's peers.

#3

All these columns help BGP to determine the best path, along with the Path and Origin Code columns. You will explore this in detail soon.

The metric indicates the cost for reaching the next hop; unlike with OSPF, this metric is not the most important criterion for selecting between routes. You might modify this to suggest how other ASes forward traffic into your AS.

You might modify Local Preference (LocPrf) to define preferred paths toward external destinations in other ASes.

You can also use weight to control route selection, but this value has local significance only. It is not advertised to other routers.

#4

The AS path attribute is included with every BGP route advertisement. A router in the AS that owns this route includes its own AS number in the path attribute, then sends it out to other ASes. A BGP router in a different AS receives the route, adds its AS to the path, and so on. In this way, the AS path indicates every AS that traffic will traverse to reach the destination. A shorter AS path is preferred.

Now you know why there is a ">" symbol next to the top entry for 1.0.0.0/8, why it is the best path. This router learned about 1.0.0.0/8 from router 192.168.1.5, which is in AS 200. You know this because only AS 200 is listed in the Path attribute. It also learned it from next-hop router 192.168.2.5 in AS 100, which learned it from AS 200. Since this is a longer AS path and all other criteria are equal, the shorter path is considered "best." The AS path attribute is also used as BGP's primary loop elimination technique. If a router receives a route with its own AS in the AS path, it drops the route.

#5

The last column is the "origin code"—where the route came from.

- "i" means the route probably came from an IGP (OSPF, RIP, and so forth) and was injected into BGP with the network command.
- "e" means the route came from the old Exterior Gateway Protocol (EGP), which is deprecated. If you see this code, you may have time traveled back to the 1980s!
- The "?" symbol means "incomplete"—there is not enough information available to determine how this route originated. This typically means the route was redistributed into BGP from a foreign routing protocol, perhaps OSPF or static.

BGP routing table versus IP routing table

The BGP routing table and the IP routing table work together to ensure efficient and accurate routing of packets within and between networks. BGP manages the selection and propagation of routes based on BGP attributes, while the IP routing table consolidates routes from various sources to make the final forwarding decisions.

The BGP table includes all routes learned via BGP. Some of these routes are valid paths; some are not. Some of these routes are the best paths; some are not. Only the valid, best paths are added to the IP routing table, which is actually used to forward packets.

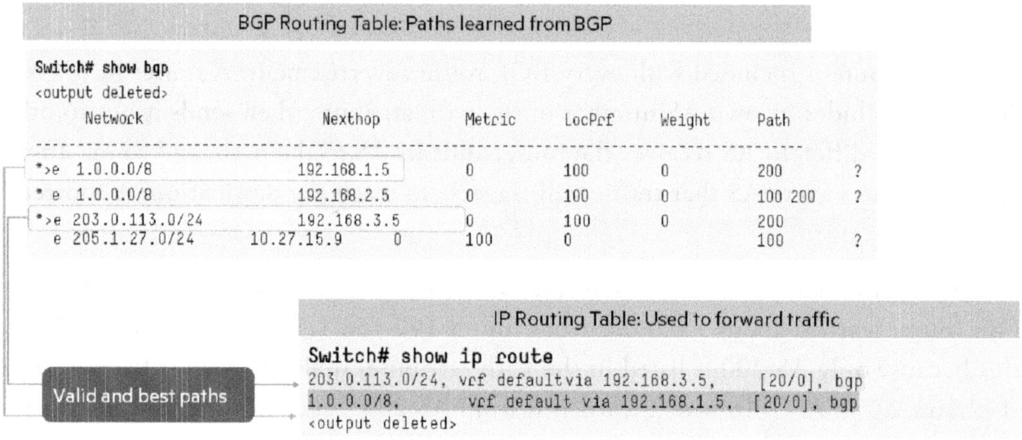

Figure 5-18: How the IP routing table is derived from the BGP routing table

The process of deriving the IP routing table from the BGP routing table involves the following steps:

1. Receiving BGP routes
2. BGP path selection
3. Injecting BGP routes into the IP routing table
4. IP routing table (RIB) update

The selected best BGP routes are installed into the IP routing table. This table now includes routes from various sources: directly connected networks, static routes, dynamic routing protocols, and selected BGP routes. The IP routing table is used by the router to make forwarding decisions. When an IP packet arrives, the router looks up the destination in the IP routing table and forwards the packet based on the next hop specified in the table.

Look at the BGP routing table in **Figure 5-18**. To see which of those paths are used to forward packets, look at the IP routing table. Two of the four BGP route table entries were added to the IP route table. The other 1.0.0.0/8 route was not added because it is not the best path—only best paths are added to the IP route table. The 205.1.27.0/24 route was not added because it is not a valid route. You can tell this because the route lacks the "*" symbol.

eBGP next hop behavior

"eBGP" next hop behavior refers to how the next hop IP address is handled when routes are advertised between BGP peers in different ASes. The following are the different scenarios in which next hop behavior changes:

Directly connected eBGP peers

When an eBGP router advertises a route to a directly connected eBGP peer, the next hop IP address is set to the IP address of the interface that the eBGP session is established on. This means the next hop is typically the IP address of the advertising router itself.

Multihop eBGP

In multihop eBGP scenarios, where eBGP peers are not directly connected (that is, they are multiple hops away from each other), the next hop is still set to the IP address of the router advertising the route. However, this requires additional configuration to ensure the next hop is reachable through intermediate routers.

Route reflectors (RR)

Route reflectors can be used within an AS to manage the propagation of routes between iBGP peers. When a route reflector advertises routes learned from an eBGP peer to other iBGP peers, the next hop remains unchanged by default. This means the next hop will still be the IP address of the eBGP peer from which the route was learned.

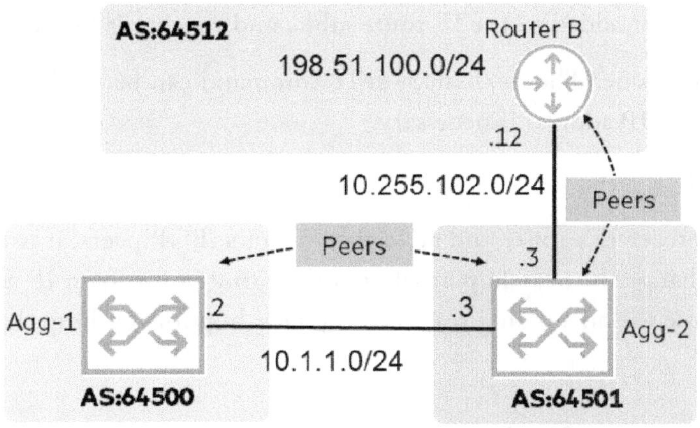

Figure 5-19: eBGP next-hop behavior

In this figure, notice that Agg-2 has two eBGP peers: 10.1.1.2 and 10.255.102.12. Router B is configured to advertise network 198.51.100.0/24. Router B advertises this route out its interface with IP 10.255.102.12.

To validate this, you go to Agg-2 and use the `show bgp ipv4 unicast paths` command. Agg-2 has this route in its BGP routing table. Agg-2 learned this route, with a next hop address of 10.255.102.12. Agg-2 is directly connected to this route, and so can reach the next hop address of 10.255.102.12. So, Agg-2 adds the route to its IP routing table; it knows it can actually reach this route and what eBGP peers share eBGP-learned routes, so Agg-2 advertises this route to its other peer, Agg-1. When Agg-1 advertises this route to an eBGP peer, it changes the next hop address to its own, 10.1.1.3. Agg-1 is directly connected to this next hop address, so it can definitely reach it. As shown in the figure, the path is added to Agg-1's IP route table.

As before, Router B advertises 198.51.100.0/24 out its interface with IP 10.255.102.12.

To validate this, you go to Agg-2 and use the `show gp ipv4 unicast paths` command. Both routers show this route in their BGP route tables; BGP has learned this route, with a next hop address of 10.255.102.12. Agg-2 is directly connected to that next hop, so it can reach it—the entry is added to its IP route table, and it routes packets to 198.51.100.0/24, all good. To verify this, you look at the IP routing table and see that, indeed, the route was added to Agg-2's route table.

However, this route is NOT in Agg-1's IP route table. In this scenario, Agg-1 and Agg-2 are iBGP peers, and iBGP peers do not change the next hop IP address of eBGP-learned routes. Router B advertises the target subnet with next hop 10.255.102.12. Agg-2 receives this and forwards it to Agg-1 with the same next hop, 10.255.102.12. Agg-1 is not directly connected to this next hop, and OSPF has not learned about this route.

So, although Agg-1 has learned about the route via BGP, it cannot use the route. The next hop is not valid, the entry is not added to the IP route table, and routing is broken for Agg-1.

How do you solve this issue? The `next-hop-self` command can be used on the router to change the next hop to its own IP address if necessary.

BGP next-hop-self

When a BGP router receives a route and advertises it to its iBGP peers, it uses the `next-hop-self` command to change the next-hop attribute of the route to its own IP address. This ensures that all iBGP peers use the advertising router as the next hop for reaching the destination network.

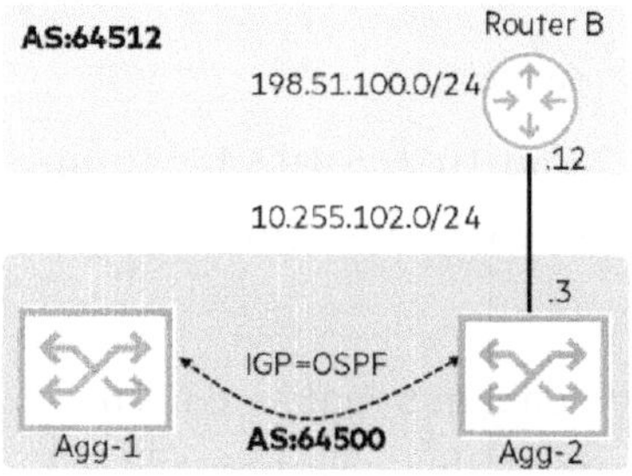


```
hostname Agg-2
router bgp 64500
  bgp router-id 10.3.0.3
  neighbor 10.3.0.2 remote-as 64500
  neighbor 10.3.0.1 update-source loop 0
  address-family ipv4 unicast
    neighbor 10.3.0.2 activate
    neighbor 10.3.0.2 next-hop-self
```

Figure 5-20: Configuring the `next-hop-self` command

The typical method to resolve the next hop issue is by configuring your iBGP peers with the `next-hop-self` parameter, as shown in **Figure 5-20**. This makes BGP act a bit more like OSPF. Agg-2 learned about the target subnet from 10.255.102.12 and forwards it on to Agg-1 with itself as the new next-hop, 10.3.0.3. This is proven by looking at the route on Agg-1; it has the expected next-hop address, as shown in the figure.

Another option is to tell Agg-2 to redistribute its directly connected network (10.255.102.0/24) into OSPF. All OSPF routers would thus know about this route, eliminating the need for the `next-hop-self` parameter. This works, but now all OSPF routers have this extra route in their route table from outside the network. This is not so bad when it is only one route, but as the network grows and changes, you must constantly keep up with all this redistribution and cluttering the route tables with external routes.

BGP route selection metrics and tuning

BGP routers typically receive multiple paths to the same destination. The BGP best path algorithm decides which is the best path to install in the IP routing table and to use for traffic forwarding. BGP assigns the first valid path as the current best path. BGP then compares the best path with the next path in the list, until BGP reaches the end of the list of valid paths.

When multiple identical BGP routes are valid and have the same administrative distance, BGP uses the criteria shown in **Table 5-2** to select the best route.

You see the BGP attributes and the preferred values for each one. As shown in **Table 5-2**, attributes are listed in the order by which the router examines them when selecting the best route. That is, the router selects the route with the better weight first. Only if those values are equal does the router move on to examine the next attribute, local preference. You can set a router's weight for specific routes. This value has meaning only on the router itself; it is not advertised. Local preference is not sent to eBGP peers, but it is one of the attributes sent in an update message to iBGP neighbors—those local to this router's AS.

Table 5-2: BGP path selection criteria

Attribute or criteria	Attribute type	Preferred values
Weight	N/A	Highest
Local preference	Discretionary	Highest
Locally originated	N/A	Prefer locally originated path, with next hop 0.0.0.0
AS path length	Mandatory	Local to this AS (learned by IGP or redistributed), then shortest path (fewest AS numbers in list)

Attribute or criteria	Attribute type	Preferred values
Origin	Mandatory	IGP < EGP < ? (i < e < ?)
MED	Optional, non-transitive	Lowest
eBGP over iBGP	Admin distance	Prefer eBGP over iBGP
IGP metric to next hop	N/A	Lowest
ORIGINATOR_ID	Optional, non-transitive	Lowest
CLUSTER_LIST	Optional, non-transitive	Smallest
Router ID	N/A	Lowest
Neighbor's IP address	N/A	Lowest

Without special manipulation of weight and local preference on your part, your router will tend to prefer routes based on the AS path. It will accept the route from the ISP that reaches the networks in question through fewer ISPs. Presumably, the shorter AS path indicates a shorter latency; however, this is not necessarily the case.

For routes with the same number of AS IDs in the AS path attribute, the origin might come into play. Routes that were originally injected into BGP are preferred to routes learned from EGP (deprecated) and redistributed routes. However, often the ISPs will have learned the route from a similar ultimate source, so the origin is identical.

Redistributed routes have an origin code of "?" (unknown) in the BGP table, as you will soon see.

The Multi-Exit Discriminator (MED) would take effect only if the ISP routers have explicitly set this attribute. They might do this if they needed to distinguish between multiple paths into your AS. This is only a consideration when you have multiple connections to the same ISP. Generally, a router only compares MED values between routes advertised from the same AS. However, in some cases, you can request different ISPs to send routes with different MED values. You must enter the `bgp always-compare-med` command to permit the AOS-CX switch to compare the MED values between the two ISPs.

However, without such special manipulation, if the AS path length is the same, your router will probably accept the route from the router to which it has a lower-cost connection. If the metric to both ISPs is the same, the choice becomes nearly arbitrary. If ECMP is enabled, in fact, the AOS-CX routing switch adds multiple routes with the same next hop cost and load balances over them.

If ECMP is disabled, the switch selects the route based on age, then router ID, and then the cluster list—but only if these checks are enabled. The final tie breaker is the advertising router ID.

In summary, here are the steps that could be involved in determining which BGP route is preferred:

1. Do not consider route if the next hop is unreachable.
2. Select the route with the highest weight.
3. Select the route with the highest LOCAL_PREF.
4. Select the route originated by the local router.
5. Select the route with the shortest AS-PATH.
6. Select the IGP, EGP, or Incomplete route in turn, referred to as the "origin" attribute.
7. Select the route with the lowest MED value.
8. Select the route learned from eBGP, confederation, or iBGP in turn.
9. Select the route with the smallest next hop metric.
10. Select the route with the smallest ORIGINATOR_ID.
11. Select the route with the shortest CLUSTER_LIST.
12. Select the route advertised by the router with the smallest Router ID.
13. Select the route advertised by the peer with the lowest IP address.

Prefix list

Prefix lists are named lists of route prefixes. They are used to match routes for inclusion in or exclusion from route policies. The sequence number determines the order of matching. The matches are performed starting from the lowest sequence number to the highest sequence number until there is a match. The sequence number is, however, optional and is autogenerated whenever it is not explicitly mentioned. All prefixes with the same prefix list name are grouped.

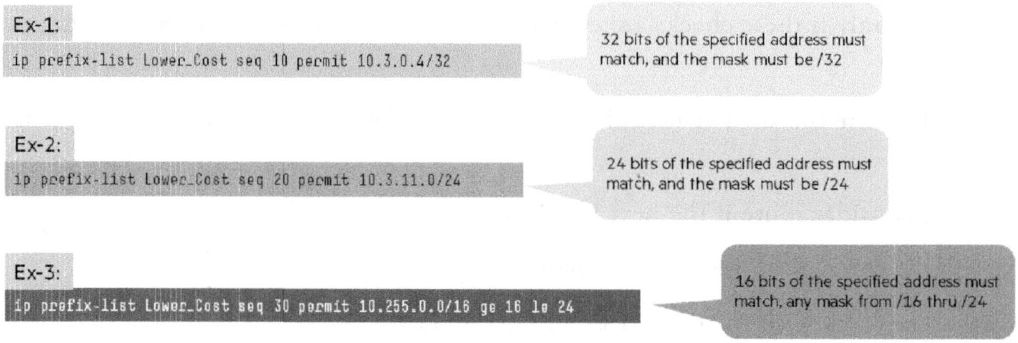

Figure 5-21: Examples of prefix lists

The auto-generated sequence number is derived by adding 10 to the highest sequence number available. This technique makes it possible to insert new prefix list sequence numbers in between.

Use the `ge` and `le` parameters to combine prefixes with a network mas rangek. For example, `10.255.0.0/16 ge 16 le 24` will match all prefixes within the 10.255.0.0/16 network that have a mask greater than or equal to 16 bits and less than or equal to 24 bits in length. For instance, subnets such as those shown at the left side of the figure would match because the first 16 bits of each address match 10.255, and because the mask is between 16 and 24 bits long, inclusive.

The examples on the right would not match, either because the first 16 bits of the address are different, or because the mask is outside of the specified range.

A prefix list is one of several ways to specify match criteria for a route map.

Route maps

You use route maps to route path selection and availability, by filtering or modifying route information. Routing policies can filter advertised, received, and redistributed routes. They can also modify attributes for specific routes. You use a three-step process to deploy route maps:

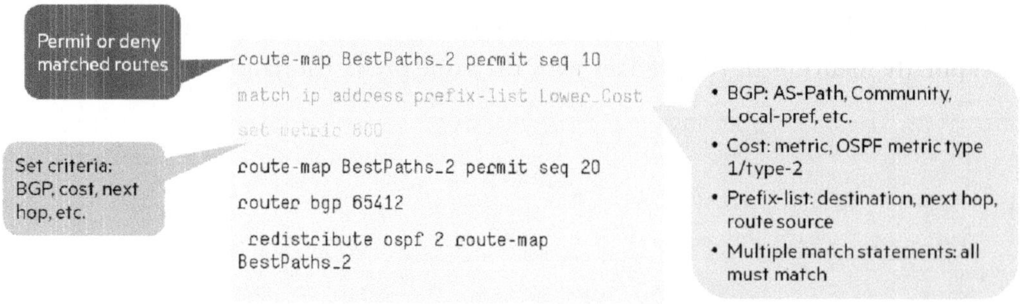

Figure 5-22: Route map example

Configure filters based on route attributes. You can use prefix lists to select a destination subnet, next hop, or route source. This figure lists other criteria you can match on, as you are about to learn.

Create route maps and apply those filters, along with appropriate set criteria as needed. Apply the route map to the redistribution statement.

The sequence numbers control the top-down processing order, with a permit statement to allow matched routes to be advertised or a deny statement to prevent their advertisement. Do you see the difference between route maps and ACLs? You often use ACLs to permit or deny user or server data traffic, and you use route maps to control the advertisement of routes.

If you specify multiple match criteria within a sequence numbered section, then all criteria must match—it is a logical AND operation. Suppose you added a second match statement under sequence 10 of `match ip route-source prefix-list Permitted_Sources`. Route advertisements are only permitted and set if they match two criteria: they must be destination subnets that match the prefix-list Lower_Cost, AND they must be sourced (advertised from) IP addresses matched in `prefix-list Permitted_Sources`.

If both criteria are not met, then there is no match for sequence 10, and sequence 20 is processed. Like an ACL, once there is a match, no more sequences are processed. Also like an ACL, there is an implicit "deny all" at the bottom. If you want unmatched routes to be advertised, simply add a line at the bottom `route-map BestPaths_2 permit sequence <XX>`. There is no need to specify any criteria or set; the statement serves to allow all previously unmatched route advertisements.

Weight

Weight is a metric that is not really part of the BGP standard, since it's not an attribute shared between BGP peers. Instead, the metric allows you to implement a policy on the local router as to which eBGP peer it prefers. The higher the weight, the more preferred the routes from the respective peer.

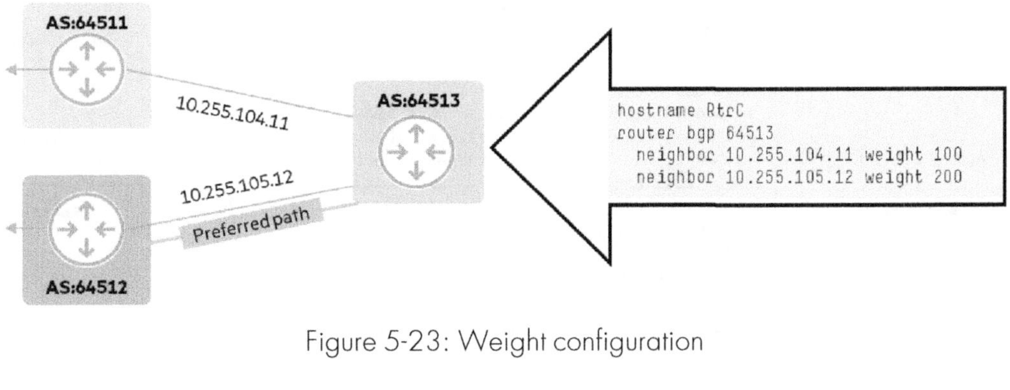

Figure 5-23: Weight configuration

shows an example of using weight to determine path selection by simply adding a weight value to the neighbor statements. You can also use route maps to adjust weight values for only specific routes from a neighbor, instead of all the routes of a neighbor, like shown in this example. If using a route map, you would use the `set weight` command to adjust the weight for matching routes.

Local preference

Local preference is used with iBGP to determine the best exit to reach a particular route. Higher values are preferred; the default is 100, and it is only shared with iBGP peers. This scenario uses local preference, such that AS 64500 prefers some paths to AS 64513 via Agg-1, and others via Agg-2.

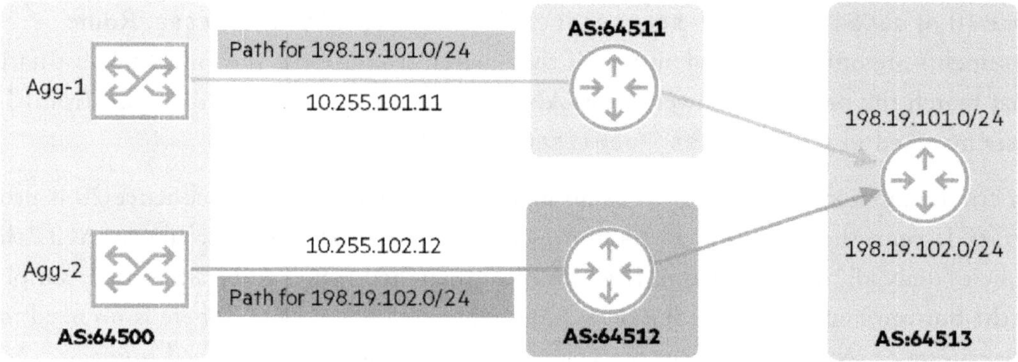

Figure 5-24: Use of local preference

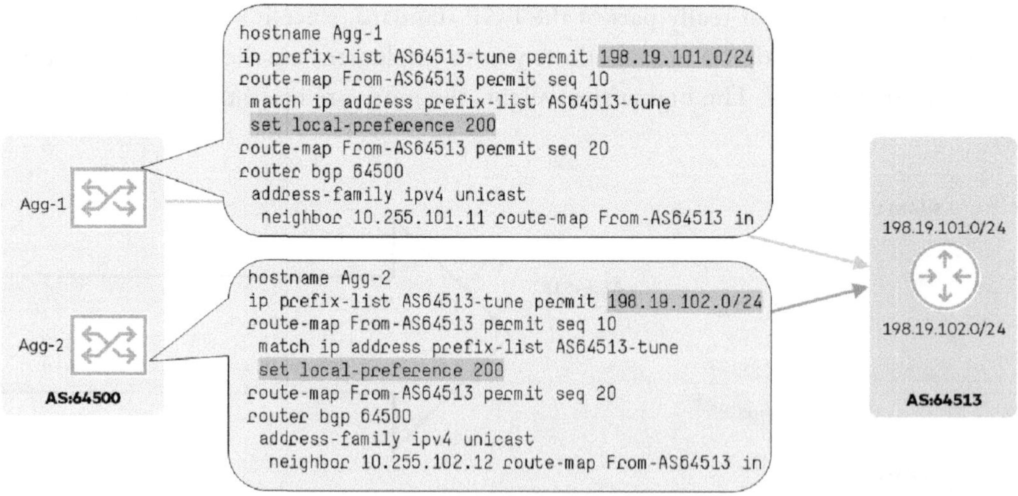

Figure 5-25: Local preference configuration

Figure 5-25 shows the configuration used to achieve the local preference scenario. The route maps for Agg-1 and Agg-2 are the same, except the prefix list matches on a different route.

AS path length

If weight or local preference is not defined, then AS path length determines the best path and also helps BGP detect route loops.

In the following example, all other things being equal, the best path to 192.0.2.0/24 is out Agg-1, directly to AS 64511, one hop away. The path out Agg-2 via AS 64512 is three hops or ASes away.

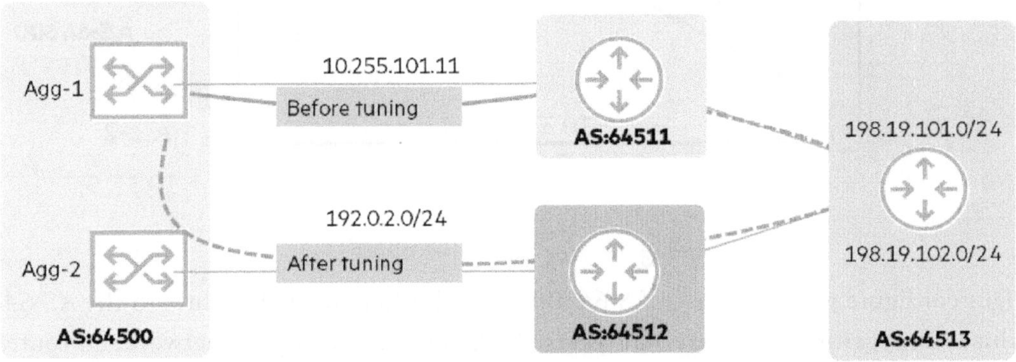

Figure 5-26: BGP AS path length metric

You can see this in the `show bgp ipv4 unicast` output, which shows all discovered BGP paths and what BGP believes to be the best paths. The greater than (>) sign indicates that, of the two possible paths, this one is the best path. Why? The weight and local preference are the same, and so "best" is determined by AS path length.

A shorter AS path is not necessarily the best path. Perhaps the link between Agg-1 and AS 64511 is very slow, and you want to tune the AS Path metric to use the other path. Let's do it.

By default, BGP does not do load sharing. The first "best" path used is what is placed in the routing table unless you enable Equal Cost Multipath (ECMP) for BGP.

Multiple exit discriminator (MED)

The MED attribute is shared with a multi-homed eBGP peer to influence how they will enter your network. **Figure 5-27** shows how AS 64500 and AS 64600 are multi-homed—one connection via the 10.3.54.0 link, and another via 10.3.44.0. This is the use case for MED.

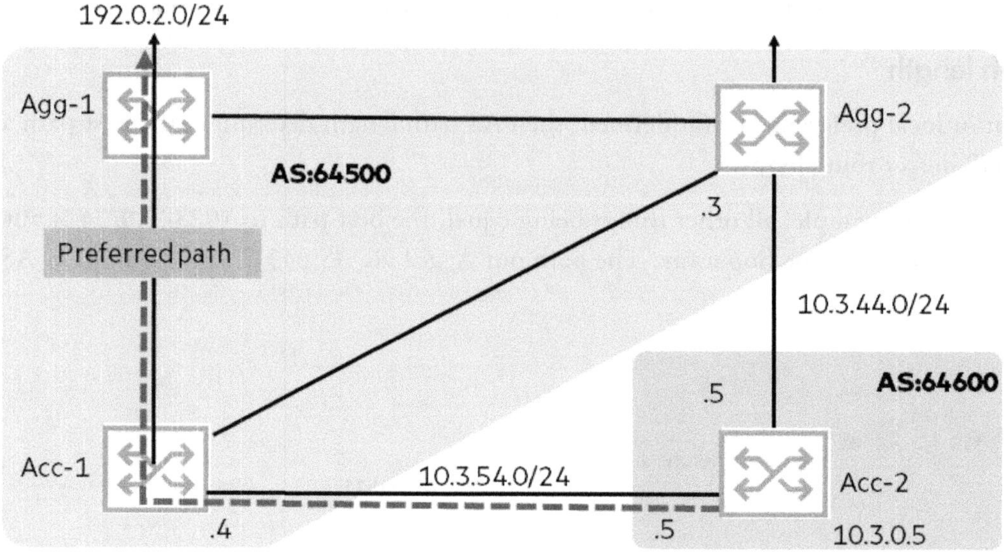

Figure 5-27: Example of MED

You might configure your routers in AS 64500 to send different MED values to the AS 64600 peer. This is a suggestion to these eBGP peers: "When packets exit your network, en route to my network, please use path A for some routes and path B for other routes."

AS 64500 might be a large cloud with thousands of subnets. You know that it is more efficient to reach some of your networks via the 10.3.44.0 link, while others should use the 10.3.54.0 link. So you make this suggestion to the external AS to forward traffic accordingly. Remember, it is called an Autonomous System because someone has autonomy (control) over it. You cannot force the AS to pay attention to your MED values. You would need to work with them as a partner, tell them of your intentions, and ask that they configure their routers to heed your request. Perhaps it makes more sense for traffic destined to 192.0.2.0 to enter AS 64500 via Access-1—maybe it is a faster or a more direct path.

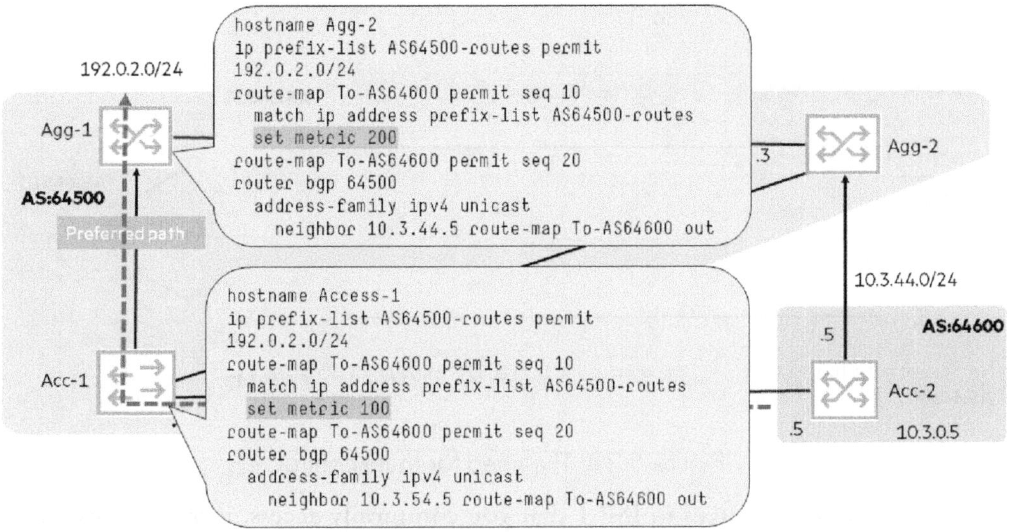

Figure 5-28: MED configuration

Create the configuration shown in this figure, in which Access-1 advertises a lower MED for that route than Agg-2.

Controlling eBGP routes

eBGP route control is essential for maintaining the stability, security, and efficiency of large-scale networks that rely on BGP. It enables administrators to implement and enforce routing policies, optimize network traffic flow, and safeguard against potential security threats. Some of the reasons are:

- Avoid incorrect routing information:

  Misconfigurations or malicious attempts to hijack routes can disrupt network traffic or direct it through unauthorized paths. Route control helps prevent these issues by implementing strict filtering and verification.

- Improve performance and reliability:

  By controlling the paths that traffic takes, administrators can optimize network performance and ensure that the most efficient routes are used. This is crucial for applications sensitive to latency or bandwidth constraints.

- Choose appropriate exit points to reach external destinations.

- Ensure that your AS doesn't become a transit AS for other ISPs.

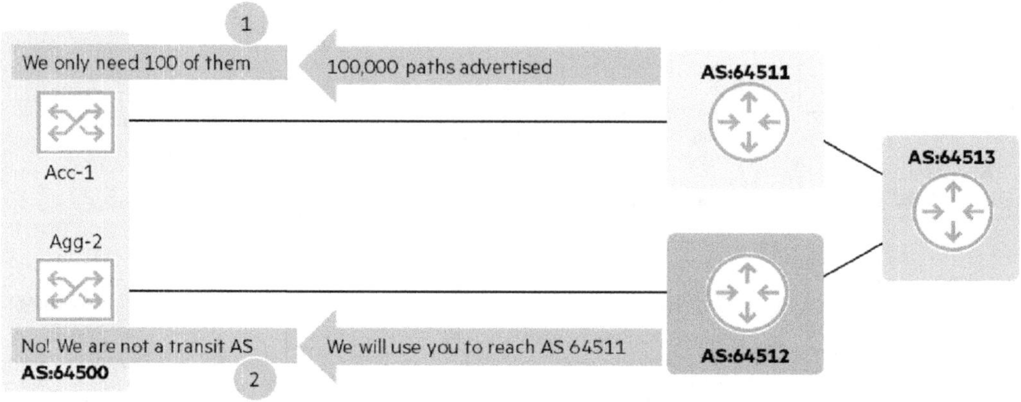

Figure 5-29: The need for route control

For example, if you have resources in ISP 1 that you commonly access, it makes sense to use the connection you have to ISP 1 to reach them versus going through another ISP to reach those resources. Second, you do not want ISPs to use your company's bandwidth in order to connect to other ISPs, which is referred to as a transit AS. Based on these two reasons, you will need to implement eBGP route control through prefix or AS path lists to control the external BGP routes you receive and/or advertise.

Use Looking Glass to verify the results of your configuration: https://www.gin.ntt.net/looking-glass-landing/.

Implementation

You can manipulate route preference using route maps, which you have already used extensively in this training. So far, you have used route maps to tune routes that already exist in the BGP table. Earlier, you used them to control OSPF route redistribution. Now you will use them to control prefixes advertised out to other ASes or accepted inbound from other ASes.

Recall that deploying a route policy is a three-step process:

1. Configure filters based on route attributes, such as the destination address and the advertising router address.
2. Create a route map and apply filters to it.
3. Apply the route map to BGP.

Avoiding transit AS for ISPs

In the multi-homed ISP scenario, you should only advertise your own public IP addresses to the ISPs. Do not announce routes that your AS learns from ISP 1 to ISP 2, nor vice versa. This can cause ISP routers to use your AS for internet traffic, and your router will quickly become overburdened by external internet traffic.

Several scenarios could cause your AS to become a transit AS.

- The ISPs advertise internet routes differently with different aggregation. Any route that your BGP routers receive from only one ISP, they could begin advertising to the other ISP.
- You connect a single AOS-CX switch or VSF fabric to both ISPs. In this case, the switch has two eBGP neighbors, and it will advertise best routes received from one to the other.

You should communicate with your ISPs to prevent issues in any of these scenarios.

Sometimes you are using BGP primarily to announce your public IP block. For traffic outbound to the internet, you do not care about finding the best path. You might want to use the connection to one ISP as the primary connection and the other as a standby only if the first fails, or you might want to load balance across both. In cases such as these, you might ask your ISPs not to send any route advertisements to you. You can then create static default routes on the BGP routers, or in some cases, you might want the ISP to send only a default route in BGP.

If you want to receive routing advertisements from the ISPs, you must ensure that your router does not pass them on to the other ISP. You can ask the ISP router to filter inbound advertisements from you, but it is best practice to filter outbound advertisements yourself.

Filtering outbound routes

Filtering outbound routes in BGP helps enforce routing policies, prevent unintended route advertisement, enhance network security, optimize traffic engineering, and support the overall scalability of the network. Here are some reasons why outbound route filtering is necessary:

Avoid route leaks

Outbound route filtering prevents the accidental advertisement of routes that should remain internal to the AS. Internal routes are not meant to be advertised to external peers, as it can lead to route leaks where traffic meant for internal networks is improperly routed through other ASes.

Enforce routing policies

Outbound route filtering allows network administrators to enforce routing policies by controlling which routes are advertised to eBGP peers. This includes enforcing specific route aggregation policies, controlling which prefixes are advertised to specific peers or regions, and ensuring that only authorized routes are propagated.

Traffic engineering

Outbound route filtering allows for effective traffic engineering by controlling the paths that traffic takes to optimize performance, reduce latency, or balance loads. This is achieved by selectively advertising routes based on traffic engineering policies.

Avoid misconfigurations

Misconfigurations can lead to unintended consequences, such as sub-optimal routing or even network instability. Outbound route filtering helps prevent these issues by providing an additional layer of protection against accidental misconfigurations.

In the following example, you are setting up route maps to restrict outbound advertisements to eBGP neighbors. There are a couple of ways that you could solve this issue:

- Use an AS Path list where the path list is empty, indicating that the route is internal to your network.
- Use a prefix list to only allow advertisement of your routes to other ISPs.

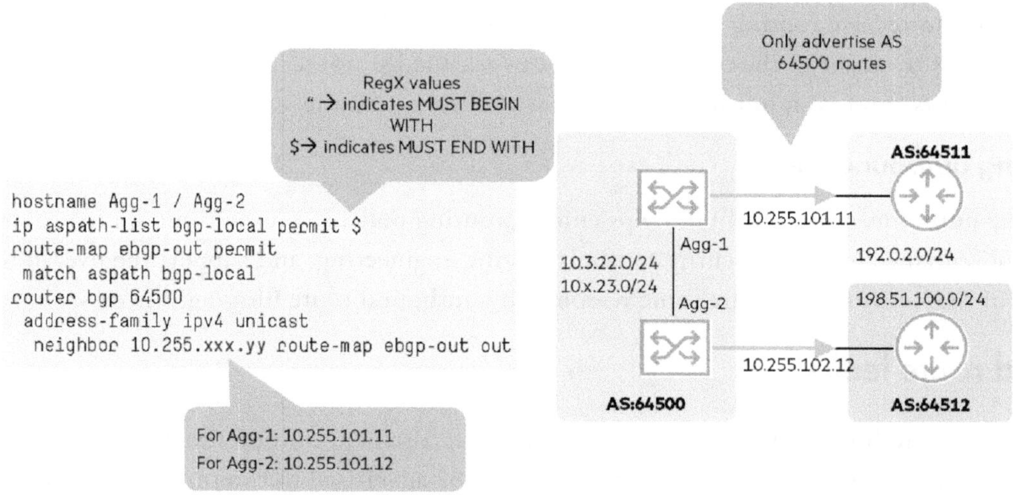

Figure 5-30: Filter outbound routes

The `ip aspath-list` command provides an elegant solution for this use case because it specifies an AS path. AS path lists are named lists of regular expression rules. They are used to match AS path attributes in the routes for inclusion in or exclusion from route policies. The sequence number is optional and is autogenerated whenever it is not explicitly mentioned. All AS path list rules with the same name are grouped together. The configuration of IP AS path lists is beyond the scope of this guide.

Filtering inbound routes

Figure 5-31 shows a route map that filters routes the BGP router receives from its eBGP neighbor. In this example, you run BGP primarily to announce your public block of IP addresses, and you care less about receiving all the BGP routes in the internet routing tables. You can filter those down in a few different ways. You may choose to just accept routes for larger ranges less than a specific prefix length. Or you might choose to accept just routes that originate in the eBGP neighbor AS because those are the ones where the company will benefit the most by sending the traffic out this specific router. You could also take a combination of these approaches.

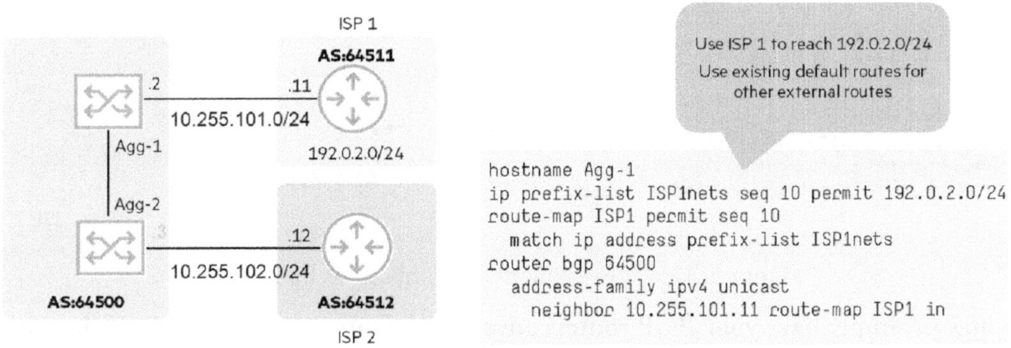

Figure 5-31: Filtering the outbound routes

In any case, you must ensure that your BGP edge routers have a default route. You could achieve this with the typical `ip route 0.0.0.0/0` command, by using the OSPF `default-information orginate` command, or by ensuring that your ISP advertises a default route, which you would then also need to permit with your route map.

In this example, you want to use ISP 1 to access resources on the 192.0.2.0/24 network. For other networks, a default route is typically sufficient. The example configuration uses an IP prefix list to select a specific route to learn from ISP 1: 192.0.2.0/24. Please note that you might need to perform similar inbound route filtering to make intelligent choices for reaching resources in ISP 2 as well.

Make sure your AS is receiving a default route if you filter inbound routes.

What would happen if your link to ISP 1 would go down? You would still be able to access 207.1.1.0/24 via a default route to leave your network via ISP 2.

Injecting external routes into OSPF (IGP)

You should consider the amount of routing information that your BGP routers receive from ISP routers, and how much of that should be available to the company IGP, typically OSPF. Although you can redistribute BGP routes into OSPF, it is not typically recommended. The large BGP routing table might overwhelm the OSPF processes on the local router and on other routers throughout the AS.

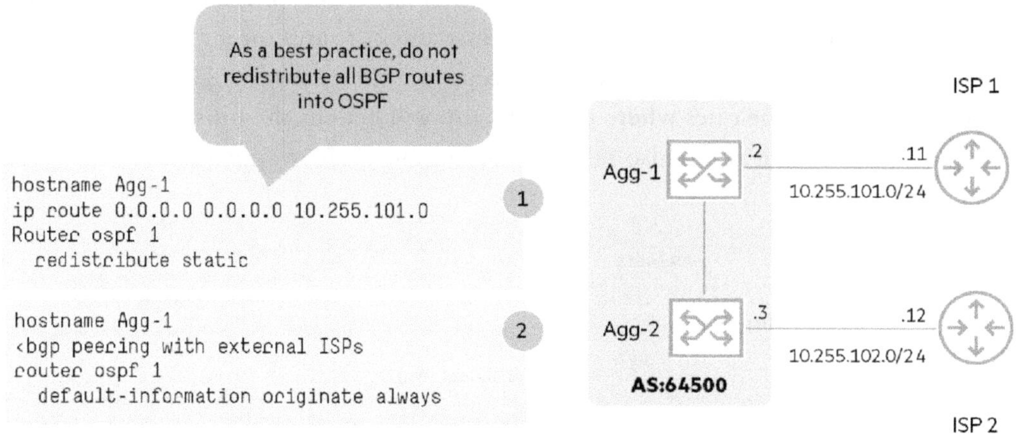

Figure 5-32: How to inject external routes into OSPF

Often, you can simply have your BGP routers advertise default routes in OSPF. If they are already acting as ABRs for stub areas, they already advertise default routes in those areas. You might need to have the BGP routers advertise default routes in the OSPF backbone, though. A simple approach is to configure a static route through its ISP connection on the BGP router, then redistribute static routes into OSPF. This is shown in **Figure 5-32** as option 1.

If you are going to redistribute BGP into OSPF, it is highly recommended that you use a route map to qualify exactly which routes you want to redistribute into OSPF. Use the techniques you learned **Advanced OSPF** chapter. The **Figure 5-32** also shows a second option, using the OSPF `default-information originate` command, with the `always` option. Without this option, OSPF only advertises default information (gateway of last resort: 0.0.0.0/0) if it actually has a valid path to 0.0.0.0. So, if the interface to 10.255.101.0 goes down, the static route goes down; thus, it will not be in OSPF.

If OSPF should have a default route, independent of the default route existence in the routing table, use the `always` option.

Your ISP might advertise a default route to your BGP router. In that case, instead of redistributing static routes, you could redistribute just the BGP default route. When you redistribute BGP into OSPF, you apply a route map that permits only that route.

Route reflectors

Route reflectors in BGP are components used to address the issue of full mesh requirements in iBGP networks. This full mesh requirement becomes impractical and difficult to scale as the number of routers increases. To fulfill the need for a full mesh, BGP route reflectors are introduced. Route reflectors help in simplifying the iBGP topology by allowing some routers to reflect BGP routes to other routers, reducing the number of iBGP sessions required.

Here are some key components of the feature.

Route Reflector (RR)

A route reflector is a router configured to reflect BGP routes between iBGP speakers (routers) within the same AS. It plays a central role in route reflection.

Route Reflector Client (RRC)

A route reflector client is a router that is configured to establish an iBGP session with a route reflector. It receives reflected routes from the route reflector.

Reflection of routes

When a route reflector receives a BGP update (route) from an iBGP client, it can reflect that route to other iBGP clients and route reflector clients within the AS.

Route reflector hierarchy

Route reflectors can be organized in a hierarchical manner to further reduce the number of iBGP sessions. In large networks, multiple route reflectors may be used, where one route reflector reflects routes to another route reflector, forming a hierarchy.

As shown in **Figure 5-33**, route reflectors reflect, or "mirror," iBGP routes to other iBGP peers. In the figure, Agg-1 and Agg-2 are configured as route reflectors. All other IBGP routers act as route reflector clients.

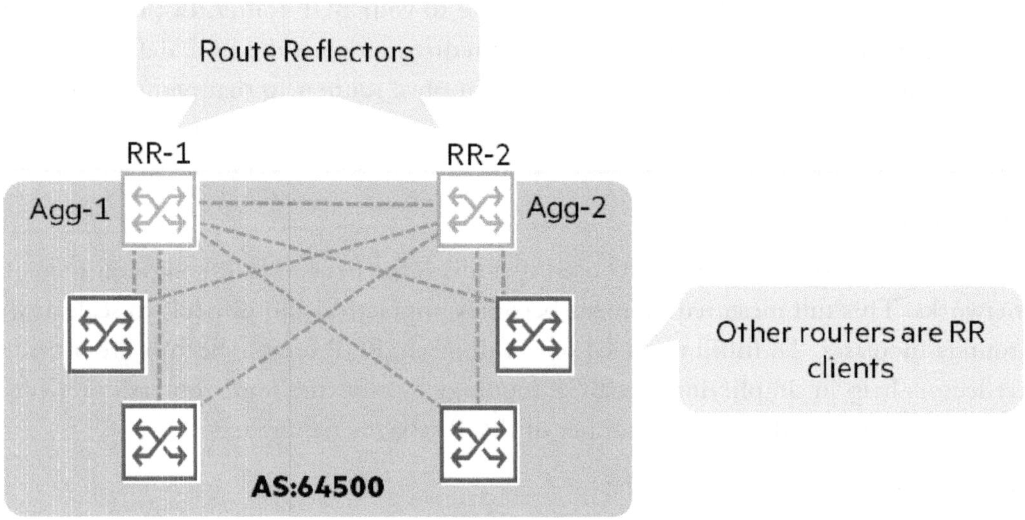

Figure 5-33: Route reflectors

Understand that a reflector client has no special configuration and is not aware that route reflectors are being used. The big difference is that instead of configuring each client with N-1 peers, you only configure them with two peers—Agg-1 and Agg-2 in this example. This greatly improves scalability.

Here's how it works:

- If a route is received from a non-client peer, the route reflector server reflects the route to EBGP peers (as normal) and to reflector clients.
- If a route is received from a client peer, the route reflector server reflects the route to all non-client peers and also to client peers, except the originator of the route, and reflects it to EBGP peers.

Confederations

At some point, the use of clusters runs into scalability issues. This is where you can then implement confederations. BGP confederations are sets of "sub-ASes" within your "real AS." Typically, the internet only sees this one real AS number. Confederations are used in very large networks where a large AS can be configured to encompass smaller, more manageable internal autonomous systems.

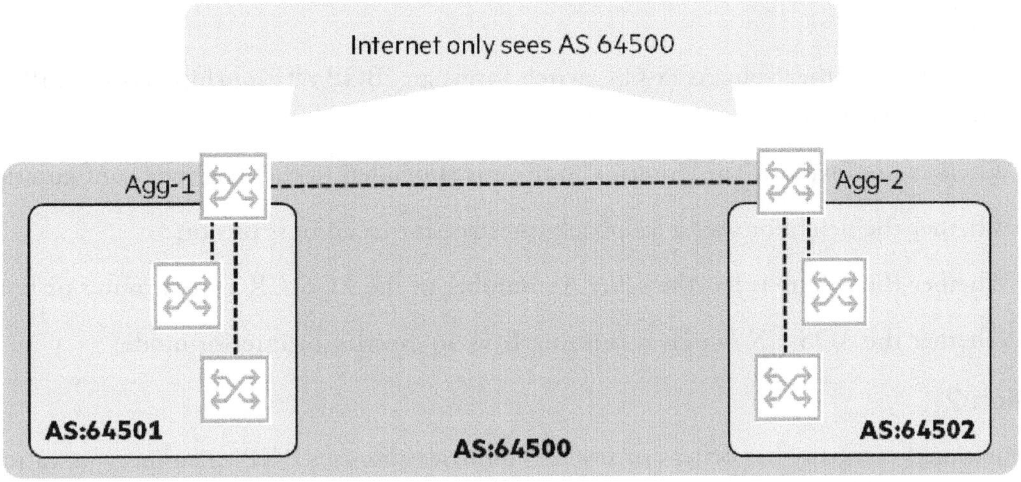

Figure 5-34: Confederation

Each confederated AS within the company has iBGP fully meshed and has connections to other autonomous systems inside the confederation. Even though these internal autonomous systems have eBGP peers to autonomous systems within the confederation, the autonomous systems exchange routing as if they used iBGP. By doing this, the confederation preserves next hop, metric, and local preference information. To the outside world, the confederation appears to be a single AS. Within this solution, iBGP transit AS problems can be resolved since iBGP requires a full mesh between all BGP routers.

Confederations can be used in conjunction with route reflectors. Both confederations and route reflectors can be subject to issues unless specific design rules, affecting both BGP and the interior routing protocol, are adhered to. Problems that can occur include:

- Route oscillation
- Sub-optimal routing
- Increase of BGP convergence time

Additionally, route reflectors, clustering, and BGP confederations were not designed to simplify BGP router configuration. However, these are common tools for experienced BGP network designers and engineers. These tools may be combined, for example, as a hierarchy of route reflectors.

The configuration of confederations is beyond the scope of this guide.

Learning check

Check your understanding of BGP routing protocol by answering the following questions.

Question 1

What determines whether your AOS-CX switch forms an eBGP relationship versus an iBGP relationship with a neighbor?

- O Whether the "exterior" or "interior" option is appended to the neighbor configuration.
- O Whether the neighbor uses a loopback interface for its address or not.
- O Whether the neighbor has the same AS number as the AOS-CX switch router or not.
- O Whether the AOS-CX switch is running BGP in exterior or interior mode.

Question 2

A multinational company has offices in several countries and uses BGP to connect its various sites over the internet. The company has recently acquired a new branch and needs to integrate its network into the existing BGP configuration. The new branch has an Autonomous System Number (ASN) that overlaps with one of the existing branches. It has a route to any subnet within this range in its IP routing table.

How should the company address the issue of overlapping ASNs to successfully integrate the new branch into the existing BGP configuration?

- O Change the ASN of the new branch to a private ASN and use BGP confederation.
- O Disable BGP at the new branch and use static routing.
- O Use the same ASN for the new branch as the existing branch and enable route aggregation.
- O Implement BGP Route Reflectors at the new branch to avoid ASN conflicts.

Question 3

A service provider is implementing BGP to manage its customer connections. One of the customers is experiencing intermittent connectivity issues and the service provider suspects a problem with BGP route propagation. The provider decides to analyze the BGP configuration and route advertisements.

Which BGP command should the service provider use to verify the advertised routes to a specific neighbor?

- O `show ip bgp summary`
- O `show ip bgp neighbors <NEIGHBOR-IP> advertised-routes`
- O `show ip route`
- O `show ip bgp`

Question 4

A company is multi-homed to two different ISPs using BGP. They want to ensure that outbound traffic primarily uses ISP1, but if ISP1 fails, the traffic should automatically switch to ISP2. How can the company configure BGP to achieve this?

- O Set a higher LOCAL_PREF for routes received from ISP2.
- O Use route maps to set a higher weight for routes received from ISP1.
- O Configure MED values to prefer routes from ISP2.
- O Announce a more specific route to ISP1.

Question 5

In BGP, what is the significance of the AS_PATH attribute?

- O It indicates the next hop IP address for the route.
- O It lists the autonomous systems a route has traversed.
- O It determines the local preference of the route.
- O It specifies the IP prefix length for the route.

6 Additional L3 features

In the previous chapters on Layer 3 technologies, you examined routing protocols such as the Open Shortest Path First (OSPF) protocol and the Border Gateway Protocol (BGP). In this chapter, you will learn some additional routing techniques that supplement those protocols. You will explore the Virtual Routing and Forwarding (VRF) feature that allows you to isolate customers and discuss the policy-based routing technique, along with some security techniques to avoid Layer 3 vulnerabilities such as DHCP or ARP attacks.

OBJECTIVES

After completing this chapter, you should be able to:

- ✓ Implement VRF to isolate routed traffic.
- ✓ Manipulate routing with policy-based routing.
- ✓ Protect networks using features such as DHCP snooping and dynamic ARP inspection.

Virtual routing and forwarding

In this section, you will become familiarized with one of the most widely used Layer 3 functionalities—VRF. This functionality facilitates the use of overlapping addresses in a single routing device. VRF sees many applications in real life as explored in this section. Along with this, you will learn to configure and validate the VRF settings.

VRF overview

You learned that VLANs allow you to create multiple virtual switches inside one physical switch. Similarly, you can create separate virtual routers inside a single physical router with the VRF feature. VRF allows you to create multiple virtual routers, as can be visualized from **Figure 6-1**.

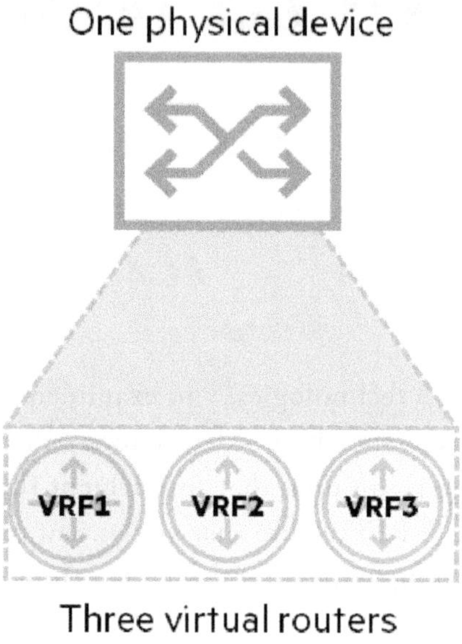

Figure 6-1: Multiple logical routers created using VRF

A router can be considered to divide the global network into smaller networks. The devices in one network can communicate with each other but need a gateway (a routing device) to communicate with those outside the network. This network is identified using a subnet mask and a range of IP addresses associated with it. Not all the addresses in this range are being used at the time. Also, there is a limit on the total number of available IP addresses. This is where the VRF feature proves to be useful. VRF divides the same network (defined by one physical router) into multiple networks, which can have the same or overlapping IP addresses. VRF allows a single physical device to host multiple logical routers with complete L3 isolation. Because the device maintains separate route processes and tables for each VRF, they can have overlapping address space. The network functionality is improved because the network paths can be segmented without requiring multiple routers.

VRF use cases

Supporting multiple customers

The scenario shown in **Figure 6-1** represents a single ISP hosting separate environments for three customers. The customers can use any address space they choose, without the fear of conflicting with other customers.

Integrating the network infrastructure of two merging companies

Another use case is when you are merging with another organization. You can physically connect the two organizations without having to worry about addressing conflicts. As you normalize the address space between the two entities, you can eliminate the VRF configuration as appropriate.

Providing logical separation of business units

Another common use case is to maintain logical separation of business units or divisions. For example, a banking company may keep a separation between their retail, lending/mortgages, investments, and internal applications departments.

Implementing EVPN-VXLAN fabric with multiple overlay

VRF supports AOS-CX switches' EVPN-VXLAN multifabric solution, which extends the basic single-fabric EVPN-VXLAN solution to support multifabric deployments.

AOS-CX switches include a specific VRF for management purposes and that can only be used in the Out-of-Band Management (OOBM) port to separate the data and control plane from the management plane.

VRF operation

Consider the example illustrated in **Figure 6-2**. It describes the operation of VRF using a simple scenario. The switch SW1 is a routing device used by a service provider. The service provider serves two customers—Cust1 and Cust2. User1 and Server1 belong to Cust1, and User2 and Server2 belong to Cust2. The IP subnets allocated to the two customers are as mentioned in **Figure 6-2**. They both use the same subnet, that is, 10.1.10.0/24. The reuse of the same subnet for different customers will make it beneficial for the service provider, yet there is an issue with such a configuration. Once the interfaces serving Cust1 are configured (in this case, the interfaces 1/1/1 and 1/1/2) with the required IP addresses, the switch will not allow you to reuse the same set of IP addresses to configure the 1/1/3 and 1/1/4 interfaces serving Cust2.

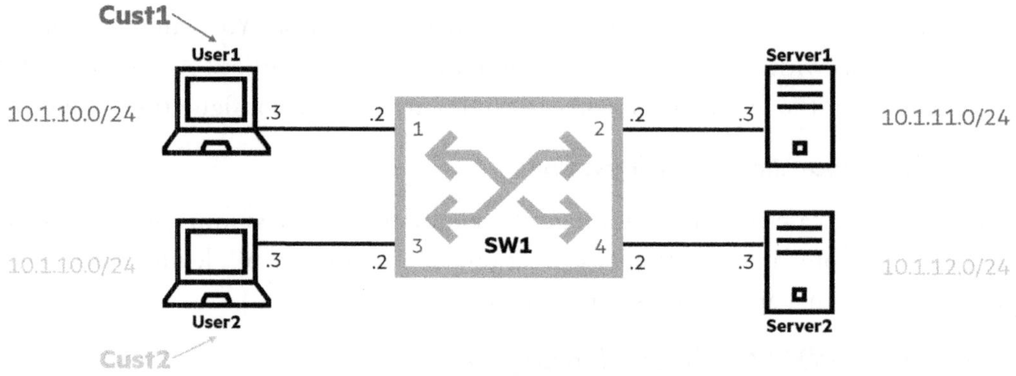

Figure 6-2: A networking device servicing two customers having overlapping addresses

This is where VRF comes into the picture. The simplest way to isolate the customer spaces using VRF is as follows:

1. Create two VRFs on SW1 with the names Cust1 and Cust2.
2. Assign the desired interfaces to each VRF, that is, 1/1/1 and 1/1/2 to VRF Cust1 and 1/1/3 and 1/1/4 to VRF Cust2.
3. Configure the IP addresses for the required interfaces in both the VRFs.

This way, you can reuse the IP addresses for different customers and still keep the communication secure.

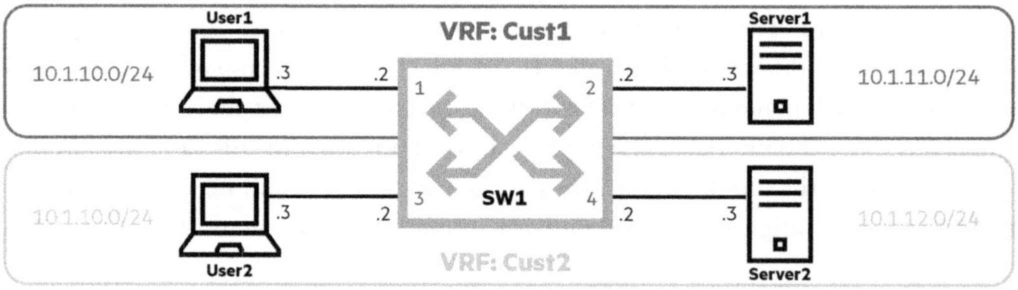

Figure 6-3: Isolation of address spaces using VRF

VRF configuration

Figure 6-4 shows a sample physical infrastructure that supports two separate OSPF networks. This type of infrastructure is particularly useful to create separate subnets for different departments in an organization—for example, one guest network and another, more secure employee network.

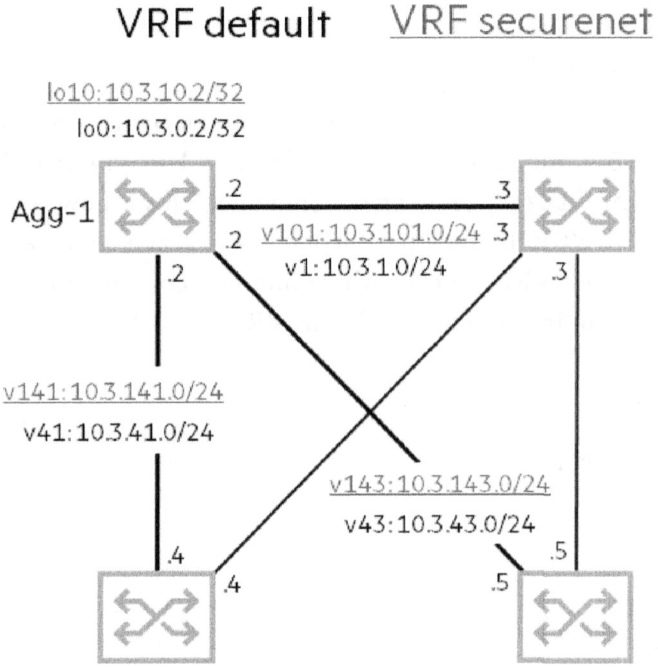

Figure 6-4: Topology for VRF configuration example

VRF configuration commands

The following configuration is for VRF "default," It is represented in the network diagram of **Figure 6-4** using non-underlined addresses.

```
hostname Agg-1
vrf default (implied)

router ospf 1
    area 0.0.0.0

interface loopback 0
    ip address 10.3.0.2/32
    ip ospf 1 area 0.0.0.0

interface vlan 1
    ip address 10.3.1.2/24
    ip ospf 1 area 0.0.0.0

interface vlan 41
    ip address 10.3.41.2/24
    ip ospf 1 area 0.0.0.0
```

```
interface vlan 43
    ip address 10.3.43.2/24
    ip ospf 1 area 0.0.0.0
```

The following configuration is for VRF "securenet." This configuration is represented in the network diagram of **Figure 6-4** using the underlined addresses. Each interface in the VRF "securenet" is configured with the `vrf attach securenet` command. This is how you tell the route which interfaces belong to which VRF. If you enter the `no vrf attach securenet` command, the interface is returned to the default VRF, and the whole interface configuration is removed. This is also true every time you move an interface to a different VRF.

```
hostname Agg-1
vrf securenet

router ospf 10 vrf securenet
 area 0

interface loopback 10
 vrf attach securenet
 ip address 10.3.10.2/32
 ip ospf 10 area 0

interface vlan 101
 vrf attach securenet
 ip address 10.3.101.2/24
 ip ospf 10 area 0

interface vlan 141
 vrf attach securenet
 ip address 10.3.141.2/24
 ip ospf 10 area 0

interface vlan 143
 vrf attach securenet
 ip address 10.3.143.2/24
 ip ospf 10 area 0
```

Validating VRF

There are different commands available to validate the VRF configuration settings that allow you to verify different aspects of the VRF configuration, such as the list of attached interfaces, the OSPF neighbor details, or the OSPF interface details. Some of these commands are described below.

Verify VRF configuration

The following code block shows how to validate which VRFs exist, and which interfaces are associated with each VRF.

```
Agg-1# show vrf
VRF Configuration:
------------------
VRF Name    : default
        Interfaces                  Status
        -----------------------------
        loopback0                     up
        vlan1                         up
        vlan41                        up
        vlan43                        up

VRF Name    : securenet
        Interfaces                  Status
        -----------------------------
        loopback10                    up
        vlan101                       up
        vlan141                       up
        vlan143                       up
```

You can use the commands `show vrf default` or `show vrf securenet` to look at each VRF separately.

Verify OSPF configuration

The following code blocks show how to verify OSPF settings for the VRF named "securenet." The syntax is identical to what you already know; just add "vrf securenet" at the end.

To display information about OSPF neighbors:

```
Agg-1# show ip ospf neighbors vrf securenet
Neighbor ID      Priority  State       Nbr Address   Interface
-------------------------------------------------------------
10.3.10.3        1         FULL/BDR    10.3.101.3    vlan101
10.3.10.4        n/a       FULL        10.3.141.4    vlan141
10.3.10.5        n/a       FULL        10.3.143.5    vlan143
```

To display information about each interface associated to a specific VRF:

```
Agg-1# show ip ospf interface brief vrf securenet
Interface    Area     IP Address/Mask  Cost  State             Status
---------------------------------------------------------------------
loopback10   0.0.0.0   10.3.10.2/32     0     Loopback          up
```

```
vlan101      0.0.0.0   10.3.101.2/24    100   Dr              up
vlan141      0.0.0.0   10.3.141.2/24    100   Point-to-point  up
vlan143      0.0.0.0   10.3.143.2/24    100   Point-to-point  up
```

> You can also use any other OSPF command to validate a VRF. For example, you could see all VRFs in a single output using the `all-vrfs` parameter.

IVRF overview

VRF allows a physical router to be partitioned into multiple virtual router forwarding instances. The control and data plane are isolated in each VRF. This isolation keeps the traffic from crossing the VRFs, and therefore multiple routing tables can coexist within the same physical L3 switch or router. Inter-Virtual Router Forwarding (IVRF) allows for routes to be distributed across multiple VRFs in a controlled manner.

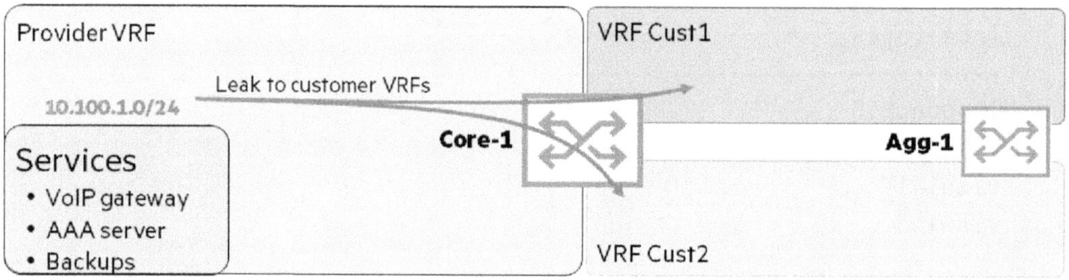

Figure 6-5: IVRF overview

Consider the multitenant use case illustrated in **Figure 6-5**. You know that VRF allows customers Cust1 and Cust2 to share the same infrastructure while enjoying logically separate, private networks, including separation from the service provider's VRF. However, suppose the provider wants to make optional services available to its clients—services such as a Voice over IP (VoIP) gateway, AAA services, or a backup service. These services exist on the service provider's VRF subnet 10.100.1.0/24. You need a way to override the isolation provided by VRFs up to some extent. This can be achieved using the IVRF functionality of AOS-CX switches. IVRF, also known as route leaking, allows you to leak routes in a controlled manner between the otherwise private VRFs. You can configure the provider's VRF subnet to leak into the customer VRFs that pay for those additional services. Mechanisms in the services themselves maintain customer isolation.

There are two methods to leak routes from one VRF to another. They are as follows.

Static route leaking: Static VRF route leaking between two VRFs is allowed using an `ip route` configuration command. To accomplish the route leak, the routes being leaked to other VRFs must be present in the source VRF. When a route leak has been accomplished and is therefore present in the source VRF, that route can then be leaked to other VRFs.

Dynamic route leaking: Dynamic VRF route leaking is allowed between two or more user-configured VRFs. Dynamic VRF is configured using export and import route targets. Multiprotocol BGP (MP-BGP) is used to leak routes between the dynamic VRFs. To configure MP-BGP, a separate address family must be configured within BGP for each VRF; then, redistribute the routes (OSPF, static, or connected) within that VRF for the routes to be leaked.

Policy-based routing

Policy-based routing, or PBR, provides a means to override normal routing decisions that occur on the basis of the system routing table. PBR lets you manipulate the path of a packet based on the various attributes of the packet. Packets that can be manipulated by PBR are packets that are already routing through the system at Layer 3, with a destination IP address that is on a network other than the packet ingress Layer 3 interface. In this section, you will explore the features of PBR and learn to configure it.

PBR overview

To understand what policy-based routing is, let us start by comparing its operation to that of the normal routing.

Normal routing

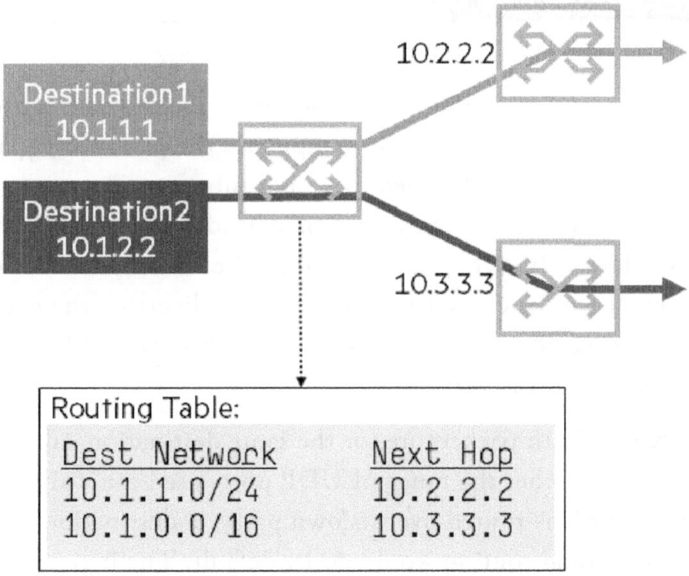

Figure 6-6: Normal routing process overview

As you know, a router (or a routing device) stores the routing information in the form of a routing table. This table is used to decide the path a data packet would traverse in an inter-network communication scenario. This is the normal routing mechanism. Normal routing is purely destination-based. As each router receives an inbound packet, it compares the packet's destination IP address to the routing table, and then chooses the entry with the longest match as the best path. In the example of **Figure 6-6**, the packet destined to 10.1.1.1 matches with both the entries in the table, but only 16 bits match with the second entry, while 24 bits match with the first entry. So, the router routes this packet via next hop 10.2.2.2, as per the first entry. Next, when a packet arrives destined to 10.1.2.2, it is routed via 10.3.3.3 because that is the only match.

Policy-based routing

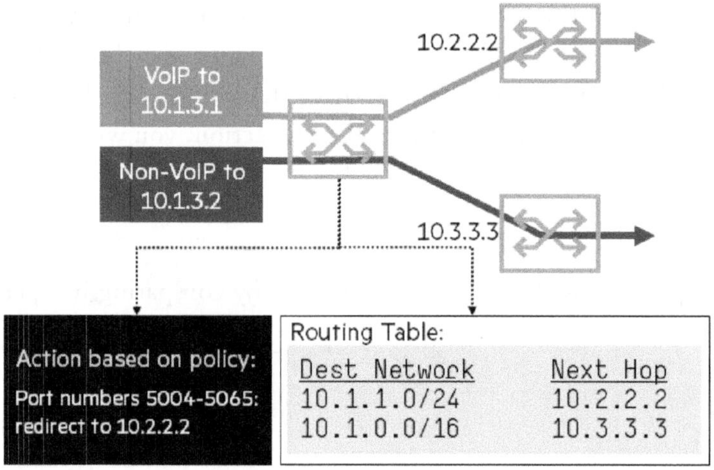

Figure 6-7: PBR process overview

PBR, as the name indicates, lets you configure policies to decide the path a packet takes when it arrives at an interface on a routing device. PBR manipulates the path of the packet based on various packet attributes. Packets must be destined to a valid subnet on your network. As a router receives an inbound packet, it compares the appropriate header and tag information to a classifier that is configured on the device. If there is a match, then the router takes an action based on a policy. You can configure actions such as dropping a packet, diverting the packet to a specific next hop, or redirecting the traffic to a tunneled interface. This way, PBR allows you to improve traffic handling, performance, and security.

In the example of **Figure 6-7**, both packets are for the same destination subnet. However, you have created a classifier that matches the range of UDP ports used for VoIP traffic—perhaps UDP ports 5004 to 5065. This traffic is routed over its own path via destination 10.2.2.2. Note that this action overrides what normal routing would do by default. The ingress router only has 10.3.3.3 as the next hop for destinations in 10.1.0.0/16. However, 10.2.2.2 is a valid path, and it is directly connected to the ingress router, as required by PBR. In other words, recursive lookups are not supported. All other, non-VoIP traffic does not match with the classifier; hence, it is routed as normal, that is, via 10.3.3.3. PBR only affects the local router's routing decision. You might have to implement the PBR policy on multiple routing devices in order to achieve a specific outcome if multiple hops exist between a source and a destination.

PBR use cases

By default, a routing device uses its routing table to determine which path to take when reaching a destination. But there are times when a company needs to override the default routing behavior based on the traffic types or network conditions. The following are some of the scenarios where PBR could prove to be beneficial.

Load sharing

Imagine that you have two links for redundancy for an OSPF network, but one link has a better cost. The higher-cost link is available, but it is not used unless the better-cost link fails. You can configure PBR to use the secondary link for certain traffic and, as a result, implement a limited form of load balancing.

Cost saving

Suppose that you have two internet connections from two different providers—ISP 1 and ISP 2. ISP 1 charges a flat rate, and ISP 2 charges a flat rate with a usage cap. Once you reach the cap, you are charged per gigabyte sent or received. You can use PBR to ensure that even if ISP 2 has better metric routes, that ISP is not used unless ISP 1 is unreachable.

Quality of service

Perhaps you have two equal-cost paths for some destination. You want to send delay and jitter-sensitive VoIP traffic on one link and all other traffic over the other link. You can maintain this redundancy while logically separating the traffic using PBR. VoIP traffic gets a link all to itself. You must configure QoS in addition to this to mitigate performance risks in case the VoIP link fails.

Source-based selection

You might need to control the traffic paths based on the source IP address, as opposed to the normal destination-based routing. Imagine that you work for a service provider that offers normal and premium-level services. The backbone has two sets of connections—10 Gbps and 100 Gbps. You can use PBR to ensure that the traffic sourced from premium-tier customers traverses the 100 Gbps connections, while the normal customer traffic uses the 10 Gbps connections.

PBR configuration

Consider the same example discussed in the previous topic for demonstrating the configuration process. The network diagram is displayed again in **Figure 6-8** for your reference.

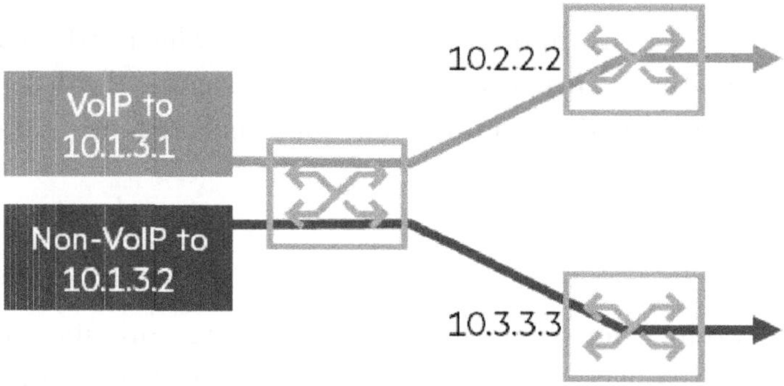

Figure 6-8: Desired policy-based routing flow

PBR configuration steps

The process to configure PBR in a routing device is described as follows.

1. **Define a classifier to match the desired traffic.**

   Define a classifier to select which traffic shall have some special actions applied to it. For the example of **Figure 6-8**, you create a class called "VoIP_Class." This class shall include the UDP traffic from any source to the destination 10.1.3.0/24, with a destination UDP port between 5004 and 5065. The commands to configure this are as follows:

   ```
   class ip VoIP_Class
     10 match udp any 10.1.3.0/24 range 5004 5065
   ```

2. **Take actions on the matched traffic.**

   Define a PBR action list to control how the desired traffic class is to be treated. In the example of **Figure 6-8**, you create a list named "VoIP_Action," which contains four possible actions in an ordered list. You can have a maximum of eight action entries in an action list and 512 total entries across the entire device. The commands to create the action list are:

   ```
   pbr-action-list VoIP_Action
     nexthop 10.2.2.2
     nexthop 10.3.3.3
     default-nexthop 10.9.9.9
     interface null
   ```

The action list entries are processed top-down, like an Access Control List (ACL). The first entry indicates that the next hop for matched traffic is 10.2.2.2. If that next hop is not available, perhaps due to a link, interface, or router-down condition, then the next hop is 10.3.3.3. If that next hop router is also down, then use a default next hop of 10.9.9.9. And finally, if that next hop is down, silently discard the packet. In other words, forward it out the "null" interface—the proverbial bit bucket.

This example describes three of the four possible PBR actions. The other one is an interface tunnel, which sends traffic over a GRE, 6in4, or 6in6 tunnel. This example does not show the default sequence numbers, which are automatically assigned based on order of entry and in increments of 10. You can configure your own sequence numbers on action list entries, if desired.

The next hop actions must point to directly connected IP addresses. PBR does not support recursive lookups to reach indirectly connected next hop addresses.

3. **Associate the class with the actions.**

   Next, create a policy to associate each class with an appropriate action list. The example of **Figure 6-8** associates the VoIP_Class with the VoIP_Action action list. As hinted at by the sequence number 10, each policy can have multiple class-action associations. Ensure that the class-action pairs are appropriate for where you intend to apply the policy. The commands to create a policy are:

```
policy pbr_policy_1
  10 class ip VoIP_Class action pbr VoIP_Action
```

4. **Apply the policy.**

   Access the `config-interface` mode on the appropriate interface and apply the policy. It is always applied as an inbound policy. In other words, the policy is applied as packets enter the router inbound. This is because the purpose of PBR is to override normal destination-based routing, the decision for which is made as soon as the packet enter the router on an interface. The commands to apply a policy are as follows:

```
interface vlan11
  ip address 10.3.11.1/24
  apply policy pbr_policy_1 routed-in
```

Validating PBR

The PBR configuration can be validated through a number of ways. Some of them are described as follows.

View interface configuration

For a quick check of the PBR configuration, the `show pbr interface vlan11` command can be used. This command displays the configuration applied to a specific interface. It shows an overview of how all PBR pieces fit together in a kind of "configuration chain."

```
SW1#show pbr interface vlan11
<output deleted>
---------------------------------------------------------------------
default
    vlan11
         pbr_policy_1
                   VoIP_Class
                           VoIP_Action
                              10  nexthop          10.2.2.2 (active)
                              20  nexthop          10.3.3.3
                              30  default_nexthop  10.9.9.9
                              40  interface        null
```

As can be seen from the output of the command displayed, the interface VLAN11 has `pbr_policy_1` applied to it, which has the classifier VoIP_Class and the associated VoIP_Action defined. This policy will apply the PBR actions defined in VoIP_Action to all traffic in class VoIP_Class. You can also see the actions listed, and the top action is currently active. Since it is a valid route, the other actions will not be used.

View details of each PBR component

For a detailed review of the PBR configuration, the `show pbr summary` command can be used. It shows all PBR configuration chains. In this example, switch SW1 only has one configuration chain. Using this command, you can verify the names of the policy, class, and action list, as can be seen in the following output.

```
SW1#show pbr summary
<output deleted>
---------------------------------------------------------------------
default
    vlan11
         pbr_policy_1
                   VoIP_Class
                           VoIP_Action
                              10  nexthop       10.2.2.2 (active)
```

These components can be further investigated by using the appropriate commands to display their details. The commands to view details of the PBR components are as follows:

Command to show PBR policy details

```
SW1#show policy pbr_policy_1
----------------------------
pbr_policy_1
       10
          VoIP_Class ipv4
                   pbr VoIP_Action
```

Command to show PBR class

```
SW1#show class ip VoIP_Class
-------------------------------------
IPv4  VoIP_Class
   10 match          udp
      any
      10.252.0.0/16    5004 - 5065
```

Command to show PBR action list

```
SW1#show pbr-action-list VoIP_Action
       Name
Seq    Type           Address/Int
---------------------------------
       VoIP_Action
10     nexthop         10.2.2.2
20     nexthop         10.3.3.3
30     default-nexthop 10.9.9.9
40     interface       null
```

PBR actions

You will now review the possible PBR actions. These are illustrated in **Figure 6-9**.

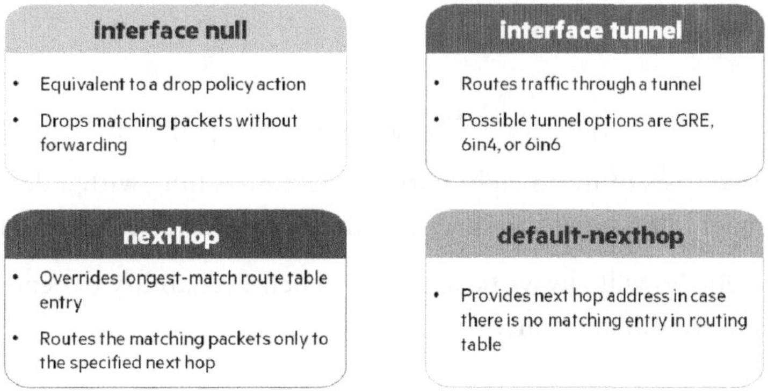

Figure 6-9: PBR actions

- **interface null:** This is equivalent to the policy drop action. Packets that match the class criteria for that policy entry are dropped and not routed further.
- **interface tunnel:** This action allows you to route the desired traffic through a GRE, 6in4, or 6in6 tunnel. You can specify the tunnel as the outbound interface for all matching packets in the PBR configuration. You have to ensure that the tunnel is configured correctly to avoid any loss of traffic. Packets sent into the tunnel interface egress at the router at the endpoint of the tunnel. If the tunnel is misconfigured or down, the traffic may be lost.
- **nexthop:** This action overrides the routing table's longest prefix match next hop router with a next hop address that you specify. If no such entry exists in the routing table for matching packets, this action still takes place.
- **default-nexthop:** This action is useful to obtain the next hop router address when there is no matching entry in the routing table. Such a default-nexthop overrides a system default route if it is already configured. In addition, it applies even if there is no system default route.

PBR routing decision process

As discussed earlier, PBR is configured by creating a criterion to match the packets to a class and apply a user-configured policy action list to it. Let us examine the routing decision-making process of PBR.

PBR decision criteria

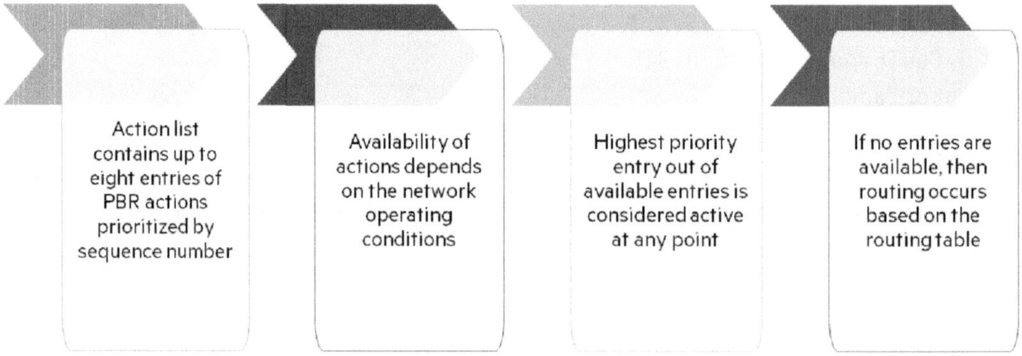

Figure 6-10: PBR decision-making criteria

- An action list consists of up to eight entries of actions along with a default or user-defined sequence number. The entries are prioritized based on these sequence numbers.
- Not all of the entries will always be available. Their availability depends on the network operating conditions at any time.

- Only one of these entries is considered active at a given time for a VRF. If multiple entries are available, then the entry with the highest priority (lowest sequence number) is considered to be the active entry.
- If no entries are available, then PBR does not impact the routing. Routing occurs based on the routing table as usual.

PBR routing example

The routing decision-making process is illustrated with an example in **Figure 6-11**.

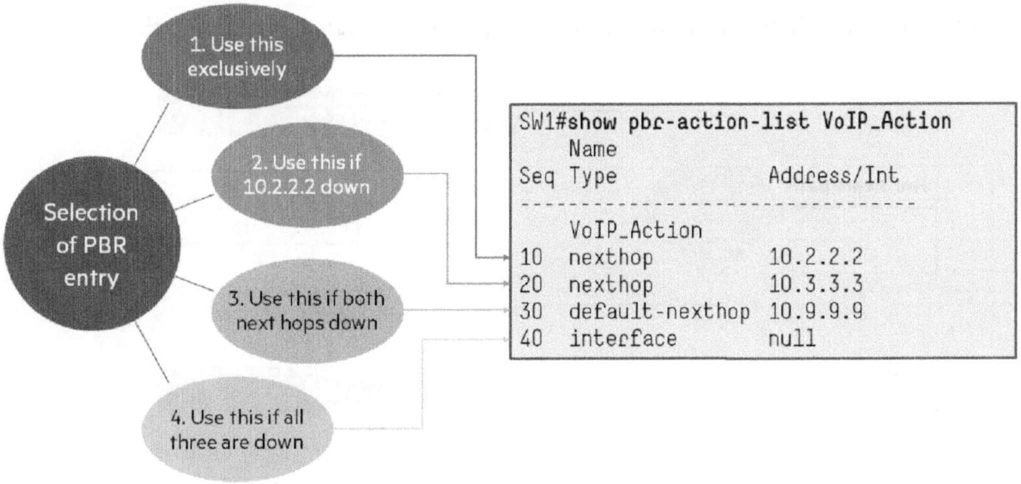

Figure 6-11: PBR routing example

The policy action list VoIP_Action consists of four entries of possible actions arranged in the order of priority.

1. It has two reachable nexthop entries. The active entry is the one with the lowest sequence number—the one entered first, by default. This entry is used exclusively unless that address becomes unavailable.
2. When you apply a PBR policy to an interface, all `nexthop` and `default-nexthop` entries in the associated action list are monitored every five seconds for reachability. If the first entry is found to be unavailable, the router uses the next entry, that is, 10.3.3.3.
3. If both the `nexthop` entries are down, then the router uses the `default-nexthop`.
4. If all three of the above entries are down, then the `interface null` action ensures that the packets are merely thrown away. If you do not add the `interface null` entry in the action list, then the router falls back to using destination-based route table entries.

Hardware vs. software path PBR

Traffic to be routed that cannot be handled by the switching ASIC, called the hardware path, includes IPv4 packets with IP options. This traffic is handled by the switch operating system kernel routing software, called the software path. The characteristics of network usage determine the proportion of PBR traffic handled by software ersus hardware path. Under normal conditions, the software path PBR is likely to be extremely low—a fraction of a percent.

There is a difference between the PBR hardware and software path operations when using the PBR default-nexthop action. **Figure 6-12** compares what happens when there is a policy match with an associated default-nexthop action for the hardware versus software path.

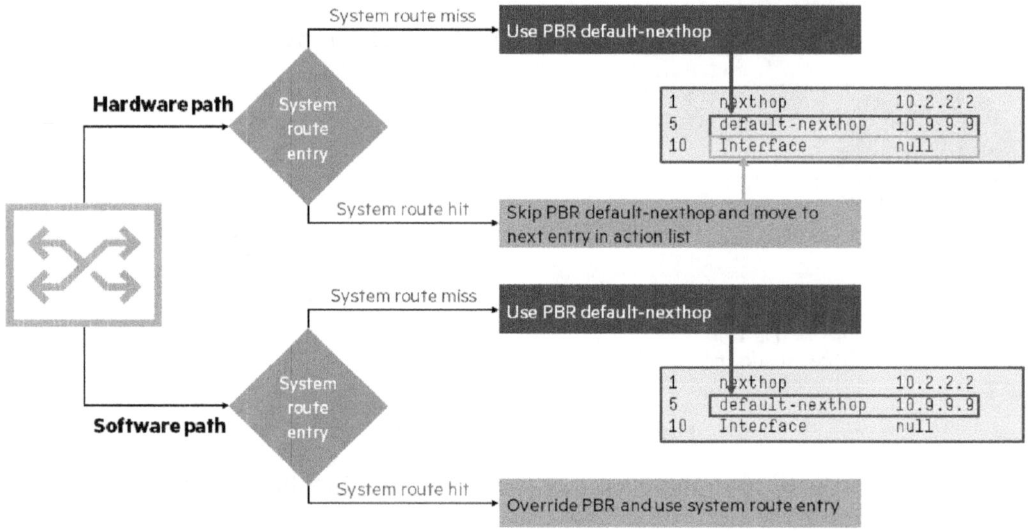

Figure 6-12: Hardware vs. software path PBR operations

Consider a policy applying two actions: `default-nexthop` and `interface null`, as shown in **Figure 6-12**.

The hardware path match criterion allows to use the default-nexthop entry from the PBR action list when there is no matching entry in the system routing table. Conversely, when there is a match in the system routing table, the default-nexthop entry from the PBR action list is skipped and the policy processing moves to the next entry in the action list (in this case, interface null).

On the other hand, in the software path, the class match with the system routing table is the preferred criterion. When there is a match with the system routing table, the packet is forwarded according to that entry, thus completely overriding the PBR.

This difference in behavior is due to a limitation in the software path matching and routing engine, relative to hardware.

Applying the same PBR to different VRFs

The PBR policies can be applied to interfaces in any VRF. Similarly, the PBR action lists can be used in different policies across different VRFs. The effect of applying the same policies/PBR action lists across different VRFs depends on the IP networks and interfaces configured in the different VRFs. Let us understand this with the help of the example illustrated in **Figure 6-13**.

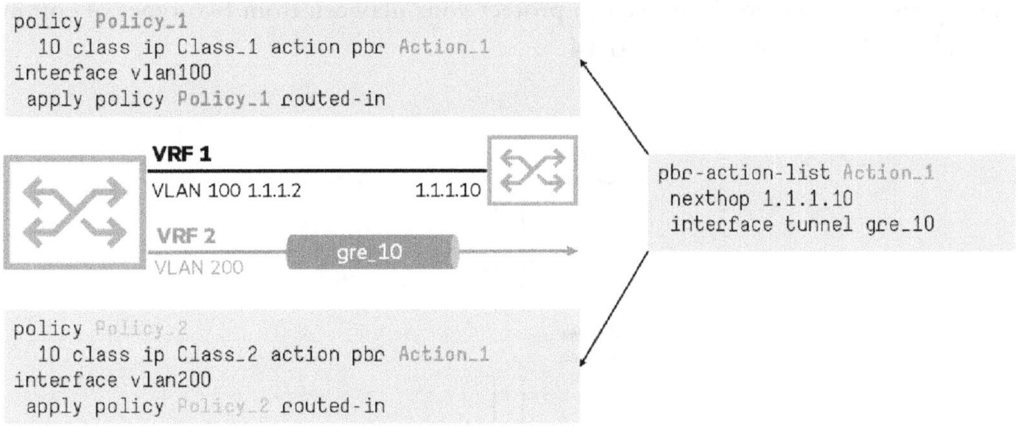

Figure 6-13: Applying one PBR action list to multiple VRFs

Figure 6-13 shows an action list with actions `nexthop 1.1.1.10` and `interface tunnel gre_10` used as the PBR action parameter for an entry in Policy_1 and also in Policy_2. If you apply Policy_1 to an interface in VRF 1, which has an interface in subnet 1.1.1.0/24 but no GRE tunnel named "gre_10," then only the "nexthop" action is relevant to VRF 1. Policy_2 contains the same action list. It is applied to an interface in VRF 2 that lacks the 1.1.1.0/24 subnet configured but does have a tunnel named "gre_10." In that case, only the interface tunnel action will apply in that VRF.

DHCP snooping

DHCP service is an integral part of most networks as it assigns IP addresses to devices. DHCP is a protocol used by DHCP servers in IP networks to dynamically allocate network configuration data to client devices (DHCP clients). Possible network configuration data includes the user IP address, subnet mask, default gateway IP address, DNS server IP address, and lease duration. This protocol enables DHCP clients to be dynamically configured with such network configuration data without any manual setup process. However, errors in assignment or misuse of this service may lead to network malfunction. DHCP snooping allows you to inspect the DHCP requests in order to avoid such problems.

DHCP vulnerabilities

Because DHCP was designed to work in the trusted internal network, it does not provide authentication or access controls. A DHCP server cannot verify if the client requesting an address is a legitimate client on the network. In addition, the DHCP client has no way of knowing if the DHCP server that offers it an address is legitimate. As a result, DHCP is vulnerable to attacks from rogue clients and servers. You need to protect your network from two types of common DHCP attacks, as illustrated in **Figure 6-14**.

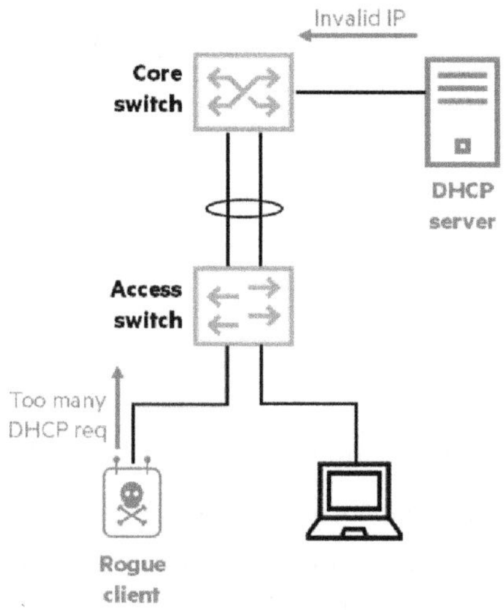

Figure 6-14: DHCP vulnerabilities

These common attacks are as follows.

Address spoofing

A rogue DHCP server on the network can assign invalid addresses to the network devices. Without valid IP addresses and network gateway addresses, these devices cannot connect to any other IP network device, which prevents users from reaching the resources they need to do their jobs. Alternatively, a rogue DHCP server might respond to a client's DHCP discover request with the rogue server's IP address as the default gateway, setting up the client for a DoS or Man in the Middle (MITM) attack and creating a potential security breach.

Address exhaustion

An attacker can access the network and request IP addresses until the DHCP server's supply of available addresses is exhausted. This prevents legitimate hosts from receiving IP addresses and accessing the network.

DHCP snooping algorithm

DHCP snooping is a security feature that helps avoid problems caused by an unauthorized DHCP server on the network that provides invalid configuration data to DHCP clients. DHCP snooping achieves this by distinguishing between trusted ports connected to legitimate DHCP servers and untrusted ports connected to general users. DHCP packets from untrusted sources are dropped, whereas DHCP packets are forwarded between trusted ports without inspection. DHCP packets received on other switch ports are inspected before being forwarded. This is illustrated in **Figure 6-15**.

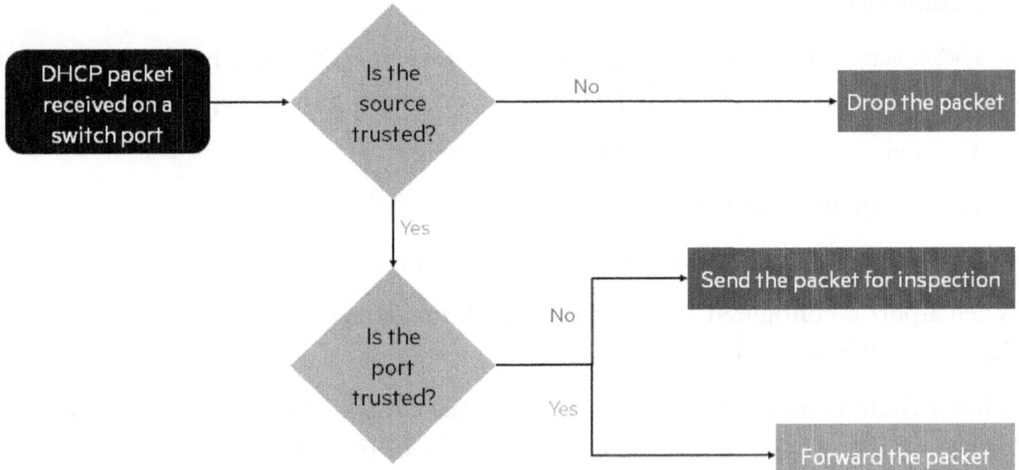

Figure 6-15: DHCP snooping algorithm

Table 6-1 lists the types of DHCP packets that may be dropped and the respective conditions for dropping the packets.

Table 6-1: Conditions for dropping DHCP packets by DHCP snooping

Packet types that are dropped	Conditions for dropping the packets
DHCPOFFER, DHCPACK, DHCPNACK	■ A packet from a DHCP server is received on an untrusted port. ■ The switch is configured with a list of authorized DHCP server addresses and a DHCP response received on a trusted port, but the source IP address is not an authorized DHCP server.
DHCPRELEASE, DHCPDECLINE	■ A broadcast packet that has a MAC address in the DHCP binding database, but the port in the DHCP binding database does not match the port on which the packet is received.

Packet types that are dropped	Conditions for dropping the packets
All DHCP packet types	■ A DHCP packet received on an untrusted port in which the DHCP client hardware MAC address does not match the source MAC address in the packet. ■ A DHCP packet containing DHCP relay information (option 82) is received on an untrusted port.

Protocol considerations

- In a VSX setup, once external storage is configured, the VSX switch role should not be changed. If there is a need to change the role, do so using these steps:
    1. Remove DHCPv4-Snooping external storage.
    2. Change the VSX role.
    3. Configure DHCPv4-Snooping external storage.
- When a port is configured as a trusted port, all the dynamic IP binding entries learned on that port will be deleted.
- When a client is connected on a trusted port, the dynamic IP binding entries will not be learned on the switch, even though the client gets an IP address.
- If DHCPv4 snooping is enabled on two back-to-back access switches, DHCP packets will be dropped, since by default option 82 is enabled on DHCPv4 snooping and the default policy is drop. The second switch with DHCPv4 snooping enabled drops the packets. In this scenario, the user should enable DHCPv4 snooping option 82 on one switch or disable it on both.
- In the HPE Aruba Networking CX 8360 Switch Series, VXLAN is trusted by default for DHCPv4 and DHCPv6 snooping over VXLAN. VXLAN operational status must be in the forward state to forward the packets over the VXLAN tunnel. DHCPv4 snooping is supported over the VXLAN IPv4 and IPv6 underlay.
- In the HPE Aruba Networking CX 8100 Switch Series, DHCPv4 snooping, DHCPv6 snooping, and ND snooping are supported over the VXLAN IPv6 underlay.

This guide only focuses on and describes the DHCPv4 snooping feature. DHCPv6 snooping has some differences in functionality.

DHCP snooping configuration

To understand the configuration, let us consider the network scenario described in **Figure 6-16**.

Configuration scenario

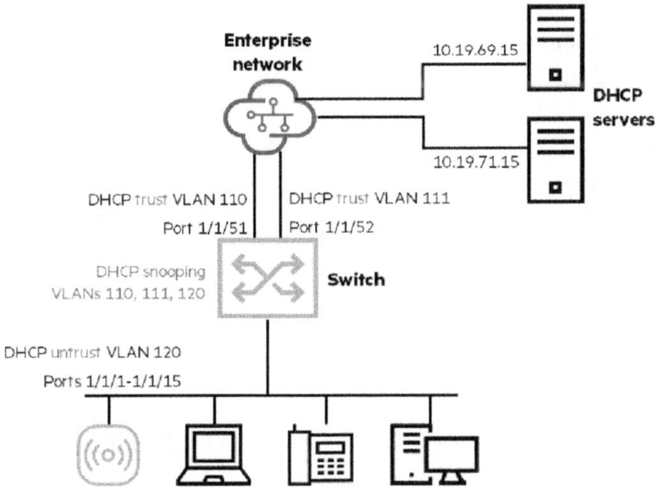

Figure 6-16: DHCP snooping configuration scenario

The switch is receiving DHCP configurations from two DHCP servers at ports 1/1/51 and 1/1/52 respectively. These are trusted networks, and hence, any communication from these must be allowed. The endpoint devices connect to the switch at ports 1/1/1 to 1/1/15. These ports must be configured as untrusted in order to inspect any DHCP requests received at these ports.

Configuration commands

The steps to achieve this configuration are:

1. Set the switch for DHCPv4 snooping.
2. Allow DHCP communication from the uplinks in the Enterprise network port 1/1/51 and 1/1/52.
3. For communication to occur, identify the authorized DHCP servers 10.19.69.15 and 10.19.71.15.
4. LAN-facing user ports 1/1/1 to 1/1/15 are not configured as trusted.

This can be configured as follows:

```
dhcpv4-snooping                                        <--enable DHCPv4 snooping
dhcpv4-snooping authorized-server 10.19.69.15          <--allow authorized servers
dhcpv4-snooping authorized-server 10.19.71.15
vlan 110
      description uplink
      dhcpv4-snooping                                  <--enable DHCPv4 snooping
on required VLANs
vlan 111
      description uplink
```

```
        dhcpv4-snooping
vlan 120
        description User
        dhcpv4-snooping
interface 1/1/1
        no shutdown
        description User
        vlan access 120
...
interface 1/1/15
        no shutdown
        description User
        vlan access 120
interface 1/1/51                                    <--Unplinks on 1/1/51
and 52 trusted
        no shutdown
        vlan trunk native 1
        vlan trunk allowed 110
        dhcpv4-snooping trust
interface 1/1/52
        vlan trunk native 1
        vlan trunk allowed 111
        dhcpv4-snooping trust
interface vlan 110
        description uplink
        ip address 172.16.69.1/30
interface vlan 111
        description uplink
        ip address 172.16.71.1/30
interface vlan 120
        description User
        ip address 172.16.10.1/254
        ip helper-address 10.19.69.15
-       ip helper-address 10.19.71.15
...
ip route 0.0.0.0/0 172.16.69.2 distance 10
ip route 0.0.0.0/0 172.16.71.2 distance 20
...
```

The key configuration commands required while setting up DHCP snooping are as follows:

Enable DHCP snooping

```
dhcpv4-snooping
```

Allow authorized servers

```
vlan <VLAN-ID>
description uplink
dhcpv4-snooping
```

Enable DHCP snooping on a VLAN

```
dhcpv4-snooping authorized-server <SERVER-IP>
```

Configure DHCP trust on interfaces

```
interface <INTERFACE-ID>
no shutdown
vlan trunk native 1
vlan trunk allowed 110
dhcpv4-snooping trust
```

Verify DHCP snooping configuration

You can use the `show dhcpv4-snooping` command to verify your DHCP snooping configuration.

Dynamic ARP inspection

Address Resolution Protocol, or ARP, maps the Layer 3 IP addresses to Layer 2 MAC addresses. An incorrect mapping can cause serious issues to a network. In this section, we will review the vulnerabilities inflicted on the ARP service and the features available in AOS-CX switches to mitigate those.

ARP vulnerabilities

ARP requests are broadcast on all devices in a local network and are not forwarded by routers. This service is enabled on the switches by default and cannot be disabled. ARP does not support any inherent security mechanism and as such depends on simple datagram exchanges for the resolution, with many of these being broadcast. This makes ARP vulnerable to attacks.

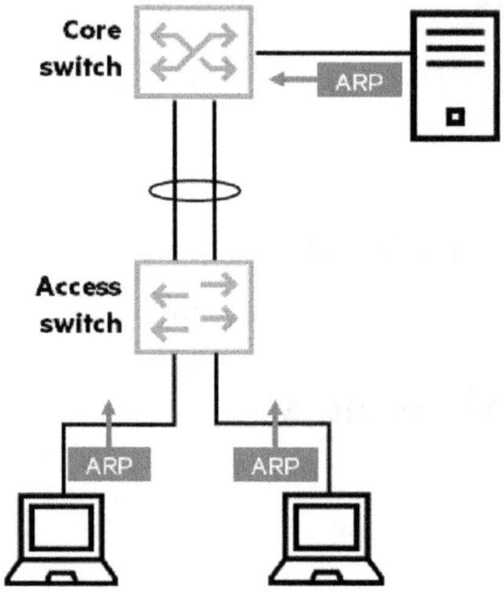

Figure 6-17: Representation of ARP requests sent in a network

The attacks primarily intend to create Denial of Service (DoS) for the other entities present in the network. Most of the attacks are carried out in one of the following three forms:

- Overwhelming the switch control plane with too many ARP packets
- Overwhelming the switch control plane with too many unresolved data packets
- Masquerading as a trusted gateway/server by wrongly advertising ARP requests

Dynamic ARP inspection feature

AOS-CX switches can use dynamic ARP inspection to protect your network against the ARP poisoning attacks. Let us go over the feature details.

Dynamic ARP inspection overview

IP-to-MAC binding table

IP	MAC
10.1.1.10	18...aa
10.1.1.11	d8...bb
10.1.1.12	d4...cc

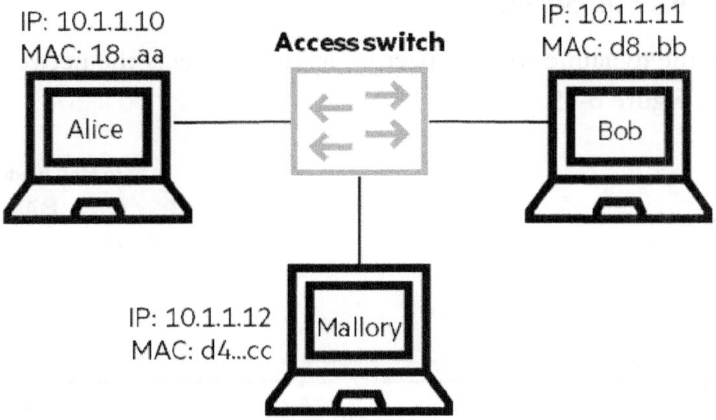

Figure 6-18: Dynamic ARP inspection feature overview

To enable the dynamic ARP inspection feature, you should first enable dynamic ARP inspection on a VLAN. You must also enable the feature globally for it to take effect on the enabled VLANs. Like the DHCP snooping feature (discussed in the next section), the dynamic ARP inspection feature allows you to designate trusted and untrusted ports. By default, all ports in an dynamic ARP inspection VLAN are untrusted. The switch checks the ARP requests and responses on all the untrusted ports. On the untrusted ports, ARP inspection takes the following actions:

- Examines all ARP requests and responses received on that port.
- Verifies that each of these packets has a valid IP-to-MAC address binding.
- Updates the switch's local ARP cache or forwards the packet to the appropriate destination, if the binding is valid. If the binding is invalid, the switch simply drops the packet, preventing other devices from receiving it and being tricked by the faulty information.

In addition to verifying IP-to-MAC address bindings, you can configure the switch to perform three further checks, described as follows. By default, none of these checks are enabled.

src-mac: The switch checks the ARP request and response packets to ensure that the source MAC address in the Ethernet header matches the sender MAC address in the body of the ARP packet. If the two addresses do not match, the switch discards the packet.

dest-mac: The switch checks each unicast ARP response packet to ensure that the destination MAC address in the Ethernet header matches the target MAC address in the body of the ARP packet. If the two addresses do not match, the switch drops the packet.

ip: The switch checks the sender and target IP addresses of the ARP packets and drops any ARP packets that have an invalid address. Invalid IP addresses include: 0.0.0.0, 255.255.255.255, all IP multicast addresses, and class E IP addresses.

Results of dynamic ARP inspection

You can analyze how the dynamic ARP inspection feature protects the network through the scenario illustrated in **Figure 6-19**.

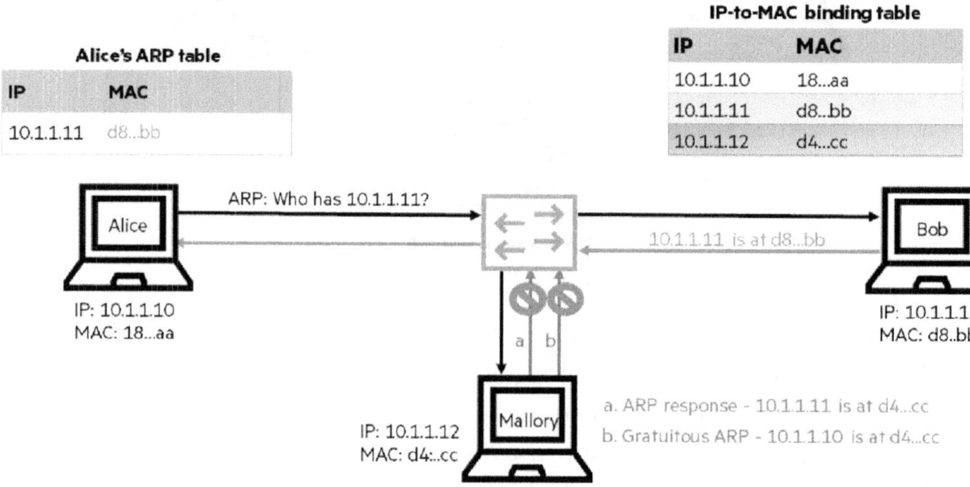

Figure 6-19: Scenario with dynamic ARP inspection enabled

The operation and decision-making occurs in the following sequence:

1. Alice broadcasts an ARP request for Bob's MAC address.
2. Mallory responds with an ARP response that maps her MAC address to Bob's IP address.
3. Mallory also sends a gratuitous ARP posing as Alice. However, the traffic arrives on an untrusted port in a VLAN on which dynamic ARP inspection is enabled. Therefore, the switch checks both ARP messages and sees that the IP-to-MAC address binding is incorrect. As a result, the switch drops the packets.
4. Bob sends his legitimate ARP response. This response also arrives on an untrusted port in an dynamic ARP inspection VLAN. However, its IP-to-MAC address binding is correct, so the switch forwards it as normal.
5. Alice and Bob can now exchange traffic.

Configuring dynamic ARP inspection

Configuration of the dynamic ARP inspection feature involves the following steps:

1. Enabling dynamic ARP inspection in the VLAN context
2. Configuring the ports as trusted or untrusted
3. Validating the ARP inspection configuration

Let us describe these steps one by one.

Enabling dynamic ARP inspection

To apply dynamic ARP inspection, you have to enable dynamic ARP inspection. As ARP acts in a local network segment, dynamic ARP inspection is configured in a VLAN. Use the `arp inspection` command, shown as follows, to enable this feature.

```
switch# configure terminal
switch(config)# vlan 1
switch(config-vlan)# arp inspection
```

Dynamic ARP inspection looks for packets received on untrusted interfaces on the VLAN and compares them with the IP-to-MAC address binding. If their ARP packet data invalidates, then the packets are dropped.

Making interfaces trusted

You can use the `arp inspection trust` command, shown as follows, to make an interface trusted.

```
switch(config-if)# arp inspection trust
```

All the interfaces are untrusted by default. If the connected port is a VXLAN tunnel, then this step can be ignored. This is because the VXLAN tunnel is always considered trusted.

Verifying ARP inspection settings

To show the current configuration of dynamic ARP inspection on all interfaces, use the following command:

```
switch# show arp inspection interface
-----------------------------------------------------------------
Interface      Trust-State
-----------------------------------------------------------------
1/1/1          Untrusted
-----------------------------------------------------------------
```

To show the current configuration on a specific interface or VLAN, use the following command:

```
switch# show arp inspection interface [<IFNAME>] or [vlan <VLAN_ID>]
```

DHCP snooping and dynamic ARP inspection

In most VLANs, devices receive their IP addresses using DHCP, and these addresses can change. The switch must discover the correct MAC-to-IP bindings automatically and dynamically. DHCP snooping fulfills this function. DHCP snooping builds and maintains a DHCP snooping-binding table, which tracks the MAC address, IP address, lease time, VLAN number, and the interface information corresponding to each DHCP lease. Using this table, the switch can learn which IP addresses should legitimately send traffic on which ports and which MAC address can legitimately use an IP address. Therefore, you must enable DHCP snooping as part of configuring dynamic ARP inspection.

If you are not using DHCP to assign IP addresses in the protected VLAN, you can configure static IP-to-MAC address bindings, and the switch will use this information to verify ARP packets. However, most networks use DHCP. Even when using DHCP and DHCP snooping, you may need to add a couple of static IP-to-MAC address bindings to the DHCP snooping table. The static bindings allow the switch to verify IP-to-MAC bindings for devices assigned static IP addresses. You must also specify the interface for the static binding.

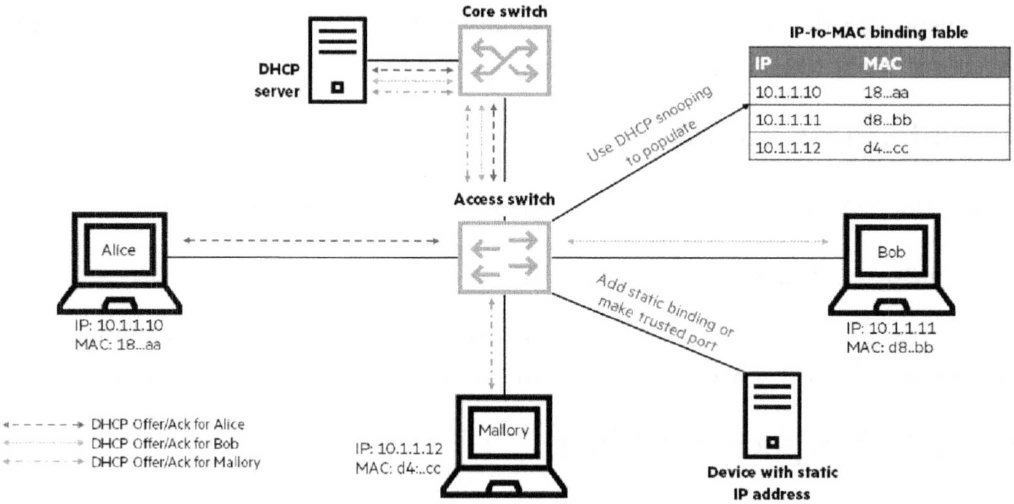

Figure 6-20: DHCP snooping with dynamic ARP inspection

In an active network, you should generally wait at least a week after enabling DHCP snooping to enable ARP. This ensures that the table is properly built up. For increased security and protection against any form of IP spoofing, you can implement dynamic IP lockdown (`ip source-lockdown`). This feature goes further than dynamic ARP inspection and enforces the correct source IP address and MAC address for all IP traffic on a port. Like dynamic ARP inspection, dynamic IP lockdown requires DHCP snooping to populate the IP-to-MAC binding table. For more details on implementing dynamic IP lockdown, refer to your switch's Access Security Guide.

Learning check

Check your understanding of this material by answering the following questions.

Question 1

Which Layer 3 technique can be used to provide isolation between customers having overlapping IP addresses?

- Network Address Translation (NAT)
- Virtual Routing and Forwarding (VRF)
- Policy-based Routing (PBR)
- DHCP snooping

Question 2

Which PBR action overrides the longest match entry from the system routing table for obtaining the next hop address?

- `interface null`
- `interface tunnel`
- `nexthop`
- `default-nexthop`

Question 3

DHCP snooping applies additional inspection on a DHCP packet received on a trusted port.

- True
- False

7 IGMP

Enterprise networks require a variety of multicast services. In this chapter, you will learn about multicast traffic and how to use Internet Group Management Protocol (IGMP) to control it. We begin with an introduction to multicasting, then move into a discussion of IGMP to finish with a lab activity.

OBJECTIVES

After completing this chapter, you should be able to:

- ✓ Explain multicast addressing.
- ✓ Identify IGMP components and messages.
- ✓ Execute multicast forwarding.
- ✓ Compare IGMP and IGMP snooping.

Multicast introduction

The purpose of multicast is to send information to a select group of devices and to avoid duplication of traffic streams. This saves precious network resources such as bandwidth and CPU utilization.

A sender device generates information and advertises it to a group. Receivers join that group and listen to the information. A multicast group is defined as a set of zero or more hosts identified by a single IP destination address. The following example shows a multicast group represented by IP address 239.2.11.1.

IP multicast messages are transmitted from one-to-many, from many-to-many, and/or from many-to-one over an IP infrastructure that may span Layer 3 boundaries. The destination nodes (receivers) send JOIN and LEAVE messages that create an on-demand community of devices interested in receiving the multicast stream. Multicast optimally uses network resources by requiring the source to send a single packet stream, even though large numbers of receivers might need to receive the messages. The replication of messages takes place at the network node or Layer 3 device and diverges the stream to the receivers.

The multicast source sends packets with its own IP address as the source and some multicast address as the destination. Multiple receivers can listen for the same address, and so switches flood the packet out all ports in the same VLAN, except the packet ingress port. Any source that is listening for that address processes the packet. Because multiple IPv4 nodes listen for the same multicast address, this address is often referred to as a *group address*.

The term "stream" is commonly used in multicast to denote traffic associated with a particular multicast destination address.

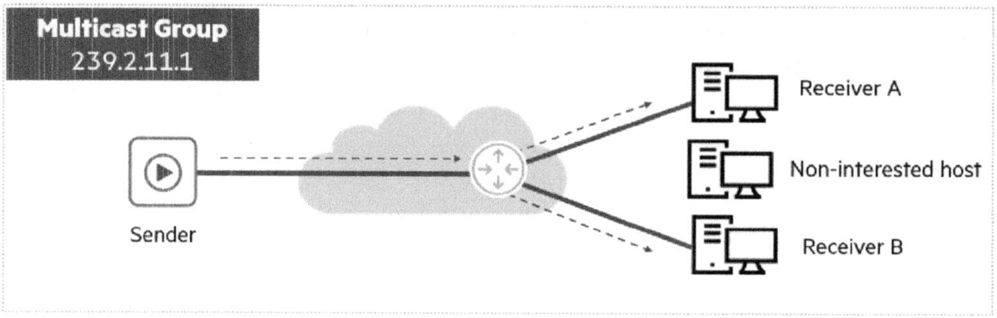

Figure 7-1: Multicast summary

Destination address review

As a review, there are three general types of destination addresses:

- **Unicast:** Traffic is forwarded to just one destination.
- **Broadcast:** Traffic is flooded throughout the VLAN/subnet (everyone will see it).
- **Multicast:** Traffic is destined to 0 or more (or all) destinations in a VLAN/subnet. It is flooded, by default, throughout the VLAN/subnet.

	Source IP	Destination IP	
Unicast	10.1.1.1	**12.2.2.2**	*"Hey you"*
Broadcast	10.1.1.1	**255.255.255.255**	*"Attention everyone"*
Multicast	10.1.1.1	**239.1.1.1**	*"Attention all members of group X"*

Figure 7-2: Three types of destination addresses

Multicast use cases

Unified communications and collaboration (UCC)

UCC applications include the following:

- IP voice conference calling
- Music or audio streaming
- Video and desktop conferencing and collaboration
- Telepresence
- Video streaming applications
- Virtual classroom applications
- Stock ticker tape streaming applications
- Network management protocols like the Network Time Protocol (NTP) and the Aruba Discovery Protocol (ADP), used by campus APs to find and connect to an HPE Aruba Networking Mobility Controller

Server management solutions

IT administrators might also assign servers or endpoints to multicast groups so that they can deliver patches, software updates, and other controls to a targeted group. Multicasting can also be used to take server or workstations snapshots—images of each system's storage drives.

Multimedia presentations

This is often the first thing one thinks about when discussing multicasting. Videos and other multimedia presentations can be large files and consume significant bandwidth. If 1000 people needed to watch the CEO's quarterly presentation for staff, you do not want 1000 individual unicast data streams. It is far better to send one single stream as a multicast, such that all staff members receive it.

Management/networking protocols

Many network-related protocols use multicasting for efficiency. This includes NTP, ADP, OSPF, and many more.

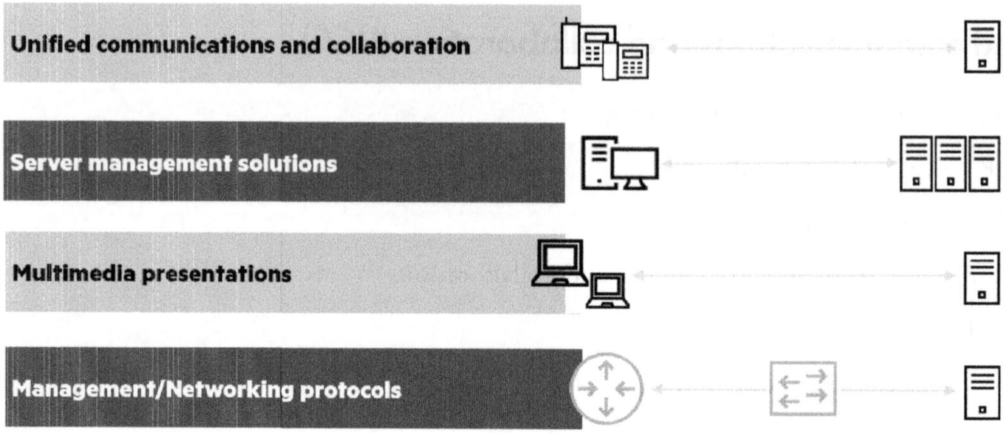

Figure 7-3: Multicast use cases

Multicast challenges

- Multicast traffic for sophisticated applications, particularly multimedia applications, can consume a great deal of bandwidth, particularly when it is copied for distribution to multiple locations. In fact, even copying multicasts can consume processing power on network infrastructure devices. To maintain a high-performing network environment, administrators must implement multicast control technologies properly to minimize unnecessary flooding and duplication. Not only do administrators need to preserve bandwidth, but they also need to protect endpoint functionality. An endpoint NIC must process the multicasts that it receives, even if it is not listening for that multicast, which can consume CPU cycles.
- Users and devices that require specific multicast traffic often connect in a variety of locations. Network administrators might need to route the traffic across multiple VLANs and even campus LAN sites.

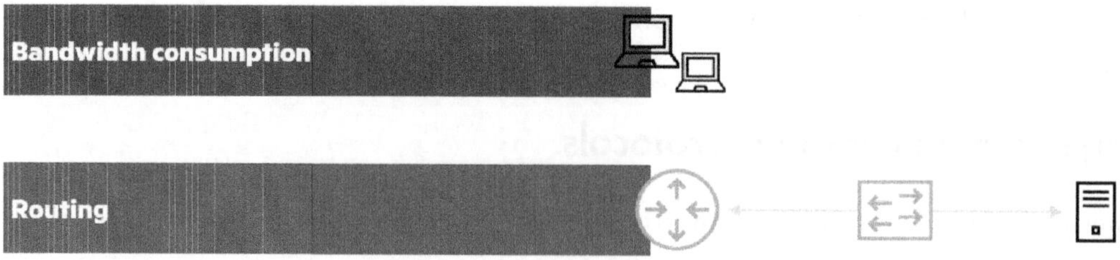

Figure 7-4: Multicast challenges

L3 multicast addressing

Multicast addresses fall within the 224.0.0.0/4 range: 224.0.0.0 to 239.255.255.255. As shown in **Figure 7-5**, the lower portion of the range is reserved for link local addresses; they are non-routable addresses used to communicate among devices on the same subnet. The remaining addresses are internal private addresses—routable multicast addresses for use across the enterprise.

The Internet Assigned Numbers Authority (IANA) has assigned specific multicast addresses to some protocols. For example, to reach all multicast host devices, use 224.0.0.1. To reach all routers, use 224.0.0.2. OSPF routers communicate with each other on multicast addresses 224.0.0.5 and 224.0.0.6. 224.0.0.13 is used for the Protocol Independent Multicast (PIM) routing protocol.

Depending on the type of application, the multicast source and multicast receivers might be statically configured to stream to, and listen for, a specific multicast address. Or they might negotiate the address dynamically. Determining the multicast address to use is the application's responsibility; the network infrastructure must simply ensure that the traffic reaches its destination.

As shown in the table below, IANA has assigned other addresses to specific organizations. Some applications, such as applications that use Session Announcement Protocol (SAP), can select multicast addresses dynamically from a certain range. Finally, individual organizations can define administratively scoped multicast addresses in the 239.0.0.0/8 range as they want; these multicasts should not leave the administrative domain.

Table 7-1: Layer 3 multicast addresses

IP multicast address range	Description	Routable
224.0.0.0 to 224.0.0.255	Local subnetwork	No
224.0.1.0 to 224.0.1.255	Internetwork control	Yes
224.0.2.0 to 224.0.255.255	AD-HOC block 1	Yes
224.3.0.0 to 224.4.255.255	AD-HOC block 2	Yes
232.0.0.0 to 232.255.255.255	Source-specific multicast	Yes
233.0.0.0 to 233.251.255.255	GLOP addressing	Yes
233.252.0.0 to 233.255.255.255	AD-HOC block 3	Yes
234.0.0.0 to 234.255.255.255	Unicast-prefix-based	Yes
239.0.0.0 to 239.255.255.255	Administratively scoped	Yes

You can refer to IANA for the latest assignments: https://www.iana.org/assignments/multicast-addresses/multicast-addresses.xhtml.

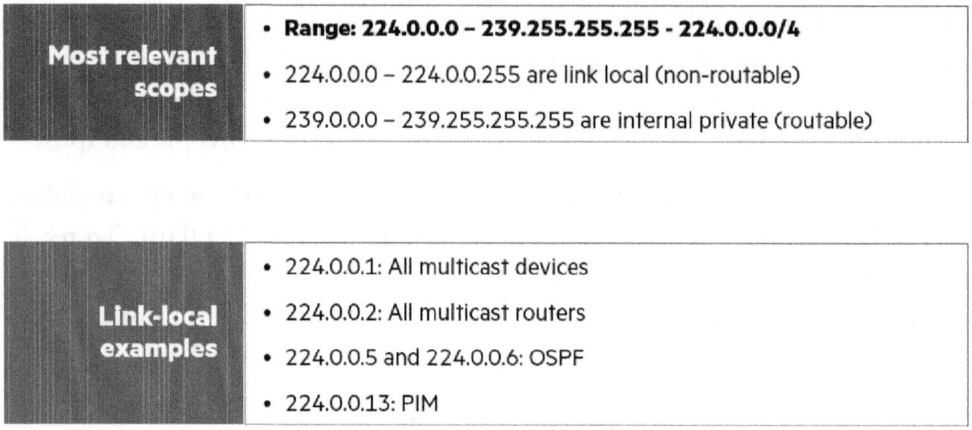

Figure 7-5: L3 multicast addressing summary

L3 multicast address assignments

As shown in **Figure 7-6**, IANA has assigned other addresses to specific organizations and uses. Some applications, such as applications that use SAP, can dynamically select multicast addresses from a certain range. Finally, individual organizations can define administratively scoped multicast addresses in the 239.0.0.0/8 range as they want; these multicasts should not leave the administrative domain.

The group address simply identifies a group of nodes interested in a flow. The group address combined with the source IP address identifies the multicast flow.

The table in **Figure 7-6** identifies the multicast address assignment:

- **Local Network Control block (224.0.0/24):** The Local Network Control block is used for specific protocol control traffic. Router interfaces listen to, but do not forward, local control multicasts—for example, OSPF "all routers" (224.0.0.5). Assignments in this block are publicly controlled by IANA. You can find a complete list of Local Network Control address assignments at the IANA website (www.iana.org).
- **Internetwork Control block (224.0.1/24):** The Internetwork Control block is for protocol control traffic that router interfaces may forward through the ASN or through the internet. Examples include 224.0.1.1 NTP, defined in RFC 4330, and 224.0.1.68 mdhcpdiscover, defined in RFC 2730. Internetwork Control group assignments are also publicly controlled by IANA.

- **AD-HOC blocks (I: 224.0.2.0–224.0.255.255, II: 224.3.0.0–224.4.255.255, and III:233.252.0.0–233.255.255.255):** Traditionally assigned to applications that do not fit in either the Local or Internetwork Control blocks. Router interfaces may forward AD-HOC packets globally. Most applications using AD-HOC blocks require few group addresses (such as, for example, less than a /24 space). IANA controls any public AD-HOC block assignments.
- **SDP/SAP block (224.2.0.0/16):** The Session Description Protocol/Session Announcement Protocol (SDP/SAP) block is assigned to applications that receive addresses through the SAP, as described in RFC 2974.
- **Source-Specific Multicast (SSM) block (232.0.0.0/8):** SSM addressing is defined by RFC 4607. SSM is a group model of IP Multicast in which multicast traffic is forwarded to receivers from only those multicast sources for which the receivers have explicitly expressed interest. SSM is mostly used in one-to-many applications. No official assignment from IANA is required to use the SSM block because the application is local to the host; however, according to IANA policy, the block is explicitly reserved for SSM applications and must not be used for any other purpose.
- **GLOP block (233.0.0.0/8):** These addresses are statically assigned with a global scope. Each GLOP static assignment corresponds to a domain with a public 16-bit ASN, which is issued by IANA. The ASN is inserted in dotted-decimal into the middle two octets of the multicast group address (X.Y). An example GLOP assignment for an ASN of X.Y would be 233.X.Y.0/24. Domains using an assigned 32-bit ASN should apply for group assignments from the AD-HOC III block. Another alternative is to use IPv6 multicast group addressing. Because the ASN is public, IANA does not need to assign the actual GLOP groups. The GLOP block is intended for use by public content, network, and internet service providers. IANA considers GLOP addressing to be experimental, and 233.252–255.0.0 is reserved.
- **Administratively scoped block (239.0.0.0/8):** Administratively scoped addresses are intended for local use within a private domain as described by RFC 2365. These group addresses serve a similar function as the RFC 1918 private IP address block (such as, for example, 10.0.0.0/8 or 172.16–31.0.0/16 blocks). Network architects can create an address schema using this block that best suits the needs of the private domain and can further split scoping into specific geographies, applications, or networks.

"GLOP," surprisingly, is not an acronym. It does not stand for anything. The authors of the RFC needed a better way to refer to "that addressing method where you put your AS in the middle of two octets." David Meyers started calling it

---

"GLOP," and the name stuck.

You can refer to IANA for the latest assignments: https://www.iana.org/assignments/multicast-addresses/multicast-addresses.xhtml

Address range	Description	Routable
224.0.0-224.0.0.225	Local Network Control	No
224.0.1.0-224.0.1.255	Internetwork Control	Yes
224.0.2.0 - 224.0.255.255	AD-HOC I	Yes
224.2.0.0 - 224.2.255.255	SDP/SAP	Yes
224.3.0.0 - 224.4.255.255	AD-HOC II	Yes
232.0.0.0 - 232.255.255.255	Source-Specific Multicast	Yes
233.0.0.0 - 233.251.255.255	GLOP	Yes
233.252.0.0 - 233.255.255.255	AD-HOC III	Yes
239.0.0.0 - 239.255.255.255	Administratively Scoped	Yes

Figure 7-6: Multicast address assignment

Switching multicast frames

The major difference between a unicast and multicast network is that in a unicast network, all traffic is fully permitted to the host, assuming no filter is applied. In a multicast network, the host must advertise its intention to join the network.

Without multicast-aware Layer 2 protocols, all hosts on a given Layer 2 segment will receive multicast packets for any groups joined by a host on that segment. This is not efficient, since non-interested endpoints must process each frame, only to drop it. When Layer 2 devices are multicast-aware, they properly send multicast traffic to only the ports where we find interested hosts.

The gateway is the network demarcation between Layers 2 and 3 and is the most appropriate device to manage host group membership for the larger network. This device receives these management messages and adds host segment interfaces to the local multicast table. A gateway communicates group interest for a multicast group to other Layer 3 devices using multicast routing (PIM).

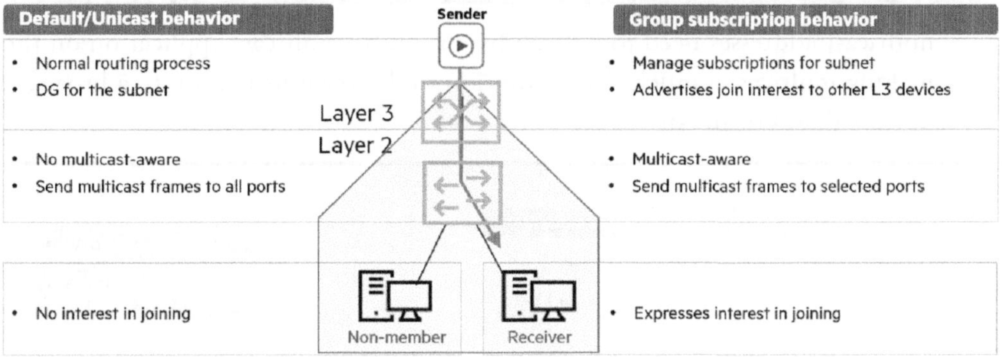

Figure 7-7: Differences between a unicast and multicast network

L2 multicast addressing

To optimize network resources, an Ethernet switch must understand multicast addressing. This is where the magic happens. The sending device must convert the destination IP multicast address into a special MAC address as follows:

- The high-order 25 bits are the official reserved multicast MAC address range from 0100.5E00.0000 to 0100.5E7F.FFFF (RFC-1112). These bits are part of the Organizational Unit Identifiers (OUI).
- The lower-order 23 bits of the destination IP multicast address are mapped to the lower-order 23 bits of the MAC address.

A switch can use this calculated multicast MAC address to distinguish a frame as a multicast and make efficient forwarding decisions.

We miss five bits of mapping information: 25 = 32. This means we will map **32 multicast IP addresses to one multicast MAC address**.

The IEEE has allocated the address block 01-80-C2-00-00-00 to 01-80-C2-FF-FF-FF for group addresses for use by standard protocols implementing multicast at Layer 2. Of these, the MAC group addresses in the range of 01-80-C2-00-00-00 to 01-80-C2-00-00-0F are not relayed by switches conforming to 802.1D; for example, they are not flooded. Examples of these would include LACP, STP BPDUs, LLDP, CDP, and others.

With a broadcast, all devices in a subnet receive it. Their respective NICs then have the local device's CPU process it. With a multicast, the NIC can make an intelligent decision as to whether the CPU should process the frame or if the NIC should drop the frame based on if the multicast MAC address is in use by a multicast-based application on the device. Therefore, the NIC can conserve CPU cycles based on what

---

multicast addresses need to be seen by a running multicast application on the device, making multicast a much better solution to deliver information to a broader range of devices than broadcasts.

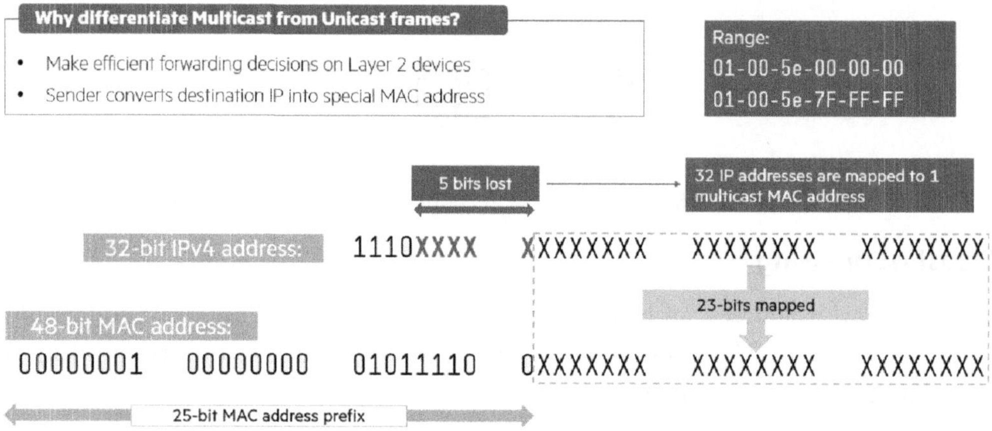

Figure 7-8: L2 multicast addressing

Overlapping addresses

All the multicast IP addresses shown in **Figure 7-9** are mapped to the same multicast MAC address (01-00-5E-00-00-01), as they share the same last 23 bits on the IP address.

This can cause some network problems. For example, a host that listens to the 239.0.0.1 multicast IP address will configure its network card to listen to MAC address 01-00-5E-00-00-01. At the same time, any device could generate traffic to 224.0.0.1 (all nodes in the link), which is mapped to the same MAC address. Hosts will have to look at the IP address (process Layer 3 header) of the received frame to see if it's for 239.0.0.1 and discard frames that are meant for 224.0.0.1. This process is not efficient.

As a recommendation, do not use x.0.0.x or x.128.0.x, since these addresses will overlay with the Link-Local Multicast address scope.

Some network devices do IGMP filtering based on MAC address. This implies that you should be careful, when picking your multicast groups, that you do not overlay the MAC addresses as to avoid odd interactions at Layer 2.

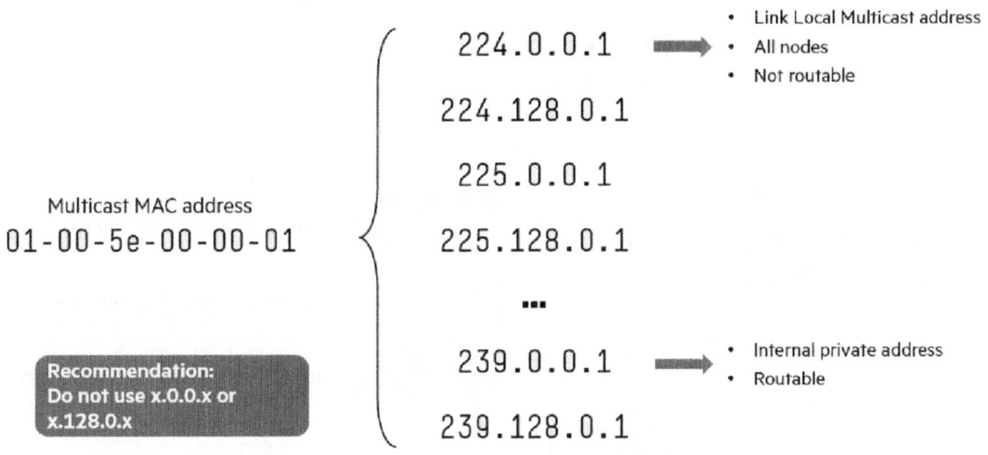

Figure 7-9: Overlapping addresses summary

IGMP messages

IGMP manages multicast group memberships based on a query and response mechanism. The multicast group uses four fundamental types of messages to communicate:

- **Query:** A message sent from the querier asking for a response from each host belonging to the multicast group. If a multicast router supporting IGMP is not present, the switch must assume this function to elicit group membership information from the hosts on the network.
- **JOIN:** A message sent by a host to the querier to indicate that the host wants to be or is a member of a given group indicated in the JOIN message.
- **LEAVE:** A message sent by a host to the querier to indicate that the host has ceased to be a member of a specific multicast group.
- **Report:** A message sent by a host, in response to a multicast router query, identifying the particular multicast stream or streams the host wishes to receive.

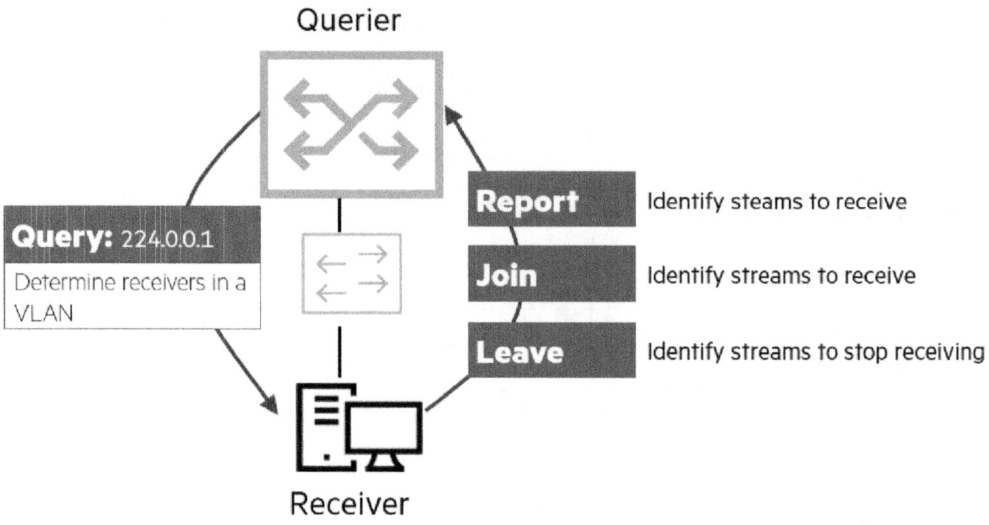

Figure 7-10: IGMP messages summary

Multicast flooding within VLANs

By default, Ethernet switches flood multicasts on all ports in a VLAN, or broadcast domain. Thus, although multicasts should target only the specific devices that require them, they act like broadcasts, creating congestion for an entire subnet. This can waste bandwidth and other resources. If all edge ports must carry traffic for all multicast groups within the VLAN, overall performance quickly suffers.

Flooding multicast packets can also increase security risk, as every endpoint in a subnet receives traffic that is only meant for a particular group. Even if most users would not know how to eavesdrop on the multicast traffic from a neighboring meeting room's videoconference that shows up unnecessarily on their ports, some users might—which could violate privacy regulations. Finally, service providers that provide for-pay services must limit multicast traffic to the properly registered receivers so that they can track who uses the services.

In summary, to deploy multicasting applications effectively, you must enable the switches to suppress specific multicasts on ports that do not require them.

Applications and processing

Applications are defined to use Layer 3 IP multicast addresses, but the actual hardware, like NICs or switches, is processing the Layer 2 addresses. By default, devices treat multicasts like broadcasts (with few exceptions). However, once multicasting support is enabled, devices can typically process the multicast traffic intelligently. For example, if a device did not have multicast enabled, the

NIC would treat it as a broadcast and forward it up the protocol stack for the CPU to process. However, with multicasting enabled, as shown in **Figure 7-11**, the NIC listens only for multicast addresses that are relevant for upper-layer applications.

The second problem is that switches flood multicast traffic. In the figure, Clients 1 and 2 receive multicast streams, even though they aren't running a multicast application. This wastes CPU cycles on these clients to basically process and drop the multicast traffic. The switches need an intelligent method of forwarding/filtering multicasts instead of treating them as a broadcast.

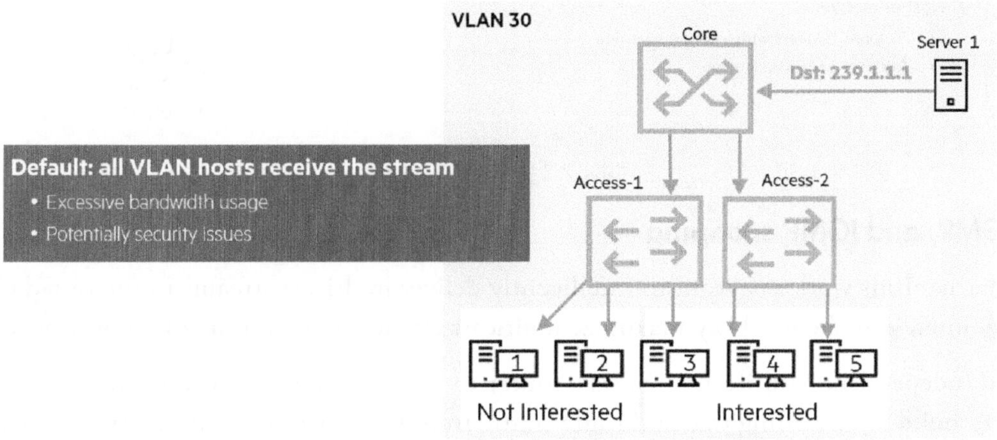

Figure 7-11: Multicast flooding within VLANs

IGMP overview

Multicast protocols are focused on "group membership." This simply indicates which multicast streams are desired. If a host wishes to receive a multicast stream destined to address 239.1.1.1, they join group 239.1.1.1.

IPv4 systems (hosts and routers) use IGMP to report their IP multicast group memberships to any neighboring multicast routers. IGMP messages are encapsulated in IP datagrams, with IP protocol number 2.

IGMP establishes two device types—Querier and Hosts or receivers.

- A querier is a Layer 3 switch that runs IGMP. It sends periodic queries, asking hosts in the subnet which multicast group traffic they wish to receive. A multicast group is commonly referred to as a multicast *stream*—the application information, like a video stream.
- A multicast receiver is a host that wishes to receive a multicast stream or streams, like a videoconference unit or a user's desktop.

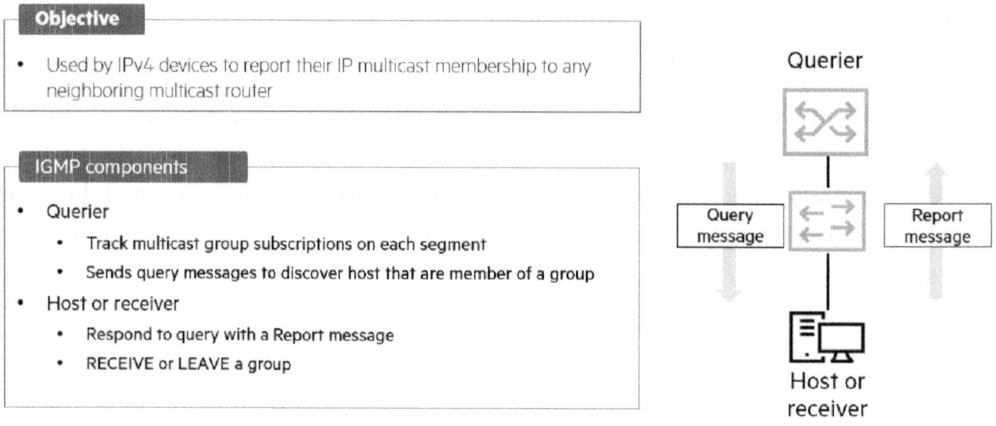

Figure 7-12: IGMP summary

PIM, IGMP, and IGMP snooping

Three mechanisms work as a team to intelligently deliver multicast streams to interested clients, and only interested clients. They maximize multicast efficiency by minimizing multicast overhead.

Protocol Independent Multicast (PIM) is a router-to-router communications protocol used by routers to build a forwarding tree and route traffic from the source toward interested clients.

The Internet Group Message Protocol (IGMP) is a host-to-router communications protocol used by IPv4 devices to report their IP multicast membership to any neighboring multicast router. It helps to determine who is interested in receiving a multicast stream. Because of IGMP, router Core-1 knows that zero clients in VLAN 40 are interested in receiving the multicast stream, and more than zero clients in VLAN 30 are interested in the stream. Thus, Core-1 only forwards the stream out its VLAN 30 interface—no wasted bandwidth on VLAN 40!

However, by default, L2 switches flood all Broadcast, Unknown unicast, and Multicast (BUM) traffic out all ports in the VLAN (other than the ingress port). This means that hosts 1 – 3 would receive the stream, even though only host 2 is interested. When you enable IGMP snooping, Layer 2 switches listen in (snoop) on IGMP traffic between hosts and routers. They learn the MAC addresses of hosts that are interested in the stream and modify their default behavior accordingly. They only forward the stream to interested hosts in the VLAN—no wasted bandwidth in VLAN 30!

Notice that PIM and IGMP are protocols, while IGMP snooping is simply a mechanism that you enable on switches. The focus of this chapter is the IGMP protocol and IGMP snooping.

The IGMP functions fulfilled by switches are sometimes called IGMP snooping because switches traditionally did not generate IGMP messages, but only listened for the ones sent by routers and hosts. However, routing AOS-CX switches can fulfill all IGMP functions. They can run the IGMP protocol as a Layer 3 router and perform IGMP snooping as a Layer 2 switch. You enable IGMP on a routed L3 interface, while you enable IGMP snooping on a Layer 2 VLAN. Whether the VLAN has an IP address has no effect on this snooping functionality.

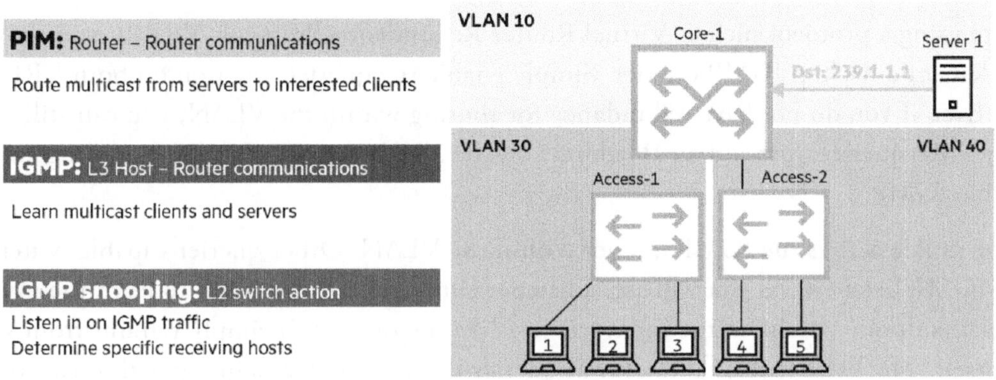

Figure 7-13: PIM, IGMP, and IGMP snooping

IGMP operations

To determine group membership, the querier (router) sends a message to every host on the subnet. In the following figure, router Core-1 is the IGMP querier, and Access-1 and Access-2 are Layer 2 switches. The querier must maintain a list of hosts in the subnet interested in multicast flows. To do this, the query message is multicast to the "all-hosts" multicast address of 224.0.0.1. This message is created with a TTL of 1, which indicates its non-routable nature.

Generally, the default router for a VLAN should act as the IGMP querier. The figure shows how to enable IGMP on an AOS-CX routing switch, both for a VLAN interface and for a physical Layer 3 interface. IGMPv3 is the default implementation. Use the `disable` parameter to disable IGMP on a Layer 3 interface. Note that if you disable IGMP on an interface and then later re-enable IGMP on that VLAN, the switch restores the last-saved IGMP configuration for that interface. Use the `no ip igmp` command to completely remove it from an interface.

```
interface VLAN 30
ip igmp enable
interface 1/1/1
ip igmp enable
```

You should plan redundancy for the IGMP querier role by implementing either VSX or VSF on your AOS-CX switches.

Multiple IGMP queriers and elections

Just as you need to consider redundancy for the default gateway, you should consider how you will provide redundancy for the IGMP querier role. In the previous example, the core is a VSF fabric, which has built-in redundancy.

If you are using a protocol such as Virtual Router Redundancy Protocol (VRRP), you should usually implement a backup IGMP querier. Simply enable the querier function on both VRRP devices. Even if you do not have redundancy for routing within the VLAN, you can still create a backup IGMP querier. Just add an IP address on the VLAN to one of the non-routing switches and enable IGMP.

Only one device acts as the IGMP querier within the VLAN. Other querier-capable switches become IGMP listeners. As you will see, a listener continues to use IGMP to learn the multicast traffic that endpoints desire in the interface, just like the querier. It simply refrains from sending queries itself. The listener also takes over as querier if it fails to hear a query within two times the querier interval.

The AOS-CX switch behavior in determining whether to act as IGMP querier or listener differs based on whether the switch is also implementing a multicast routing protocol such as PIM. The behavior with PIM enabled is described as follows for your reference.

Behavior without PIM enabled

When you enable IGMP on a switch VLAN with an IP address, the querier capability is enabled by default. However, based on the presumption that a multicast router will act as IGMP querier, the switch does not attempt to become querier immediately. Instead, it listens for IGMP queries for a waiting period equal to twice the IGMP querier interval. If it hears any queries in this period, it becomes an IGMP listener. If it does not, it becomes the querier. However, if the switch later hears a query from another device, it disables the querier function and defers to that device.

If you enable IGMP on two AOS-CX switches that are not running PIM, the first switch on which you enable IGMP becomes the querier.

Behavior with PIM enabled

When an AOS-CX switch has PIM and IGMP enabled on a VLAN, it uses the IGMPv2 rules for its querier functions. As soon as IGMP is enabled, the switch begins to send IGMP queries at the designated interval. If it ever hears a query from another querier with a lower IP address, it stops sending queries and becomes an IGMP listener. In other words, the switch running IGMP and PIM with the lowest IP address becomes querier, possibly preempting the role from another switch.

IGMP operates separately from VRRP. That is, if two routing switches implement VRRP and IGMP in VLAN 20, either the VRRP standby router or the VRRP active router could be elected as the IGMP querier regardless of its VRRP role. This is because the switches use their actual IP addresses for the election (the VRRP active router does not use its virtual address).

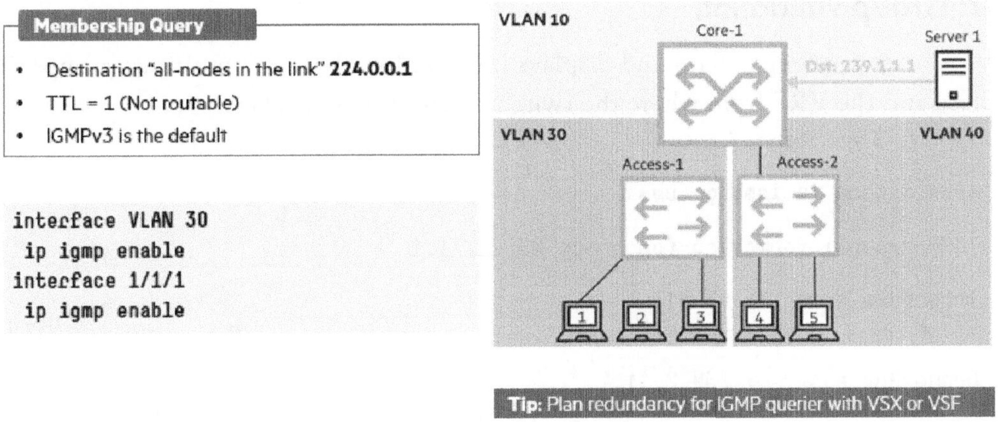

Figure 7-14: IGMP overview

Querier validation

The `show ip igmp` command displays information about the querier. Notice that the output includes which device is running the querier role for the segment.

This command includes information such as:

- VRF Name
- Layer 3 Interface
- IGMP Configured Version (v3 by default)
- Querier State
- Querier IP address

```
AGG-1# show ip igmp

VRF Name    : default
Interface   : vlan11
IGMP Configured Version  : 3
IGMP Operating Version   : 3
Querier State             : Querier
Querier IP [this switch]  : 10.2.11.2
Querier Uptime            : 12m 17s
Querier Expiration Time   : 1m 23s
IGMP Snoop Enabled on VLAN : False

Active Group Address   Vers Mode Uptime    Expires
---------------------- ---- ---- --------- ---------
239.2.11.1             3    EXC  0m 12s    4m 8s
```

Querier information

Active group

Figure 7-15: Command display information

Querier: Group validation

The `show ip igmp groups` command displays information about the multicast groups, including the interface and the VRF from where the switch received these packets.

```
AGG-1# show ip igmp groups

IGMP group information for group 239.2.11.1

Interface Name    : vlan11
VRF Name          : default

Group Address     : 239.2.11.1
Last Reporter     : 10.2.11.10

                                  V1        V2        Sources   Sources
Vers Mode Uptime    Expires   Timer     Timer     Forwarded Blocked
---- ---- --------- --------- --------- --------- --------- ---------
3    EXC  5m 9s     3m 0s
```

Interface and VRF

Host that sent the report

Figure 7-16: Command display information

Querier: Interface validation

You could also use the command `show ip igmp interface` to see IGMP interface-specific configuration and operation information.

```
switch# show ip igmp interface vlan 30
IGMP Configured Version : 3
IGMP Operating Version : 3
Querier State : Querier
Querier IP [this switch] : 20.1.1.1
Querier Uptime : 1m 46s
Querier Expiration Time : 0m 1s
```

```
Snoop Enabled on VLAN : True
```

Querier: Joined group validation

Validate IGMP joined group information with the `show ip igmp group <GROUP-IP> [source <SOURCE-IP>] [vrf <VRF-NAME> | all-vrfs] [vsx-peer]` command.

You can also use the `show ip igmp groups` command, which will list all IGMP groups the querier knows about. Many other `show` commands are available to verify the operation of IGMP.

```
switch# show ip igmp group 239.1.1.10
IGMP group information for group 239.1.1.10
Interface Name : vlan2
VRF Name : default
Group Address : 239.1.1.10
Last Reporter : 100.1.1.10
                                V1        V2        Sources   Sources
Vers Mode Uptime    Expires   Timer     Timer     Forwarded Blocked
---- ---- --------- --------- --------- --------- --------- --------
3    EXC  16m 34s   2m 27s
```

IGMPv3 behavior

As you should consider redundancy for the default gateway, you should also provide redundancy for the IGMP querier role. IGMPv3 elects a single querier per subnet. Other querier-capable switches become IGMP listeners. A listener continues to use IGMP to learn the multicast traffic that endpoints desire in the interface, as the querier. It simply refrains from sending queries itself. The listener also takes over as querier if it does not hear a query within two times the querier interval.

When a host receives an IGMP query packet, it kicks off a timer that begins with a random value that is less than the maximum response time. If no other host responds with a membership report before this random timer expires, the host will then reply with a report. This decreases the number of total IGMP reports needed to maintain the group state. This preserves local bandwidth because the host suppresses its own reports unless necessary.

The AOS-CX switch behavior, in determining whether to act as IGMP querier or listener, differs based on whether the switch is also implementing a multicast routing protocol such as PIM.

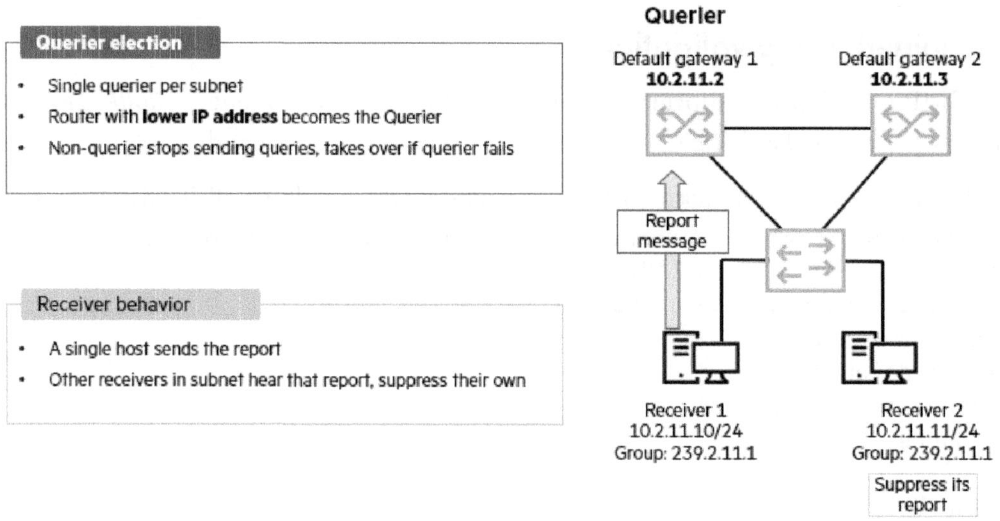

Figure 7-17: IGMPv3 behavior

Membership reports

Membership report messages are sent by IP endpoints to report (to neighboring routers) interest in joining or leaving a multicast stream. These messages are sent either in response to a membership query message or when the multicast application in the endpoint starts. There are two subtypes of membership messages:

- **JOIN:** Sent from host to querier to identify the multicast stream the host wishes to receive.
- **LEAVE:** Sent from host to querier to indicate that the host shall no longer be a member of a multicast group.

In **Figure 7-18**, a server sends a multicast stream to 239.1.1.1 and another stream to 239.2.2.2. Depending on the PIM configuration, these streams may not be forwarded anywhere—Core-1 discards the frames because nobody is interested in receiving the streams—there are no group members.

Then host 2 in VLAN 30 sends an IGMP group membership report: "I wish to join 239.1.1.1." Meanwhile, host 5 in VLAN 40 sends a group membership report: "I wish to join 239.2.2.2." Assuming all devices run IGMPv3, the messages are sent to 224.0.0.22. For IGMPv1 and v2, messages are sent to the group address—239.1.1.1 or 239.2.2.2.

When Core-1 receives the first report for a particular multicast group on a VLAN interface, it adds a table for that group to the VLAN. Each VLAN that implements IGMP has its own forwarding tables. Because the switch also performs IGMP snooping at Layer 2, the physical interface on which the switch received the report becomes a forwarding interface for that group. If the switch receives another membership report on a different physical interface, it adds that interface to the forwarding table as well.

IGMPv1 group membership entries do not persist indefinitely. The switch also sets a timer for the group membership, which it resets every time that its query produces another report. If the timer expires, the switch removes the group membership from the interface.

IGMPv2 and v3 have a faster way for an interface to leave a group. When a host no longer requires multicast traffic in a particular group, it sends an IGMP leave group message. The IGMP querier instantly sends one or more queries for the multicast group in question. This allows another IGMP host connected to the same interface to send a membership report indicating that it still wants to receive the multicasts.

Supplemental information on how AOS-CX switches handle reports

In the following example, only one IGMP host in VLAN 30 wants to join the 239.1.1.1 group. As you see, the core VSF fabric receives only one membership report. Even if multiple hosts on Access-1 wanted to join this group, though, they would all receive the membership report from the first host to submit the report because this report is a multicast. Those hosts would then suppress their own reports, and a single report still flows to the core. (This behavior is a bit different when Access-1 also implements IGMP.)

IGMP defines this behavior because it was originally designed to inform multicast routers if at least one host in the subnet requires multicasts in the subnet. How many hosts required the multicasts does not really matter. Because routing AOS-CX switches also implement IGMP at Layer 2 and need to discover on which physical interfaces to forward traffic, they alter this behavior a bit. The core VSF switch, which implements IGMP, suppresses the membership reports that it receives on each physical interface and does not forward them on other physical interfaces in the VLAN. This behavior prevents hosts on Access-2 from hearing the membership reports from hosts on Access-1. In this way, the core fabric could receive a report on the Access-2 link—for example, if any hosts connected to that switch wanted to listen on 239.1.1.1.

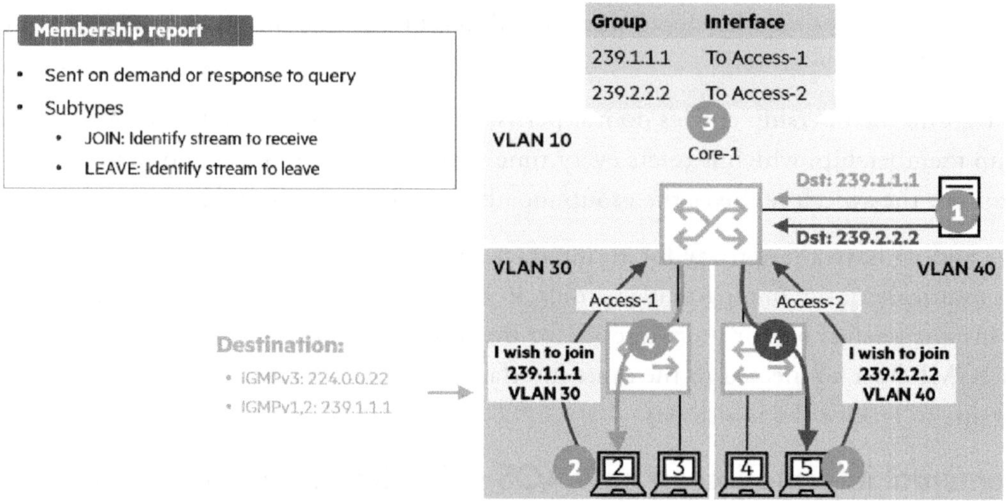

Figure 7-18: Membership reports summary

Challenge: IGMPv1 query and report issue

The original IGMPv1 only supported two message types: queries and reports. There are two problems with the original implementation and these message types:

- The host must wait for a query before sending a report; this slows down access to the stream in which the user is interested.
- When a user is done with the multicast application, the host would still receive the stream until a corresponding query would resolve the issue, wasting bandwidth.

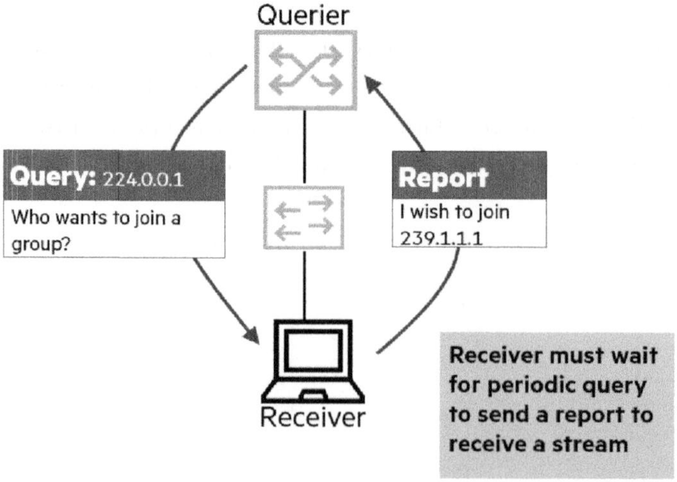

Figure 7-19: Challenge: IGMPv1 query and report issue

Solution: IGMPv2/3 JOIN and LEAVE messages

IGMPv2 and v3 introduce two new message types: JOINs and LEAVEs. A JOIN message allows a multicast host to immediately signal a querier of a multicast group that the host needs to receive, expediting the receipt of the stream. And a LEAVE message allows a host to immediately notify a querier when the host no longer needs the stream for the multicast group.

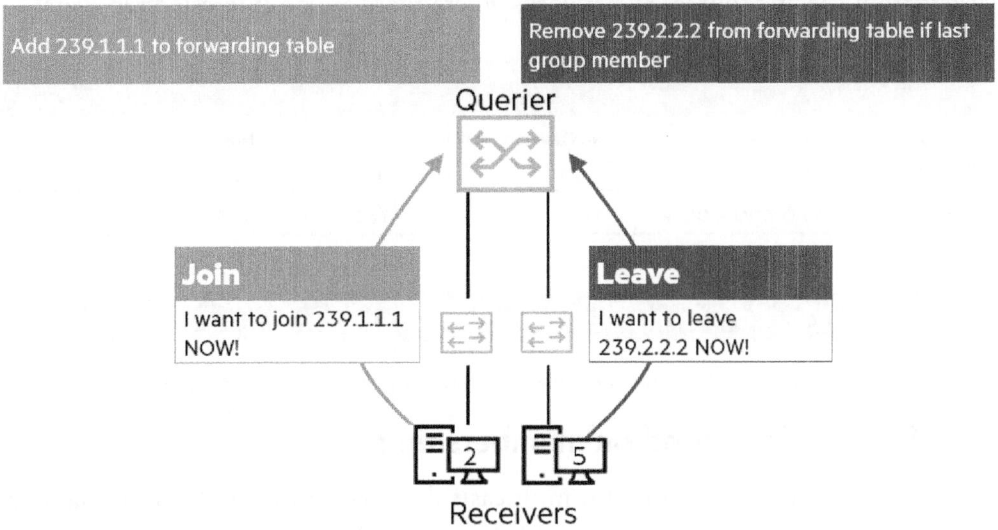

Figure 7-20: Solution: IGMPv2/3 JOIN and LEAVE messages

Summary review

The following figure summarizes the differences between IGMP versions, as described below.

- **IGMPv1:** Defined in RFC 1112, IGMPv1 offers a basic query-and-response mechanism to determine which multicast streams should be sent to a particular network segment. IGMPv1 has no mechanism for a host to signal that it wants to leave a group. When a host using IGMPv1 leaves a group, the router will continue to send the multicast stream until the group times out. As you can imagine, this can create a large amount of multicast traffic on a subnet if a host joins groups very quickly. IGMPv1 also does not elect a querier. If there are multiple queriers (routers) on the subnet, a designated router (DR) is elected using PIM to avoid sending duplicate multicast packets. The elected querier is the router with the highest IP address. IGMPv1 is rarely used in modern networks.
- **IGMPv2:** Defined in RFC 2236, this newer version made improvements over IGMPv1. One of the most significant changes was the addition of a LEAVE process. A host using IGMPv2 can send a leave-group message to the querier indicating that it is no longer interested in receiving a particular multicast stream. IGMPv2 also added group-specific-queries. This feature allows the querier to send a message to the hosts belonging to a specific

multicast group. The querier election process determines the querier based on a priority.

- **IGMPv3:** Defined in RFCs 3376 and 4604, this version 3 adds support for "source filtering"—the ability for a system to report interest in receiving packets **only** from specific source addresses, or from **all but** specific source addresses, sent to a particular multicast address. Multicast routing protocols can use this information to avoid delivering multicast packets from specific sources to networks where there are no interested receivers.

Version	RFC	Querier election	Leave process	Source-specific query
IGMPv1	1112	No (Use higher IP)	No	No
IGMPv2	2236	Yes	Yes	No
IGMPv3	3376 and 4604	Yes	Yes	Yes

Default version in AOS-CX, compatible with IGMPv1 and v2

Figure 7-21: Summary of the differences between IGMP versions

Challenge: Forwarding based on membership tables

The multicast group table filters traffic for multicasts destined to that address. In **Figure 7-22**, Core-1 is the only switch running IGMP, and it is the IGMP querier. Based on the membership reports that it has received in VLAN 30, it has created two forwarding tables: one for 239.1.1.1 and one for 239.2.2.2. Bandwidth is saved—the 239.1.1.1 stream is only sent toward Access-1, and 239.2.2.2 is only sent toward Access-2.

However, recall the default L2 switch behavior for BUM traffic—multicast traffic received by Access-1 and Access-2 is flooded out all ports in the VLAN. This includes the ports connected to hosts that are not a member of any group and have no need for this potentially high-bandwidth video stream.

In essence, the core filters multicasts, but the access layer switches do not.

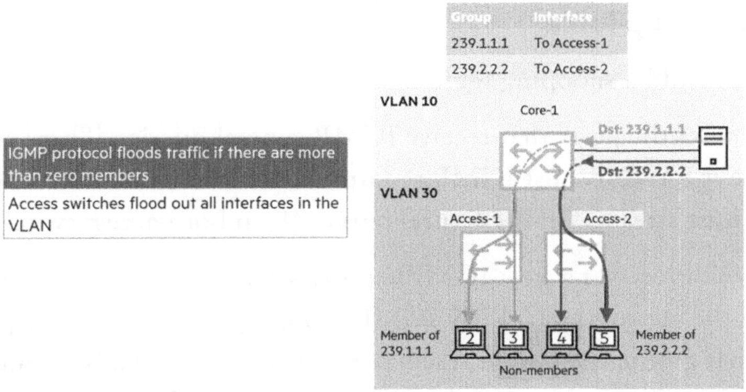

Figure 7-22: Challenge: Forwarding based on membership tables

Solution: IGMP snooping

IGMP snooping solves this issue.

- You configure IGMP snooping on Access-1 and Access-2. These Layer 2 devices are transparent to the L2 communications between routers and hosts. They do not run the IGMP protocol, and so act as neither queriers nor hosts; they just quietly listen and learn.

It listens in on (snoops) the normal IGMP packets between querier routers and receiver hosts.

- It learns the MAC addresses of group members and creates Layer 2 multicast forwarding entries, one table for each VLAN interface on which IGMP snooping is enabled. No more traffic to non-member hosts!

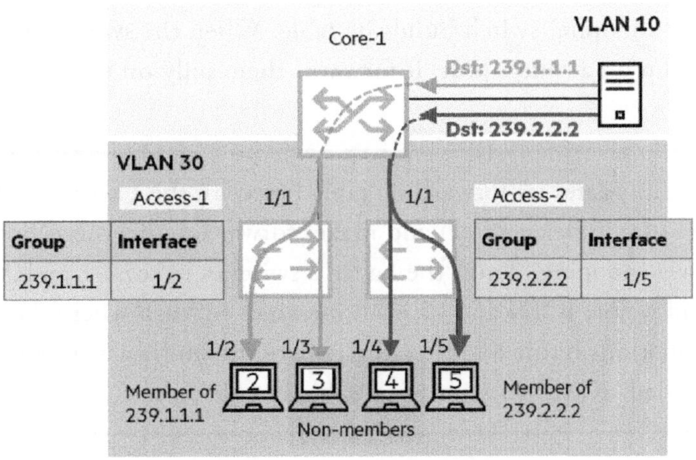


```
vlan 20
  ip igmp snooping enable
```

Figure 7-23: Solution: IGMP snooping

Building the IGMP snoop table

Here is how switches build a snooping table:

1. The switch tracks the ports that receive IGMP general queries. These are called querier ports—where it forwards host IGMP reports, JOIN and LEAVE messages, and it forwards querier multicast streams. In the figure, ports 1/1 on both access switches are querier ports.
2. When the switch receives a host IGMP report, it maps the incoming port to the group in the report in the appropriate VLAN IGMP snooping table. Or, if the port is already mapped to that group, the switch resets the dynamic port member aging timer for that port and group. In the figure, port 1/2 maps to 239.1.1.1, and port 1/5 maps to group 239.2.2.2
3. When hosts hear IGMP reports from other hosts in the same group, they typically suppress their own reports to minimize floods or reports. However, the IGMP snooping switch requires more granular information about the receivers on each port. Therefore, the switch only forwards IGMP reports on router or querier ports. Other hosts, not hearing the report, send their own reports, and the switch can learn their group memberships.
4. The snooping switch minimizes IGMP report floods—it only forwards the initial entry for a multicast group in its IGMP snooping table (rather than every time it adds a new port to the entry).
5. Inactive group memberships time out. How the switch handles LEAVE messages depends on the fast-leave setting, which you will examine in a moment.

This is how the IGMP snooping switch builds its table. When the switch receives multicast packets for a particular group on a router port, it forwards them only on the ports mapped to the group.

The switch updates its forwarding table based on the Layer 2 multicast MAC addresses (01:00:5e:xx:xx:xx) and makes forwarding decisions based on what it learned via snooping. Thus, the Layer 2 switches (Access-1 and Access-2) spy on the query and report JOIN and LEAVE messages to learn where the multicast sources and destinations reside and therefore where the multicast stream (239.1.1.1) needs to be forwarded. IGMP snooping is disabled by default.

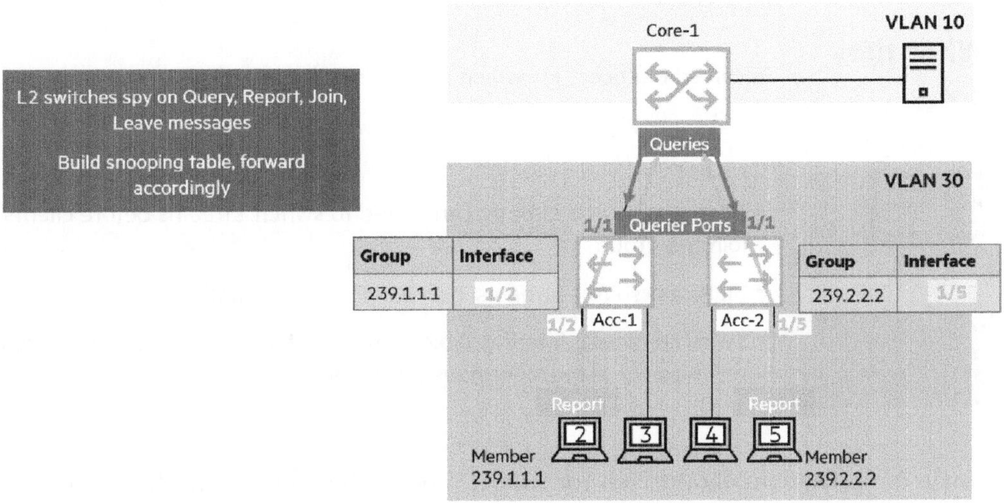

Figure 7-24: How to enable IGMP snooping

Unknown multicast

Sometimes an IGMP or IGMP snooping switch receives a multicast for a group for which the switch does not have a filtering table. Called "unknown multicasts," these multicasts can occur when:

- The multicast source is directly connected to this switch, and it starts streaming before any clients join the group.
- A multicast source is streaming, and another switch in the path between the source and this switch does not support IGMP.
- The switch has removed a multicast group filtering table because the last member left the group. However, a multicast router forwarding the traffic has not yet received the leave, so it continues to forward the multicasts.

What they are	Received multicast for which there is no filtering table
When they occur	• Multicast source directly connected to switch, streams before clients join the group • Multicast source streams, non-IGMP switch in path • Switch removed filtering table – last member has left. Router has not received the leave, continues forwarding.

AOS-CX switches operate in data-driven (or smart) mode. They filter all multicasts for which they do not have a multicast table. If no host has joined the group, but the switch receives multicasts, the switch drops them.

One exception applies: an IGMP snooping switch DOES forward both unknown and known multicasts on any port on which it has heard IGMP queries—port 1/1 in . This behavior is required so that a multicast stream can reach the multicast router. In fact, by default, the switch forwards unknown multicasts that arrive on one VLAN on any port on which it has heard queries in any VLAN.

To change this behavior and restrict forwarding to querier ports for that specific VLAN, enter the `ip igmp snooping drop-unknown vlan-exclusive` command.

The default mode is **vlan-shared**. It is important that you understand that the `igmp filter-unknown-mcast` command has no effect on how the data-driven switch behaves with non-querier ports. The switch always filters unknown multicasts on these ports.

Data-driven mode and fast learn

Configuring Fast Learn on a port enables a faster response to Topology Change Notifications (TCN). When spanning tree changes the port state from Blocked to Forwarding, the device acting as querier will immediately send a general query on the Fast Learn-enabled port. Then the device acting as a non-querier will replay the JOINs. This will help in the faster convergence of multicast flows.

```
Switch(config)# vlan <ID>
Switch(config-vlan)# ip igmp snooping fastlearn <port_ID_list>
```

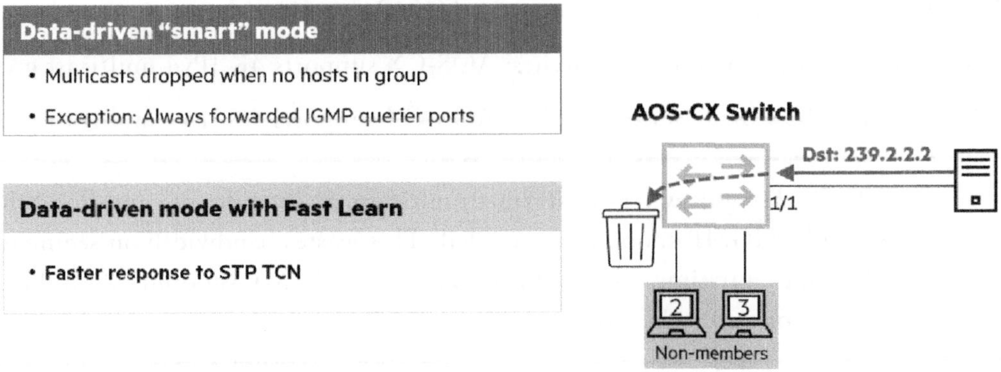

Figure 7-25: Unknown multicast

AOS-CX

With the factory default setting, multicast data transmitted from the sources will be flooded on all ports in the VLAN. Configuring IGMP snooping avoids flooding and causes the switch to forward data only to the receivers.

IGMP configuration considerations include the following:

- For IGMP to be operational, the interface must be administratively up. For interface VLANs, the L2 VLAN must be up, and one of the ports in the VLAN must be up.
- The IP address must be assigned for the interface to become a querier. Without an IP address, the device will remain in a non-querier state.
- A querier is required for proper IGMP operation. For this reason, you must enable IGMP on the L3 Interface. If the querier functionality is not configured or is disabled, you must ensure that there is an IGMP querier in the same VLAN.
- For IGMP snooping to be operational on a VLAN, the VLAN must be administratively up, and at least one port in the VLAN must be up.
- If the switch becomes the querier for a particular interface, then subsequently detects queries transmitted from another device on the same VLAN, the switch ceases to operate as the querier for that interface.
- The switch automatically ceases querier operation in an IGMP-enabled interface if it detects another querier on the interface. You can also use the switch CLI to disable the querier capability.

Each of the AOS-CX switches have different capabilities regarding the number of multicast streams supported by the switch. Here's a summary:

- CX 8400 Series switches: AOS-CX supports 16K IPv4 multicast groups.
- CX 8320 Series/CX 8325 Series switches: AOS-CX supports 4K IPv4 multicast groups.
- CX 6300 Series/CX 6400 Series switches: AOS-CX supports 4K IPv4 multicast groups.

By default, though, switches will treat multicasts as "broadcasts" and flood them in a VLAN unless IGMP snooping is enabled. This wastes bandwidth on segments in a VLAN. Thus, any time multicast is implemented, IGMP snooping should be enabled for VLANs on AOS-CX switches.

Configuration task	Context	Default
IGMP querier	Layer 3 interface	Disabled
IGMP snooping	Layer 2 VLAN	Disabled

Figure 7-26: AOS-CX capabilities

VSX and IGMPv3

You should also understand how VSX affects the forwarding of multicast traffic. Each VSX switch constructs its own IGMP group database, which is identical to its peer's IGMP group database. Both switches hear JOIN/LEAVE messages they receive from the downstream VSX LAGs because the ISL is always included as a forwarding port for IGMP. For example, SW1 has a VSX LAG to the aggregation switches.

In the following example, the switch chooses the link to Agg-1 for forwarding the IGMP JOIN, and Agg-1 learns the group on the VSX LAG. The IGMP JOIN crosses the ISL to Agg-2. The VSX IGMP process translates the IGMP JOIN learned on the ISL into an IGMP JOIN message learned on the VSX LAG, just as if the JOIN had arrived on the link from SW1. Therefore, Agg-2 also learns the group on the VSX LAG to SW1.

In this way, each VSX switch can determine on which interfaces to forward or prune multicast traffic for particular IGMP groups, based on its own IGMP group database. Because the IGMP database construction relies on this dataplane-based process, ISLP does synchronize IGMP groups between VSX peers. If a VSX switch reboots, it needs to relearn all the IGMP groups. For about

120 seconds after rebooting, the VSX switch floods multicast traffic within VLANs that have active, forwarding physical ports, which prevents disruption of multicast forwarding during the learning process. The VSX switch also sends an All-Hosts Query (AHQ) message. Clients send JOIN messages in response, so the VSX switch can relearn the groups and re-create the IGMP group database.

You can configure IGMP snooping on access VLANs. It enables VSX switches operating at Layer 2 on the VLAN to similarly construct identical IGMP group databases.

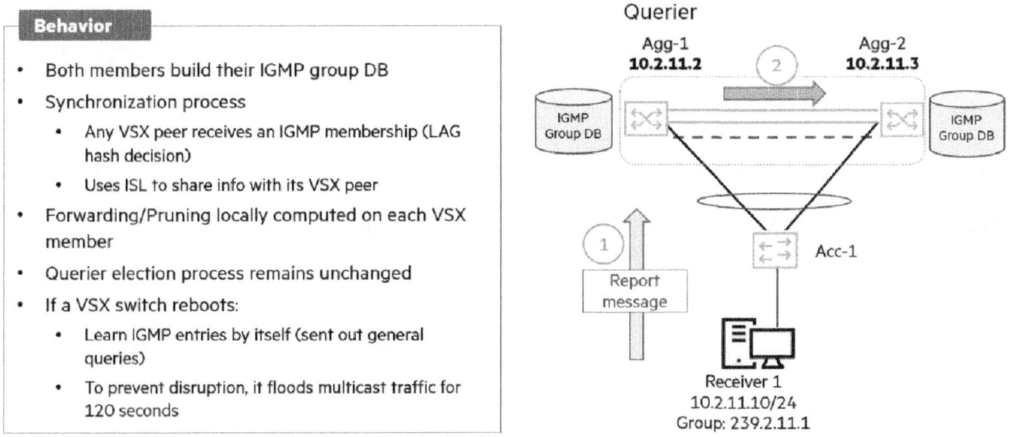

Figure 7-27: VSX and IGMPv3 example and summary

Learning check

Check your understanding of IGMP by answering the following questions.

Question 1

How do AOS-CX switches treat multicast traffic at the factory default settings?

- ○ They forward the multicasts on interfaces that connect to other switches but drop them on edge ports.
- ○ They drop the multicasts.
- ○ They forward the multicasts only on interfaces that have received IGMP reports for destination group.
- ○ They flood traffic throughout the VLAN in the same manner as a broadcast.

Question 2

Which features should be configured to ensure proper treatment and delivery of multicast traffic within a VLAN? (Select two.)

- ☐ IGMP on every switch within the VLAN
- ☐ IGMP on L3 switches acting as gateways for the VLAN
- ☐ IGMP snooping on every switch within the VLAN
- ☐ IGMP snooping on every switch acting as L2 switches within the VLAN

Question 3

By default, IGMP snooping is enabled.

- O True
- O False

Question 4

For an IGMP querier, IGMP is enabled on a Layer 3 interface.

- O True
- O False

8 Multicast routing

In this chapter, you will be introduced to Protocol Independent Multicast (PIM), a multicast routing protocol that enables switches to determine which interfaces should receive multicast traffic in a Layer 3 network, like a campus.

In the last , you learned about how AOS-CX switches can learn where the multicast source (typically a server) and the multicast receivers (typically user devices) reside. The problem in a larger campus network is that the multicast source and destinations are commonly on different subnets, which requires a routing function. Unicast routing protocols like OSPF will not route multicast traffic. You need a multicast routing protocol, and PIM is the de facto standard. This chapter introduces the PIM routing protocol and the two modes of PIM: dense mode and sparse mode.

OBJECTIVES

After completing this chapter, you should be able to:

- ✓ Distinguish between PIM-DM and PIM-SM.
- ✓ Implement PIM-DM and PIM-SM to route multicast traffic.

PIM introduction: Multicast routing use cases

For small networks with only one router or a single VLAN, there is no need for a multicast routing protocol like PIM. You just studied scenarios like this in the previous chapter about IGMP.

However, if you have a more complex, enterprise-class network with multiple routers, you need PIM.

In the following figure, the multicast source is in VLAN 10. This traffic must traverse two routers to reach the receivers in VLANs 20 and 30. Multicast routing enables a source in one VLAN to stream to receivers in multiple VLANs.

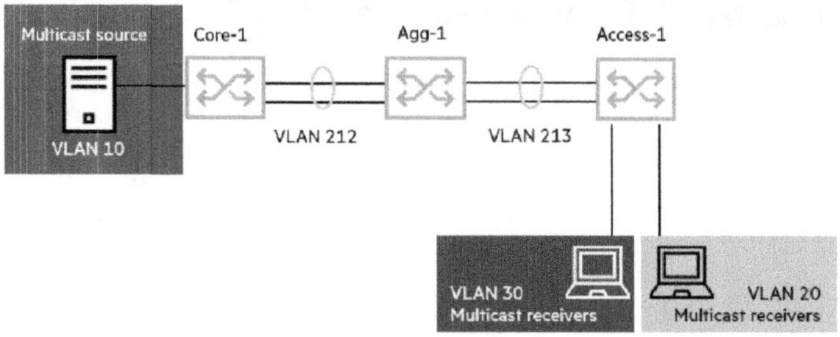

Figure 8-1: Multicast routing example

Multicast routing

You know that regular unicast routing protocols like OSPF send packets *toward a destination*. The focus is on each packet's destination IP address. The objective of multicast routing is to send multicast messages *away from the source* toward all interested receivers. Multicast packets must always flow away from the source, and never back on a segment from where the transmission originated. This means that rather than tracking only destinations, multicast routers must also track the location of sources, the inverse of unicast routing. Even though multicast routing uses the exact inverse logic of unicast routing protocols, you can leverage the information obtained by those protocols (unicast routing—OSPF and BGP for example) for multicast forwarding.

This allows a network to avoid using another full routing protocol, minimizing the use of memory space. All the functionality that is built into the unicast routing protocol, including loop prevention, path selection, failure detection and so on, can be used for multicast. Modern IP multicast routing uses PIM, which leverages unicast routing information, to forward messages to receivers. The process of determining where the multicast messages are to be sent is referred to as *building the tree.*

With IPv4, IGMP determines who needs to receive a multicast stream, and PIM helps determine how to get the stream to those destinations. It relies on IGMP to request the traffic. IPv6 uses MLD instead of IGMP.

If you want to route multicast traffic into a VLAN, it is recommended that you avoid multinetting on that VLAN (configuring more than one subnet).

Multicast tree

A network tree is essentially a graph, or directed graph (digraph), that represents a hierarchical logical structure. The primary purpose of a multicast tree is to build an efficient loop-free forwarding path from the source of the multicast stream toward all the receivers. A multicast tree is composed of the following components:

- **Root:** Router closest to the source.
- **Leaf:** Router with an attached receiver.
- **Branch:** Intermediate router that performs replication to connect the root to the leaves.

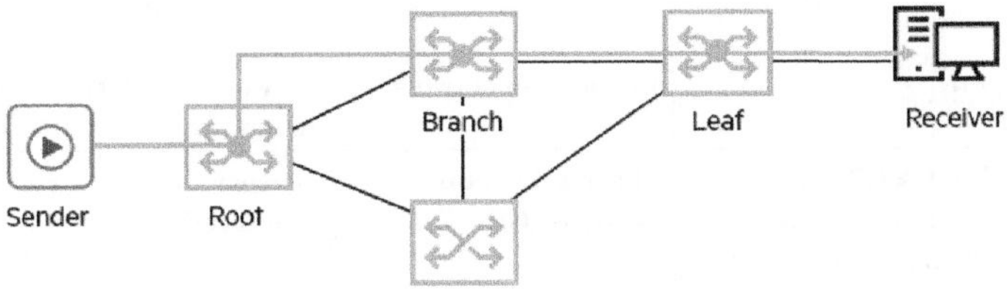

Figure 8-2: Multicast routing

Multicast versus unicast routing

The following figure summarizes the difference between unicast and multicast routing. While unicast routing forwards any traffic toward some destination, multicast traffic forwards traffic away from a specific source, with a multicast group address as its destination.

Unicast routing protocols like OSPF do not care about the source address of packets. They compare the destination IP address of each packet to their route table and forward packets along the best path.

For multicast routing, the source matters. A multicast route must track the upstream interface that is closest to the source so that the router can verify that the multicasts are arriving on the correct interface. It can then drop multicasts that arrive on the wrong interface, preventing multicasts from looping through the network.

The following figure shows the output of the `show ip mroute brief` command—the multicast route (mroute) table. The router has learned that it should forward all packets for group 239.1.1.1 efficiently away from the source of this stream, 10.101.10.10, toward its neighbor at 10.101.121.1.

An active multicast route is called an (S,G) route because it includes both the source IP address (S) and the destination multicast group address (G). Multicast routing devices create an (S,G) route after they receive a multicast from a source to a group. For example, the (10.101.10.10, 239.1.1.1) route refers to traffic from 10.101.10.10 to 239.1.1.1.

You will soon learn about (*,G) entries, which act as a kind of placeholder until the multicast source is discovered.

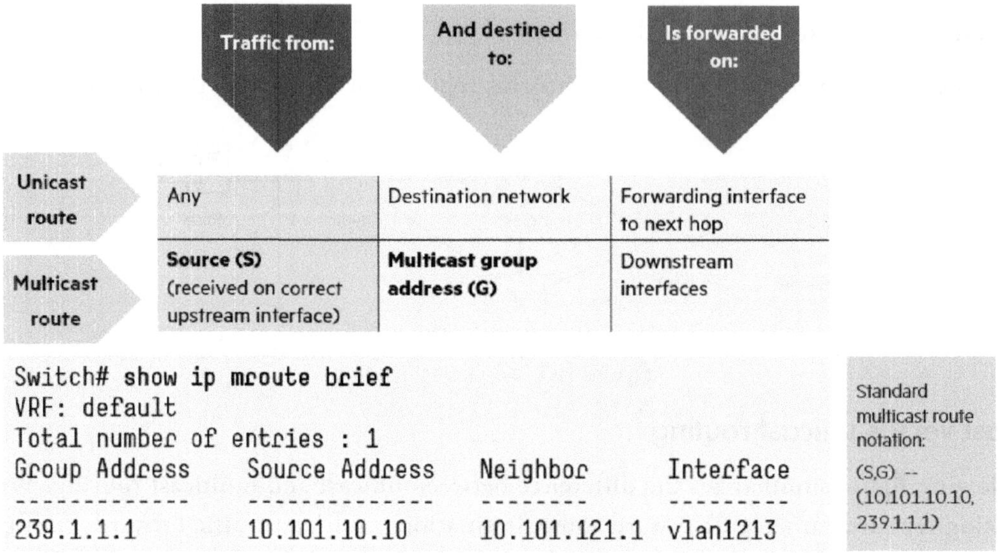

Figure 8-3: Difference between unicast and multicast routing

Reverse path forwarding

PIM uses a so-called Reverse Path Forwarding (RPF) technique to save bandwidth and maximize efficiency. Here's the scenario. A multicast source originates a stream with source address 10.1.1.1 and multicast destination 239.1.1.1. Multicast router MR1 receives this stream and forwards it to its two neighbors, MR2 and MR3. MR3 receives this stream and forwards it to its only other neighbor, MR2.

Thus, MR2 receives two copies of the multicast stream. This is a waste of bandwidth! Remember, the advantage of multicasting is that a *single stream* can reach multiple destinations. One of these streams must be eliminated, so the router looks in the unicast routing table: "What is the best path back to source 10.1.1.1?"

MR2's OSPF process has learned that it can reach 10.1.1.1 via 10.2.2.2 (MR1) with a cost of 100 and can reach it via 10.3.3.3 (MR3) with a cost of 200. Thus, 10.2.2.2 is the best path back to 10.1.1.1. Of course, if the best "reverse path" back to 10.1.1.1 is via 10.2.2.2, then that is the best path for receiving multicast traffic from 10.1.1.1.

So MR2 sends a PIM message to MR3: "Hey MR3, I don't need you to send that stream to me. Please stop." MR3 complies with this request. In other words, the 10.3.3.3 interface is "pruned from the tree." Bandwidth is no longer wasted sending the stream over that interface. If MR3 has no receivers directly attached to it (as determined by IGMP) and hosts no other downstream routers with receivers, then MR3 has no reason to receive the stream. In that case, MR3 sends a message to MR1: "Please stop sending the stream to me."

MR1 complies and prunes that interface from the multicast distribution tree. In a large, complex network with dozens or hundreds of routers, every router uses this RPF process. The result is that multicast traffic is only sent where needed. Nice! And, like spanning-tree, if a link failure occurs, the PIM routers quickly discover the issue and determine an alternative, loop-free path.

We have an efficient multicast tree for the current situation, but networks are dynamic. New users join a multicast group; existing users leave. How is this handled?

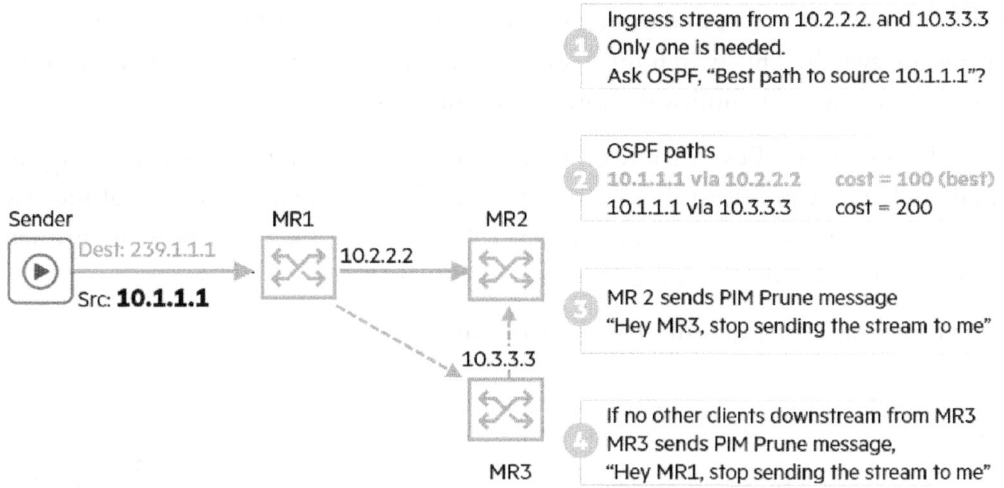

Figure 8-4: Reverse path forwarding example

IGMP and PIM

In the previous example, we had an efficient multicast forwarding tree that only forwarded the stream out needed interfaces. Then things change.

Host 1 is the last remaining receiver downstream of multicast router MR2, and it sends an IGMP Leave message: "I no longer need the stream."

In response, MR2 sends a PIM message to MR1: "I no longer need the stream." MR1 sees that there are no more downstream neighbors out that interface, and so prunes that interface from the multicast distribution tree.

Meanwhile, an employee sits down at their desk and runs some application that requires the stream. The host sends an IGMP Join message to MR3: "I need that stream." In response, MR3 sends a PIM message to MR1: "I need that stream." MR1 add the MR3-connected interface back to the distribution tree, and the stream again flows to MR3 and on to Host 2.

This confirms what you learned earlier:

- IGMP is host-to-router communications, so routers can discover which receivers need a particular stream and which receivers do not.
- PIM is router-to-router communications, so multiple routers can build a shortest path tree (SPT) for a multicast stream, leveraging the internal routing protocol (OSPF, and so forth) for reverse path forwarding.

It is important that you understand that IGMP alone is never enough to provide multicast routing, even if the multicast source and multicast receivers are connected to VLANs on the same multicast routing switch. This switch must still have multicast routing enabled and run PIM on the VLAN interfaces to route multicast traffic from one subnet to another.

In other words, IGMP and PIM are a team that enable efficient multicast routing. They build and maintain a dynamic multicast distribution tree. Each device maintains a list of interfaces that are part of the tree, out which a multicast stream is forwarded. Other interfaces are pruned from the tree—removed from this Outgoing Interface List (OIL).

There are two methods used to build this distribution tree.

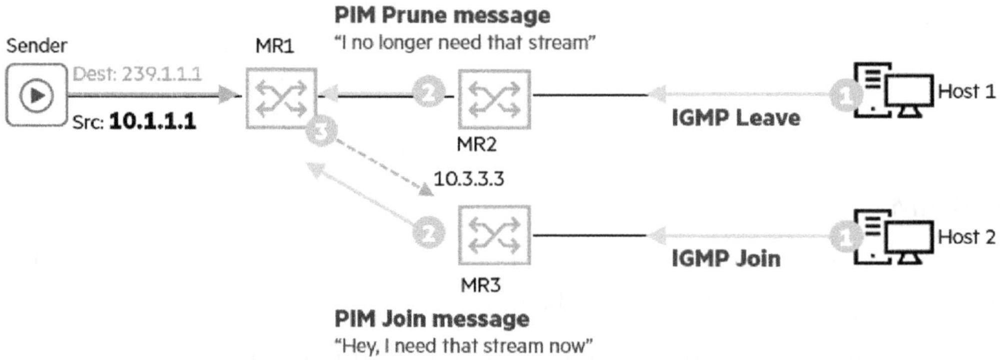

Figure 8-5: IGMP and PIM example

PIM-DM versus PIM-SM

The routing switches can construct the multicast routing tree in one of two modes.

PIM Dense Mode (DM)

PIM-DM (RFC 3973) assumes everyone wants to receive the stream—a push concept. When a source begins streaming traffic, PIM-DM-enabled VLAN interfaces flood multicasts to all neighbors. This continues until neighbors send a Prune message: "Stop sending that stream to me." Upstream neighbors set a timer for the pruned interface. When the timer expires, the interface reverts to the default behavior: forwarding multicasts until explicitly requested not to. You will soon learn how to prevent the timer from expiring.

This mode works best in environments with high bandwidth and high tolerance for congestion, such as a high-speed Ethernet network. Because the PIM-DM routers automatically create the multicast routing tree for a particular stream when that stream begins, PIM-DM is quite simple to set up.

A VRF can run PIM-DM or PIM-SM—not both.

PIM Sparse Mode (SM)

PIM-SM (RFC 4601) assumes nobody wants to receive the stream—more of a pull concept. It builds the multicast routing tree for a group in advance and then forwards multicasts on that tree when a source begins to stream to that group. PIM-SM-enabled routing interfaces must specifically indicate that they want to receive multicast traffic in a particular group by sending a PIM Join message. The interface must then periodically refresh its join. If its upstream neighbor does not receive joins, it reverts to its default behavior: not forwarding traffic.

Because PIM-SM establishes efficient forwarding paths, it is well suited for a lower-bandwidth environment, such as one built on WAN links. PIM-SM also gives you more control over which multicast groups are routed.

PIM-SM has more configuration options and tends to be more complex to set up. PIM-DM can waste bandwidth during initial tree establishment and may not be suitable for networks with many lower-speed WAN links.

Your AOS-CX switch supports both PIM-DM and PIM-SM; however, it can only run one mode concurrently (within a VSF). All multicast routing switches in the network (in their respective VRF) must also agree and run the same type of PIM.

PIM-DM interoperates with IGMP/MLD and the switch's routing protocols. It does not matter which unicast routing protocol is in use—PIM can use it for RFP. That is why it is called the Protocol Independent Multicast routing protocol. PIM-DM can be used with RIP, OSPF, BGP, or static routes configured.

PIM Hello message

Routers and L3 switches running PIM send periodic Hello messages on PIM-enabled interfaces to discover neighbors. These messages are sourced with the IP address of the outgoing interface (In AOS-CX, you can specify the interface that should be used to source those packets) and are sent to all PIM routers with the multicast destination address of 224.0.0.13. They contain information such as the hello period time, hello hold time, LAN prune delay time, and capabilities. The hold time signifies how long the neighbor will keep the neighbor relationship without receiving an update. The LAN prune delay is used on multi-access networks for propagation delay.

The packet capture in **Figure 8-6** shows a Hello message. Pay special attention to the areas: the destination multicast addresses at Layer 3, the PIM version (v2), and the PIM message type (Hello).

The PIM process begins when two or more routers establish a PIM neighbor adjacency on a given segment. To begin the adjacency process, any interfaces participating in multicast routing must be configured for PIM communications. Routers discover PIM neighbors by sending PIM hello messages to the link-local multicast address 224.0.0.13 out each PIM-enabled interface.

Hello messages are used to elect a Designated Router (DR)—a single PIM router that acts on behalf of directly connected hosts in a shared media, like Ethernet. Hello messages are also the way that option negotiation takes place in PIM, so that additional functionality can be enabled or parameters tuned.

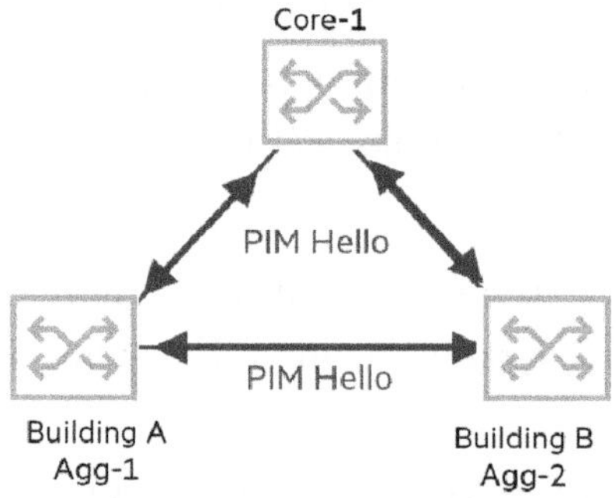

Figure 8-6: PIM Hello message topology

AOS-CX defaults and restrictions

PIM (DM or SM) is disabled by default. Either PIM-SM or PIM-DM can be configured within a VRF at a time. All the interfaces within the VRF must run with the same mode.

PIM uses unicast routing information from any of the routing protocols that are running on the system, such as OSPFv2, OSPFv3, or BGP. Static routes are also supported with next-hop IP addresses.

PIM can be enabled on a ROP (routed port) and SVI (interface VLAN) interfaces. PIM can be enabled across all VRFs on a maximum of 1000 interfaces with an upper limit of 128 per VRF.

Although up to 128 PIM-DM enabled interfaces can be configured, when configuring trunk interfaces with multiple dense enabled SVIs, the trunk interfaces must have sufficient bandwidth or have only the required number of trunks they can support. This ensures that the link utilization is not exceeded due to the initial flooding nature of the protocol.

PIM-DM is compatible with IGMP version 2 and version 3, MLD version 1 and version 2, and is fully interoperable with IGMP/MLD for determining multicast flows. PIM-DM is fully interoperable with VRRP to quickly transition multicast routes in a failover. PIM-DM can run on multiple VRF instances in parallel. It is supported on all VRFs supported in the system.

PIM-DM currently does not support the following:

- VXLAN, 6in4, 6in6, and GRE interfaces.
- PIM-DM can be enabled on VSX deployments; however, active-active functionality is not available.

Since active-active is not supported, a slightly higher traffic recovery time can be expected compared to sparse mode, in the event of failovers. Hence, it is recommended to use PIM-SM on VSX. Ensure that the ISL is not oversubscribed while using PIM-DM with VSX. In case of numerous PIM-enabled SVIs, ensure that ISL is configured to handle the flooding on all VLANs since ISL is member of all SVIs.

If BFD is enabled globally, it will be enabled by default on all PIM interfaces as well. The only exception is when it is disabled specifically on an interface using the `ip pim-dense bfd disable` interface command. If BFD is disabled globally, it will be disabled by default on all interfaces. The only exception is when it is enabled specifically on an interface using the `ip pim-dense bfd` interface command.

PIM-DM

PIM Dense Mode is a great option for many deployments. Let's take a look.

Configure PIM-DM

PIM-DM must be enabled on all VLAN interfaces between the multicast sources and the multicast receivers:

- Interfaces on which multicast sources reside
- Interfaces on which multicast receivers reside (IGMP-enabled interfaces)
- Interfaces between all routers and routing switches that connect sources and receivers

The following figure shows where to enable PIM-DM and IGMP. Note that you should also enable IGMP on the server VLAN; although IGMP is not strictly required in the source VLAN, it enables Core-1 to filter multicasts within that VLAN.

First, validate that unicast routing is in place and fully functional. Next, enable PIM-DM multicast routing globally. Then enable PIM-DM on appropriate VLAN interfaces, as shown in the figure.

A PIM router uses hello packets to automatically discover neighbors. They are sent to 224.0.0.13 every 30 seconds by default. You can change the default within the range of 5–300 seconds. The router uses this setting to compute the hello hold time, which is included in hello packets sent to neighbor routers. The hello hold time tells neighbor routers how long to wait for the next hello packet from the routing switch. If another packet does not arrive within that time, the router removes the neighbor adjacency on that interface from the PIM adjacency table, which removes any flows running on that interface. Shortening the hello interval reduces the hello hold time. If they do not receive a new hello packet when expected, it changes how quickly other routers stop sending traffic to the routing switch.

BFD can be used to speed up the convergence of PIM. If BFD is globally enabled, BFD is automatically enabled for PIM on interfaces where PIM is also enabled. BFD was previously covered.

MLD and PM-DM are not currently supported for IPv6 in AOS-CX; however, MLD and PIM-SM are.

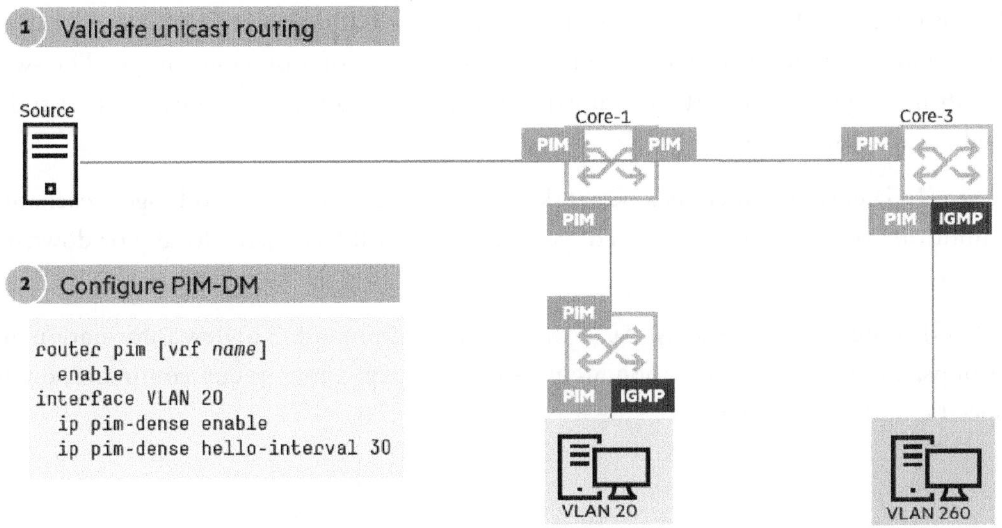

Figure 8-7: Configure PIM-DM

Interpreting PIM-DM (*,G) entries

When you are monitoring and troubleshooting a multicast routing solution, you do not have an illustration tracing multicasts for you. It can be helpful for you to understand more about the multicast route entries, how to interpret them, and how the routing switch uses them to forward traffic.

Let's begin with (*,G) entries.

When you enable PIM-DM on a VLAN interface, the switch creates a (*,G) entry for each group membership discovered by IGMP on that VLAN. This entry specifies the downstream interfaces from which a tree can be formed when multicasts begin to arrive from a particular source.

When you use PIM-DM, the (*,G) entry essentially acts as a placeholder, listing the downstream interfaces that require multicasts in this group, as discovered by IGMP. The framework of the tree that delivers these multicasts is not filled in. For example, no upstream interface is specified (the routing switch does not know where multicasts will arrive).

The (*,G) entry exists only as a pending route; it is not an active route forwarding multicasts. The (*,G) entry remains in the PIM routing table for as long as an IGMP membership for the group exists on at least one interface. This signals to the routing switch that it should stay part of the multicast routing tree for any multicast stream to this group, regardless of whether it needs to forward the traffic to any downstream PIM-DM neighbors.

Typically, you will see (*,G) entries when there has been a purge. A client wants to receive a stream now, but the switch isn't receiving the stream because of a previous purge. The switch generates a graft message and sends it to its upstream neighbors to find the source so that traffic can be flooded back to the switch.

You also see (*,G) entries if there is a network issue where the source is no longer reachable and thus no multicast stream is being received, yet the switch still has clients locally or downstream that desire to receive the stream.

To view (*,G) routes, enter the `show ip pim pending` command. To view information about IGMP groups, enter the `show ip igmp groups` or `show ip igmp group` command, qualifying it with a specific multicast address.

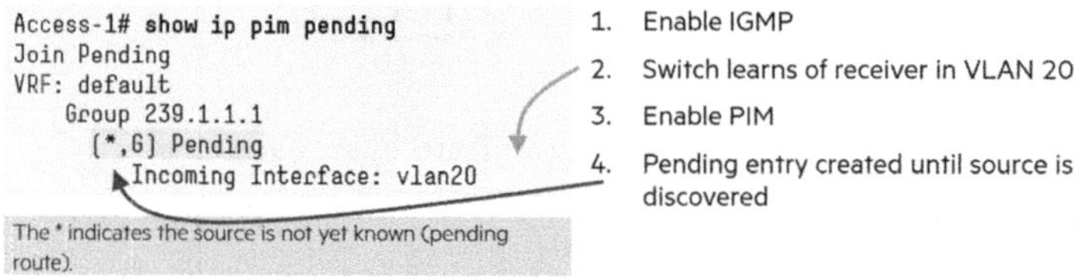

Figure 8-8: Interpreting PIM-DM (*,G) entries

See IGMP groups

To view (*,G) routes, enter the `show ip pim pending` command. To view information about IGMP groups, enter the `show ip igmp groups` or `show ip igmp group` command, qualifying it with a specific multicast address.

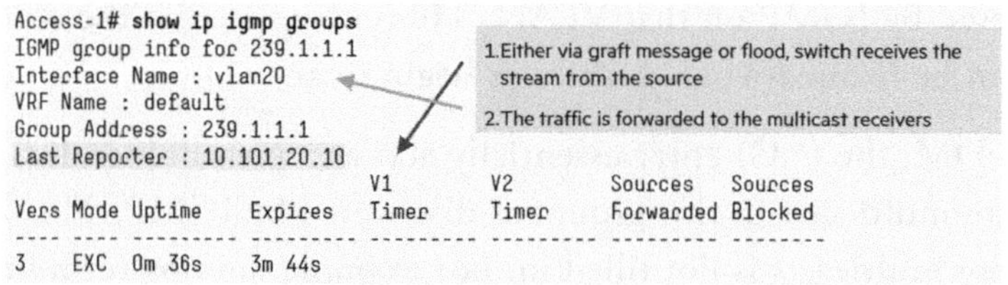

Figure 8-9: Example with commands to view information about IGMP groups

See IGMP routes pruned

When the switch receives a multicast on the proper upstream interface for the source, it creates an (S,G) entry. It uses this entry to route the multicast traffic on the correct interface or interfaces.

The (S,G) entry includes:

- The source for the incoming multicasts.

- The upstream interface on which the multicast arrived. You will learn in a moment how the switch checks the validity of this interface. It will not add the route if the upstream interface is invalid.
- Initially, the downstream interface includes all PIM-DM interfaces set to a forwarding state. The switch maintains the forwarding state unless a neighbor sends a prune on that interface. The switch also always maintains the forwarding state for interfaces discovered by IGMP if they are in the (*,G) entry associated with this group. As you see, if the interface receives a prune, the switch maintains the interface in the list but places it in the pruned state and does not forward multicasts on it.

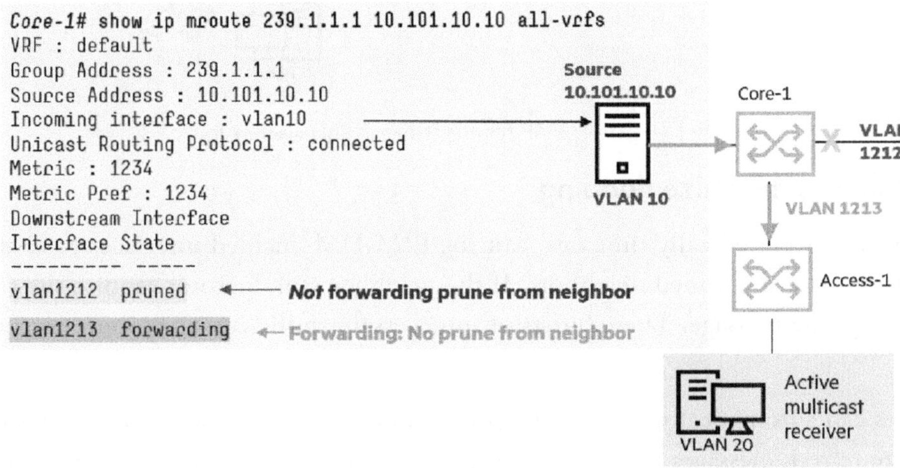

Figure 8-10: See IGMP routes pruned example

See IGMP routes forwarding

If the (S,G) entry has downstream interfaces or if it the source is connected directly to the routing switch, the routing switch adds the (S,G) entry to its active multicast routing table. The routing switch then uses the (S,G) entry to forward multicasts.

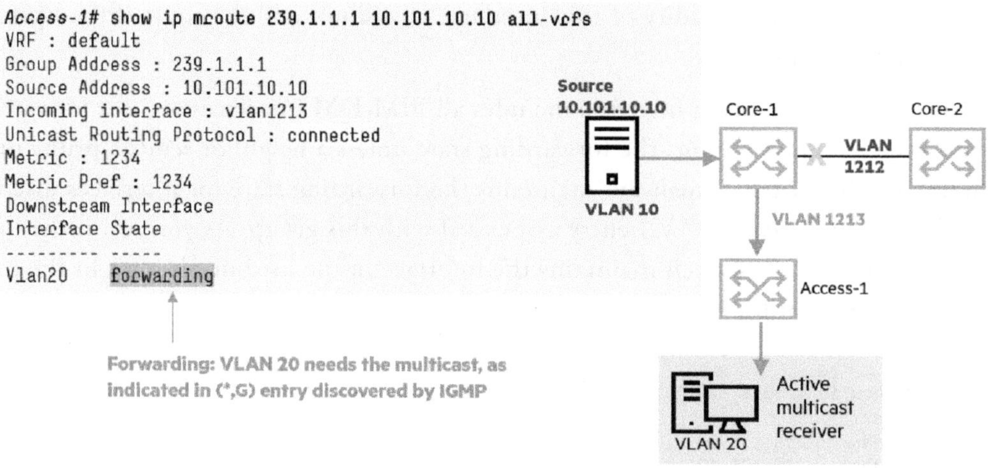

Figure 8-11: See IGMP routes forwarding example

Use state refreshes to minimize flooding

You learned that prunes periodically time out, causing PIM-DM-enabled interfaces to resume the multicast flood to previously pruned neighbors. If the neighbor still does not require the traffic, it must send another prune message. These bursts of unnecessary traffic can contribute to congestion.

AOS-CX switches can automatically prevent the prune state time-out—the source-connected switch floods state refresh messages on all PIM-DM interfaces. This indicates state information about the transmitting interface, whether pruned or forwarding. It propagates throughout the system, triggering all PIM-DM routers to send joins and prunes only if they need to change the current state of the upstream interface. Routers can then also refresh their prune states with confidence without needing to flood multicast traffic.

In more detail, receiving a state refresh message triggers the PIM-DM interface to take these actions:

- Drop the message unless it has arrived from the RPF upstream neighbor.
- For messages from the proper upstream neighbor:
    - Check the indicated state of the upstream interface and determine whether it matches this routing switch's requirements.
    - If the indicated state is pruned and this routing switch has no downstream interfaces for the (S,G) entry, the indicated state is correct. Similarly, if the indicated state is forwarding and this routing switch does have downstream interfaces on which it

must forward traffic, the state is correct.

- If the indicated state is pruned and this switch requires the multicasts, the switch sends an (S,G) join upstream to alter the state. Similarly, if the indicated state is forwarding, but this switch does not have an active entry forwarding multicast traffic downstream, the switch sends an (S,G) prune upstream.
- Reset all pruned interface timers. The switch can safely reset all timers because its downstream neighbors will correct any mistakes on receipt of the state refresh message, just as this routing switch did.
- Forward the state refresh message (with one less value in the Time To Leave) on all interfaces with PIM-DM neighbors except the interface on which it received the message. The TTL prevents the state refresh message from endlessly looping through the PIM-DM domain.

AOS-CX switches automatically support state refresh messages, but only the routing switch connected directly to the unicast source initiates state refresh messages. All other PIM routers in the network only propagate these state refresh messages.

You can configure the interval in seconds (10 to 100) between successive state refresh messages. The default setting is 60 seconds. You should keep the interval below the PIM-DM prune expiry time to prevent periodic unnecessary floods of traffic.

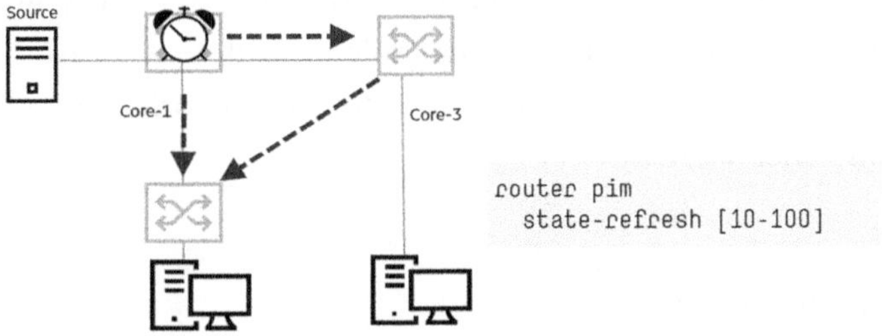

Figure 8-12: Using state refreshes to minimize flooding

Multicast forwarder election

In the following scenario, routers Access-1 and Access-2 have interfaces in VLAN 20, connected to a downstream switch with multicast receivers. They have chosen this solution rather than a VSF fabric, which is typically recommended. The switches run VRRP and PIM-DM. Just as only one of the redundant switches can act as the active VRRP master, only one of the switches can forward multicasts. Otherwise, endpoints could receive duplicate copies of the multicasts, causing errors.

When Access-1 and Access-2 receive multicasts from Core-2, they flood the multicasts to each other. As you learned earlier, RPF ensures that only the best path continues to forward frames. Let's look at this in more detail.

The PIM routers send an assert message to elect a forwarder, then examine each other's asserts to elect a multicast forwarder for the interface. The assert includes the (S,G) entry under dispute and information about the PIM routing switch's unicast path to the source. The routing switch with the more favorable path becomes the multicast forwarder, as determined first by administrative distance (preference) and then, if the administrative distance ties, by metric or cost. If the metric is the same, the router with the highest priority value is elected. If that matches, the interface with the higher IP address is elected.

Suppose that both routers have the same administrative distance, cost, and priority for the route. However, Access-2 has the higher IP address. Therefore, Access-2 is elected the multicast forwarder for VLAN 20.

Access-1 removes this VLAN from its list of downstream interfaces in its (S,G) entry. It also sends a prune to Access-2. Access-2 continues to forward multicasts in VLAN 20 because multicast receivers connect in this VLAN. Of course, if all receivers send an IGMP leave message, there would be no receivers in VLAN 20. In that case, Access-2 would send a prune message upstream to stop the unneeded multicast flow.

In a design like this, you should pay attention to the IP addresses that you assign to routers. To load balance the primary and multicast forwarder role on different VLANs, assign one switch a higher IP address and VRRP priority on some VLANs and the other switch a higher IP address and VRRP priority on other VLANs.

The election winner:

1. Has OSPF route to source
2. Admin distance
3. Best OSPF metric
4. Highest priority
5. Higher IP address

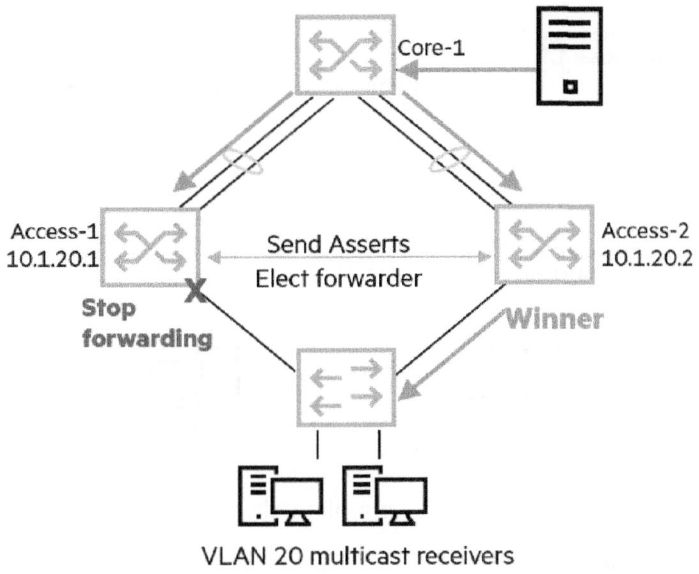

Figure 8-13: Multicast forwarder election

PIM-DM operation: Flooding and pruning

A multicast source begins to transmit, and its default router receives the multicasts. The VLAN interface on which the multicasts arrive should be enabled for PIM-DM. The router then floods the multicasts on every PIM-DM interface, including:

- Interfaces that connect to PIM-DM neighbors: Core-2 and Access-1 in **Figure 8-14**.
- Interfaces that connect to receivers that have requested the multicast using IGMP. In this example, Core-1 has no directly attached multicast receivers.

Downstream multicast routers follow the same steps to flood the multicast. In **Figure 8-14**, Access-1 and Core-2 are only neighbors with Core-1. Access-1 has learned that VLAN 20 is a downstream interface for this group from IGMP, and so continues to accept and forward traffic from Core-1.

Neither Core-2 or Core-3 have multicast receivers, so they send prune messages to stop the stream from flowing to them. Core-2 still maintains the multicast route (the (S,G) entry), which it can use to receive multicasts later, if necessary.

To summarize, for a multicast group "X" on a given interface, when the last host belonging to group "X" leaves the group, PIM places that interface in a prune state. Multicast traffic from group "X" is now blocked to that interface. The prune state remains until a host on the same interface issues a join for group "X," in which case the router cancels the prune state and changes the flow to the forwarding state.

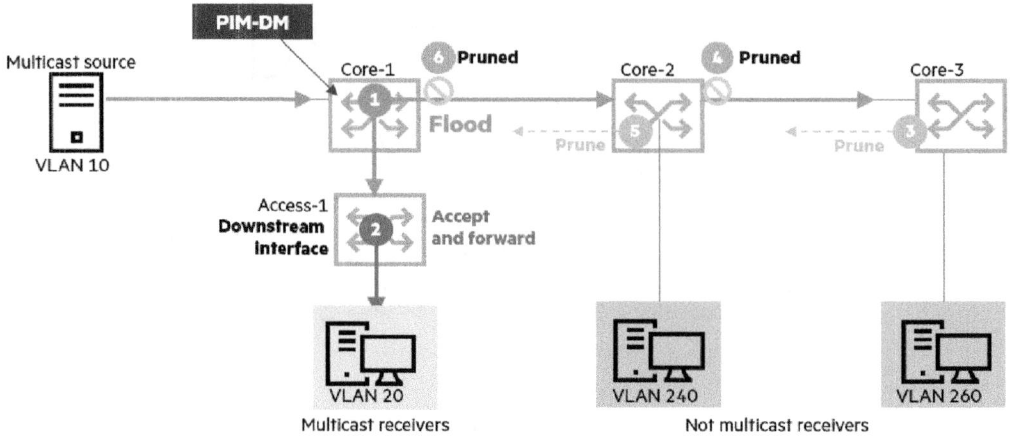

Figure 8-14: PIM-DM operation: flooding and pruning

PIM-DM operation: Grafts

After pruning, Core-1 only forwards multicast traffic to Access-1, which forwards the traffic to receivers in VLAN 20.

As group memberships change, the SPT must change to accommodate. Recall that PIM-DM prefers to forward unnecessary multicast traffic over failing to forward necessary traffic. Therefore, prune messages expire, and the routing switch that had received the prune message begins forwarding multicast traffic again.

In addition, group members leave groups. For example, if all multicast receivers in VLAN 20 left the group, then Access-1 would send a prune message to Core-1, stopping the flow of traffic.

Figure 8-15 illustrates a new host that wishes to receive the multicast stream, probably because a user sat at this host and opened an application that required the stream. The host sends an IGMP report or join message for this group, and Core-2 adds VLAN 240 as a downstream interface for its (S,G) multicast route for this multicast stream. Eventually, the prune message would expire, and Core-2 would receive traffic again. However, PIM-DM provides a faster method of pulling the multicast traffic down. Core-2 sends a graft message on the upstream interface, requesting to be added to the tree.

Core-1 adds the interface that connects to Core-2 back to the (S,G) multicast route's list of downstream interfaces. Core-2 begins to receive the traffic again, and it forwards it to the receiver.

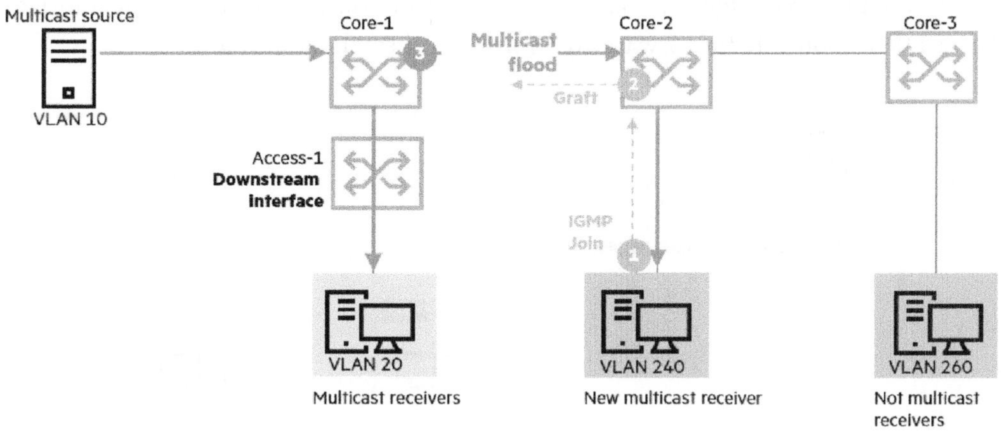

Figure 8-15: PIM-DM operation: grafts

PIM-SM

PIM-SM involves a little more setup than PIM-DM because PIM-SM routing switches cannot assume that they are part of the initial multicast routing tree for all traffic. Instead, PIM-SM routing switches must explicitly join that tree, and you must configure them to do so.

Recall that PIM-SM assumes that most hosts do not want to receive multicast traffic. It uses a non-flooding multicast model to direct traffic from the source to the interface when there are multicast receivers in the group. As a result, this model sends traffic only to the routers that specifically request it.

PIM-SM can be configured on physical ports, VLAN interfaces, LAG interfaces, and loopback interfaces.

Configuring PIM-SM is like configuring PIM-DM in some ways. For example, you must enable PIM-SM on every interface between multicast sources and receivers, including:

- Layer 3 interfaces on which multicast sources reside
- Layer 3 interfaces on which multicast receivers reside (IGMP-enabled interfaces)
- Layer 3 interfaces between all routers and routing switches that connect sources and receivers

As with PIM-DM, PIM SM-enabled interfaces exchange hellos on 224.0.0.13. This helps explain why you cannot implement PIM-SM and PIM-DM in the same domain. PIM-enabled routing switches operating in one mode would expect certain behavior from neighbors operating in another, leading to failed establishment of the multicast forwarding tree.

In addition to establishing neighbor relationships, PIM interfaces in broadcast or multiaccess networks elect a DR responsible for forwarding multicasts. Let's look at that.

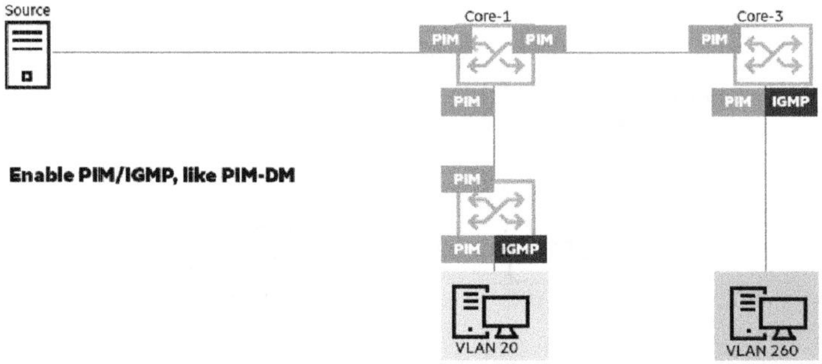

Figure 8-16: PIM-SM overview

PIM-SM: Designated router

A single network interface and IP address may connect to a multipoint or broadcast network, such as Ethernet. If the network is PIM-enabled, many neighbor adjacencies can occur, one for each PIM-enabled IP neighbor. If every PIM router on the segment were to attempt to manage the tree state, confusion and excess traffic could occur. Much like an OSPF DR, the PIM DR manages state information and PIM updates for all the routers on the segment.

The DR selection process occurs when all neighbors on a segment have replied to PIM hello messages. Each router on a segment selects one router to function as the DR. The DR router is chosen using the PIM DR priority parameter, where the highest priority router is selected as the DR. The DR priority is configurable and sent in the PIM hello message. If the DR priority value is not supplied by all routers, or the priorities match, the router with the highest IP address wins—the IP address of the interface sending the PIM message. By default, all interfaces have a PIM DR priority of 1. All routers on the segment should select the same DR router to avoid conflicts.

When the DR receives an IGMP membership report message from a receiver for a new group or source, the DR creates a tree to connect the receiver to the source. To do this, it sends a PIM join message out the interface toward the agreed-upon root of the multicast tree, the Rendezvous Point. When the DR determines that the last host has left a group or source, it sends a PIM prune message to remove the path from the distribution tree.

From the RFC 4601 section 4.3, DR election occurs on all interface types, point-to-point or broadcast: "Because the distinction between LANs and point-to-point interfaces can sometimes be blurred, and because routers may also have multicast host functionality, the PIM-SM specification makes no distinction between the two. Thus, DR election will happen on all interfaces, LAN or otherwise."

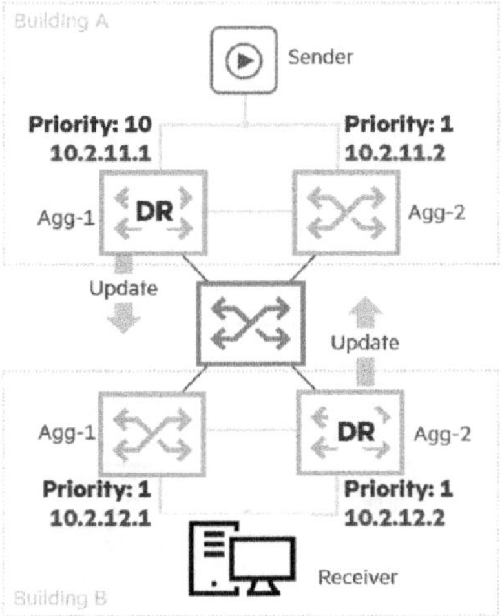

Figure 8-17: Designated router

PIM-SM: Rendezvous point

A Rendezvous Point (RP) is a router that has been configured to be used as the root of the non-source-specific distribution tree (shared tree) for a multicast group. Join messages from receivers for a group are sent toward the RP, and data from senders is sent to the RP. Thus, receivers discover senders and start to receive traffic destined for the group. The RP router is not a packet source but simply a location from which to root and calculate a loop-free topology. The RP can also offload the processing of network joins and leaves during the initial setup of a multicast stream, saving network resources by concentrating them in a single location.

You should place the RP close to the source if possible. This is only applicable to a few sources that are of key importance to the business. Enterprise-wide deployments for multicast normally use RPs in the data center for the enterprise scope. The RP might not be in the path between sender and receivers. PIM-SM uses the RP to build a **shared tree** as an *initial* state, but it can move to an optimized **shortest path tree**. This means that traffic is not sent to RP all the time and so RP location has minimal impact on network performance.

Let's compare shared versus shortest path trees.

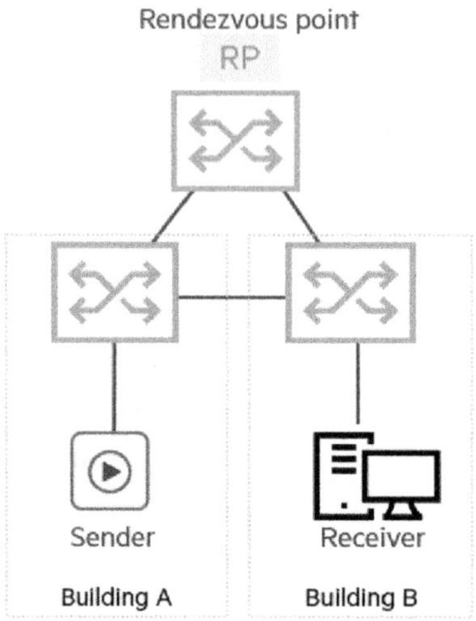

Figure 8-18: Rendezvous point

PIM SM: Multicast trees types

The two types of trees used in multicast networks to control the distribution of traffic are source trees and shared trees.

- **Source trees (shortest path trees):** This is the most common type of multicast tree. From a network perspective, the root of the source tree is the router closest to the source. The tree extends from the root router in a direct path toward all the group members. This creates a tree in which the path might look like the one shown in the figure. Each router in the network must calculate the tree independently based on the information it has received, whether by dynamic protocol updates (PIM) or by configuration. Using RPF, the router checks for loops and creates the Outgoing Interface List (OIL) from the best (sometimes called shortest) paths from the unicast routing table. For this reason, a source tree is also called a shortest path tree.
- **Shared trees (rendezvous point tree):** The shared tree uses a shared point in the network from which to build the distribution tree. This location is called an RP and is therefore the root of the tree. Each router still uses RPF checks to keep the topology loop-free, but the RPF check is in the direction of the RP, not the source. Notice the RP placement has no relation to the IP multicast source. The RP may or may not be in the direct path (shortest path) between the group source and the group client. In fact, it can be placed outside of the path entirely. For this reason, a shared tree is also sometimes called a Rendezvous Point Tree (RPT). The advantage to using this type of tree is that all non-RP routers can conserve control plane resources during stream establishment or during host maintenance. In

addition, the network path is predetermined and predictable, which leads to consistent network operations.

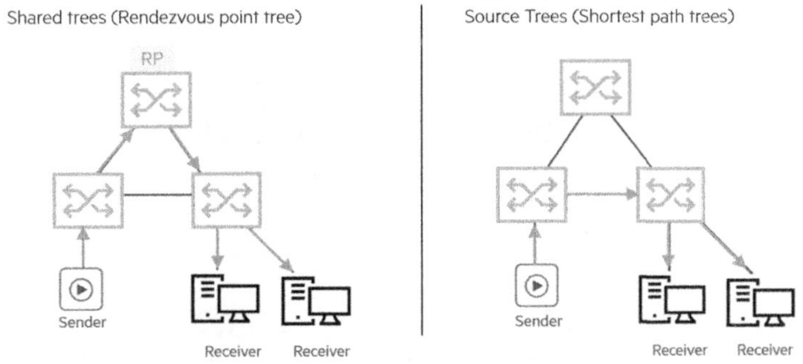

Figure 8-19: Multicast trees types

PIM-SM: Group states

Entries in the multicast routing table (mroute table) have the form (source, group).

A **(*,G)** entry in an mroute table represents a router's relationship to the leaves of a tree. All the leaf-facing interfaces are added to an outgoing interface list. This type of entry is used in an RPT because routers have a preconfigured meeting point—the RP. This means that routers need to only get one interested receiver for a particular multicast group to build the tree—the source is not really needed. The RP requests this information from the source.

The **(S,G)** entry in the mroute table represents the router's relationship to the source of the multicast stream. This type of entry is used in shortest path trees. In this case, not only the multicast group is needed, but also the source IP address.

Routers can forward packets using either the (*,G) or the (S,G) entries, if the paths are determined to be loop-free.

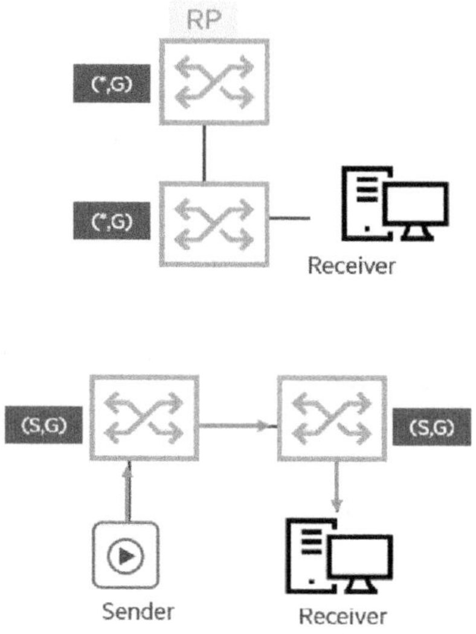

Figure 8-20: Group states

PIM-SM build-up process

RFC 4601 defines three phases for the PIM-SM build-up process.

Phase 1: Rendezvous point tree

1. The receiver endpoint sends an IGMP Join request asking for a specific multicast group to the Last-Hop Router (LHR). This message includes only the Multicast group, therefore multicast state (*,G) is created on the First Hop Router's (FHR) multicast routing table. This action also adds the receiver-facing interface to the outgoing interface list.
2. LHR Sends a PIM Join message to the RP:
    a. LHR uses RPF to look in the unicast routing table and determine the next-hop to reach the RP.
    b. PIM Join messages are sent periodically to maintain the multicast state.
3. If intermediate routers exist between the LHR and RP, those devices add a (*,G) state on their multicast routing table and forward the PIM Join message to the RP. In the example, this step is not needed.

Eventually, the packet reaches the RP, which adds a (*,G) state in its multicast routing table and adds the receiver-facing interface in the OIL.

4. The source sends the multicast stream to the FHR, which adds an (S,G) state to its multicast routing table.

5. The FHR must send the multicast traffic to the RP, but there is a problem. Intermediate routers (between the FHR and the RP path) and the RP do not have an (S,G) state in their multicast routing table. This means that routers will drop the multicast traffic coming from the FHR. To resolve this problem, the router closest to the source must encapsulate the multicast packet inside of a unicast packet with the IP address of the RP as the IP destination. In other words, the multicast stream must be tunneled from the router at the source to the RP. The encapsulation packets are known as PIM register packets.

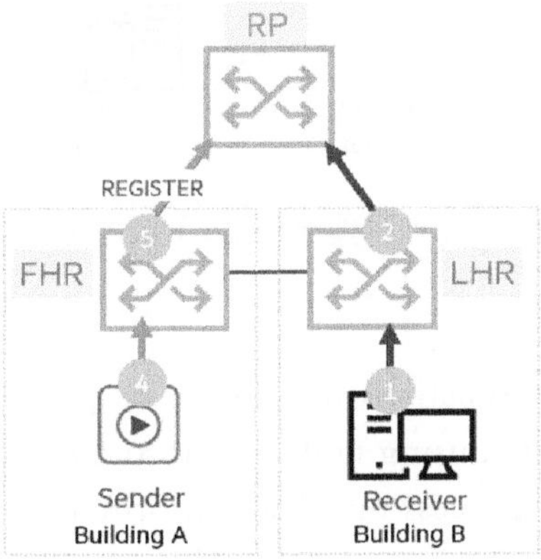

Figure 8-21: Phase 1: rendezvous point tree

Phase 1 conclusion

6. The RP receives the PIM Register message (encapsulated traffic) and checks for an interested receivers present (*,G) entry in its multicast routing table.
    a. If there is a receiver present, then it decapsulates the traffic and forwards the multicast packets onto the shared tree.
    b. If there is no reciver present, then the RP sends a Register Stop message to the FHR. Each router (between the RP and source) maintains a register suppression timer that is initiated by the Register Stop message. Upon expiration of this timer, the FHR sends a register packet to the RP again to verify if there are any active receivers for that group.

At the end of phase 1, multicast traffic flows encapsulated to the RP, and then natively over the RP tree to the multicast receivers. During this time, the flow is not efficient.

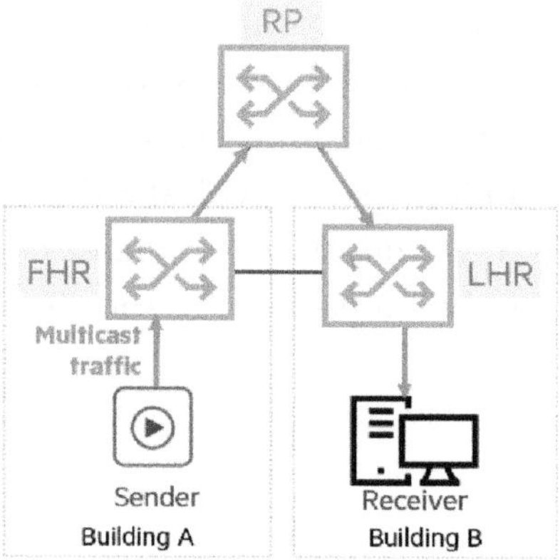

Figure 8-22: Phase 1 conclusion

Phase 2: Register-stop

1. The FHR receives a source-specific PIM Join message from the RP. This adds an (S,G) entry to the FHR multicast routing table and to the intermediate routers in the path. This state permits all devices to process multicast information that is received from the source, meaning that the encapsulation process is no longer required.
2. The FHR sends the multicast traffic natively (non-encapsulated) to the RP. All intermediate devices in the path can route the traffic.
3. At this moment, the RP receives two copies of the same traffic, one encapsulated in the Register message and other natively received. The RP discards the encapsulated copy.
4. The RP sends a PIM Register Stop message to the FHR.
5. The RP receives the multicast traffic from the source with no encapsulation and forwards this traffic to the LHR. This action creates a new (S,G) state in the multicast routing table on all routers between the RP and the LHR.

At the end of phase 2, encapsulation overhead has been removed, but the multicast traffic might travel a long distance to reach receivers.

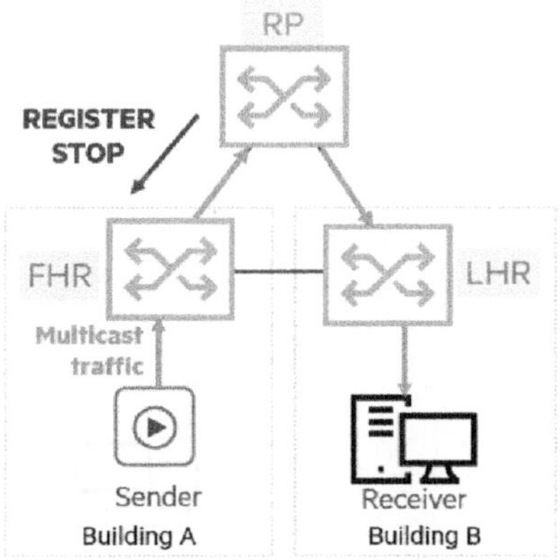

Figure 8-23: Phase 2: Register-stop

Phase 3: Shortest path

1. Now, the LHR knows the multicast traffic source IP address and asks itself if there a best path to that source, using the RPF process. If there is not a new best path available, then the LHR keeps using the shared tree path. But if there is a new best path, then a path optimization process must be performed—Step 2.
2. The LHR initiates a transfer from the shared tree to the SPT. The LHR sends a PIM Join message (S,G) to the FHR.
3. When the FHR receives the PIM message, it starts forwarding multicast packets to the LHR via the best path.
4. The LHR receives two copies of the traffic, one from the RP using the shared tree and the other from the source using the SPT.
5. The LHR sends a PIM Prune message toward the RP.

At the end of phase 3, receivers get the multicast traffic efficiently.

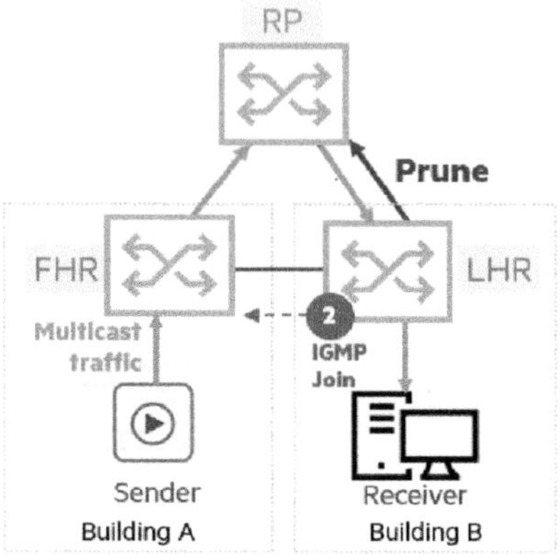

Figure 8-24: Shortest path

Static RP configuration

Enabling PIM sparse mode using a static RP is perhaps the simplest way to configure a multicast domain. You must manually configure this, telling every device the IP address of the RP—including the desired RP. You are mapping a device to function as the RP for a particular group address. This configuration must be defined identically on every router. Otherwise, the domain will be incomplete.

The configuration shown in **Figure 8-26** statically configures the router as the RP for a specified multicast group or range of multicast groups. This must be configured on all PIM-SM routers in the domain. If the group address is not specified, it applies to all IPv4 multicast addresses (224.0.0.0–239.255.255.255). PIM-SM supports a maximum of eight static RPs per VRF.

Why is it important to configure the main network loopback interface with PIM-SM? This is a recommended practice for any multicast overlay. This ensures that the router can fully participate in the multicast domain, even if errors occur on leaf-facing interfaces. It also allows the loopback interfaces to be used as RP addresses or mapping agents in dynamic RP propagation, making them more reliable.

This static configuration option can be suitable in small deployments with few multicast group mappings. However, for large deployments with many overlapping domains or many multicast applications and many rendezvous points, a consistent group mapping through static commands may become extremely difficult to manage. You may need a more scalable solution than static configuration—the Bootstrap Router (BSR) mechanism.

Same configuration on all devices

```
router pim
 enable
 rp-address 10.2.0.2 239.2.0.0/16
```

Figure 8-25: Static RP configuration

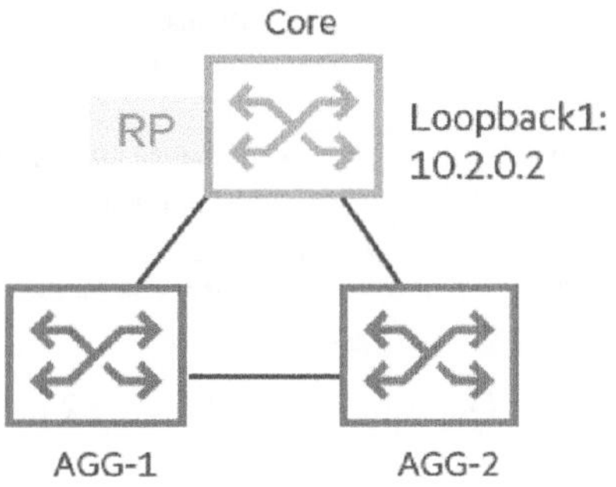

Figure 8-26: Static RP configuration topology

Dynamic RP and the BSR

The BSR mechanism for PIM is defined in RFC 5059.

BSR is an RP high-availability mechanism that provides active/standby functionality and automatic downstream RP information propagation. This mechanism dynamically establishes a method to share group-to-RP mappings (also known as RP-Set) to all PIM routers in the domain. This mechanism avoids wrong mappings that can cause unexpected problems.

With BSR, you configure some routers within a PIM domain to be potential RPs—so-called Candidate-RPs (C-RPs). A subset of the C-RPs will eventually be used as the actual RPs for the domain. You also configure some PIM routers in the domain to be candidate bootstrap routers, or Candidate-BSRs (C-BSRs). One C-BSR is elected to be the BSR for the domain. All PIM routers in the domain learn the result of this election through bootstrap messages. The C-RPs then report their candidacy to the elected BSR, which chooses a subset of these C-RPs and distributes corresponding group-to-RP mappings to all the routers in the domain through bootstrap messages.

RP-Set elements shared by bootstrap are:

- Multicast group range, expressed as an address and prefix length
- RP priority
- RP address
- Hash mask length
- SM/BIDIR flag

Router type	Description
Bootstrap Router (BSR)	• Receive messages from C-RPs • Distributes corresponding group-to-RP mappings to all routers in the domain
Candidate RP (C-RP)	• Potential RP for a specific Group • Report their candidacy to BSR
Candidate BSR (C-BSR)	Potential BSR for the PIM domain

Figure 8-27: Dynamic RP and the BSR mechanism

BSR mechanism

The bootstrap process has four phases, as defined in RFC 5059.

Phase 1: BSR election

The function of the BSR is to collect and broadcast the RP set to all routers in the domain. Each C-BSR originates BootStrap Messages (BSMs) announcing their capacity to become a BSR. Every BSM contains a BSR Priority field. Routers within the domain flood the BSMs throughout the domain. A Candidate or C-BSR that hears about a higher-priority C-BSR than itself stops sending BSMs for some period of time. The single remaining C-BSR becomes the elected BSR, and its BSMs inform all the other routers in the domain that it is the elected BSR. If the priority value is the same, the router with the highest IP address is elected.

There are three states during the BSR election:

- **Candidate-BSR (C-BSR):** The router is a candidate to be the BSR for the scope zone, but currently, another router is the preferred BSR.
- **Pending-BSR (P-BSR):** The router is a candidate to be the BSR for the scope zone. Currently, no other router is the preferred BSR, but this router is not yet the elected BSR. This is a temporary state that prevents rapid thrashing of the choice of BSR during BSR election.
- **Elected-BSR (E-BSR):** The router is the elected BSR for the scope zone, and it must perform all the BSR functions.

A key part of the election mechanism is that we associate a weight with each BSR. The weight of a BSR is defined to be the concatenation in the fixed-precision unsigned arithmetic of the BSR Priority field from the bootstrap message and the IP address of the BSR from the bootstrap message (with the BSR Priority taking the most significant bits and the IP address taking the least significant bits).

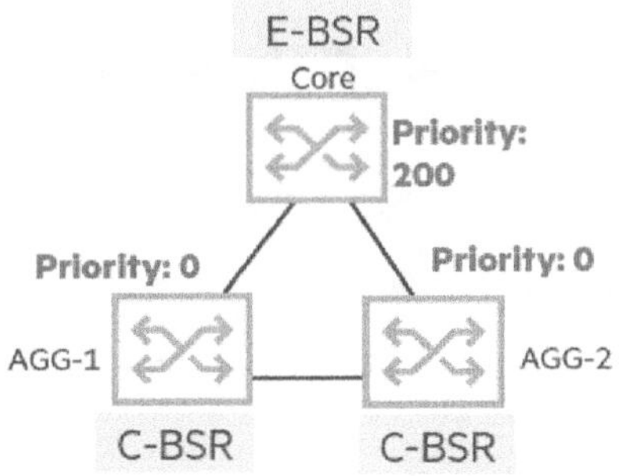

Figure 8-28: Phase 1: BSR election

Phase 2: C-RP advertisement

Each Candidate-RP within a domain sends periodic Candidate-RP-Advertisement (C-RP-Adv) messages to the elected BSR. A C-RP-Adv message includes the priority of the advertising C-RP, as well as a list of group ranges for which the candidacy is advertised. In this way, the BSR learns about possible RPs that are currently up and reachable.

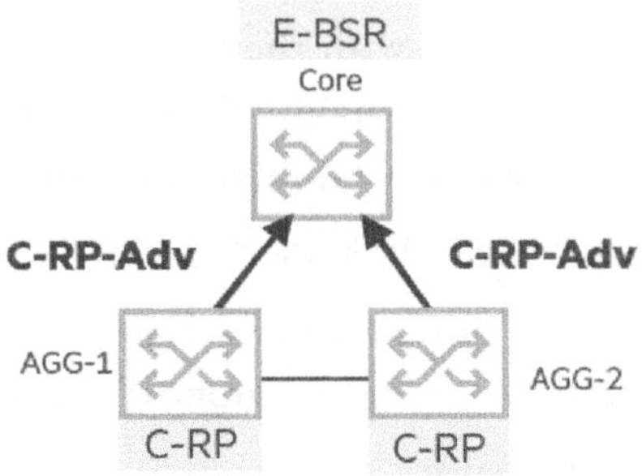

Figure 8-29: Phase 2: C-RP advertisement

Phase 3: RP-Set formation

To form the RP-Set, the BSR selects a subset of the C-RPs from which it has received C-RP-Adv messages. In general, it should do this in such a way that the RP-Set is neither so large that all the routers in the domain cannot be informed about it, nor so small that the load is overly concentrated on some RPs. It should also attempt to produce an RP-Set that does not change frequently.

For security reasons, there should be a way to restrict which IP addresses the BSR accepts C-RP-Adv messages from—such as an access list. For use of a scoped BSR, it may also be useful to specify which group ranges should be accepted.

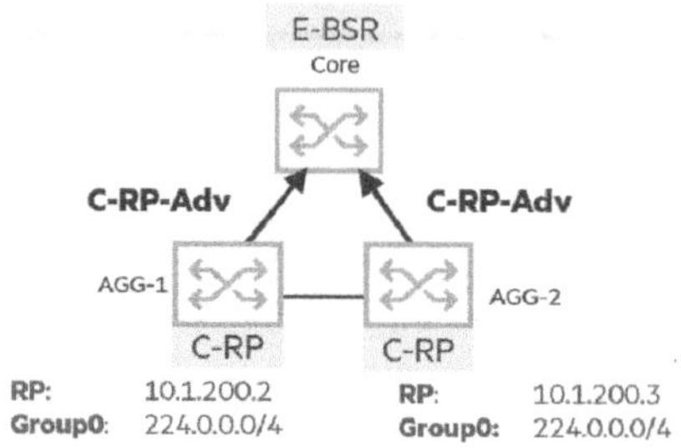

Figure 8-30: RP-Set formation

Phase 4: RP-Set flooding

When a PIM router receives a BSM, it adds the group-to-RP mappings contained therein to its pool of mappings obtained from other sources (for example: static configuration). It calculates the final mappings of group addresses to RP addresses using the following criteria:

1. Select the RP that advertises the most specific range of multicast addresses that includes the desired address.
2. If more than one C-RP advertises a range of the same specificity, select the C-RP configured with the highest priority.
3. If more than one C-RP is configured with the highest priority, a hash function generates a series of mask length values that are individually assigned to the set of eligible C-RPs. If the hash function matches a single RP candidate to a longer mask length than the other candidates, that candidate is selected to support the group.
4. If the hash function matches the longest mask length to multiple RP candidates, the C-RP that has the highest IP address is selected to support the group.

In future BSMs, the BSR includes the RP-Set information. BSMs are flooded through the domain, which ensures that the RP-Set rapidly reaches all the routers in the domain. BSMs are originated periodically to ensure consistency after failure restoration.

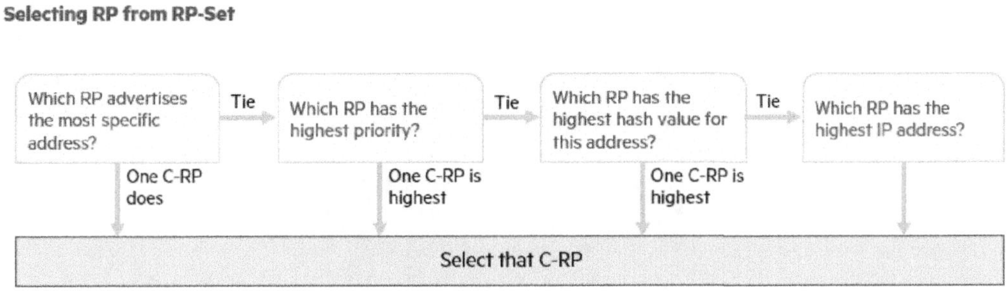

Figure 8-31: RP-Set flooding

VSX and PIM

AOS-CX supports dual DR functionality for VSX. One VSX member is elected as PIM DR for each IP subnet, using the typical PIM election process. The other VSX member becomes a PIM proxy DR.

VSX and PIM: Behavior

Both VSX peers have the same control plane information. This means that both members will be able to establish PIM neighborships, send PIM JOIN/LEAVE messages to the RP, populate (S,G) entries on their multicast routing table, and build an SPT. However, multicast traffic (data plane) is only routed from the VSX peer that acts as the PIM DR. Database route entries for downstream ports in the DR and proxy DR are always in sync but are written differently on each device; multicast entries in the DR are written as route entries and the entries in the proxy DR as bridge entries.

You can implement VLAN-based load balancing so that for each VSX pair, one switch can be the DR for a group of VLANs and the other switch the DR for the other group of VLANs.

> This solution works for IPv4 and IPv6.

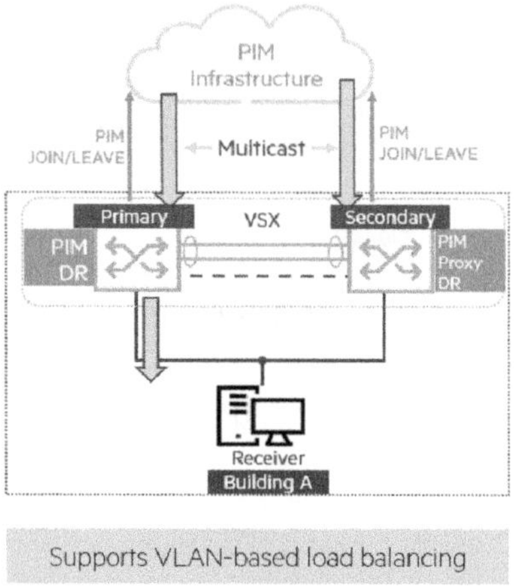

Figure 8-32: VSX and PIM: behavior

VSX and PIM: DR failure

The mechanism to have a pre-established control plane on both VSX peers permits the VSX cluster to achieve a fast failover in case the PIM DR fails.

Once the VSX peer that is acting as DR does down, the proxy DR detects the ISL status and removes the PIM neighborship of the VSX peer immediately.

The proxy DR changes its role to DR. This action turns the bridge entries for downstream VSX ports to route entries.

Traffic is now forwarded to the downstream multicast devices using the new DR.

The conversion from bridge to route entries happens in the Configuration and State Database, and the change propagates to hardware. This approach takes much less time compared to forming the SPT for the multicast traffic.

When the VSX peer is back online, it becomes the proxy DR and starts to learn about multicast routes. This implies building PIM neighbour relationships, populating (S,G) entries in their multicast routing table, and building an SPT. To prevent the new proxy DR from becoming the DR again after the reboot (because of a higher IP address), the current DR sends hello packets with the highest DR priority (4294967295). Any user-configured DR priority is ignored.

Keep the PIM hello-interval timer at the default value: 30 seconds. This timer no longer has the effect of improving VSX PIM fast fail-over. Consider configuring PIM-Sparse DR-Priority such that each VSX node has an equal distribution of

---

VLANs, as relates to the actual DR role. This helps to further minimize the impact during fast fail-over.

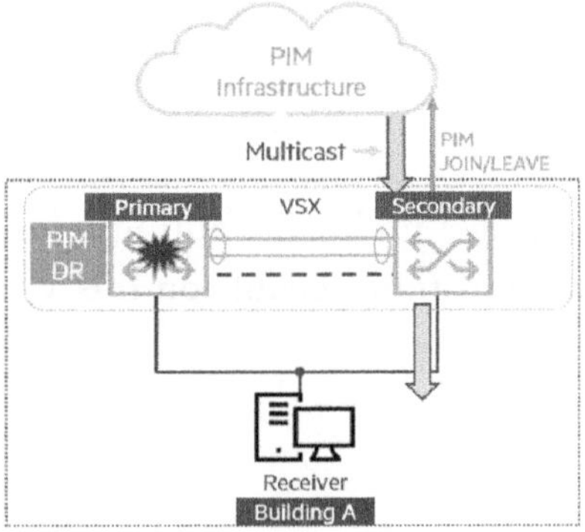

Figure 8-33: VSX and PIM: DR failure

VSX and PIM: ISL failure

If the ISL goes down with the VSX keepalive present and both VSX peers are up, then the VSX primary stays as the DR. The VSX secondary becomes the proxy DR when its downstream links are cut by VSX infrastructure. This takes at least five seconds. During this time, duplicated traffic is expected as both VSX peers act as DR.

If the ISL goes down when the keepalive is not configured, both VSX peers act independently and the downstream VSX ports are not brought down. This results in duplicate traffic since both VSX peers start acting as DR.

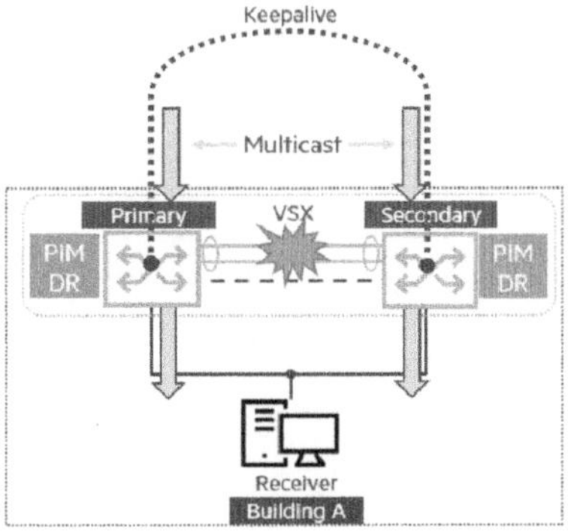

Figure 8-34: VSX and PIM: ISL failure

Learning check

Check your understanding of Multicast Routing by answering the following questions.

Question 1

When using PIM-SM, AOS-CX switches will automatically join the tree?

- O True
- O False

Question 2

A network administrator is configuring a multicast routing solution. Potential multicast receivers are on VLAN 5. What must the administrator enable on this VLAN?

- O PIM only
- O PIM and IGMP
- O IGMP only
- O PIM-DM and PIM-SM

9 Access Control Lists

With internal threats so common, you must secure your network's internal infrastructure and IT resources. One of the basic mechanisms for controlling traffic is Access Control Lists (ACLs). ACLs let network administrators control the passage of traffic based on various characteristics of the traffic. ACLs see an array of applications, from the simple restricting of traffic arriving at or destined to a particular address to a more complex prevention of a host from accessing a particular application.

OBJECTIVES

After completing this chapter, you should be able to:

- ✓ Define and create ACLs.
- ✓ Apply ACLs to different types of interfaces.
- ✓ Leverage object groups in ACLs for improved efficiency.
- ✓ Implement classifier policies.
- ✓ Analyze restrictions and resource utilization in implementing ACLs.

ACL introduction and creation

You will explore the creation, usage, and implementation processes of ACLs in this section.

ACL overview

ACLs allow you to configure networks to permit and deny traffic based on the packet characteristics. In other words, you can use ACLs to control traffic based on a list of rules. Each rule matches on the fields in an IP or Ethernet header to permit or deny traffic. Each rule consists of specific criteria, and a firewall or a routing switch compares ingress or egress packets against the criteria. If it finds a match, some action is taken—permit the traffic to enter or exit a router interface, or simply deny it. ACLs are composed of one or more Access Control Entries (ACEs), also referred to as rules in an ACL. These rules consist of criteria and an action.

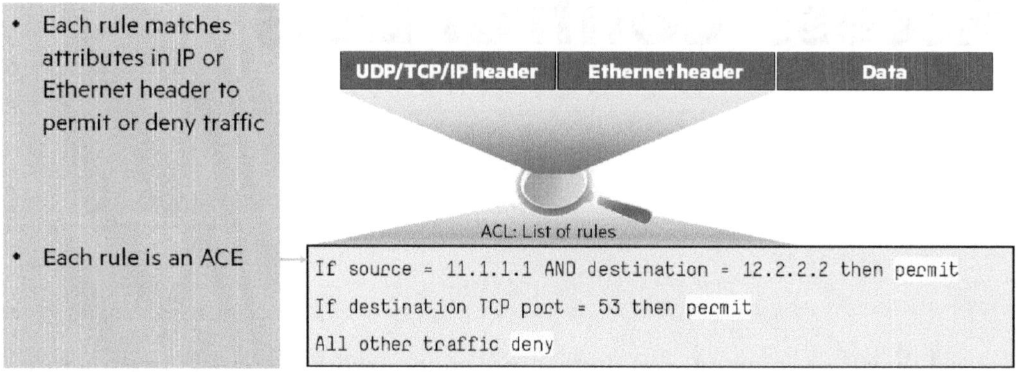

Figure 9-1: ACL overview

The sample ACL shown in **Figure 9-1** does not follow proper syntax; it is represented in a simplified manner to help visualize how an ACL works. This ACL consists of three rules. The first rule allows the passage of traffic between a pair of source and destination addresses, the second rule allows everyone access to the application at port 53, and the last rule denies all other traffic.

ACLs have several use cases, but this chapter focuses on packet-filtering ACLs—ACLs that are applied to routing switch interfaces to filter incoming and outgoing traffic.

Benefits of ACLs

Some of the benefits of using ACLs are:

- Blocks unwanted traffic or users
- Classifies traffic for Quality of Service (QoS) features
- Restricts routing update content
- Mitigates Denial of Service (DoS) attacks
- Controls switch management access
- Is useful with Policy Based Routing (PBR)

ACL use cases

ACLs can be implemented for a variety of reasons. They can be used at different layers of a network infrastructure and perform different functions according to the layer at which they are applied. ACL functions in different application scenarios are described below.

At the network perimeter

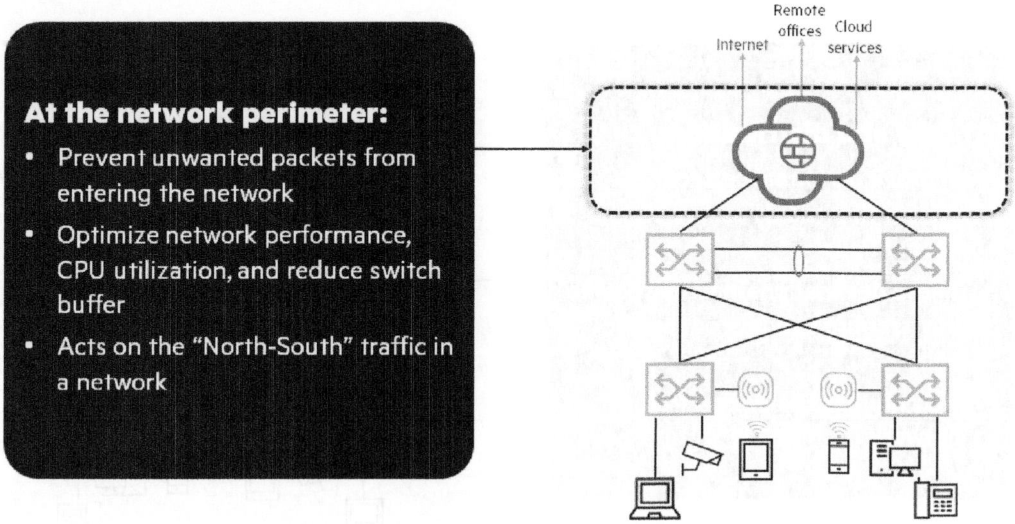

Figure 9-2: ACL functions at the network perimeter

At the network edge, ACLs can prevent unwanted IP packets from entering the network infrastructure. You can take a restrictive approach in which you permit the required traffic and deny all other traffic, or you can take a permissive approach in which you deny specific traffic and permit all other traffic. By combining permit and deny rules, you can create highly flexible policies.

Implementing ACLs at the network edge can also improve the network performance and reduce the switch buffer and CPU utilization by reducing the volume of packets that the upstream switches and routers handle.

At the core and distribution layer

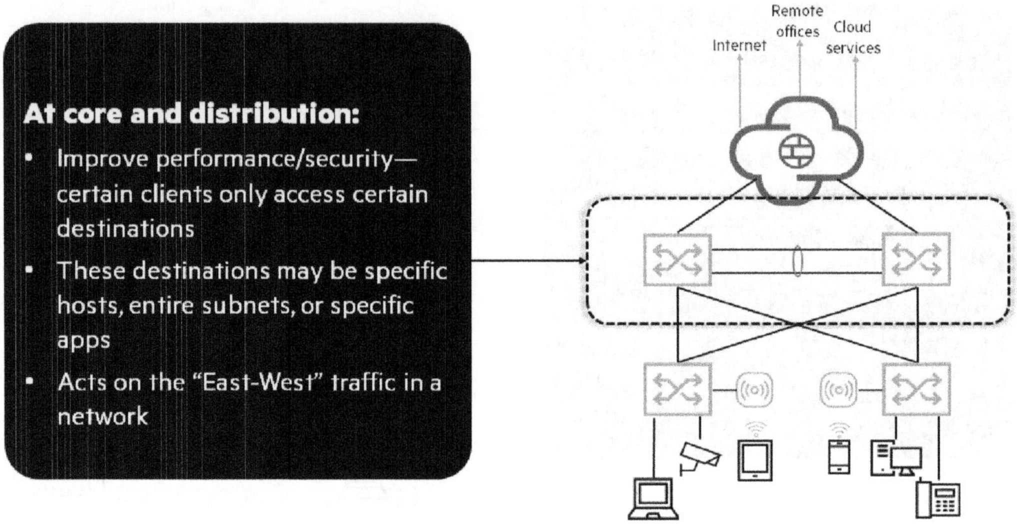

Figure 9-3: ACL functions at the network core and distribution layer

At the network core and distribution layer, ACLs can improve performance and security. You can use ACLs to ensure different groups of clients only have access to specific destinations. These destinations may be specific hosts, entire subnets, or even particular applications. For example, you may want to ensure that all hosts and servers in a given VLAN are only allowed to communicate within that VLAN or with a limited number of other specific VLANs.

At the management layer

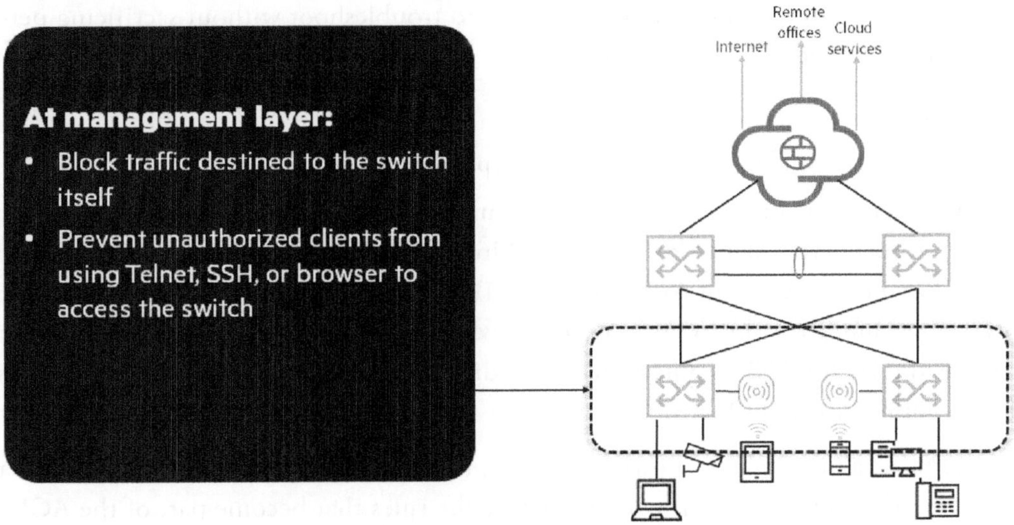

Figure 9-4: ACL functions at the management layer

You can use ACLs to block IP traffic that has the switch as the destination address. For example, ACLs can deny unauthorized clients attempting to access the switch using functions such as Telnet, Secure Shell (SSH), or a web browser.

For performance and troubleshooting

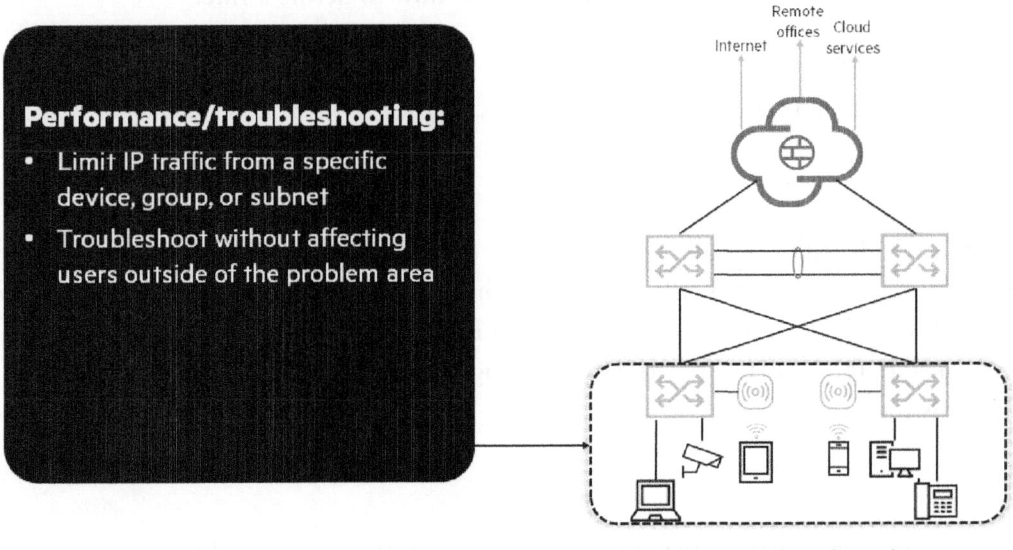

Figure 9-5: ACL functions in performance and troubleshooting related applications

In addition to securing the network infrastructure, you can also use ACLs to control a performance problem quickly by limiting the IP traffic generated by a specific device, group of devices, or a subnet. This gives you an opportunity to troubleshoot without sacrificing performance for users outside of the problem area.

ACL creation

AOS-CX switches support three types of ACLs—IPv4, IPv6, and MAC. Based on the type of ACL, relevant packet attributes can be used to define the criteria for matching. These attributes could be the protocol type, source/destination addresses, or source/destination ports. When you create an ACL, you give it a valid ACL identifier (ID). The ID groups the rules in the ACL, so you can view and modify it. Every ACL ID on a switch must be unique. With AOS-CX, you can use a name as the ID for ACLs, something that indicates its intended purpose. The name is a case-sensitive alphanumeric value.

Use the `access-list` command, shown as follows, to create an ACL. This command will take you inside the ACL context. At this context, define the rules that become part of the ACL.

```
switch(config)# access-list ip/ipv6/mac <ACL_NAME>
switch(config-acl-ip/ipv6/mac)# <ACL_RULE>
```

You can reuse an ID for an IPv4 and an IPv6 list. An ACL's ID must be unique in IPv4, and it must be unique in IPv6, but two ACLs in IPv4 or IPv6 cannot use the same ID.

Next, you will examine the composition of a rule and how to define a rule.

ACL rule composition

An ACL rule consists of three basic components, as follows:

Sequence ID

The sequence ID identifies the rule within the list, enabling you to identify and, if necessary, remove it. The sequence ID also tells the switch when to process this rule in relation to other rules. The ID with the lowest number is processed first. Specifying a sequence ID is optional. If you do not specify the sequence ID, the switch assigns an ID automatically, placing the rule at the end of the existing list.

Action

ACLs have the following actions that are performed when there is a match:

- **permit:** Matching packets will be allowed to progress through the switch.
- **deny:** Matching packets will be dropped.

Optionally, you can have the switch log or count the packet matching events.

- **count:** A specific rule matching event increments hit counts for that entry.
- **log:** Rule matching events are logged and displayed on the specified logging destination.

Matching criteria

At the end, you define the criteria involving a packet attribute, which will be used to filter the packets. The main attributes that can be used to filter traffic are protocol type, source and/or destination address, or source and/or destination port.

ACL rule creation

To define the criteria, start by specifying the protocol, such as IP, TCP, UDP, ICMP, and so forth. Next, specify the source and destination addresses or both. The address is specified like any standard IP address—a host address or a subnet/prefix length. In case you do not want to restrict specific source or destination addresses, use the keyword "any" to allow all traffic from that source or to that destination. At the end, provide the additional relevant characteristics, if any. These include TCP/UDP port numbers, ICMP type, and more such characteristics. This is illustrated in **Figure 9-6** with a few examples.

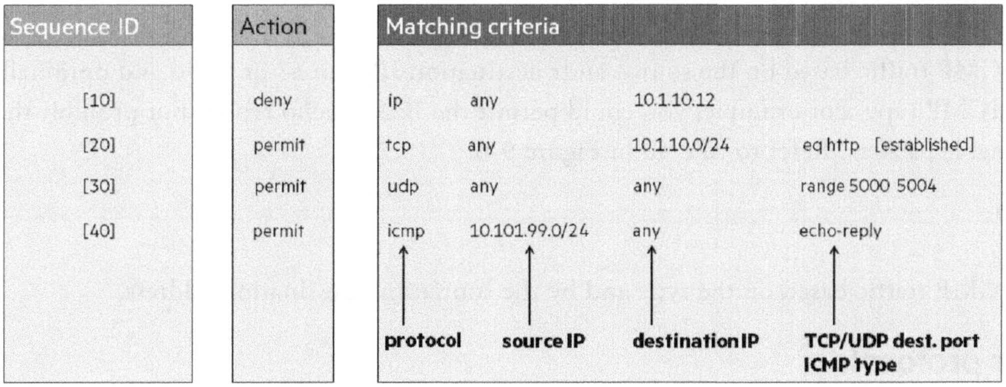

Figure 9-6: Examples of ACL rule creation

You can categorize IPv4 ACL rules by the IP protocol type, which is the first option that you specify after "permit" or "deny." The following are some of the common protocols and their respective ACL features.

IP

Filter traffic based on the source and destination IP address or both. Refer to the sample rule with sequence ID 10 in **Figure 9-6**.

TCP or UDP

Filter traffic based on the source andr destination IP address or both, plus the source andr destination TCP or UDP port numbers or both. When you specify the port, you must also include a keyword that indicates whether to select that specific port, a range of ports, or any port except that port. Instead of using the port number, you can use a keyword for the well-known ports. For example, you can specify a port as "http," in which case the switch automatically interprets the port number as 80.

For TCP, you can also filter the traffic based on the TCP flag. For example, you might want to drop TCP RST packets from untrusted IP addresses. In addition, you can include the "established" keyword to indicate that the rule applies only to sessions that are already established. Assume that an ACL is filtering traffic from an untrusted zone to a trusted zone. A rule with the "established" keyword can prevent untrusted endpoints from establishing certain types of TCP sessions, but allow sessions that were established by trusted endpoints to flow. Refer to rules 20 and 30 in **Figure 9-6**.

ICMP

Filter ICMP traffic based on the source andr destination IP address or both, and optionally, based on the ICMP type. For example, you could permit the ICMP echo replies but prohibit the ICMP unreachable packets. Refer to rule 40 in **Figure 9-6**.

IGMP

Filter IGMP traffic based on the type and by the source and destination address.

Other protocols

Permit or deny specific IP protocols such as OSPF.

The CLI provides keywords for specifying several protocols. If you do not see a keyword for the protocol that you want to control, use the IP protocol number. The syntax of the command depends on the type of protocol in the rule. Refer to *ACLs and Classifiers Policy Guide* for the syntax details.

Unlike with TCP, UDP, ICMP, or IGMP, you can only specify the IP protocol and the source and destination IP addresses, and no other options.

ACL processing order

The switch processes the rules in an ACL in a top-down manner—from the lowest to the highest sequence ID. The processing takes place as described in the flowchart in **Figure 9-7**. The switch first compares a packet against the first rule's matching criteria. If the packet matches, the switch performs the specified permit or deny action, and stops processing the ACL. If it does not match, then the switch compares the packet against the next rule in the sequence. It repeats this process until it finds a match. If a packet matches with none of the rules, then the switch uses the "implicit deny" rule. This rule does not have to be defined by the user (implicit). This is the default action of an ACL, in case none of the user-defined rules provide a match.

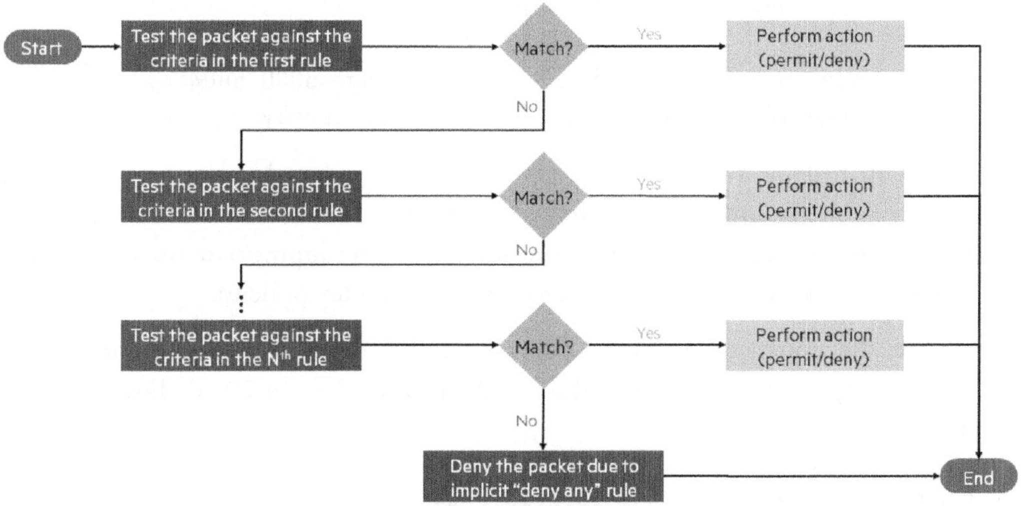

Figure 9-7: ACL processing order

Implicit deny rule

There is an "implicit deny" rule any at the end of all ACLs—a `deny any any` statement. If you apply an ACL to an interface and the switch finds no matches between an inbound packet and the ACL rules, the packet is dropped. Its an "implicit" rule because you can't see it when viewing an ACL.

Some network administrators like to explicitly define the rule at the end of an ACL to clarify the behavior. If you do so, make sure to assign the rule a very high sequence ID, so that you can add as many rules to the ACL as you want.

If you need to override the "implicit deny" so that a packet that does not have a match is permitted, then you should create a rule that permits all traffic as the last rule in the ACL. Again, assign this rule a high sequence ID.

ACL sequence IDs

Sequence IDs are useful to identify a rule in an ACL, as well as to decide the order in which the rules will be executed. You will now assess the significance of sequence IDs and learn how to assign them to a rule.

Impact of sequence of rules

Consider the following snippet of code representing an ACL.

```
access-list ip ACL-1
  [10] deny    ip   any            10.1.10.12
  [20] permit  ip   any            10.1.10.0/24
  [30] permit  udp  any            any           range 5000 5004
  [40] permit  icmp 10.1.99.0/24 any           echo-reply
```

Observe how the first (deny) rule prevents everyone from accessing the server at 10.1.10.12. The second (permit) rule then allows everyone to access all other addresses in that subnet. This way, you can restrict access to a specific server in a subnet. With the appropriate use of sequence IDs, you can combine permit and deny rules to create highly complex policies.

At the same time, the sequence IDs should be set carefully. Incorrect sequencing may result in an undesired result. Suppose that you swapped the order of rules 10 and 20, as shown in the following example.

```
access-list ip ACL-1
  [10] permit  ip   any            10.1.10.0/24
  [20] deny    ip   any            10.1.10.12
  [30] permit  udp  any            any           range 5000 5004
  [40] permit  icmp 10.1.99.0/24 any           echo-reply
```

Destination 10.1.10.12 would match 10.1.10.0/24, and so rule 20 would never be processed. As a result, anyone would be able to access Server 2.

Sequence ID assignment

The assignment of sequence IDs could be done manually, or it can happen automatically.

Automatic assignment

If you do not enter a sequence ID manually, the switch automatically adds the IDs in order, leaving spaces. This means that the rules are processed in the order in which you entered them. New rules are added to the end of the list, even if spaces exist earlier in the list. For example, if the first rule that you entered does not have a manual ID, an AOS-CX switch numbers the rule as 10. Subsequent rules without a sequence ID are numbered in increments of 10, that is, 10, 20, 30, and so on. An ACL can have a combination of automatic and manually assigned rules. Consider an ACL consisting of rules with manual ID assignment in which the last rule has an ID that is not divisible by 10, such as 45. If a new rule is added to this ACL using automatic assignment, then the switch adds the new rule with the next ID that is divisible by 10, that is, 50.

For example, the following ACL is created without providing the sequence IDs for the rules. The sequence IDs assigned automatically can be seen using the `do show access-list` command, as shown below it.

```
switch(config)# access-list ip ACL-2
switch(config-acl-ip)# permit udp any 172.16.1.0/24
switch(config-acl-ip)# permit tcp 172.16.2.0/24 gt 1023 any
switch(config)# do show access-list
<outout omitted>
10      permit                              udp
        any
        172.16.1.0/255.255.255.0
20      permit                              tcp
        172.26.1.0/255.255.0.0   > 1023
        any
        syn
        ack
```

Manual assignment

Manual assignment happens when you define a rule and specify the sequence ID along with it. When you enter the IDs manually, it is best practice for you to space the IDs, preferably by tens, so that you can insert additional rules at a later point, if necessary. An example is shown in the following code snippet.

```
switch(config)# access-list ip ACL-2
switch(config-acl-ip)# 100 permit udp any 172.16.1.0/24
switch(config-acl-ip)# 110 permit tcp 172.16.2.0/24 gt 1023 any
```

Manual or automatic, you must plan the ACL carefully and configure rules for processing in the correct order, generally entering more specific rules with a lower ID followed by the less specific rules with higher IDs.

Sequence ID renumbering

Sometimes, despite your best planning, you need to change the sequence IDs to place rules in the correct order. This can be done using the following command.

```
switch(config)# access-list {ip | mac} <ACL-name> resequence <start-sequence>
<increment>
```

The `start-sequence` is the sequence number you want to assign to the first rule, and `increment` is the incrementing value by which subsequent rules will be numbered. This command leaves the rules in the same relative order. However, with more spaces between the rules, you can remove rules and add them in the appropriate location as necessary. To remove a rule, use the `no <sequence-number>` command. The sequence renumbering command is described using the following example.

```
switch(config)# access-list ip ACL-2 resequence 1 1
switch(config)# do show access-list
<outout omitted>
1       permit                          udp
        any
        172.16.1.0/255.255.255.0
2       permit                          tcp
        172.16.2.0/255.255.0.0   > 1023
        any
```

IP address format in ACL rules

You can use a network address and a prefix length to define a range of IP addresses in an ACL rule. This option works well to permit or deny an entire subnet. When you view an ACL using the `show access-list` command, you can see the range denoted as the IP address and a subnet mask. You can also use either format to specify the IP addresses while creating an ACL rule. A sample result of the `show access-list` command is shown as follows.

```
switch# show access-list ip ACL-2
Type     Name
Sequence Comment
         Action                   L3 Protocol
         Source IP Address        Source L4 Port(s)
         Destination IP Address   Destination L4 Port(s)
         Additional Parameters
-----------------------------------------------------------
```

```
IPv4      ACL-2
       10 permit                udp
          any
          10.1.101.0/255.255.255.0
```

AOS-CX switches do not support wildcard masks—only prefixes or subnet masks are supported—when creating ACL rules.

Log and count options using ACL rules

In addition to the options discussed until now, an ACL rule on an AOS-CX switch can include a log or a count option at the end.

Logging

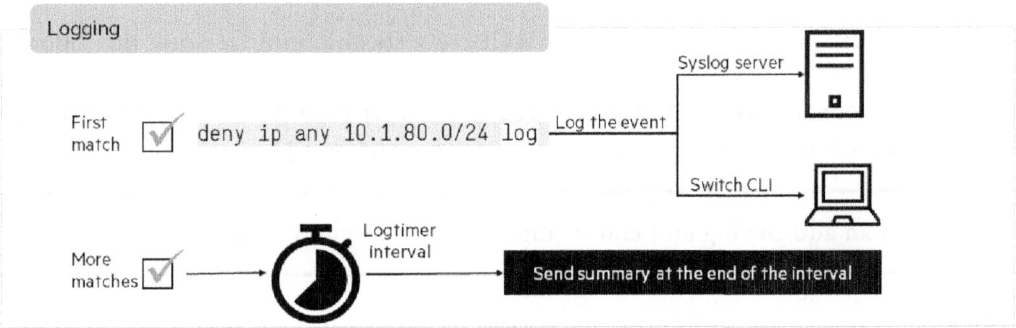

Figure 9-8: Logging using ACLs

The "log" keyword tells the switch to generate an "event log" message when a packet matches a rule. The log is created irrespective of the action—whether it is "deny" or "permit"—if the logging is enabled. Refer to the following example command used to enable logging in a rule.

```
deny ip any 10.1.80.0/24 log
```

Depending on the platform, the "log" keyword also increments hit counts for the specified entries. The first packet that hits any log entry is copied to the CPU and logged to the operator's specified destination; it could either be the switch's CLI or the syslog server. This first packet starts the five-minute ACL "log-timer." When the timer expires, the summary of all hit counts per ACL entry is sent to the specified logging destination. This capability allows throttling of logging ACL hits.

When using multiple ACL types such as IPv4, IPv6, or MAC, with logging on the same interface, the first packet that matches a rule with the log option is logged. Any packets matching other ACL types do not create a log until the "log-timer" is over. At the end of the timer, a summary log is created from all the ACLs that were matched, regardless of the type.

To change the default logging period, use the following command:

```
Switch(config)# access-list log-timer {default|<INTERVAL>}
```

Counting

Figure 9-9: Counting using ACLs

The "count" keyword keeps track of each match on a rule. Refer to the following example of a command used to enable counting in a rule.

```
deny ip any 10.1.80.0/24 count
```

To reset the hit counts for all ACLs, a specific ACL, or a specific interface, use the following command:

```
switch# clear access-list hitcounts {all | [{ip|ipv6|mac} <ACL-NAME>] [interface
<IFNAME>| vlan <VLAN-ID>] [in|out|routed-in|routed-out]}
```

> You can add the log and count functions in a single ACL rule statement.

Apply an ACL

When you create an ACL, it does nothing until you apply it to one or more interfaces or VLANs, as shown in **Figure 9-10**. ACLs configured on an interface using this syntax are called static ACLs—they remain on the interface permanently until you modify or remove them. AOS-CX switches can also apply dynamic port-based ACLs as part of an Authentication, Authorization, and Accounting (AAA) solution. This chapter focuses on static ACLs.

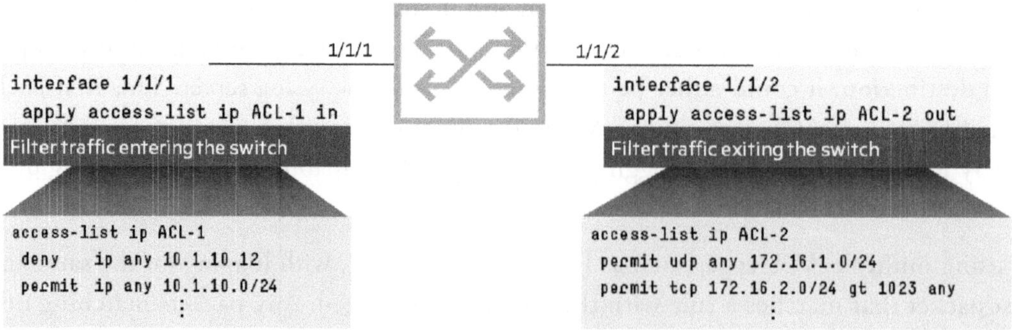

Figure 9-10: Applying ACLs

You have three choices while applying an ACL:

- Choose the interface (Layer 2 or Layer 3) on which to apply the ACL.
- Choose the direction, which determines whether the ACL filters the traffic that arrives on that interface or that is destined out that interface.
- Choose the ACL type (IPv4/IPv6/MAC) and apply the ACL using the appropriated syntax based on the type.

ACLs can be applied to L2 or L3 ports or LAGs, VLANs, or interface VLANs, as well as management and control plane interfaces. **Table 9-1** illustrates the options available for the application of an ACL, depending on the type and direction.

Table 9-1: ACL application options

ACL type	IPv4+6	IPv4	IPv6	MAC
Direction	**In**	**Out**	**Out**	**In**
L2 interface (port)	Yes			Yes
L2 LAG	Yes			Yes
L3 interface (port)	Yes	Yes		Yes
L3 LAG	Yes	Yes		Yes
L3 interface (port) subinterface[1]	Yes			Yes
L3 LAG subinterface[2]	Yes			Yes
VLAN	Yes	Yes	Yes	Yes
Interface VLAN	Yes (routed)	Yes (routed)	Yes (routed)	
Management interface	Yes			
Control plane (per VRF)	Yes			

You can apply an ACL to multiple different interfaces, but each interface supports a maximum of one ACL in each direction.

On VLAN interfaces, you are allowed to use the same ACL for both the inbound and outbound directions. Often, however, you should not do so because the source and destination addresses for your rules might not be valid in both directions.

1Not applicable to CX 8320 Series and CX 9300 Series

2Not applicable to CX 8320 Series and CX 9300 Series

Stateless vs. stateful ACLs

To fully understand the operational details of ACLs, it is important to know how the ACLs offer the firewall abilities. Some firewalls operate based on the network layer whereas some others use both the network and transport layers. Let us discuss these two types of firewalls and then see the features supported by the AOS-CX switches.

Stateless firewalls

Stateless firewalls operate at the network layer, making decisions based on static information such as IP addresses, ports, and protocol types. They inspect each packet independently, without considering the context of previous packets. The operation of these firewalls is typically faster because they do not need to maintain or track the state of network connections. In addition, they are easier to configure and manage due to their straightforward rules.

Stateless firewalls apply predefined rules to each incoming or outgoing packet. For example, a rule could allow all traffic on port 80 (HTTP) but block traffic on port 25 (SMTP). Moreover, since they do not track the state of connections, they may be less effective at blocking certain types of attacks, such as TCP SYN flood attacks.

Stateful firewalls

Stateful firewalls operate at both the network and transport layers, meaning they can track the state of active connections. They keep a state table that records the details about active sessions, such as IP addresses, ports, and the current state of the connection. Moreover, they can inspect packet headers and payloads, making them more effective against sophisticated attacks. As a result, their operation is slightly slower due to the overhead of maintaining state information, but generally more secure.

In contrast with stateless firewalls, the stateful firewalls dynamically adjust their rules based on the state of network connections. For example, they can allow packets for an established connection while blocking unsolicited packets. They achieve this by maintaining a table that tracks the state of each connection, for example, SYN_SENT, ESTABLISHED, and FIN_WAIT.

Stateless vs. stateful firewall operation example: TCP session establishment

Let us look at how stateless and stateful firewalls handle a TCP session establishment, which involves a three-way handshake (SYN, SYN-ACK, and ACK).

Stateless firewall

- Initial SYN packet: When an external client sends a SYN packet to initiate a TCP connection, the stateless firewall checks its rules to see if the traffic is allowed on the specified port, as illustrated in **Figure 9-11**. In this example, there is a rule allowing internal clients to access HTTP applications (port 80) on remote destinations.
- SYN-ACK packet: The internal server responds with a SYN-ACK packet using another port (socket 10378). As the return packet is using another port (socket), the stateless policy is not aware that this is an answer to a previous request. Unless there is a rule allowing that port to be used as a socket, the packet will be dropped. Remember that the socket assignment is random, making it almost impossible for the network administrator to open each one individually. Moreover, opening all ports is a security breach.

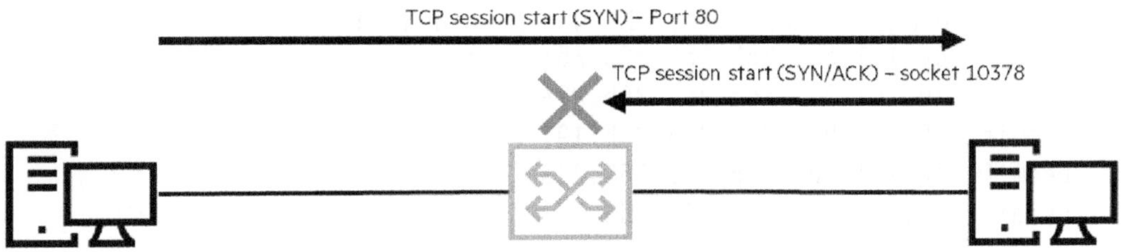

Figure 9-11: Stateless firewalls

Stateful firewall

- Initial SYN packet: When an SYN packet arrives, the stateful firewall checks its rules and records the state of this connection in its state table. If allowed, it forwards the packet and records the state as SYN_SENT.
- SYN-ACK packet: The internal server responds with a SYN-ACK packet. The firewall recognizes this packet as part of an existing connection (SYN_SENT state) and updates the state to SYN_RECEIVED. It forwards the packet to the client, as shown in **Figure 9-12**.
- ACK packet: When the client sends an ACK packet, the firewall sees it as a continuation of the existing connection. It updates the state to ESTABLISHED and forwards the packet.

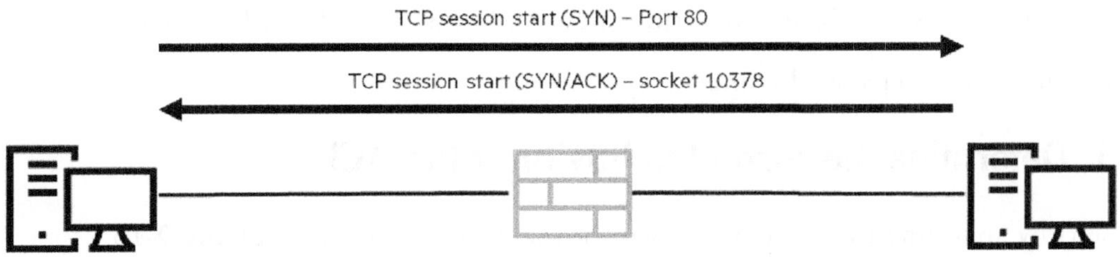

Figure 9-12: Stateful firewalls

In summary, while stateless firewalls make decisions based solely on predefined rules for each individual packet, stateful firewalls make more informed decisions by tracking the state of network connections. This allows stateful firewalls to provide more robust security, especially for protocols like TCP that rely on connection states.

All AOS-CX switches support stateless ACLs for traffic filtering, except CX 10000 Series switches which are able to act as stateful firewalls.

ACL application scenario: analyze and modify an ACL

In this section, you will start analyze an existing ACL and then modify it to perform the desired behavior, if required.

Assume that the following ACL is applied outbound to a VLAN interface, and a switch is now comparing packets being routed out that VLAN against the ACL.

```
10    deny    10.1.1.31  255.255.255.255   any
20    deny    10.1.1.32  255.255.255.224   any
30    permit  10.1.1.0   255.255.255.192   any
40    permit  10.1.1.40  255.255.255.248   any
50    permit  10.1.1.48  255.255.255.248   any
```

You want the switch to permit only the packets with the following source IP addresses:

- 10.1.1.1 to 10.1.1.30
- 10.1.1.40 to 10.1.1.55

10.1.1.0 is the network address in the case of the first address range specified above, so it is okay if the rule permits packets from this address. The switch will not route packets from it in any case.

This task can be performed by taking the following steps:

1. Determine the current behavior of the ACL.
2. Compare it with the desired behavior to check if any modifications are required.
3. Make the necessary changes or write a new ACL to get the desired behavior.

Let us perform these steps one by one.

Step 1: Determine the current behavior of the ACL

Let us start by analyzing the current behavior of the ACL, as shown in **Figure 9-13**.

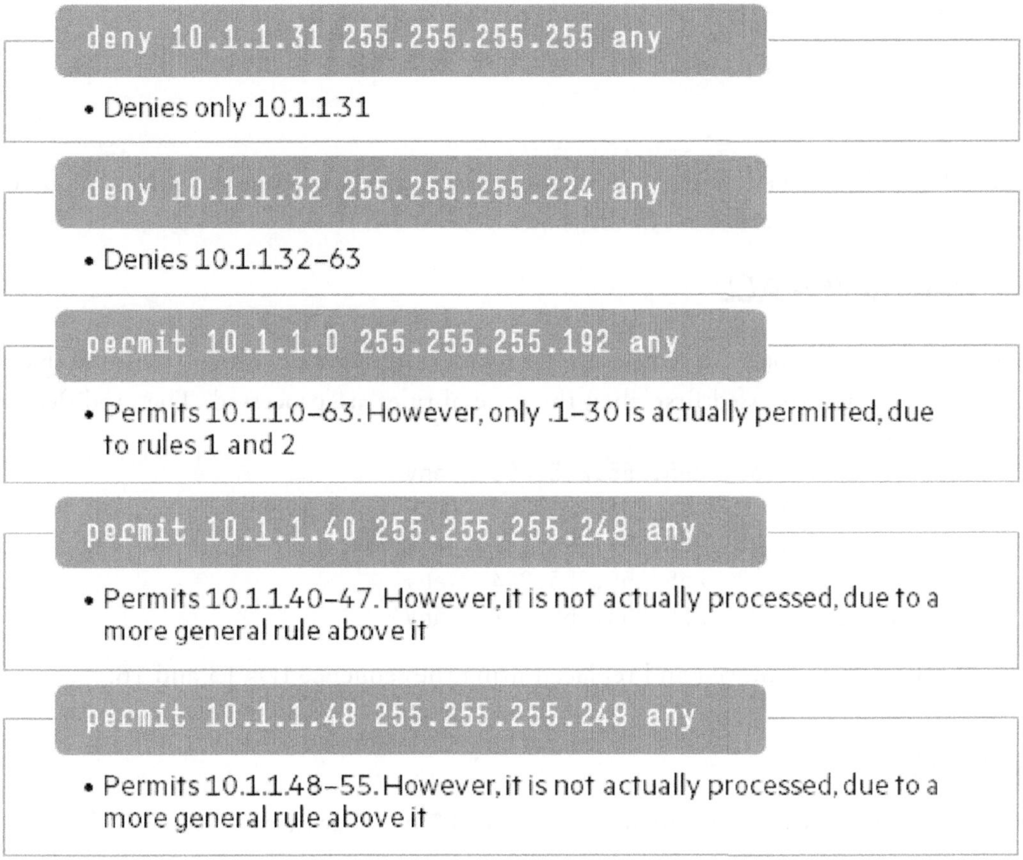

Figure 9-13: Analyzing the current behavior of the ACL

- Rule 1 is `deny 10.1.1.31 255.255.255.255 any`. It only denies access to source IP 10.1.1.31.
- Rule 2 is `deny 10.1.1.32 255.255.255.224 any`. On conversion, this provides an address range from 10.1.1.32 to 10.1.1.63. And as per the rule, all the traffic from these source IPs is denied.
- Rule 3 is `permit 10.1.1.0 255.255.255.192 any`. This rule permits all IP addresses from 10.1.1.1 to 10.1.1.63. But, because of the first two rules, the effective range of addresses that are permitted is from 10.1.1.1 to 10.1.1.30.
- Rules 4 and 5 act on a subset of addresses already covered in the more general rules listed previously. Thus, these two rules are not processed.

Step 2: Compare the current behavior with the desired behavior

The desired behavior is:

- Permit 10.1.1.1–10.1.1.30
- Permit 10.1.1.40–10.1.1.55
- Deny all others

Although rules 2, 4, and 5 support the desired behavior, they are being overshadowed by the more general rules that occur before them in the sequence.

Step 3: Rewrite the ACL

The desired behavior can be achieved by simply reordering the rules in the ACL such that the more specific rules are processed first, then the general rules are processed. The modified ordering is shown as follows.

```
10    deny    10.1.1.31  255.255.255.255   any
13    permit  10.1.1.40  255.255.255.248   any
16    permit  10.1.1.48  255.255.255.248   any
20    deny    10.1.1.32  255.255.255.224   any
30    permit  10.1.1.0   255.255.255.192   any
```

Rules 4 and 5 have been removed and replaced using the sequence IDs 13 and 16.

Applying ACLs

As you have explored earlier, you can apply ACLs to L2 or L3 ports or LAGs, VLANs, or as interface VLANs. You can apply ACLs inbound to filter the traffic entering the switch, outbound to filter the traffic leaving the switch or both. An outbound ACL cannot filter the traffic sourced from the switch itself. In this section, you will explore some scenarios that focus on each of these ACL types—both inbound and outbound.

Inbound and outbound ACL application

You know that ACLs are created globally and applied on an interface. An ACL does not take effect until it is applied on an interface. An ACL can be applied inbound or outbound on an interface. Inbound application acts on the traffic entering the switch and outbound application acts on the exiting traffic.

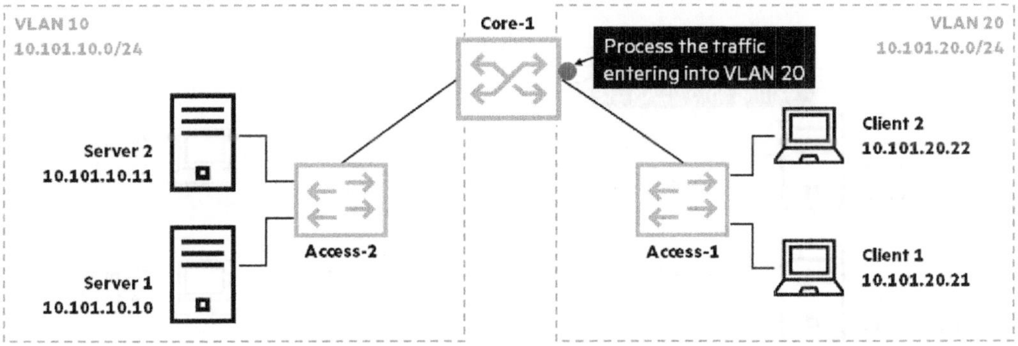

Figure 9-14: Inbound ACL application

Assume that an ACL is applied on an interface of Core-1 of **Figure 9-14**. There is going to be traffic entering and exiting different interfaces of this device. When deciding whether to apply the rules of the ACL on a particular traffic, always consider the direction of the traffic with respect to the device at that interface. For example, to process the traffic from Clients 1 or 2—the devices in VLAN 20, destined towards the devices in VLAN 10—there are two options as follows:

- An ACL applied inbound on VLAN 20
- An ACL applied outbound on VLAN 10

The decision of whether to apply an ACL inbound or outbound depends on multiple factors. If the aim is to restrict the traffic coming from a specific VLAN, the ACL should be applied inbound on that VLAN. If there are multiple VLANs and the restrictions have to be imposed only on a specific VLAN, then it is advantageous to apply the ACL closer to the source. This would limit the amount of traffic that needs to be processed, resulting in lower resource consumption.

In the example in **Figure 9-14**, the goal is to only allow Client 1 from VLAN 20 to reach other hosts/servers outside the VLAN. Therefore, it is apt to apply an ACL inbound to VLAN 20. Such an ACL will process the traffic coming from Clients 1 and 2, but only allow the desired traffic to pass.

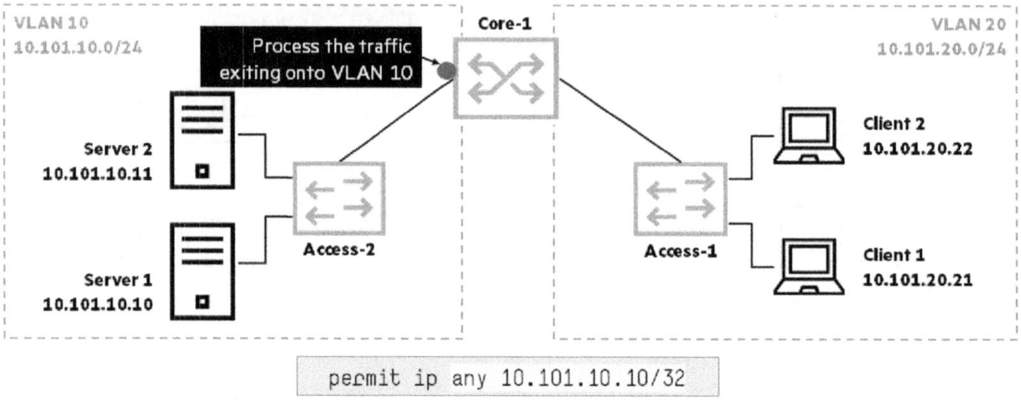

Figure 9-15: Outbound ACL application

In some scenarios, there could be a need to restrict access to a specific server or a host. For example, as illustrated in **Figure 9-15**, consider that you want outside devices to only access Server 1 but not Server 2. One way to achieve this is applying an inbound ACL on all the VLANs apart from VLAN 10, and permitting all the devices access to Server 1 and denying access to Server 2. These ACLs would process all the traffic generated from the clients from those VLANs irrespective of whether it is targeted towards VLAN 10 or not. This will result in unnecessary processing of traffic by the switch and additional resource consumption.

It would be more beneficial to just apply an outbound ACL to VLAN 10 instead. This ACL would process any traffic that exists onto VLAN 10; that is the traffic which is targeted towards Server 1 or Server 2. And then, based on the permit and deny rules created in the ACL, only access to Server 1 will be permitted, as desired.

In summary, you should consider the following factors to determine whether to use an inbound or an outbound ACL solution. Analyze the possible traffic that would get processed by an ACL, and if it can be minimized, when deciding whether to apply the ACL inbound or outbound. The advantage of applying an ACL inbound is that you are dropping the traffic closer to the source, thus less processing has to be done by the switch. However, if you have many interfaces but only one exit interface, like an uplink, it might be easier to manage the application of the ACL by applying it outbound on the uplink interface. In addition, applying an ACL outbound on an uplink will consume fewer resources than applying it inbound on a multitude of ports.

In addition to appropriately choosing the direction for ACL application, it is important to specify the source and destination addresses correctly. Swapping the positions of the addresses or including/excluding any host address may result in an undesired output. In the example in **Figure 9-15**, the rule must specify the addresses as follows to achieve the desired goal of allowing traffic from "any" source to "Server 1":

```
permit ip any 10.101.10.1.0/32
```

ACLs on VLANs

As discussed earlier, the ACLs applied to VLAN interfaces do not filter the traffic that flows within the same subnet. When you want to control the traffic switched within a subnet, you must apply a different type of ACL. An ACL applied to a port filters both the routed and the switched traffic and meets this requirement. However, when you want to control all the devices in a VLAN in the same way, an ACL applied to the VLAN is the most efficient mechanism. One such ACL applies to all the ports that are members of the VLAN. It filters all the IP traffic entering the switch on the specified VLAN interface. In a multinetted VLAN—a VLAN with multiple subnets—this includes the inbound IP traffic from any subnet.

The switch does not require an ACL rule to permit the DHCP traffic that it itself relayed. However, if the ACL was applied to a switch operating at Layer 2 and not implementing DHCP, then you have to permit DHCP by allowing ports 67 and 68 to and from any address.

Inbound application

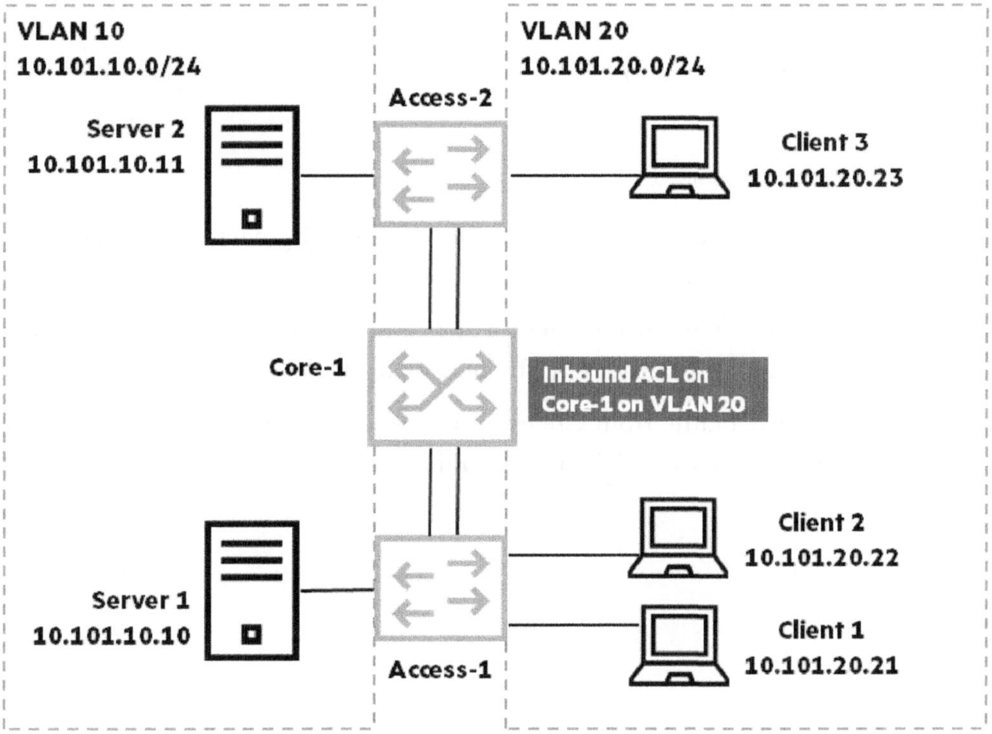

Figure 9-16: Inbound ACL on a VLAN

An inbound ACL applied on a VLAN processes the following traffic:

- Traffic received on this VLAN interface and routed out another subnet. In the example in **Figure 9-16**, this is the traffic from clients to servers.
- Traffic received on this VLAN interface and forwarded in the same subnet. This is the traffic between the clients.
- Traffic received on this VLAN interface and destined to this network infrastructure device.

An inbound ACL applied on a VLAN does not process the following traffic:

- Traffic received on another interface and routed to this VLAN interface
- Traffic generated by the switch itself using its IP address on this VLAN and routed out another VLAN

An example ACL configuration for this scenario applied inbound for VLAN 20 on Core-1 is as follows.

```
access-list ip example
 permit ip 10.101.20.23/32 any
 exit
vlan 20
 apply access-list ip example in
```

This ACL is applied inbound on VLAN 20 and allows traffic originated from Client 3 to any destination. The effect of applying this ACL is as follows:

- Intra-VLAN access: Traffic from Client 3 and destined to any of the servers is allowed. No other client of VLAN 20 can send traffic outside of the VLAN.
- Inter-VLAN access: Traffic from Client 3 to other clients of the same VLAN is allowed. On the contrary, Clients 1 and 2 cannot send any traffic to Client 3.

Outbound application

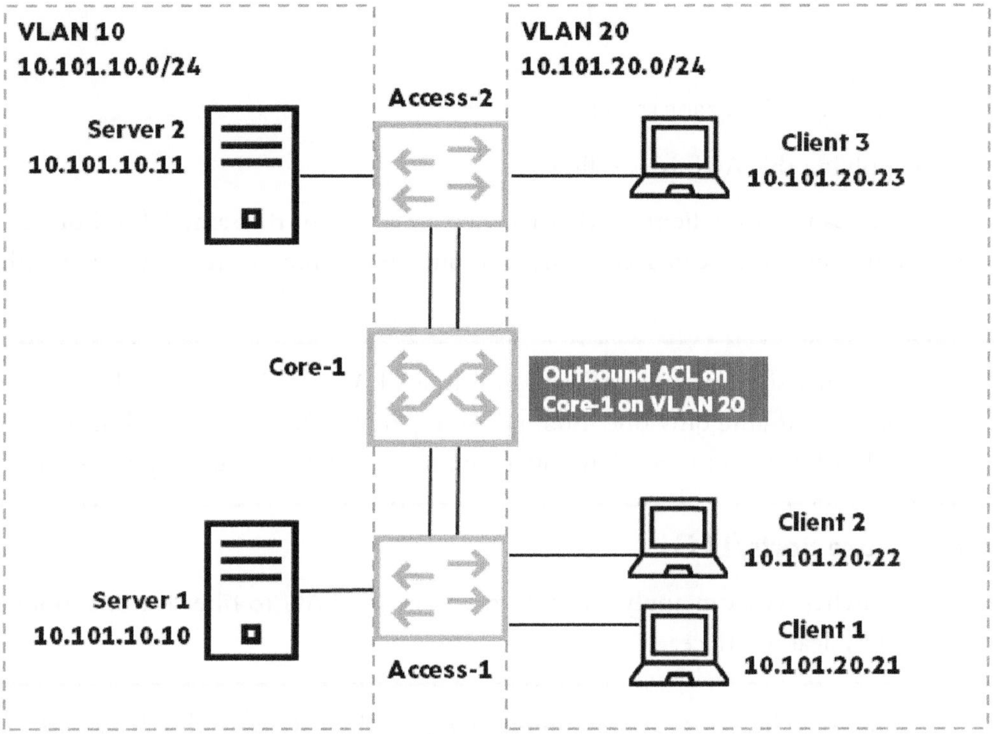

Figure 9-17: Outbound ACL on a VLAN

An outbound ACL applied on a VLAN processes the following traffic:

- Traffic received on another VLAN interface and routed out this VLAN. In the example, this the server-to-client traffic.
- Traffic received on this VLAN interface and switched out this VLAN—traffic between the clients. The ACL also processes traffic from the clients to Core-1's own IP address on VLAN 20.
- If multinetting applies, then the traffic received on one subnet in this VLAN and routed out another subnet on this VLAN.

An outbound ACL applied on a VLAN does not process the following traffic:

- Traffic received on this VLAN interface and routed out another VLAN. Here, this includes the traffic from the clients to the servers.

An example ACL configuration for this scenario applied outbound for VLAN 20 on Core-1 is as follows.

```
access-list ip example
 permit ip any 10.101.20.23/32
 exit
vlan 20
 apply access-list ip example out
```

The effect of applying this ACL is as follows:

- Any device can access Client 3, whether they are a part of the same VLAN or not. As a result, both the servers can access Client 3, but they cannot send any traffic to Clients 1 or 2.

You can assign the same ACL to multiple VLANs. AOS-CX switches, however, allow you to configure only one inbound ACL per VLAN. This is in addition to any other ACLs of different types that you assign to the VLAN or to the ports in the VLAN.

ACLs on interface ports/LAGs

On AOS-CX switches, you can apply an ACL to a port or a LAG to filter traffic inbound or outbound on that physical interface.

Some switches support only inbound application. Refer to the *ACLs and Classifiers Policy Guide* of the respective switch for details.

An inbound ACL applied to a port/LAG filters all inbound traffic regardless of whether it is routed or switched. The ACL also applies to traffic inbound on the interface with a destination on the switch itself.

An outbound ACL applied to a port/LAG filters all outbound traffic on the port regardless of whether it is routed or switched.

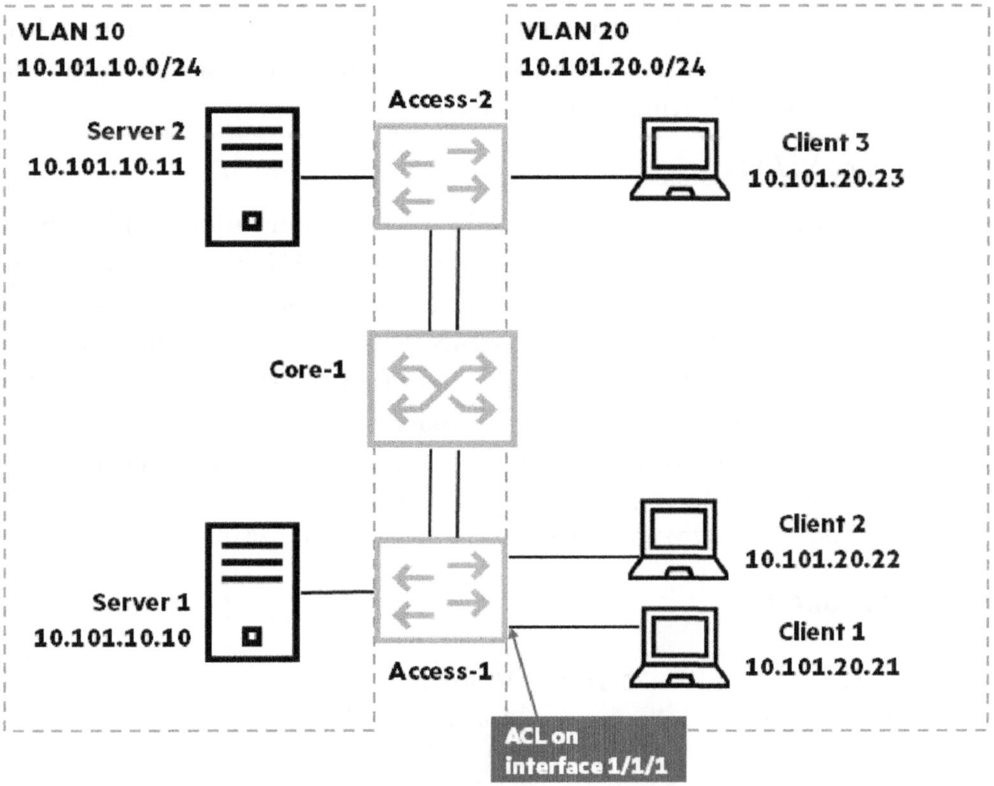

Figure 9-18: ACL applied on an interface port

The following code provides an example configuration for this scenario, applied inbound to port 1/1/1 on Access-1. The network diagram is shown in **Figure 9-18**.

```
access-list ip example
 deny ip 10.101.20.21/32 10.101.20.22/32
 permit ip 10.101.20.21/32 any
 exit
interface 1/1/1
 no routing
 apply access-list ip example in
```

This ACL filters all the traffic from the connected client, Client 1, sent to any device or address on the network. Similarly, an outbound ACL applied to port 1/1/1 would filter all the traffic destined to Client 1.

If you apply an ACL to a switch-to-switch link, such as a LAG, the ACL affects all the incoming traffic from that link. For example, an ACL applied inbound on the LAG between Access-1 and Core-1 of **Figure 9-18** would filter the traffic from Server 1, Client 1, and Client 2. When applying an ACL to the inter-switch link, you must apply the ACL to the LAG (trunk) rather than to individual ports within the LAG. It then automatically applies to the member ports. Similarly, removing an ACL from a LAG removes it from the member ports.

ACLs on interface VLANs

You should configure an ACL to control the traffic at VLAN boundaries and filter the routed traffic between the subnets. Such ACLs function at Layer 3 VLAN interfaces, and you should typically apply them on a routing switch that handles the routing for that VLAN. For example, in **Figure 9-19**, Core-1 is the default gateway for VLAN 20, which is also configured on several access layer switches. To control the traffic routed to or from VLAN 20, apply an ACL to VLAN 20 on Core-1. You can apply ACLs inbound or outbound—one ACL per direction, per physical interface (Layer 2 or Layer 3). To apply an ACL to an interface, write a command in the interface VLAN context by specifying the direction shown as follows:

```
apply access-list {ip|ipv6} <ACL-NAME> {routed-in|routed-out}
```

Inbound application

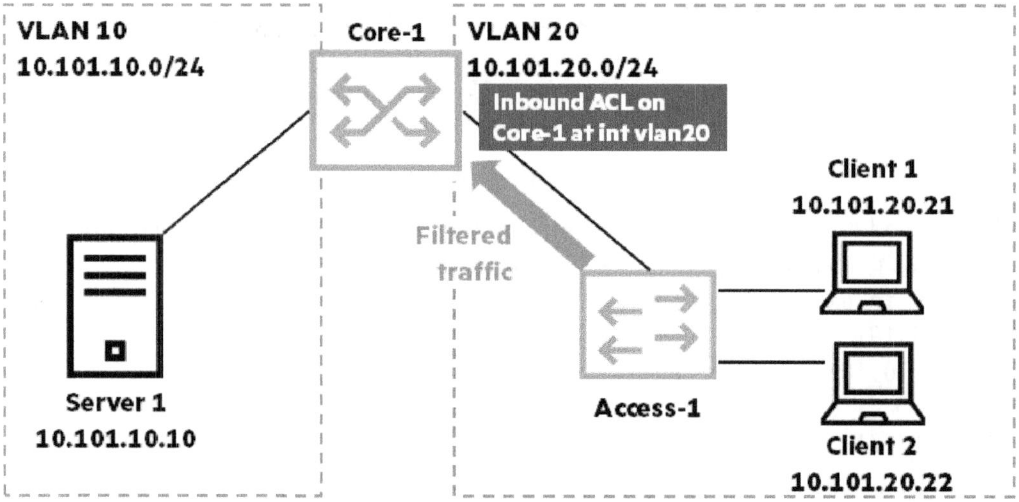

Figure 9-19: Inbound ACL on a VLAN interface

An inbound ACL on an interface VLAN processes the following traffic:

- **Traffic received on this physical interface and routed out another subnet:** Assume that you applied the ACL inbound to routed interface VLAN 20, which is the 10.101.20.0/24 subnet, as shown in **Figure 9-19**. The ACL filters packets that ingress this interface and are routed to other subnets. This includes traffic from Client 1 or Client 2 to Server 1.

- **Traffic received on this physical interface and destined to this switch:** This ACL does not filter switched IP traffic unless the switch's own IP address on the interface is the destination. For example, as shown in **Figure 9-19**, the ACL processes the traffic from Client 1 or Client 2 to all IP addresses on Core-1, including the VLAN 20 IP address. This helps you to protect the access to the switch itself.

> The ACL does not filter traffic generated on the interface by the switch itself.

An inbound ACL on an interface VLAN does not process the following traffic:

- **Traffic received on the VLAN and switched on the same subnet, except to the switch itself:** This ACL does not process the traffic between Clients 1 and 2.
- **Traffic received on other VLANs and routed out this VLAN:** This ACL does not process the traffic from Server 1 to either of the clients. That would require the ACL to be applied outbound.

An example ACL configuration for this scenario is shown below.

```
access-list ip example
 permit ip 10.101.20.21/32 any
 exit
interface vlan 20
 apply access-list ip example routed-in
```

We need to consider the specification of the rule in the ACL that acts on this traffic. In this case, the rule is `permit ip 10.101.20.21/32 any`. Considering this, out of the traffic that is processed, only the following traffic is allowed, and the rest is denied.

- The following traffic would be processed by the ACL and allowed:
    - Client 1 to Server 1
    - Client 1 to Core-1
- The following traffic would also be processed by the ACL but denied:
    - Client 2 to Server 1
    - Client 2 to Core-1
- The following traffic is not processed by the ACL:
    - Client 1 or Client 2 to any other client in VLAN 20
    - Server 1 to Client 1 or Client 2

Outbound application

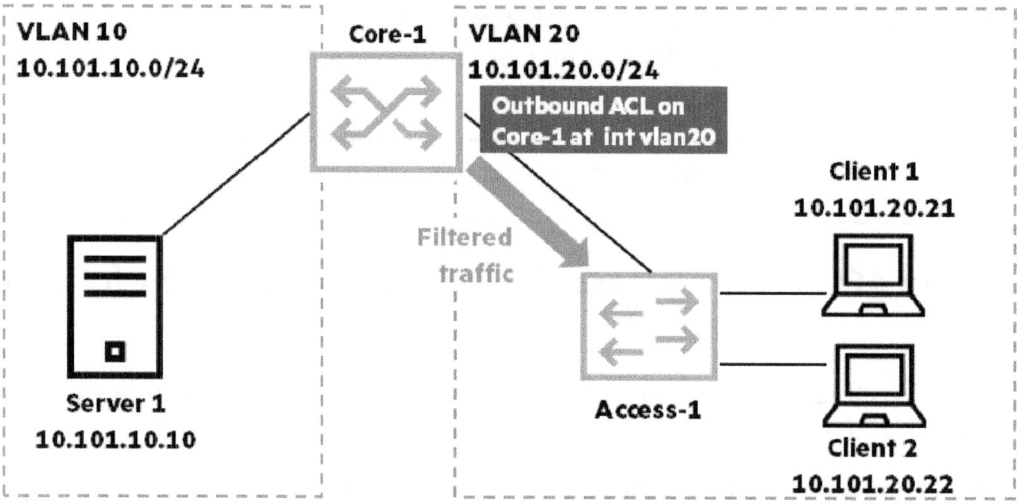

Figure 9-20: Outbound ACL on a VLAN interface

The outbound ACL on an interface VLAN filters the traffic received from any subnet and the traffic routed to a subnet on this VLAN. Assume that you have applied an ACL outbound to Core-1's routed interface VLAN 20 that connects to the subnet 10.101.20.0/24.

An outbound ACL on an interface VLAN processes the following traffic:

- The switch receives a packet from Server 1 on VLAN interface 10, and the packet is destined to Client 1 at 10.101.20.21. This ACL processes that packet because the switch routes the packet to VLAN 20.

An outbound ACL on an interface VLAN does not process the following traffic:

- Traffic generated by the switch and routed out this VLAN interface.
- Traffic received on this VLAN and routed out another subnet. That would require an inbound ACL.

> You can apply only one ACL per interface to the inbound traffic and one ACL per interface to the outbound traffic. Plan the ACL rules accordingly.

An example ACL configuration for this scenario is shown as follows.

```
access-list ip example
 permit ip any 10.101.20.21/32
 exit
```

```
interface vlan 20
 apply access-list ip example routed-out
```

Following is a summary of the traffic flow applicable to the example in **Figure 9-20**.

- The following traffic will be processed by the ACL and allowed:
  - Server 1 to Client 1
- The following traffic would be processed by the ACL but denied:
  - Server 1 to Client 2
- The following traffic is not processed by the ACL:
  - Client 1 or Client 2 to any other client in VLAN 20
  - Client 1 or Client 2 to Server 1 or Core-1

Object groups

Object groups provide a mechanism to optimize the creation and management of ACLs. In this section, you will explore the advantages of using object groups and how to implement them.

Object group overview

Object groups are useful for defining groups of IP addresses and Layer 4 ports for use exclusively in the ACL rules. Often, common groups of addresses and ports or port ranges are used repeatedly in many ACL rule definitions. With address and port object groups, the IP addresses and ports can be defined once and reused easily wherever required.

There are two types of object groups:

- IP addresses
- Layer 4 ports

Creating an object group is done using the `object-group` command. You can include many objects in a group, but they must be of the same type—either addresses and/or networks for an IP address object, or ports for a port object. Once the object groups are created, you can reference them in a rule creation command at the appropriate place. They can be used as the source and/or destination IP address or the source and/or destination Layer 4 port.

The following two code snippets show an example of creating both types of object groups and then applying them in the creation of an ACL.

Creating object groups

```
object-group port DNS_Ports
  eq 53
```

```
object-group ip address DNS_Svrs
  10.1.1.10
  10.1.1.11
```

Applying object groups

```
access-list permit_DNS
 permit tcp any DNS_Svrs group DNS_ports
 permit udp any DNS_Svrs group DNS_ports
```

A port group is preceded in the ACL with the keyword "group."

Object group configuration example

An organization wishes to allow access to its servers to all the users connecting through its internal network. In this case, it is advisable to use object groups considering the possibility of the service ports or the internal network subnets changing in the future. The current service port numbers are 80 and 443, and the internal IP networks are addressed as 172.16.0.0/16 and 10.1.0.0/16. The following code shows the required configuration.

```
object-group port Web-Servers
  eq 80
  eq 443
object-group ip address Internal-Network
  network 172.16.0.0/16
  network 10.1.0.0/16
  exit
access-list ip Web-Internal
  permit tcp any Internal-Network group Web-Servers
```

Benefits of using object groups

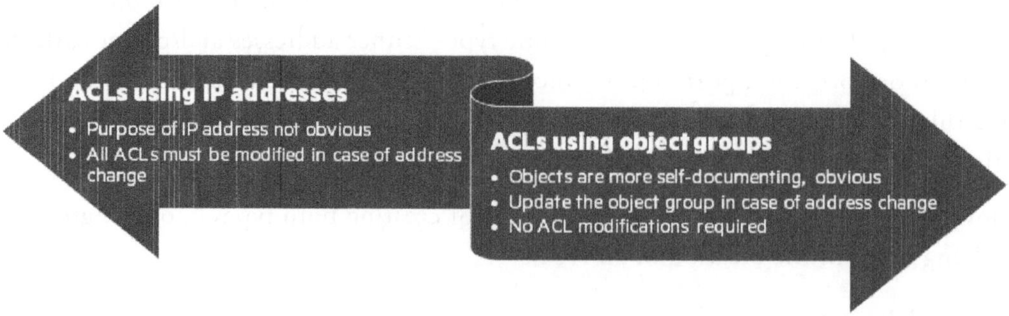

Figure 9-21: Benefits of using object groups

Figure 9-21 compares ACLs that use direct IP addresses to ACLs that use object groups. This can be further elaborated through the following example. As shown in the following code snippet, the employees in various VLANs have an ACE entry that allows them access to the internal network, which is the 172.16.0.0/16 network. Without prior knowledge, the purpose and use of subnet 172.16.0.0/16 is not obvious. You must reference other documents to learn that it is the organization's internal network.

```
access-list ip finance
  10 permit ip any 172.16.0.0/16
access-list ip sales
  10 permit ip any 172.16.0.0/16
```

Now, suppose the network grows, and you must add an additional subnet—10.1.0.0/16—to the list of allowed internal addresses. With the traditional method, you will have to modify every ACL in the organization. There could be dozens, or even hundreds, of such ACLs spread across various Layer 3 switches.

Object groups, on the other hand, are self-documenting. You can create an object group named "Internal-Net" with network 172.16.0.0/16 assigned, as shown in the following code snippet. This descriptive name makes the use more obvious to the network administrators. Later, when the network grows, you can simply add the new network—10.1.0.0/16—to the object group, and it instantly becomes active in any ACL where that object group is used. No individual ACL modifications are required.

```
object-group ip address Internal-Net
  network 172.16.0.0/16
  network 10.1.0.0/16
  exit
access-list ip finance
  10 permit ip any Internal-Net
access-list ip sales
  10 permit ip any Internal-Net
```

Classifier policies

This section describes how to use classes and policies to define very specific QoS and other traffic polices for protocols and connections at the global or interface level.

Classifier policy overview

AOS-CX switches support two feature sets related to traffic policy implementation: ACLs and classifier policies. ACLs are primarily used to filter traffic. On the other hand, classifier policies are used to implement a multitude of policies related to filtering traffic, rate limiting the traffic, changing QoS markings for DSCP or 802.1P, and traffic mirroring.

Classifier policy use cases

Use case

- **Policing traffic entering the ISP network** – classify the traffic and take predefined actions on each packet:
  - Forward the packet if the result is conforming
  - Drop the packet if the result is excess
  - Forward the packet with its precedence remarked when the result is conforming

Figure 9-22: Classifier policy use case

A typical application of classifier policies is for traffic policing applied to inspect the traffic entering the ISP networks. It can classify the traffic and take a predefined action of either forwarding or dropping a packet.

Other use cases include:

- Supervising the specification of traffic entering a network and limiting it within the specified range
- Restricting extra traffic to prevent the aggressive use of network resources by an application

Classifier policy configuration

Configuring classifier policies involves the following three steps:

1. Defining a class
2. Configuring a policy
3. Applying the policy

Let us explore them one by one.

Defining a class

The following code snippet shows how to create classes, also called "class maps," to identify the traffic to be applied to a policy. Traffic to include is indicated by a `match` command, whereas the `ignore` command indicates which traffic to ignore. The entries are processed in numerical order, so the sequencing of the entries is important.

```
switch(config)# class ip <CLASS-NAME>
switch(config-class)# [<seq-#>] match any
switch(config-class)# [<seq-#>] match tcp <protocol-criteria>
```

```
switch(config-class)# [<seq-#>] match udp <protocol-criteria>
switch(config-class)# [<seq-#>] match icmp <protocol-criteria>
switch(config-class)# [<seq-#>] match <protocol-name-or-number>
switch(config-class)# [<seq-#>] ignore <protocol-information>
switch(config-class)# [<seq-#>] comment <comment>
```

Use the `show class` command, shown as follows, to verify your configuration. Refer to the *Command-Line Interface Guide* for a more detailed description of the configuration of classes.

```
class ip MY_IP_CLASS
 10 match icmp any any
 20 ignore udp any any
 30 match tcp 192.168.0.1 192.168.0.2 eq 80
 exit
```

Configuring a policy

The following code shows how to reference classes in a policy and assign a policy action, or multiple actions, to a class. Use the `dscp` or `local-priority` parameters to define the policy. You can specify multiple classes within a policy. They are processed numerically based on each entry's sequence number.

```
switch(config)# policy <POLICY-NAME>
switch(config-policy)# [<seq-#>] class ip <CLASS-NAME> action {<REMARK-ACTIONS> |
<POLICE-ACTIONS> | <OTHER-ACTIONS>}
```

Use the `show policy` command to verify your configuration.

```
policy MY_Policy
 class ip MY_IP_CLASS action dscp EF
```

Applying a policy

Policies can be applied globally, to an interface or VLAN.

To apply a policy globally, use the following command:

```
switch(config)# apply policy <POLICY-NAME> in
```

Globally, policies can only be applied inbound. Only one policy can be globally applied at a time. Applying a policy globally again replaces the previous policy.

To apply a policy to a VLAN or an interface, use the `apply` command in the respective context, shown as follows:

- To a VLAN:

```
vlan 10
 apply policy vlan_Policy in
```

- To an interface:

```
interface 0/0
 apply policy int_Policy in
```

Classifier policy configuration example

In the example illustrated in the following code snippet, the traffic is affected when it enters inbound on interface 1/1/1. UDP traffic from 172.16.0.0/16, with a source port greater than 1023 and destined to 172.16.1.0/24, will be re-marked. All other traffic will keep its assigned DSCP value, if any. Also, any ARP packet from 00-14-22-01-23-45 is dropped, but all other ARP packets are allowed.

```
class ip IP_CLASS
  10 match udp 172.16.0.0/16 gt 1023 172.16.1.0/24
  exit
class mac MAC_CLASS
  10 match 00-14-22-01-23-45 any arp
  exit
policy EXAMPLE_POLICY
  10 class ip IP_CLASS action dscp EF
  20 class mac MAC_CLASS action drop
  exit
interface 1/1/1
  apply policy EXAMPLE_POLICY in
```

There are no implicit denies in policies. If you want to apply a policy on traffic, it must match a permit. Matching on a deny statement exempts the traffic from the policy.

Restrictions and resource utilization

This section introduces the rules and restrictions regarding the use of ACLs and the software and hardware resources that they consume.

Rules and restrictions

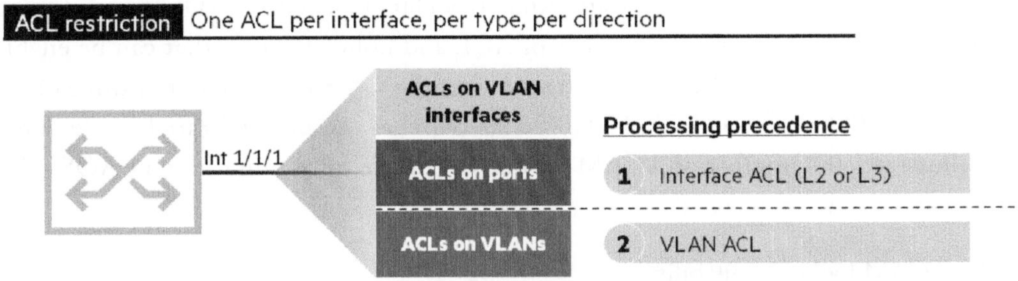

Figure 9-23: ACL rules and restrictions

Only one access list of each type (IPv4, IPv6, or MAC) can be applied on an interface or VLAN in the same direction at a time. As shown in **Figure 9-23**, you can have one—and only one—ACL inbound on a VLAN interface, and one outbound. You can also have one inbound and one outbound ACL on the physical interface (port/LAG). In addition, you can apply one inbound and one outbound ACL on a VLAN. If you attempt to apply more than one ACL on an interface, the new list replaces the old one.

For the precedence of ACLs, if the traffic is to be processed by both an interface (Layer 2 or Layer 3) ACL and a VLAN ACL, based on the VLAN the interface is associated with, then the ACLs would be processed in the following order:

1. Interface ACL
2. VLAN ACL

TCAM overview

Switching/routing ASICs have different types of tables where they store the MAC addresses, ARP or Neighbor Discovery Protocol (ND) entries, IPv4/IPv6 routes, ACLs, and policies. The switch uses Ternary Content Addressable Memory (TCAM) to perform lookups at the hardware speed and to provide an index corresponding to a given search key. All AOS-CX switches have "internal TCAMs," which are responsible for lookups inside the switch or line card ASIC.

When processing access lists, the first ACE match is performed by the CPU using a software process. However, this is cached in the ASIC's TCAM, so that all the subsequent matches can be performed quicker and with less CPU utilization in the hardware ASICs. Therefore, you must have ample switch resources to support the number of ACLs you need. When applying ACLs or policies, hardware resources are consumed to enforce the following rules on the data plane.

- Management plane resources are used when an ACL is created.
- ASIC/TCAM data plane resources are used when an ACL is applied.

TCAM lookups are a finite hardware resource used in the application of ACLs and policies to packets being processed in switch hardware. Analytics Data Collection (ADC) also consumes TCAM lookups. There are a limited number of ACL and policy features that can be enabled at the same time. At the most, eight classifier features on the CX 8320 Series, and three on the CX 8325 Series, CX 9300 Series, and CX 10000 Series switches can be configured at the same time. **Table 9-2** illustrates the number of TCAM lookups available for use in different AOS-CX switch series.

Table 9-2: TCAM lookup availability

Switch series	# of TCAM lookups available
CX 6300 Series, CX 6400 Series, CX 8325 Series, CX 9300 Series, and CX 10000 Series	9
CX 8100 Series, CX 8320 Series, and CX 8360 Series	8
CX 8400 Series	2

Resource validation

Let us look at the commands used to validate the hardware as well as software resources.

TCAM resource validation

To verify allocated TCAM resources, the `show resources` command can be used. The following code snippet shows an example of the `show resources` command, used to troubleshoot the installation of ACLs.

```
switch# show resources
Resource Usage:
Mod Description
       Resource                        Width    Used  Reserved       Free
-----------------------------------------------------------------------
1/1 Ingress Control Plane Policing
       Ingress TCAM Entries                3     231      2304
    Egress Control Plane Policing
       Egress TCAM Entries                 2      84       512
    Total
       Policers                                    0                 6144
       Ingress TCAM Entries                      231      2304       6912
       Egress TCAM Entries                        84       512       1536
       Ingress L4 Port Ranges                      0                   32
Resource data is updated every 10 seconds.
```

Troubleshooting TCAM allocation issue

In some cases, the configured ACL might not be activated due to an issue with the TCAM resource allocation. To identify the issue, follow these steps:

1. Use the `show access-list commands` command to see the active switch configuration—the configuration that is currently being used in the RAM and TCAM.

   The active configuration provides details on:

   - The ACLs that have been configured and accepted by the system
   - The interfaces on which the ACLs have successfully been programmed in the hardware

2. Use the `show access-list commands configuration` command to see the configured, but not necessarily accepted, ACLs.

   This command's output may not match what was programmed in the hardware or what is active on the switch. This may be because the configured parameters or the specified applications are unsupported, or applying the ACL was unsuccessful due to lack of hardware resources.

   - If the active ACLs and configured ACLs are not the same, the switch shows the following warning message in the output:

     ```
     ! access-list ip MY_IP_ACL user configuration does not match active
     configuration.
     ! run 'access-list TYPE NAME reset' to reset access-list to match
     active configuration.
     ```

   - If the configured ACL is processing, the switch shows the following in-progress warning:

     ```
     ! access-list ip MY_IP_ACL user configuration currently being
     processed
     ! run 'access-list TYPE NAME reset' to reset access-list to match
     active configuration.
     ```

3. If the switch shows a warning message or an in-progress message, additional changes can be made until the error message is no longer shown in the output, or you can run the `access-list {all|ip <ACL-NAME>|ipv6 <ACL-NAME>|mac <ACL-NAME>} reset` command. This command changes the user-specified configuration to match the active configuration.

Software resource validation

ACL, class, and policy definitions consume software-based resources on the management plane of a switch. The switch resource capabilities can be seen with the `show capacities` and `show capacities-status` commands. The following codes show an example of these commands for a CX 6300 Series switch.

```
6300# show capacities classifier
System Capacities: Filter Classifier
Capacities Name                                                       Value
Maximum number of Access Control Entries  configurable
in a system                                                           64000
Maximum number of Object Group Entries configurable
in a system                                                           1000
Maximum number of Object Groups configurable
in a system                                                           1000
Maximum number of Access Control Lists configurable
in a system                                                           4000
Maximum number of class entries configurable
in a system                                                           8000
Maximum number of classes configurable in a system                    1000
Maximum number of entries in an Access Control List                   8000
Maximum number of entries in a class                                  1000
Maximum number of entries in an Object Group                          64
Maximum number of entries in a policy                                 128
Maximum number of classifier policies configurable
in a system                                                           4000
Maximum number of policy entries configurable
in a system                                                           8000
Maximum number of PBR Action Lists configurable
in a system                                                           64
Maximum PBR Action List entries configurable
per PBR Action List                                                   8
```

Here is an example of the `show capacities-status classifier` command for a CX 6300 Series switch:

```
6300# show capacities-status classifier
System Capacities Status: Filter Classifier
Capacities Status Name                                          Value    Maximum
----------------------------------------------------------------------------------
Number of Access Control Entries currently configured               0      64000
Number of Object Group Entries currently configured                 0       1000
Number of Object Groups currently configured                        0       1000
```

```
Number of Access Control Lists currently configured          0       4000
Number of class entries currently configured                 0       8000
Number of classes currently configured                       0       1000
Number of policies currently configured                      0       4000
Number of policy entries currently configured                0       8000
Number of PBR Action Lists currently configured              0         64
```

The number of resources available is dependent on the switch model and will typically vary. Also, the available resource limits might increase based on the version of AOS-CX code running on the switch.

Refer to the *Command-Line Interface Guide* for details on these commands.

Learning check

Check your understanding of this material by answering the following questions.

Question 1

What are the three general components of an ACL rule?

- ○ Prefix length, IP or MAC address, and rule action
- ○ ACL type, ACL application, and rule
- ○ ACL type, subnet mask, and prefix length
- ○ Sequence number, action, and matching criteria

Question 2

An administrator wants to configure an IPv4 ACL to deny one device within a subnet (10.1.10.10) but permit every other device in the complete subnet (10.1.10.0/24). Which set of ACL entries meets these requirements?

- ○ 10 permit 10.1.10.0/255.255.255.0 any , 20 deny 10.1.10.10/255.255.255.255 any
- ○ 10 deny 10.1.10.10/255.255.255.255 any
- ○ 10 deny 10.1.10.10/255.255.255.255 any, 20 permit 10.1.10.0/255.255.255.0 any
- ○ 10 permit 10.1.10.0/255.255.255.255 any, 20 deny 10.1.10.10/255.255.255.255 any

Question 3

The following ACL is applied to Core-1. Which of the following interpretations regarding the traffic flow are valid based on the adjacent topology? (Select two.)

```
access-list ip example
 permit ip any 10.101.10.11/32
 exit
vlan 10
 apply access-list ip example out
```

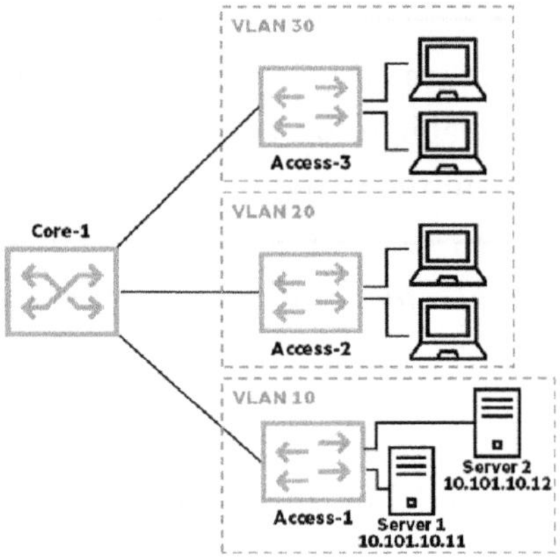

- ☐ Devices from VLAN 20 can reach Server 1
- ☐ Devices from VLAN 30 can reach Server 1
- ☐ Devices from VLAN 20 can reach Server 2
- ☐ Devices from VLAN 30 can reach Server 2

Question 4

The classifier policies contain an implicit deny which drops all the traffic which is not explicitly allowed in the policy definition.

- O True
- O False

Question 5

When the traffic matches two ACLs applied on a port and VLAN respectively, which ACL will take precedence?

- O Port
- O VLAN

10 802.1X authentication

Authentication is the process of verifying the identity of a user or a device before granting access to a network. It ensures that only authorized individuals or devices can access protected network resources, thereby enhancing security and preventing unauthorized access.

802.1X authentication is a network access control protocol widely used to provide secure network access. It is a part of the IEEE 802.1X standard for Port-based Network Access Control (PNAC) and operates primarily in wired and wireless LANs. The protocol uses the Extensible Authentication Protocol (EAP) for passing authentication messages between the client, the network device, and the authentication server.

802.1X authentication on AOS-CX switches provides a robust and secure mechanism for controlling network access. The protocol's flexibility allows for dynamic policy application, making it suitable for both static and dynamic network environments. Configuring 802.1X is simple and integrates seamlessly with existing authentication infrastructure, providing a comprehensive solution for network access control and policy enforcement.

OBJECTIVES

After completing this chapter, you should be able to:

- ✓ Understand the roles of supplicant, authenticator, and authentication server in 802.1X authentication.
- ✓ Understand different authentication services for 802.1X.
- ✓ Implement 802.1X on AOS-CX switch ports.
- ✓ Understand different user roles.
- ✓ Integrate HPE Aruba Networking ClearPass with AOS-CX switches as an external authentication server.
- ✓ Understand device fingerprinting.

Authentication overview

User authentication for network access is a security process used to verify the identity of users attempting to access a network. This process ensures that only authorized users can connect to network resources, providing a layer of protection against unauthorized access. Common uses of user authentication are as follows.

- Verifies the user identity before allowing network access
- Differentiate between interactive and non-interactive types
- Assign a role to control network access
- Optionally, generate encryption keys

The most common method is called interactive authentication, where users enter a username and password to gain access. The other method is called non-interactive authentication, where users use cryptographic certificates (digital certificates) to authenticate. These certificates are typically issued by a trusted certificate authority (CA).

After a successful user authentication, network access control (NAC) should be enforced. NAC defines rules and policies (roles) to decide who can access what resources and under what conditions. This includes Role Based Access Control (RBAC), Attribute-Based Access Control (ABAC), and other policy-based access control mechanisms.

There are different protocols and technologies available to manage authentication. Remote Authentication Dial-In User Service (RADIUS) is a protocol for centralized authentication, authorization, and accounting (AAA), commonly used in network access servers, VPNs, and wireless networks. Terminal Access Controller Access-Control System Plus (TACACS+) is similar to RADIUS but provides more granular control over authentication and authorization processes. Lightweight Directory Access Protocol (LDAP) is used for accessing and maintaining distributed directory information services, such as Microsoft Active Directory. Kerberos is a network authentication protocol that uses tickets to allow nodes to prove their identity securely. 802.1X is a standard for network access control, commonly used in conjunction with RADIUS to provide port-based access control on wired and wireless networks.

Controlling network access

When a user joins the organization, they are given an account. It holds details about their relationship to the organization and their authentication credentials. As you just learned, these credentials are often stored in a Microsoft AD or LDAP database. **Figure 10-1** shows the different steps involved in network access control.

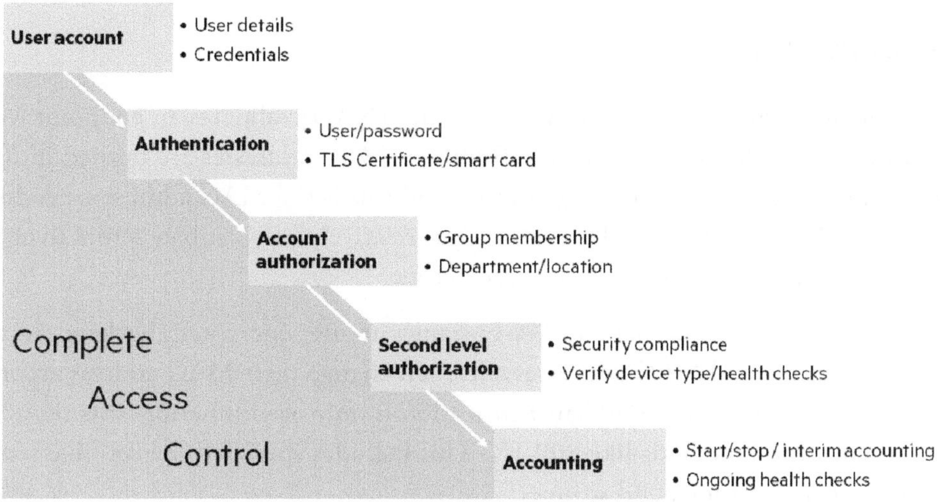

Figure 10-1: Steps involved in network access control

1. During authentication, the user provides credentials, which the network validates against the data store. This is typically done via an AAA server like ClearPass.
2. Then the network authorizes the user's account details and client device. The network may grant or deny access based on a complete picture of the client.
3. If access is granted, the system continues to monitor client activity using start/stop and interim accounting messages, with ongoing compliance checks.
4. By leveraging all available information and implementing sound access rules, ClearPass AAA-based deployment provides a complete access control system, with three common authentication methods.

Authentication methods

The authentication process validates the user or device identity to permit or deny network access. Typically, you only wish to permit authorized users with proper network credentials. A MAC address, passphrase, password, or certificate can all serve to authenticate users, with credentials preconfigured in some LDAP or AD data stores. Commonly used authentication methods are as follows:

- MAC authentication
- 802.1X authentication
- Captive portal authentication

MAC authentication

With MAC authentication, the Network Access Device (NAD) validates an endpoint MAC address against a list of valid MAC addresses. Endpoint MAC addresses are burned in ROM at the factory but are copied to RAM during bootup, and this is the MAC address an endpoint adds to Ethernet or 802.11 Wi-Fi frames. This means it is relatively easy for bad actors to change or "spoof" their MAC address to gain unauthorized access.

Since the MAC addresses in Ethernet and Wi-Fi frames are not encrypted, hackers can use packet capture software to find valid MAC addresses and then change their MAC address accordingly. Due to these security weaknesses, MAC authentication is only used for endpoints that cannot support better authentication methods like 802.1X. This includes specialty devices, like some older medical equipment, IoT sensors, and scanners.

To improve overall security with MAC authentication, use ACLs to strictly limit what devices can access. You can also use device fingerprinting to examine device protocol information like DHCP and HTTP payload information. Then use this information to identify additional information about the device, like the product, operating system, and other information.

802.1X authentication

802.1X is the most common method used to validate clients on enterprise-class networks, used in conjunction with an EAP. There are many flavors of EAP, including PEAP, EAP-TLS, EAP-TTLS, and others. These EAP types are primarily intended for use with open media such as wireless.

You can think of 802.1X and EAP as "sister protocols" that work as a team. EAP uses 802.1X to transport the user credentials between the user device and the NAD. 802.1X happens at Layer 2, where users are authenticated before an IP address is assigned. This is independent of the Layer 3 protocol. Most NAD solutions will then forward the EAP information onward to a centralized AAA server using the RADIUS protocol.

Captive portal / web portal authentication

Some systems use Layer 3 authentication, such as captive portal web authentication or VPN software. Captive portal is commonly used in guest networks to register and restrict their access. Captive portal is web-based, where users are authenticated after obtaining the IP address. Thus only requires a web browser on the user device. Given that the web is based on the HTTP protocol, only minimal Layer 3 access is required—DHCP, DNS, and restricted HTTP to perform the captive portal process.

AAA services

This is a common NAC solution, based around running AAA services on the authentication server.

Authentication: Who are you? The RADIUS or TACACS+ server permits or denies access based on your credentials. Perhaps your username is Maria, with password Secret123; if those are valid credentials, access is granted. The user might alternatively submit a digital certificate that includes a valid username and is signed by a trusted CA.

Authorization: What can you do? That is authorization. User Maria is a member of the marketing group, so she can access all standard business applications, plus the marketing application and database. Other resources are off-limits for Maria. You might define a policy to authorize Maria to be on a certain VLAN, with certain Quality of Service (QoS) settings, which the switch port dynamically applies.

Accounting: What did they do? So, we have auditing, reporting, and tracking of user activity. When and from where they logged in, when they logged out, and what resources they accessed during that time. One thing we are trying to eliminate here is "plausible deniability." If the log file indicates that you accessed an HR database at 2:00 a.m., it is difficult to deny that fact.

802.1X authentication

You have endpoints that must connect to an access switch or AP to gain appropriate network access. For tight security, these endpoints often run a "supplicant"—software that enables use of 802.1X network authentication protocol—and for wireless, in conjunction with an EAP. In this context, the switch or AP is called the Authenticator, the device that controls initial network access. In many solutions, the AP forwards these authentication messages on to its Gateway (GW), which acts as the authenticator.

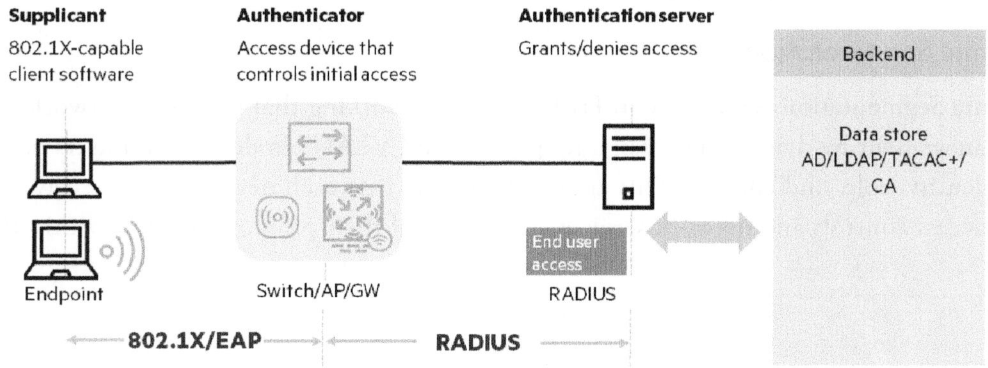

Figure 10-2: 802.1X authentication process

The switch or AP/GW harvests credentials carried in the 802.1X message, creates a RADIUS message, and sends it to the RADIUS server. RADIUS is both an authentication service running on some server and a protocol used to communicate between the authenticator and authentication server, used specifically for end user access to network services and apps.

Now the RADIUS server might have a local data store of usernames, passwords, and group memberships, especially in a smaller deployment. In this case, the server checks this data store, validates credentials, and sends "access granted(access-accept)" or "access denied(access-reject)" messages to the authenticator, which in turn notifies the endpoint.

However, for larger, more scalable deployments, you have several authentication servers all accessing a common set of backend servers that provide a common data store. This is often a Microsoft server running AD. This sets you up for single sign-on; the same credentials you use to access AD are also used to access the network itself. Nice! Some deployments might use an LDAP database, and you might also have a CA service for even tighter endpoint and end user authentication.

Port-based authentication

Port-based authentication secures the edge. You must prevent unauthorized users from connecting to open ports; this may be required for compliance with corporate security policies and government regulations. Port-based security provisions a switchport dynamically based on user/device identity.

The old way is to manually configure specific ports for specific VLAN, access lists, and QoS. It is tedious, and you must ensure that everyone connects each device to the designated port—it is error-prone. Maybe the orange ports are for accounting users, red ports are for IT administration, green is for sales, blue is for access points, purple is for printers, and so on. Instead, consider minimally configuring a collection of colorless ports that are ready to be defined base on a role provided by 802.1X port authentication.

Dynamic Segmentation

Dynamic Segmentation is a feature in HPE Aruba Networking that enhances network security and management by dynamically assigning policies and VLANs to devices and users based on their identity, role, and context. This approach ensures that each device or user receives appropriate access controls and network services, regardless of their physical location within the network.

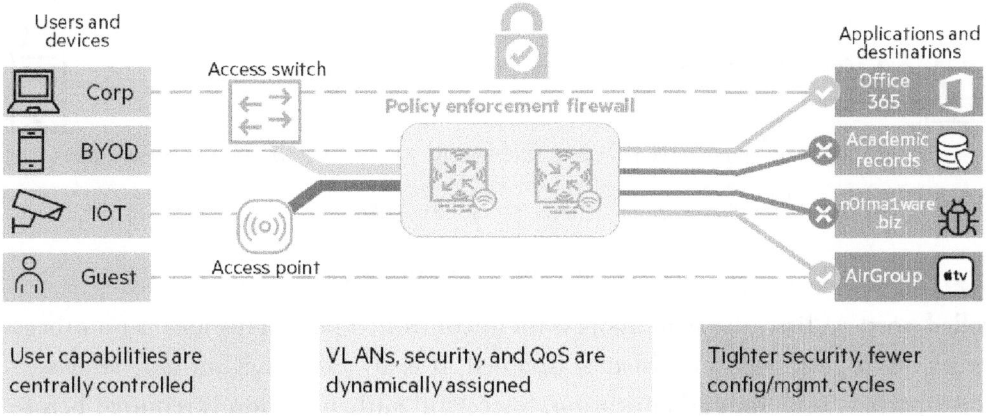

Figure 10-3: Dynamic Segmentation

Key Benefits of Dynamic Segmentation are

- Enhanced Security. Users and devices are authenticated via protocols like 802.1X or MAC authentication.
- Network policies are assigned by ClearPass based on predefined roles and contexts.
- Traffic is dynamically segmented into appropriate VLANs or tunneled to the HPE Aruba Networking Mobility Controller for additional policy enforcement.

That policy is pushed down to the NAD (switch), and a tunnel forms between that gateway device and the user's connected switch or AP, which permits and denies access to appropriate resources. Thus, user capabilities are centrally controlled, and VLAN connectivity, security, and QoS are dynamically assigned. Your network is much more secure, with far less initial configuration and fewer management cycles.

802.1X authentication overview

802.1X forces users to authenticate as soon as the connection between the switch port and the client comes up. To restrict an unauthenticated user so that the client can send only authentication messages, 802.1X divides the port into two logical ports.

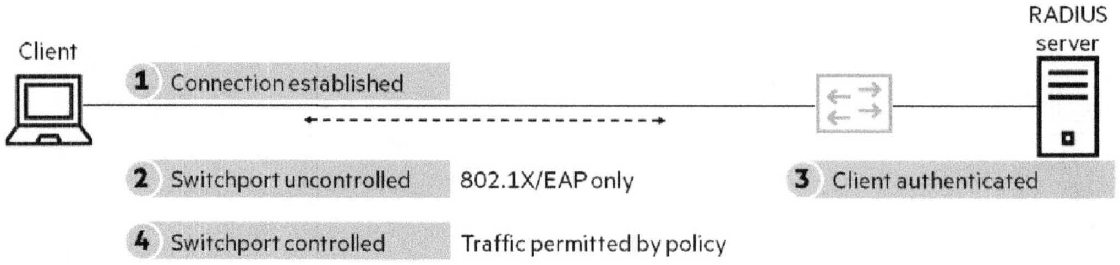

Figure 10-4: 802.1X logical ports

Uncontrolled port: At first, the switchport is an uncontrolled port. This means no process controls whether this logical port is enabled or disabled. It is always active but only allows 802.1X/EAP packets used for authentication. Successful authentication is required before the controlled port is enabled.

Controlled port: The controlled port is typically only enabled upon successful authentication. After authentication, the controlled port can accept all types of traffic, but you can control this with locally defined access lists or a centrally defined policy. Both sides of the connection control the port based on their assessment of the authentication state. In other words, no unauthorized traffic flows across the link.

This is merely a formal way to ensure that no ingress switch traffic is allowed until you authenticate, except for the 802.1X/EAP traffic required to become authenticated. The uncontrolled (non-authenticated) port does not even allow DHCP or DNS traffic. Authentication happens, and then IP addressing information is assigned.

802.1X device roles

Supplicant: This is a device or software seeking access to the network and supports the EAP. It can be a computer, smartphone, tablet, or any other network-capable device. The suplicant initiates the authentication process by providing its credentials to the network.

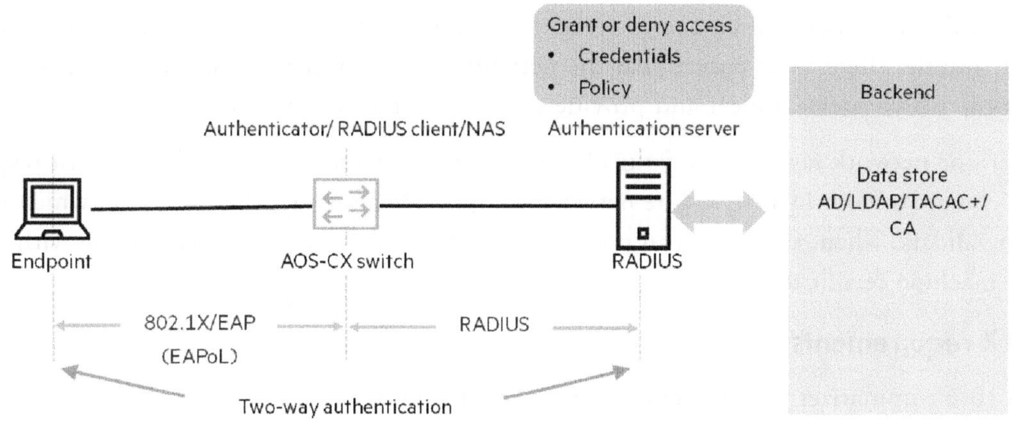

Figure 10-5: 802.1X device roles

As shown in **Figure 10-5**, the supplicant is the client's 802.1X/EAP component, authenticating in response to a challenge from the switch port. However, endpoints could receive a challenge before they are fully booted, causing authentication to time out before the user can submit their credentials. If this happens, the endpoint can initiate authentication by sending an EAP-Start packet.

Authenticator: This is the device to which endpoints connect directly. Since these access switches, access points, or firewall devices ultimately control access, they are also referred to as the RADIUS clients or Network Access Servers (NAS). 802.1X EAP messages that traverse a LAN, whether wired or wireless, are also referred to as EAP over LAN (EAPoL).

The NAS authenticator encapsulates these messages in RADIUS packets and sends them to the RADIUS authentication server. Understand that the actual authentication is between the endpoint and the RADIUS server. Think of the NAS as a mediator or translator; the endpoint speaks 802.1X/EAP, the server speaks RADIUS, and the NAS passes packets between the two.

Authentication server: This performs the actual authentication of the supplicant. This is typically a RADIUS server that validates the supplicant's credentials and informs the authenticator whether the supplicant is allowed to access the network. The authentication server primarily manages user credentials, policies, and access permissions. It communicates with the authenticator using RADIUS protocol and with the supplicant using EAP.

The authentication server can grant or deny access based on several factors:

- Credentials: username/password, user certificate, and/or machine certificate
- Policy and context: Who, what, how, where, when

For example, username John with password Secret123 attempts to log in using his corporate laptop via Ethernet connection from Corporate HQ during business hours. He is granted full access. But if he attempts to log in using his personal tablet via Wi-Fi at a café near his home at 1:00 a.m. on Saturday, he may only be granted limited access—or no access at all.

Understand that authentication can be in both directions: the network must ensure that the client is valid, and the client must ensure that the network is valid. What if a bad actor sets up a rogue network to which users connect and provide their credentials to the hacker? Bad news!

So when the network says, "Prove your identity—give me your credentials," the client responds, "First, you prove your identity!" The network can provide a server-side certificate, which the client can validate. Then, the client can provide its username/password, perhaps along with a user and/or machine certificate.

802.1X requirements

Figure 10-6 summarizes requirements for 802.1X on clients, servers, and switches.

The endpoint must have an 802.1X supplicant, which comes standard in most modern operating systems like Windows, MacOS, iOS, and others. Some organizations may prefer a third-party supplicant for additional features or personal preference.

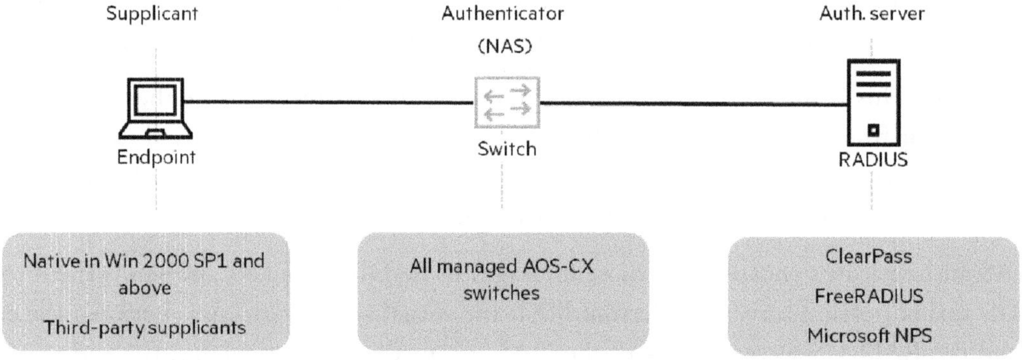

Figure 10-6: 802.1X authentication requirements

You must have an EAP-compliant RADIUS server.

ClearPass provides a RADIUS server, as well as other capabilities for monitoring and managing user access. This includes the powerful context features, device profiling, centralized policy management, and more.

You can alternatively use a third-party RADIUS server, such as Microsoft Network Policy Server (NPS) or an open-source server such as FreeRADIUS.

Authenticator (NAS) to authentication server (RADIUS) communication

Let us look at how the authenticator (NAS) relays messages between the client supplicant and RADIUS server. You must configure the NAS and RADIUS server to work together. You must configure the NAS with the RADIUS server's IP address so it knows where to send packets, and you must configure the RADIUS server with the IP addresses of acceptable NAS devices so it knows who to accept requests from. Each must be configured with a matching shared secret: a simple alphanumeric string of your choice.

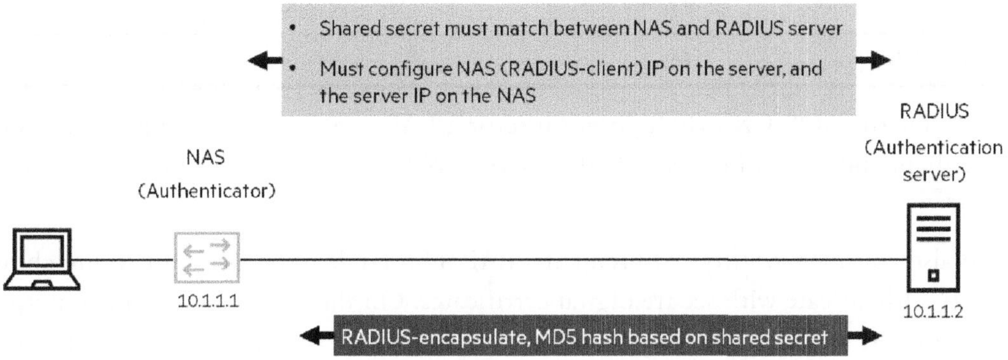

Figure 10-7: Communication between NAS and RADIUS

The authenticator encapsulates a supplicant's EAP message in a RADIUS Access-Request packet. It also hashes the packet with MD5, using the RADIUS shared secret as the key, and adds the hash to a message authenticator field. The authenticator then transmits the packet to the RADIUS server. This communication is between the NAS IP and the server IP address; the client does not need an IP address or any network access.

The RADIUS server checks that the message has arrived from a valid network device (sometimes called a NAS or RADIUS client); that is, the switch IP address must match a network device address configured on the server. The RADIUS server also checks the message authenticator field by hashing the packet with its shared secret. If the request is not from a known network device with a correct shared secret, the RADIUS server does not respond to the message—it silently discards the request.

If the switch is known and the secret is correct, the server decapsulates the EAP messages and formulates the appropriate responses, which it encapsulates in RADIUS challenge messages.

EAP methods

When choosing an EAP method, you must always consider if both the supplicant and the server support the method. EAP-TTLS tends to be less widely supported. For example, Microsoft Windows does not support TTLS natively; you must install additional software to provide this support. For this and other reasons, EAP-TLS and PEAP MS-CHAPv2 are the most popular choices.

Table 10-1: Different EAP methods

EAP method	RADIUS server certificate	Supplicant certificate
EAP-TLS	Yes	Yes

EAP method	RADIUS server certificate	Supplicant certificate
EAP-TTLS	Yes	No
PEAP	Yes	No
PEAP MS-CHAPv2	Yes	No

NAS devices like the AOS-CX switch do not interpret EAP messages; they simply encapsulate and decapsulate them and forward to the RADIUS server. NAS devices can do this for any valid method.

You should also consider security requirements. EAP-TLS tends to be the most secure choice because users authenticate with secure digital certificates. On the other hand, many enterprises prefer PEAP or EAP-TTLS because user names and passwords provide enough security for their purposes. The value of PEAP's ease of implementation is more important, especially when used with an inner method such as MS-CHAPv2, as is typical.

EAP-TLS, EAP-TTLS, and PEAP all require that the RADIUS server supply a server-side certificate, so the client can ensure that the network is valid. Many PEAP supplicants do allow you to disable this requirement, but it creates a security risk and is not recommended.

EAP-TLS also requires that each client supply a certificate. Historically, this is the reason why EAP-TLS is both considered more secure and more difficult to configure, making PEAP a popular alternative. You only need to distribute server-side certificates for a few servers, as opposed to client-side certificates for hundreds or thousands of clients.

However, modern networks now support more automated ways to distribute certificates, and so the difficulty of implementing EAP-TLS is far lower than in the past. The server uses this client-side certificate to authenticate to clients and generate key material for the secure TLS tunnel. You must install a valid certificate, signed by a trusted CA, on the RADIUS server. If you choose to use a self-signed certificate on the RADIUS server, you should install that certificate as trusted on client. Or, you could configure clients to prompt users to validate an untrusted certificate on the first connection.

You can configure local authentication for 802.1X and have the switch act as the authentication server. In this case, you would need to consider the EAP methods that the switch supports. However, this use case is less common and is not covered in this guide.

EAP-TTLS and PEAP could use TLS as an inner method, which would require a supplicant certificate. However, this is not a typical implementation.

802.1X authentication process

Let us look at the 802.1X authentication process in detail to improve your theoretical knowledge and real-world diagnostic abilities.

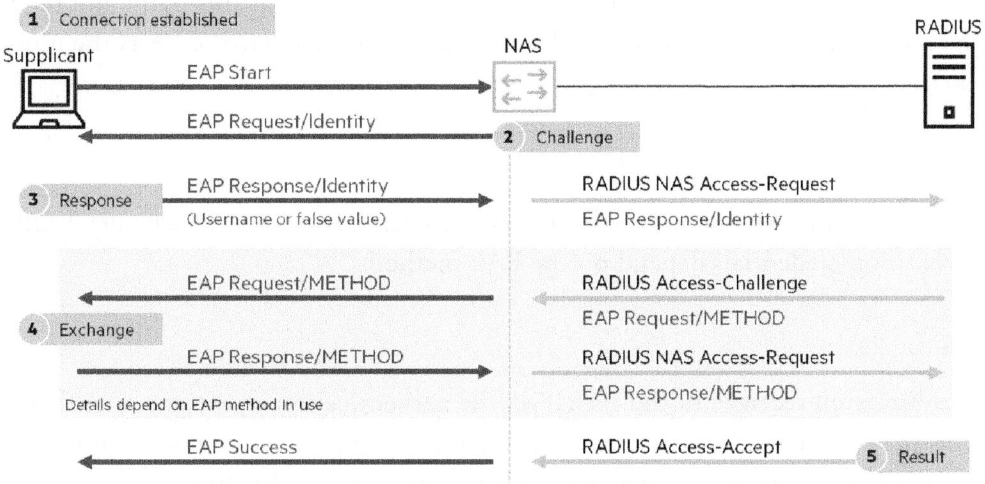

Figure 10-8: 802.1X authentication process

802.1X carries EAP messages, which are used for authentication. EAP defines a flexible framework that supports several authentication methods to meet your particular environment and policy needs. **Figure 10-8** shows this general framework. As you follow the process, remember that the NAS authenticator relays all messages from the client to the RADIUS server, encapsulating them with RADIUS as necessary. The vertical line underneath the switch shows the point at which the authenticator encapsulates the messages.

Step 1

The client and switch port establish an Ethernet connection.

Step 2

The switchport is controlled, blocking all traffic except EAP, and sends an EAP Request/Identity challenge. This initiates authentication and asks the client only to identify itself—do not send any other information.

Step 3

The client responds with an EAP Response/Identity packet, which typically includes a username. This is encapsulated in a RADIUS message and forwarded to the RADIUS authentication server. The NAS authenticator includes the user's identity in all future frames so that the server can keep track of which EAP messages belong to which user. The user's identity also marks any accounting frames for the connection, important for companies with rigorous auditing requirements. Depending on the EAP method, though, the username submitted at this point might not be the actual username. Some methods have the supplicant submit a false name at this point and the true name later in the process after a secure, encrypted session is established. This protects the username from hackers.

Step 4

The server initiates an EAP method for exchanging credentials. Message exchange details, flow, and authentication credentials depend on the EAP method.

Step 5

Based on information received in the exchange, the authentication server grants or denies access and informs the switch. The switch then either activates the connection and transmits an EAP-Success or leaves the connection deactivated and transmits an EAP-Failure.

802.1X with EAP-TLS

As mentioned earlier, step 4 in the EAP/RADIUS process is where users prove their identity. The server's EAP Request/METHOD packet starts the process and indicates the method that the server requires. You can configure most RADIUS servers to select methods according to conditions such as user identity and location.

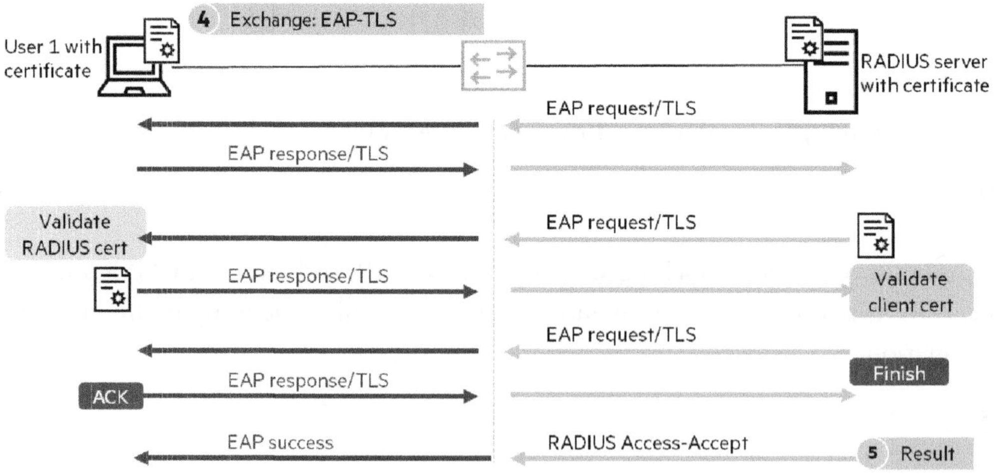

Figure 10-9: EAP-TLS process

If the station supports the requested method, it continues the exchange. Otherwise, the station sends an EAP NAK packet, which can suggest a different method. The server, if it supports the alternative method, may then initiate the exchange with the new method. The station and the RADIUS server must agree upon the method for the process to continue.

The server sends their certificate to the user, which validates it against a local store of acceptable certificates. The EAP request/TLS and EAP response/TLS packets include information such as:

- Digital certificates and certificate verifications
- Supported encryption suites
- Values for generating encryption keys (not the keys themselves)

The digital certificate includes the device's identity, and it is signed by a CA. The RADIUS server validates the certificate submitted by the user, and the supplicant validates the RADIUS server's certificate. Validation occurs through asymmetric encryption with public and private keys, which is beyond the scope of this guide. The validation process also includes checking the CA signature, so the certificate must be signed by a CA that the device is configured to trust.

After the server validates the user certificate, it sends a message to indicate a successful finish, and the client responds. Recall that after the EAP exchange successfully completes, step 5 occurs: the RADIUS server sends a RADIUS Access-Accept.

Advantages of EAP-TLS method

EAP-TLS is one of the most secure EAP methods because it provides mutual authentication with Public Key Infrastructure (PKI) digital certificates, ensuring both parties are authenticated before communication. It encrypts the authentication process, protecting the credentials and data exchanged during authentication.

Digital certificates are verified by reliable CAs using asymmetric keys that are incredibly strong. It is always preferable to use digital certificates for authentication rather than static passwords, which have a greater chance of being compromised. The digital certificate is typically stored on a laptop or a smart card, providing stronger security. Although you can set up requirements to force users to create "stronger," and therefore more secure, passwords, you cannot prevent users from telling people their password or writing it on a paper displayed in plain sight.

As a widely accepted and standards-based protocol, EAP-TLS ensures compatibility across various devices and network equipment from different vendors.

Hackers often run dictionary attacks on passwords, phish for passwords, or use other social engineering techniques to get users to leak their passwords. Although no credentials can be totally secure—malware might be able to steal certificates, for example—digital certificates are more resistant to many common attacks.

Concerns of EAP-TLS method

EAP-TLS is not vulnerable to eavesdropping and replay attacks that affect less secure EAP methods such as EAP-MD5. Some concerns include the following:

- EAP-TLS requires PKI to issue, distribute, and manage digital certificates for all clients and servers, which can be complex and resource-intensive.
- Setting up a PKI and configuring EAP-TLS can be more challenging compared to simpler authentication methods. Security comes at the cost of purchasing and managing the digital certificates—substantially more expensive than managing passwords.
- Maintaining a large number of certificates requires specialized software and trained IT staff.

In addition, in environments that have multiple operating systems, installing certificates on all clients can be difficult. Bring Your Own Device (BYOD) environments can make managing certificates even more difficult. However, ClearPass BYOD onboarding capabilities can help companies surmount these challenges. ClearPass Onboard can provide new user devices with limited access and helps users.

802.1X authentication with EAP-PEAP

The following figure starts at step 4 of the 802.1X authentication process, this time showing a PEAP MS-CHAPv2 exchange.

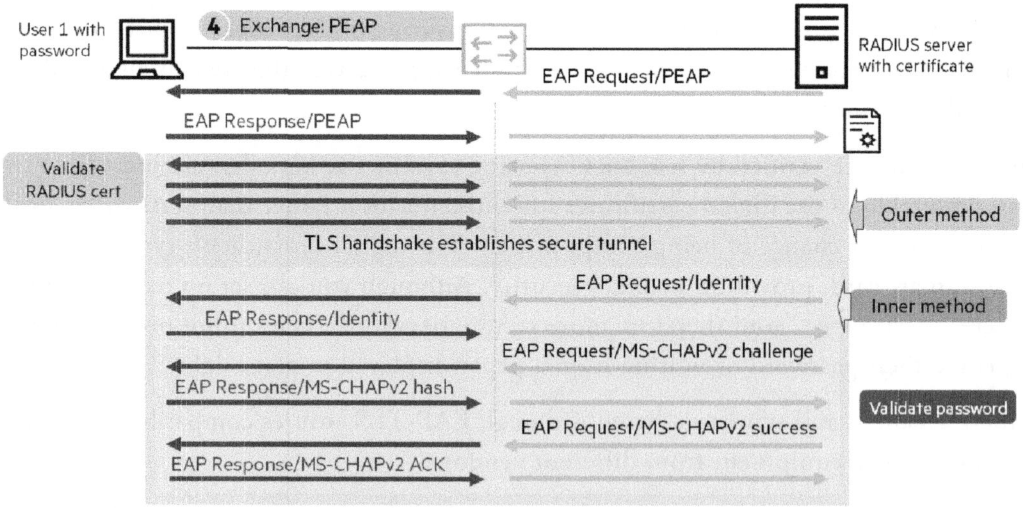

Figure 10-10: The EAP-PEAP process

Step 1: outer method

Like EAP-TLS, TTLS and PEAP use a TLS handshake to generate encryption keys and negotiate a tunnel secured by those keys. Only the server authenticates itself with a digital certificate during this exchange, but you must ensure that the clients trust the CA that signed the server certificate. The encryption keys for the tunnel are derived from the public key in this certificate.

Step 2: inner method

The client authenticates itself in the second step. It uses a weaker—and so more easily implemented—authentication method, protected by the secure tunnel established during the outer method exchange. The RADIUS server sends a new EAP Request/Identity packet, and the supplicant responds with their identity. The server then begins the inner authentication method.

For the inner authentication method, TTLS can use a weaker EAP method, such as EAP-GTC, or a legacy RADIUS method, such as CHAP, PAP, or Microsoft CHAP variants (MS-CHAP v1 or MS-CHAP v2). PEAP supports methods such as MS-CHAP v2, EAP-GTC, and TLS. Because Windows native supplicants support PEAPv0 with MS-CHAP v2, this is by far the most prevalent EAP method explained in this guide.

The RADIUS server sends a challenge. The supplicant responds with a hash of the password. The server validates that the hash is correct, and, if so, it sends a message indicating the success. The supplicant responds. The figure shows only the PEAP part of the complete 802.1X authentication process. After these messages, the tunnel is taken down, and the RADIUS server sends a RADIUS Access-Accept to the switch.

For 802.1X on wireless connections, whether implemented through EAP-TLS, EAP-TTLS, or PEAP, the authentication server sends the authenticator the encryption material. The wireless client and authenticator continue to use encryption to secure the connection. On Ethernet connections, however, the client and switch port to do not encrypt traffic in this way.

EAP-PEAP strikes a balance between security and ease of deployment. It offers robust security through TLS encryption without the complexity of managing client-side certificates, making it a popular choice for many enterprise environments. Like EAP-TLS and EAP-TTLS, PEAP provides a strong and mutual authentication.

Advantages of EAP-PEAP method

EAP-TLS creates a secure tunnel between the client and the server, protecting the authentication process from eavesdropping and man-in-the-middle attacks. It creates an inner authentication tunnel and supports a variety of inner authentication methods like EAP-MSCHAPv2, and EAP-GTC, allowing for flexibility in the choice of authentication protocols.

Unlike EAP-TLS, EAP-PEAP only requires a certificate on the server side, which simplifies the deployment and management of certificates. There is no need for a client-side PKI, reducing the cost and complexity associated with managing client certificates. EAP-PEAP is widely supported by network devices and client operating systems, ensuring broad compatibility and ease of integration into existing networks. It allows the use of password-based authentication methods within the secure TLS tunnel, making it user-friendly and easier to deploy in environments where users are accustomed to using passwords. The ability to use various inner EAP methods (like EAP-MSCHAPv2) provides flexibility to adapt to different security requirements and user environments.

Concerns of EAP-PEAP method

The security of EAP-PEAP is dependent on the strength of the inner authentication method. For example, EAP-MSCHAPv2, commonly used within PEAP, has known vulnerabilities if not properly managed. If the server certificate is compromised, the entire authentication process is at risk.

While simpler than EAP-TLS, EAP-PEAP still requires the deployment of a server certificate, which involves some level of administrative effort and initial configuration for ensuring all clients are correctly configured to trust the server certificate, and using the appropriate inner authentication method can be challenging.

The TLS handshake process introduces some performance overhead, which might be noticeable in environments with a large number of simultaneous authentication requests. While widely supported, there can be compatibility issues with certain clients or network devices, particularly older or non-standard implementations.

Configuring 802.1X authentication on the switch

Before you enable 802.1X on a switch, you must tell the switch who to send RADIUS messages to and, optionally, how to send them. Use the syntax shown in **Figure 10-11**, which refers to the RADIUS server by IP address. You can also use the Fully Qualified Domain Name (FQDN) if preferred.

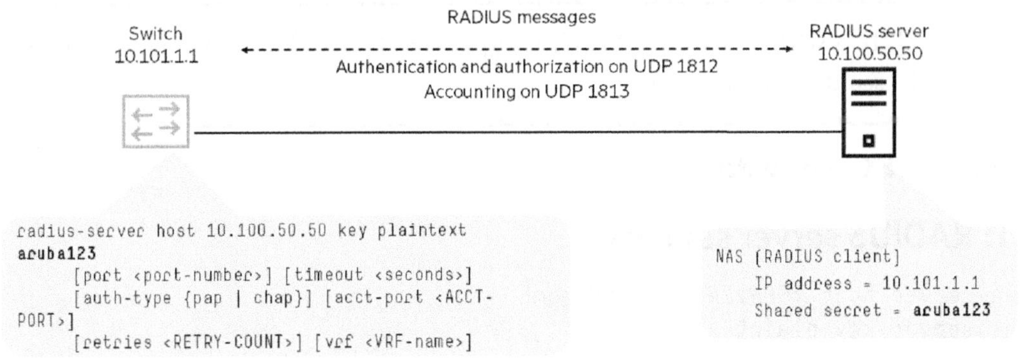

Figure 10-11: 802.1x configuration

If you define a FQDN name for the RADIUS server, you will have to define a DNS server to resolve the name to an IP address.

The switch sends a RADIUS Access-Request to this IP address whenever it needs to complete any type of RADIUS authentication. By default, the switch uses the RADIUS standard UDP port 1812 for the authentication messages. After the RADIUS server has authorized a user for access, the switch can continue to communicate information about the user session in accounting packets sent to port 1813.

You can also include the auth-port and acct-port options in the `radius-server host` command to change the authentication and accounting ports. Other options are documented in the switch *Security Guide.*

Global key or shared secret

The key option in the `radius-server host` command sets an individual shared secret for communicating with the RADIUS server. The RADIUS server has the switch IP address configured on it as a RADIUS client (sometimes called Network Access Server (NAS) or, on ClearPass, a network device). The shared secret configured for the switch on the RADIUS server must match the key configured for the RADIUS server on the switch. Otherwise, the RADIUS server will not respond to the switch's messages.

Setting the key for the individual server is optional if you set a global key. An individual key configured for a host overrides the global key, but if you do not set an individual key, the RADIUS server uses the global key. You might choose to use the global key if you have multiple RADIUS servers that use the same shared secret. Use the `radius-server key plaintext <SECRET-KEY>` command for setting a global key.

Defining a RADIUS host with a key will override the global key setting.

RADIUS with IPSec (RADSEC) is supported by AOS-CX switches. To configure this, see the *Security Guide.*

Enabling 802.1X in a switch

Step 1: RADIUS server settings

```
radius-server host <SERVER-IP> vrf mgmt
radius-server key plaintext <KEY>
```

Step 2: 802.1X global configuration

```
aaa authentication port-access dot1x authenticator auth-method eap-radius
aaa authentication port-access dot1x authenticator radius server-group radius
aaa authentication port-access dot1x authenticator enable
```

Step 3: 802.1X on an interface

```
interface <IFACE-ID>
aaa authentication port-access dot1x authenticator
enable
aaa authentication port-access allow-lldp-bpdu
aaa authentication port-access allow-cdp-bpdu
```

Add the optional configuration If you need to allow LLDP/CDP or BPDU packets on a secure port. This configuration might be necessary for devices like a phone or camera where LLDP (or Cisco's CDP) must be exchanged for the device to operate correctly prior to authentication occurring on the port.

Now the switch will act as described earlier. It will keep the port closed to all non-EAP traffic and send EAP messages to trigger an 802.1X supplicant on the connected device to authenticate. If the user succeeds in authenticating to the RADIUS server, the switch opens the port for normal traffic. If the user fails, the port remains closed.

Dynamic authorization

The following examples illustrate situations where a user has authenticated, but their authorization information must change based on what is happening in the network.

In the first example, the user authenticates and is assigned their normal VLAN. However, once the user's session times out, they should be placed in a restricted VLAN until they reauthenticate.

In the second example, a company originally deployed PEAP for 802.1X but is migrating to certificates. If a user connects and uses PEAP, they should be placed in a restricted VLAN that allows them to perform onboarding. Once the user is onboarded and has downloaded a root certificate and a new identity certificate, they should reauthenticate using EAP-TLS and be placed in their normal VLAN.

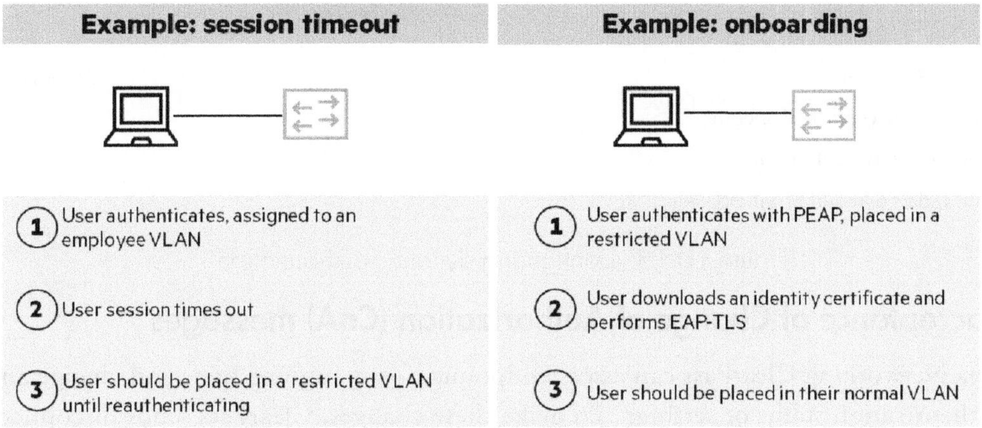

Figure 10-12: Examples for dynamic authorization

There are many examples where a company needs to change authorization information based on what devices are doing in the network; using the authorization information originally assigned to the user when they authenticated can create problems based on these network changes. RADIUS supports a Change of Authorization (CoA) that allows a RADIUS server to push new authorization information to a network device and used for a user's session.

Most networking devices ignore RADIUS CoA messages from an AAA server unless you explicitly enable it. This is true of both AOS-S and AOS-CX switches.

When you use HPE Aruba Networking ClearPass as the RADIUS server, it is recommended that you add a few settings to ensure correct operation between the AOS-CX and the ClearPass RADIUS server.

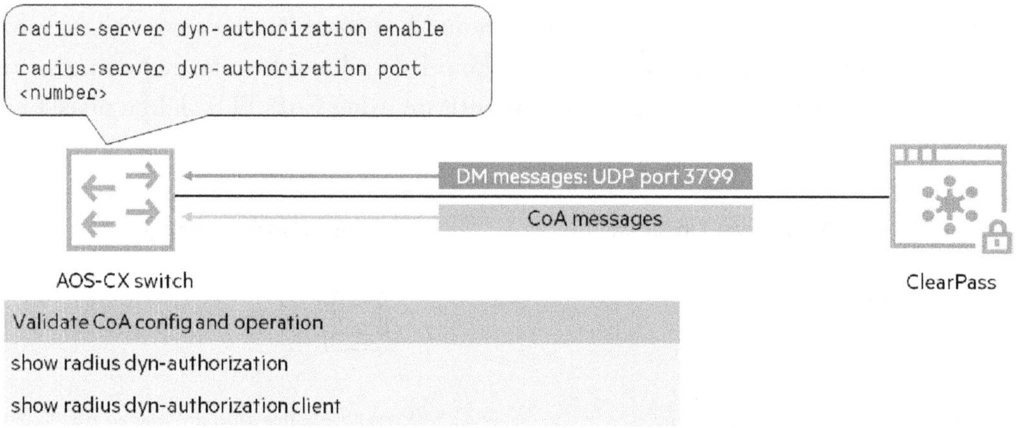

Figure 10-13: Configuring dynamic authorization

Enable acceptance of Change of Authorization (CoA) messages

HPE Aruba Networking ClearPass can assess endpoints on an ongoing basis and change an endpoint's authentication status or settings. To make these changes, ClearPass sends disconnect messages (DMs) and CoA messages. For the switch to accept these messages, you must enable dynamic authorization for the RADIUS globally on the switch, as shown in the previous figure. Without this option, some solution components might work, but others will fail.

If `dyn-authorization` is enabled, the switch accepts these messages on the default UDP port 3799, which is also the default port on ClearPass. You can change that setting, if necessary, in your environment, as shown in the previous figure.

If you are using a different RADIUS server than ClearPass, you should check whether the server needs to send DM and CoA messages.

Use the `show radius dyn-authorization` command to verify the configuration and operation of CoA for all servers, and restrict the view to one server with the `show radius dyn-authorization client` command.

Dynamic authorization time settings

When an AOS-CX switch is configured to allow CoA and DM messages, it checks these messages to determine if they are current. The message should include a time stamp for the event that caused the RADIUS server to change the user status. The switch only accepts the message if the time between the time stamp and the current time on the switch falls within its RADIUS time window.

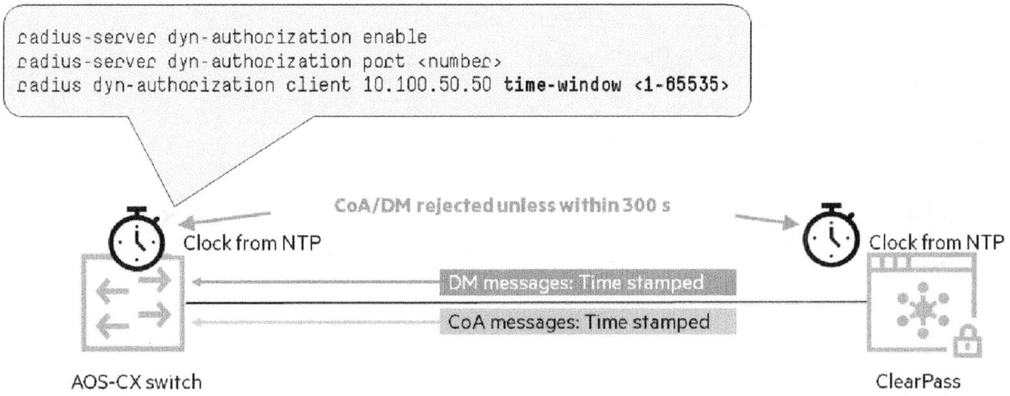

Figure 10-14: Dynamic authorization time settings

The default time window is 300 seconds. You can adjust the time window as shown in the figure, optionally specifying a VRF.

To ensure that the switch can accept the CoA and DM messages, make sure that the switch has the correct time. It is a best practice to configure Network Time Protocol (NTP) on the switch and the RADIUS server and have them take their time from the same clock.

You must also determine whether the RADIUS server always includes a time stamp with its DM and CoA messages. ClearPass does not.

Unlike the older HPE Aruba Networking switches, you cannot disable the time window check. Instead, set up NTP, as shown in the previous figure. For the explanation of these commands, see the *Fundamentals Guide.*

NTP settings on a switch

To ensure that the switch can accept the CoA and DM messages, make sure that the switch has the correct time. It is best practice to configure NTP on the switch and the RADIUS server and have them take their time from the same clock.

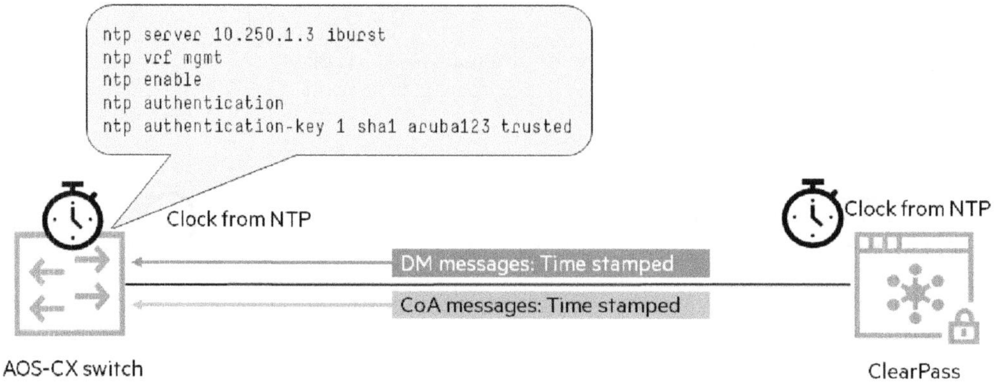

Figure 10-15: NTP settings on a switch

NTP is essential in network environments to synchronize the clocks of various devices on a network. Consistent time across devices helps in the proper functioning of network management protocols and systems, ensuring accurate monitoring and reporting.

Configuring NTP on an AOS-CX switch

Add the NTP servers that the switch should use for time synchronization, and enable the authentication.

```
ntp server 10.250.1.3 iburst
ntp vrf mgmt
ntp enable
ntp authentication
ntp authentication-key 1 sha1 aruba123 trusted
```

By following these steps, you can ensure that your AOS-CX switch is accurately synchronized with your NTP server, enhancing the reliability and security of your network.

Configuring RADIUS accounting on an AOS-CX switch

```
aaa accounting port-access start-stop group {radius | <GROUP-NAME>}
aaa accounting port-access start-stop interim <minutes> {radius | <GROUP-NAME>}
```

RADIUS server groups

Companies deploy multiple RADIUS servers to provide redundancy. The servers that you add are automatically added to the global RADIUS server group, which is called **radius**, and referred to as the **global group**. The switch uses this group, by default, for all types of authentication for which you have enabled RADIUS. Servers are used in the order they are defined.

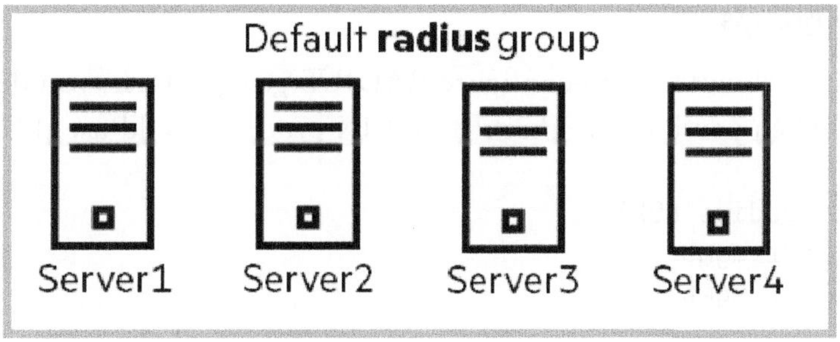

Figure 10-16: Default/global server group

You might want to use only a subset of servers for a particular task, instead of all the globally defined ones. You can do this by creating RADIUS groups—maximum 28 groups per switch. As shown in the following figure, you might create a group of servers for end user authentication using 802.1X, and another group used for administrative access to the network devices' CLI and GUI.

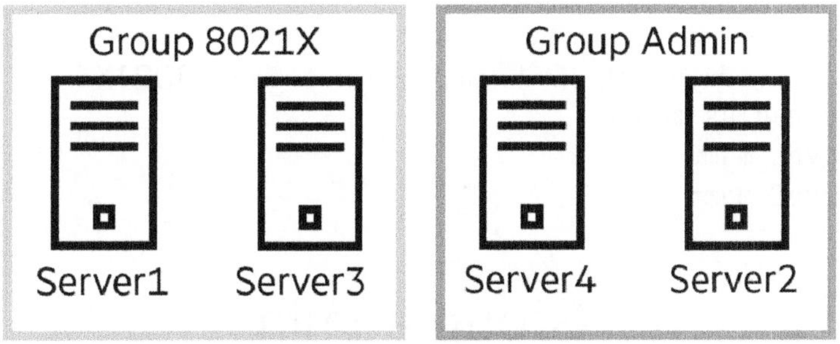

Figure 10-17: Customized server group

Configuring server group

Global server group

The following example in the figure shows how to configure RADIUS with 802.1X using the global default. The switch uses the global radius group to authenticate users who connect to ports with 802.1X enabled on them.

```
aaa authentication port-access dot1x authenticator radius server-group radius
```

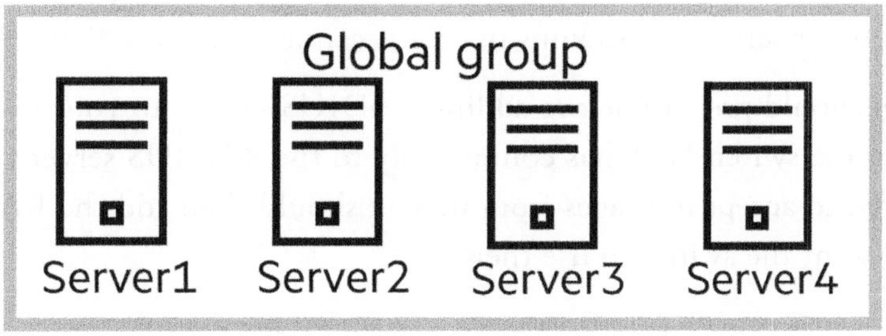

Figure 10-18: Global server group

Customized server groups

The following example shows how to configure RADIUS groups. First, you define the group and its member servers. Then you indicate that the 802.1X process should use this group to authenticate users.

```
aaa group server radius 8021X
server 10.1.1.1 vrf mgmt
server 10.1.1.3 vrf mgmt
```

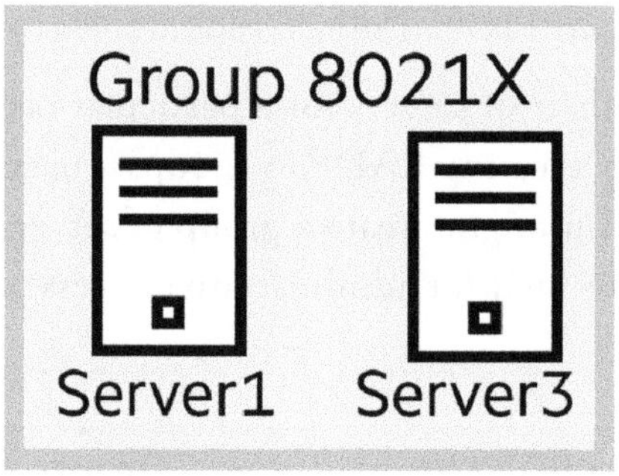

Figure 10-19: Example—server group **8021X**

```
aaa group server radius admin
server 10.1.1.2 vrf mgmt
server 10.1.1.4 vrf mgmt
```

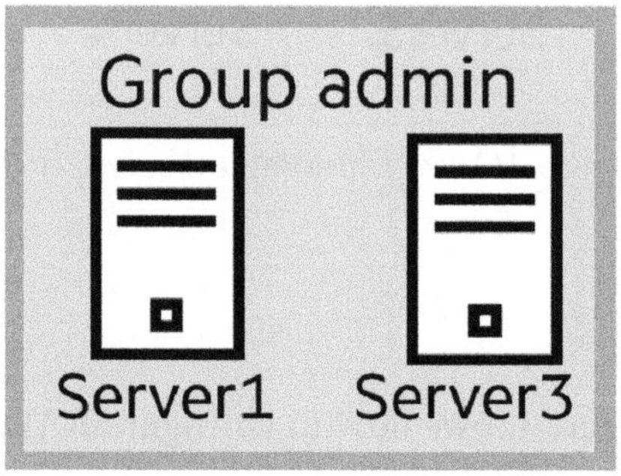

Figure 10-20: Example—server group **admin**

When a RADIUS server group has multiple servers, the switch does not load balance requests to different servers. Instead, it sends all requests to the first server in the group that it successfully accesses. It uses the other servers as backups in case it cannot reach the first server.

For this reason, you should prepare before adding RADIUS servers to your switch and enable 802.1X. Ensure that the switch NAS has connectivity to the RADIUS servers, and the RADIUS servers are configured to accept messages from it. You should then add the RADIUS servers in the order in which you want the switch to use them.

However, you can configure the switch to wait a set period after the last failed retransmit to a particular server before attempting to communicate with that server again. You might do this to prevent a switch from endlessly attempting to send RADIUS packets during a network outage. This period is called dead time.

These are global settings that apply to all RADIUS servers and groups. You can adjust the timeout (in seconds—5 is the default) and the number of retransmits (1 is the default), and the dead time (in minutes) with these commands:

```
Switch(config)# radius-server timeout <1-60>
Switch(config)# radius-server retries <0-5> fail-through
```

By default, if a RADIUS server is reachable and the user fails authentication, the switch uses this result. However, if the user is a visitor from a different location, like a neighboring campus, and their credentials are not stored on the local campus server but the remote campus server, the user would be denied access by default.

You can enable the fail-through feature, so that if the first RADIUS server is reachable and returns a "failed" result, the switch can proceed and try other RADIUS servers in the list or group. This feature is disabled by default but can be enabled with this command:

```
Switch(config)# aaa authentication allow-fail-through
```

Load sharing

For rudimentary load sharing, specify different servers first on different switches, as shown in the following figure. So, the default timeout is five seconds and the default retransmit setting is 3. After 20 seconds, the switch determines that the RADIUS server is not available. The switch then tries to reach the next RADIUS server in the list. If the switch cannot reach that RADIUS server either, after 20 seconds, it tries the next server on the list. If the switch cannot reach any of the servers, it tries the first server in the list again.

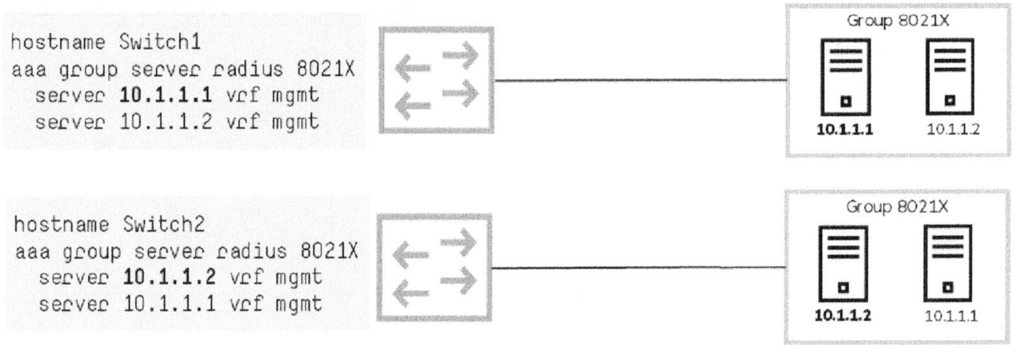

Figure 10-21: Load sharing example

However, as you will learn in other chapters, AOS-CX switches can use RADIUS with other types of authentication such as captive portal, MAC authentication, and the authentication of Telnet, SSH, or web UI managers.

To use different RADIUS servers for different types of authentication, you must place those servers in different RADIUS server groups; each group can have up to three RADIUS servers.

A RADIUS server can be part of more than one group. If, for example, you add a server that is part of the radius server group to a new group, the server remains part of the **radius** group, as well. Therefore, you should define the servers that you want to use for the **global** group before adding the servers for the other groups. Otherwise, the servers that you intend for other groups will take up the slots in the global radius group.

Remember, to use the RADIUS server group, you must choose the type of authentication to which it applies:

```
Switch(config)# aaa authentication port-access eap-radius server-group <GROUP-NAME>
```

RADIUS attributes for the dynamic settings

RADIUS attributes play a crucial role in the authentication, authorization, and accounting processes. They enable granular control over user access, and enforce policies dynamically. By leveraging RADIUS attributes, organizations can apply dynamic settings on a switch, enhance their network security, optimize resource allocation, and ensure compliance with usage policies.

Colored ports versus colorless ports

In the context of AOS-CX switches, "colored" and "colorless" ports refer to specific configurations that determine how the switch handles port settings, such as VLAN, ACL, QoS, and so forth, for the network endpoints, particularly in environments with dynamic configurations and automation needs.

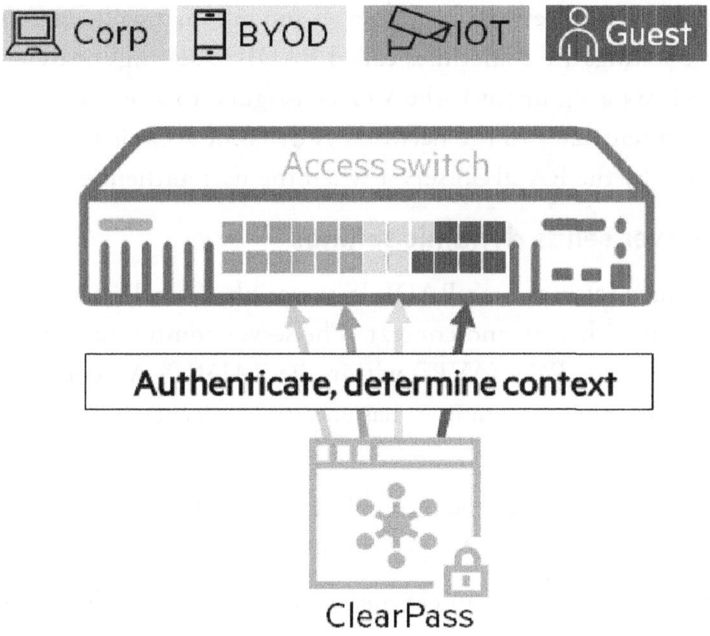

Figure 10-22: Example of colored ports

Colored ports have specific, predefined configurations applied to them. This configuration includes VLAN assignments, QoS policies, security settings, and so forth. These ports are typically used when the network administrator wants to ensure that certain ports always have a consistent configuration regardless of the connected device. They are particularly useful in static network environments where the configuration does not change frequently.

Colorless ports, on the other hand, do not have predefined configurations. Instead, their configuration is dynamically assigned based on the device or user connecting to the port. This is achieved through mechanisms such as authentication methods (for example, 802.1X, MAC authentication), User-Based Tunneling (UBT), and dynamic VLAN assignment.

Beyond the security benefits, port-based 802.1X authentication helps you to provision the network edge more quickly. You learned about the old days, when you had to configure each port with the appropriate settings for the intended user—tedious and error–prone.

With 802.1X, the RADIUS server and NAS switch automatically determine user identity as well as other context characteristics—who, what, how, where, when. A RADIUS server policy can use this information to automatically configure appropriate settings: VLAN, ACLs, QoS/CoS, rate limits, and more. The network edge becomes more flexible and adaptable.

Because the switch applies these settings to a user session automatically, as dictated by the RADIUS server, these settings are sometimes called RADIUS settings or dynamic settings. For example, a dynamic VLAN assignment is the VLAN assigned to a user session by the RADIUS server when the user authenticates to the network. A dynamic ACL is the (port-based) ACL applied to a user session by the RADIUS server when the user authenticates, and so on.

How the RADIUS server sends dynamic settings

You define dynamic policy settings on a RADIUS server like ClearPass. These settings apply to certain users based on their identity and context. The server communicates these settings to the NAS switch in Attribute-Value Pairs (AVPs) within the RADIUS-Accept message. They are called Attribute-Value Pairs because each data item has an attribute type and an associated value. The figure shows three AVPs:

- Attribute Type 64 defines the type of medium over which this communication occurs, with a value of 6, which means an 802-type "tunnel".
- Attribute Type 65 defines the tunnel type, which is set to 13. This is used to indicate that we intend to assign a VLAN ID.
- Attribute Type 81 defines the actual VLAN ID—in this case, VLAN 30.

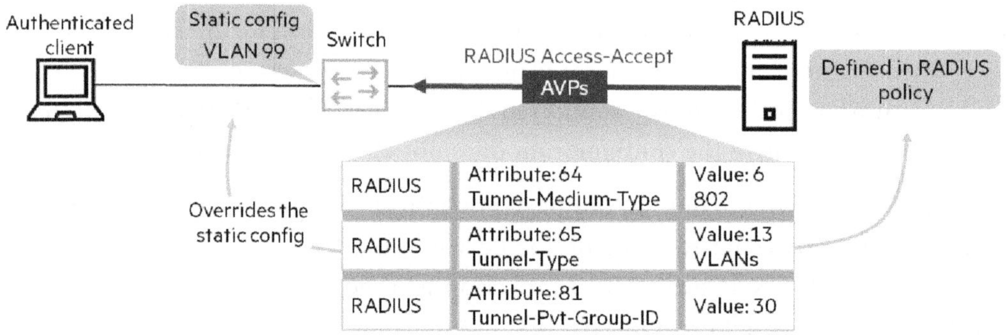

Figure 10-23: RADIUS attribute assignment

Thus, the purpose of this set of AVPs sent from RADIUS server to switch is to override the port's statically configured VLAN 99 with VLAN 30.

There are standard AVPs defined by the IETF in an RFC, and there are custom attributes that are defined specifically by a vendor like HPE. These are called Vendor-Specific Attributes (VSA). There are also extended RADIUS attributes.

When possible, you should use standard RADIUS attributes because they are supported across many vendors. Depending on the software version, AOS-CX switches might support multiple attributes for some settings. Generally, earlier software used vendor specific attributes while the more up-to-date software added standard RADIUS attributes for applying the same setting. HPE recommends that you typically use the standard RADIUS setting, and only use the vendor–specific one if you need to accommodate a legacy solution.

Guidelines for dynamic settings

- Tunnel-Private-Group-ID can be a VLAN ID or a name
- Use name to simplify policies in large deployments
- Attribute VLAN name must match actual VLAN name
- Attribute VLAN ID must match actual VLAN ID on switch
- If mismatched, authentication fails
- RADIUS server must support these AVPs

Keep a few guidelines in mind when coordinating the value for a dynamic VLAN with ClearPass administrators. The Tunnel-Private-Group-ID can be a VLAN ID or a name. You might use the name option to simplify policies in a large or multi-site network. For example, switches in one building might name VLAN 10 "Students" while switches in another building name VLAN 11 the same name. The RADIUS server could use a single policy to assign students to the "Students" VLAN with the Egress-VLAN-Name AVP. The switch through which the individual student authenticated would then assign the student's untagged traffic (native VLAN) to the correct VLAN ID associated with the "Students" name on that switch.

Be very careful that the name in the attribute matches the name of a VLAN on the switch; names are case–sensitive. Similarly, a VLAN ID must match a VLAN on the switch. If the RADIUS server sends a VLAN name or ID that does not exist on the switch, the switch does not apply the port's static VLAN to the user session. Instead, the authentication fails, much as it would if the user entered an incorrect password.

Make sure that the RADIUS server supports these AVPs. Also, make sure that the administrators know to define the vendor-specific AVPs, when required.

Device-based versus Client-based authentication

Device-based and client-based authentication are two methods used to control network access on AOS-CX switches. They differ in terms of where and how the authentication policies are applied, offering different levels of flexibility and security.

Device-based authentication

Port-based authentication controls access to the network based on the physical port to which a device is connected. This method typically uses protocols such as 802.1X, MAC authentication, or Captive Portal for authenticating devices connected to the switch ports.

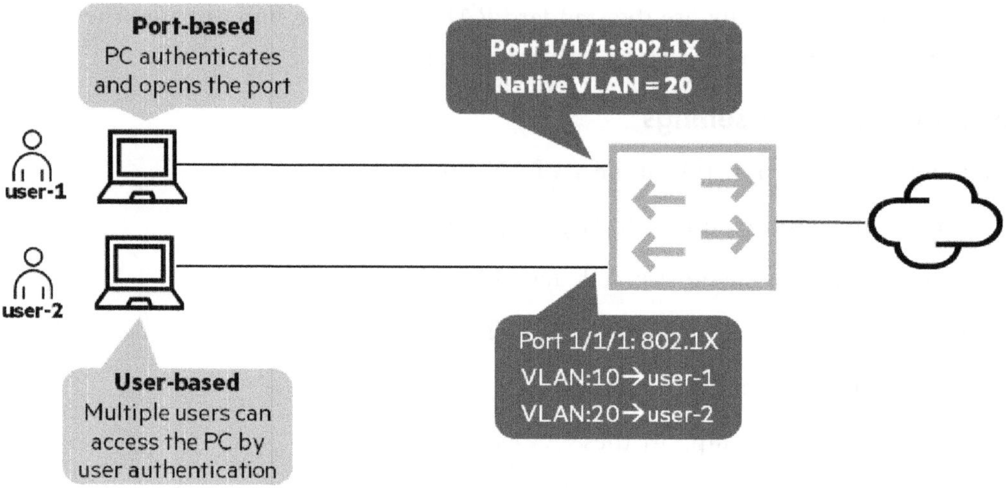

Figure 10-24: Port-based and user-based authentication

Client-based authentication

Controls access based on the identity of the user rather than the physical port. This method often involves more advanced mechanisms such as Role Based Access Control (RBAC), UBT, and integration with Identity Management Systems. The authentication can be done via 802.1X, Captive Portal, or other methods, but the policies applied are based on the user's identity and credentials.

Device-based mode

A switch applies AVP settings based on whether the port operates in 802.1X port-based or user-based mode. The mode also determines how the port establishes authentication sessions.

With **device-based** control, the switch establishes a single session for all traffic sent and received on the port. If any one user authenticates, the switch permits all traffic without referring to the source or destination MAC address. Untagged traffic (native VLAN), regardless of source MAC, is forwarded in the native VLAN assigned to the port (either statically or through RADIUS, as you have learned).

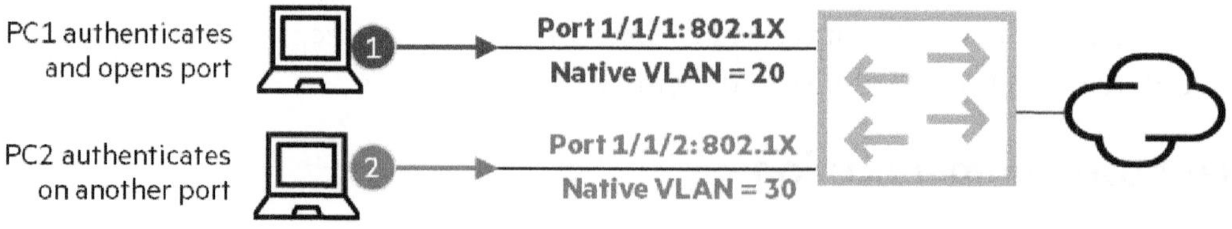

Figure 10-25: Example of the port-based authentication

Often, only a single device connects to the switch port, so all traffic does originate from a single source MAC address. This example illustrates this situation, where PC1 and PC2 are connected to two different switch ports and are assigned different VLANs based on their 802.1X credentials.

Configure the port mode for device mode using the following command.

```
interface 1/1/1 [and 1/1/2]
aaa authentication port-access auth-mode device-mode
```

> Client mode/user-mode is the default mode of the port.

Device-based mode use case

The concern with some customers is that the access switch might be in an unsecured environment. Suppose a company statically defines switch port policy settings for Instant APs (IAPs).

Someone unplugs an IAP from statically defined port 1/1/1 and connects an unauthorized device to that port. The unauthorized device would have the same policy settings applied as the IAP.

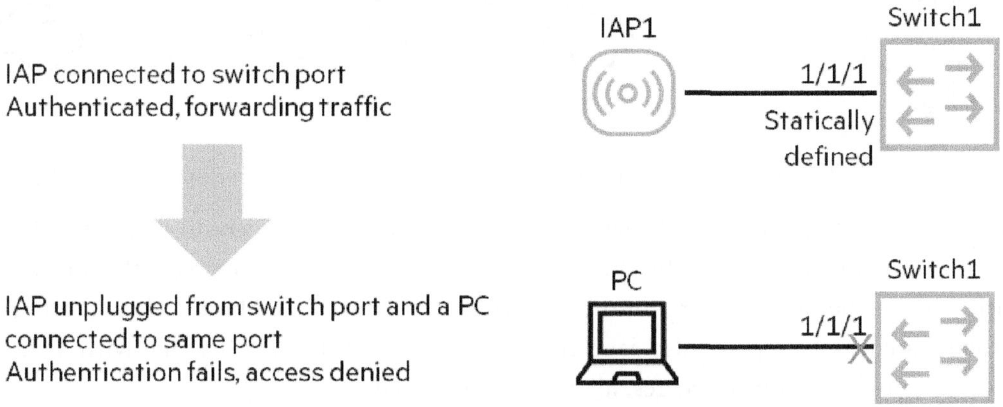

Figure 10-26: Device-based mode use case

If this were a corporate user, they would not have the correct VLAN settings; access would probably be denied. If this were a bad actor, they might be able to gain unauthorized access to corporate resources, depending on how the port is configured.

Device-based authentication mitigates this issue. If someone unplugged the IAP and plugged in an authorized device in this situation, the unauthorized device would be denied access to the network unless it successfully authenticated.

Device-based mode is useful when you know that you'll only have one device connected to a port, you need to authenticate access on the port, and you want to apply policies dynamically to the port based on authentication credentials (and possibly other information, like fingerprinting).

Issues with device-based mode

An 802.1X-enabled interface might also connect to multiple devices. You could have an old Ethernet hub or unmanaged switch that does not know how to process EAPOL frames, so the switch or hum simply floods them. The following figure shows how port-based mode interfaces handle this situation.

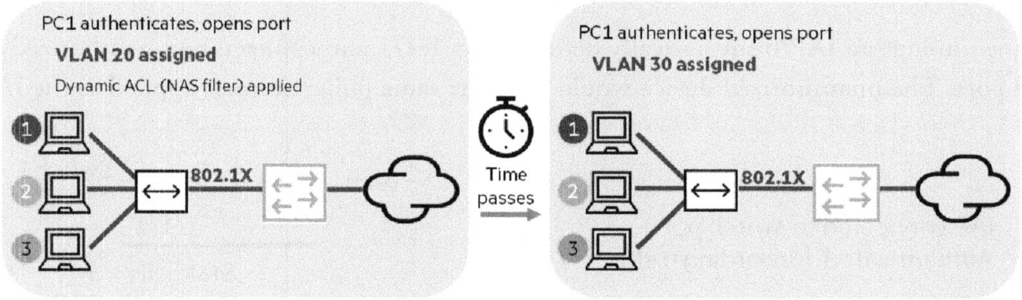

Figure 10-27: How device-based mode interfaces handle multi–device connectivity

The interface has a single untagged (native) VLAN assignment. This could be a statically assigned VLAN, or a dynamic VLAN sent by the RADIUS server when the most recently authorized user authenticated. In the figure, User 1 authenticates and assigns to VLAN 20. Users 2 and 3 have not authenticated, but they can also send traffic in VLAN 20. Remember that with device-based mode, when one device authenticates, all are authenticated.

Understand that AOS-CX switches automatically bind a dynamic ACL to a user's MAC address, even when 802.1X operates in port-based mode on the interface. Therefore, unauthenticated devices can receive access, and their traffic is not filtered by the dynamic ACL. Any static port ACL (PACL) on the port applies to both the authenticated client and the unauthenticated ones—as it does in VACLs or RACLs that would normally filter the devices' traffic.

Suppose that some time later, PC2 authenticates. The RADIUS server sends new dynamic settings, which override those from the first device. This can cause connectivity issues as the devices are moved to a new VLAN.

If only one device connects to the port at once, port-based mode can work well. But if multiple devices connect to the same port, device (port) based control introduces these potential–issues:

- Unauthorized users gain access by "piggybacking" on the same connection as an authorized device.
- Dynamic settings are inconsistent.
- Dynamic ACLs do not filter all incoming traffic.

Client-based mode

Client-based mode is recommended for scenarios in which you want the switch port itself to authenticate and control multiple clients. This is recommended when the RADIUS server might implement different dynamic settings for different clients. Often, you should also enable Client-based control even when the port should only connect to a single device to prevent unauthorized devices from "piggybacking" on the port.

In Client-based mode, the port authenticates each downstream device, as identified by source MAC address. You enable this mode when you set the client limit on the port to any value, including 1. Client mode is the default, with a default client limit of 1. See the configuration example in the following figure.

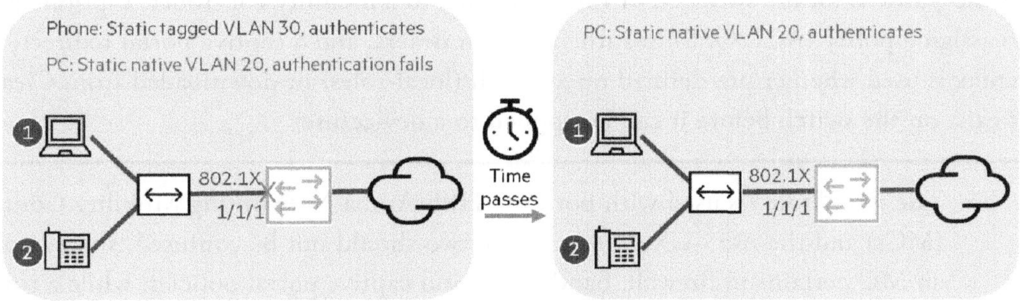

Figure 10-28: Client-based mode example

The port creates a separate session for each device, and each session has its own authorized status. In other words, the switch permits ingress traffic with a particular source MAC address only if that MAC address's session is authorized. Conversely, it permits egress traffic based on the destination MAC address's authorization status. The limit value indicates how many devices can be authorized.

After you set the limit, the port begins tracking MAC addresses and defines the authorization status and settings for each separately. For example, in the scenario with the computer and Voice over IP (VoIP) phone, the switch port sends an EAP Request/Identity to each separate MAC address detected on the port. If the VoIP phone authenticates successfully but the computer fails, the computer traffic is blocked.

Configuring client-based mode

```
interface 1/1/1
aaa authentication port-access auth-mode client-mode
```

```
aaa authentication port-access client-limit 2
```

User roles overview

The benefits of assigning "roles" to users and/or devices is well known within the wireless world. This has simply been adapted to the wired switch port. Just as with HPE Aruba Networking wireless, you simplify configuration, grouping policies into a "role" that can be referenced by many device or user types. Additionally, when ClearPass is incorporated, that who, what, how, where, when context can be used to control user access rights.

You can configure roles locally on the switch using a Local User Role (LUR). When a device connects and authenticates successfully, a User-Role name (VSA) is passed to the switch, and then the switch will apply that User-Role to the device.

Or, configure Downloadable User Roles (DURs) on a ClearPass server. The ClearPass HPE-CPPM-Role VSA is used with HTTPS to transfer the role to the switch.

A role is a role. At a minimum, they dictate the tagged or untagged VLAN assignment and whether the traffic is locally switched or tunneled back to a mobility controller. Optionally, a role can also assign a policy (ACL/QoS), reauthentication timers, and a captive portal redirect. The same syntax is used whether pre-defined on a switch (local-roles) or downloaded from ClearPass. It must exist on the switch before it can be applied to a device/user.

ⓘ The term "role" is used with both the HPE Aruba Networking Mobility Controllers (MCs) and the AOS-CX switches. The two should not be confused, since a role on an MC pertains to firewall, bandwidth, and captive portal policies, while a role on an AOS-CX switch pertains to VLAN, QoS, ACL, and other policies. In other words, roles on both devices apply policies, but the policy types differ.

Role assignment using local user role (LUR)

Remember with LURs, the RADIUS AAA server only passes a VSA that indicates the name of the locally defined role to be applied.

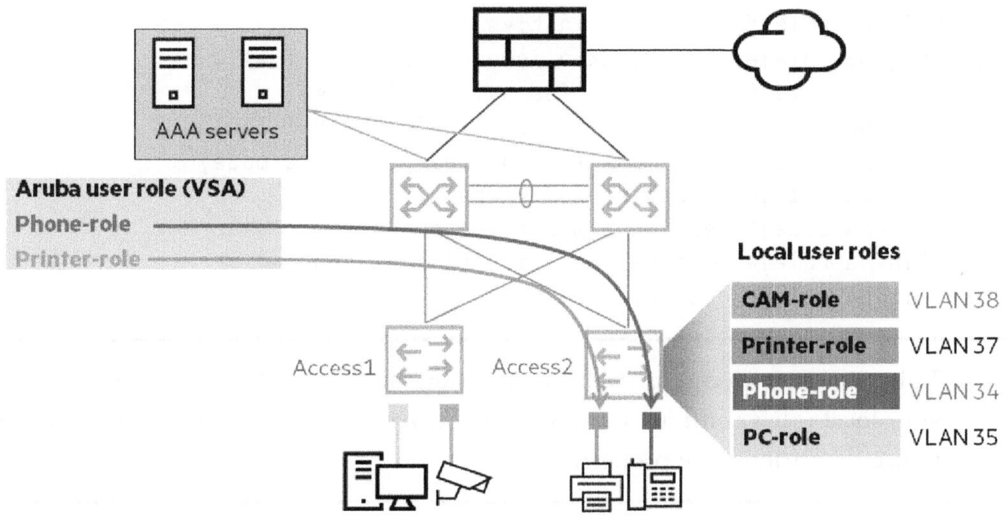

Figure 10-29: Dynamic role assignment using LUR

A user role consists of the following optional parameters:

Ingress user policy: L3 (IPv4 and/or IPv6) ordered list of classes with actions.

Captive-portal-profile: Assigns a captive portal profile for this role.

Inactivity-timeout: The inactivity timeout period in seconds with a range of 300 to 4294967295 for the authenticated client for an implicit logoff.

Reauth-period: Sets the reauthentication period in seconds, or 0 to disable.

VLAN access: Sets the untagged VLAN ID (native VLAN).

VLAN-id-tagged: Sets the tagged VLAN ID.

Auth-mode: Sets the configuration in user role to either device-mode or port-mode (client-mode).

The following are the attributes:

PoE-priority: Specifies the PoE priority for the interface.

MTU: Configures the MTU support for the client.

Tagged-VLAN: Specifies the list of tagged VLANs configured for the interface.

Trust-mode: Configures the QoS trust mode for the client.

To create a local user role, use the following configuration:

```
Switch(config)# port-access role <role-name>
Switch(config-pa-role)# description <description>
Switch(config-pa-role)# associate captive-portal-profile <profile-name>
Switch(config-pa-role)# associate policy <policy-name>
Switch(config-pa-role)# vlan access <VLAN-ID>
```

```
Switch(config-pa-role)# vlan trunk native <VLAN-ID>
Switch(config-pa-role)# vlan trunk allowed {<VLAN-ID> | all}
Switch(config-pa-role)# trust-mode {dscp | cos | none}
Switch(config-pa-role)# poe-priority {critical | high | low}
Switch(config-pa-role)# auth-mode {client-mode | device-mode}
Switch(config-pa-role)# reauth-period <seconds>
Switch(config-pa-role)# session-timeout <seconds>
Switch(config-pa-role)# client-inactivity time <seconds>
Switch(config-pa-role)# mtu <byte-size>
```

Role names are case-sensitive! Use the `show port-access role` command to verify your local role's configuration. See the *Security Guide* for more information to configure the various role properties.

Role assignment using DUR

Downloadable user roles (DUR) are downloaded from HPE Aruba Networking ClearPass Policy Manager Platform to the switch, as opposed to locally configuring roles on switches. The download enables you to set up policies and attributes for a specific user role which can then be downloaded and stored on the switch.

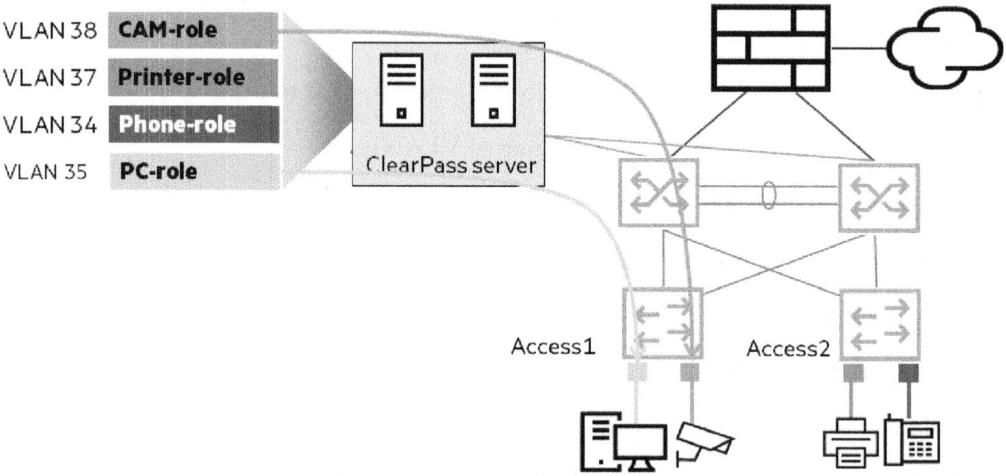

Figure 10-30: Dynamic role assignment using DUR

Configure new users and assign them the same stored version of the user role in ClearPass, saving the time to reconfigure each user individually.

Use the `radius-server host` command if the entire user role configuration is to be downloaded from the Policy Manager server, in this figure.

Because the login process from the AOS-CX switch uses HTTPS, the switch must validate the ClearPass server certificate. On older AOS-S, they would automatically download and install the ClearPass certificate. Unfortunately, in AOS-CX version 10.4, this process is not automatic. You must access the ClearPass server, download its SSL certificate, and install it on the AOS-CX switch.

Using ClearPass or third-party RADIUS

You have learned how AOS-CX switches can accept RADIUS VSAs to customize access for authenticated devices.

Some companies prefer to take a different approach. They might use ClearPass to apply user roles to authenticated users based on identities and other criteria. HPE Aruba Networking MCs, acting as RADIUS clients, use these roles to decide how to filter user traffic. Rather than create new profiles with the RADIUS VSAs for dynamic VLANs and ACLs, you might prefer to have ClearPass simply send a role to AOS-CX switches too.

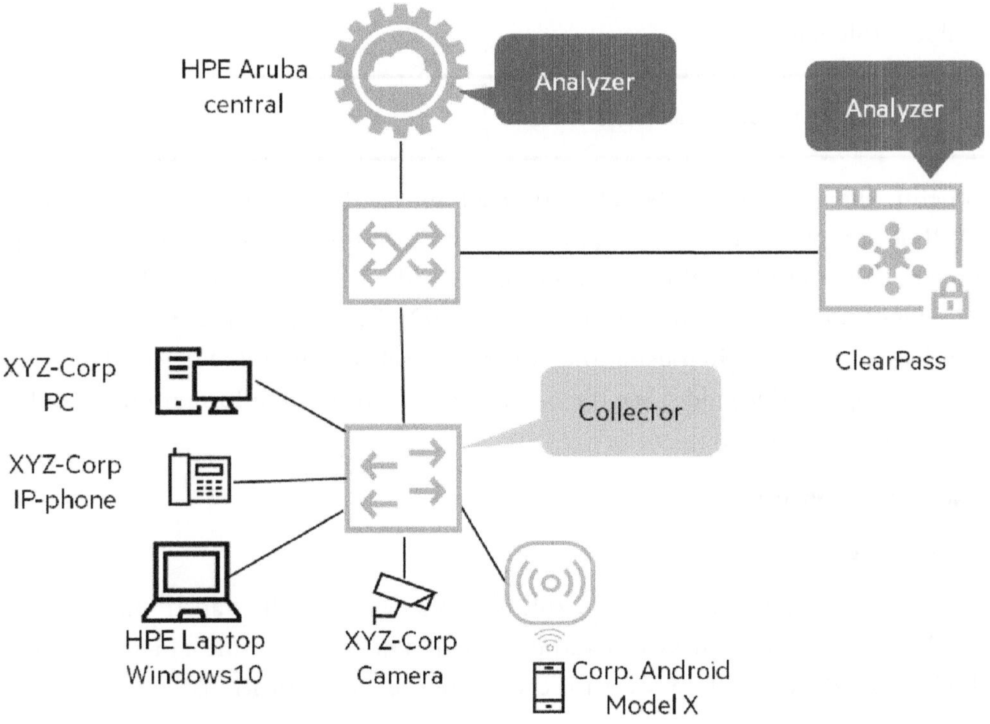

Figure 10-31: Role assignment using third party servers

Every client is associated with a role. Roles associate with a set of attributes for authenticated clients (clients with authentication configuration) and unauthenticated clients, applied to each user session. On a switch, roles are always enabled by default.

Examples of user roles are:

- **Employee**—All access
- **Contractor**—Limited access to resources
- **Guest**—Browse internet

Each user role determines the client network privileges, frequency of reauthentication, applicable bandwidth contracts, and other permissions.

The VSA that AOS-CX switches accept from RADIUS servers like ClearPass has these properties:

- Type: RADIUS vendor—specific for HPE Aruba Networking
- Name: Aruba-User-Role
- ID: 25
- Value: <role name>

ⓘ The role name is case-sensitive

You can define the user role locally. This approach moves the control over customizing settings from RADIUS server administrators to infrastructure administrators, which some companies might prefer. The User Derived Role is applied when the roles are enabled. UDR will have the same precedence order as the authentication type (802.1X, MAC authentication) and UDRs are locally configured on the switch using the port-access role command. The configuration of UDRs is beyond the scope of this guide. See the *Security Guide* for more information to configure the port-access role profile locally.

Local user-role options

- **Pre-auth role**

  The pre-authentication (pre-auth) role allows a device, such as an IP phone, to have network access before the device is authenticated. The pre-auth role can also be used to allow DHCP, DNS, and web redirect for captive portal authentications.

- **Auth-Role**

When implementing AAA, a AAA server will authenticate the user. If the AAA server does not reply with a role name, or the role name sent in the AAA request does not exist on the switch (case-sensitive), then a default role can be assigned to the user if the reject role has been configured.

- **Reject role**

  Normally, if a user fails authentication, they will have no access beyond the switch port. Optionally, you can assign a reject (failed authentication) role to the user that would give them minimal network access, perhaps to download software to become compliant in order to connect to the network, like an 802.1X supplicant.

- **Critical role**

  A critical role can be assigned to a user when the switch sends an AAA request to an AAA server, but no response is received, perhaps because there is a network issue preventing communication between the two sets of devices. By default, the authentication would fail and the user would be denied access. However, you can override this behavior by defining a critical role for the interface.

  If the AAA server is not reachable, then the critical role is applied. This allows important devices—like VoIP phones, for example—to stay connected to the network at all times.

  Once an AAA server is reachable, however, the device connected to the interface must be authenticated, at which point the critical role is not used.

- **Critical-voice role**

  A role critical-voice-role is applied when an authenticated client fails to re-authenticate because the RADIUS server is unreachable. This role is not applied when the multidomain authentication mode is not enabled and the client fails to re-authenticate because the RADIUS server is unreachable.

Either a pre-auth role VLAN (voice and/or data) or a pre-auth role can be configured for a port. However, both a VLAN and role cannot coexist for an interface. Initial traffic on the port is restricted only by ACLs configured for the port or for VLANs or ACLs in the role.

The voice client must first be authenticated successfully with the RADIUS server for the criticial-voice-role to be applied. If the client is never authenticated, then this role is not applied.

Device fingerprinting overview

Device Finger Printing (DFP) is a feature that identifies and classifies devices connecting to the network based on their unique characteristics. This capability helps network administrators to enforce policies, optimize network performance, and enhance security by understanding the types of devices that are accessing the network.

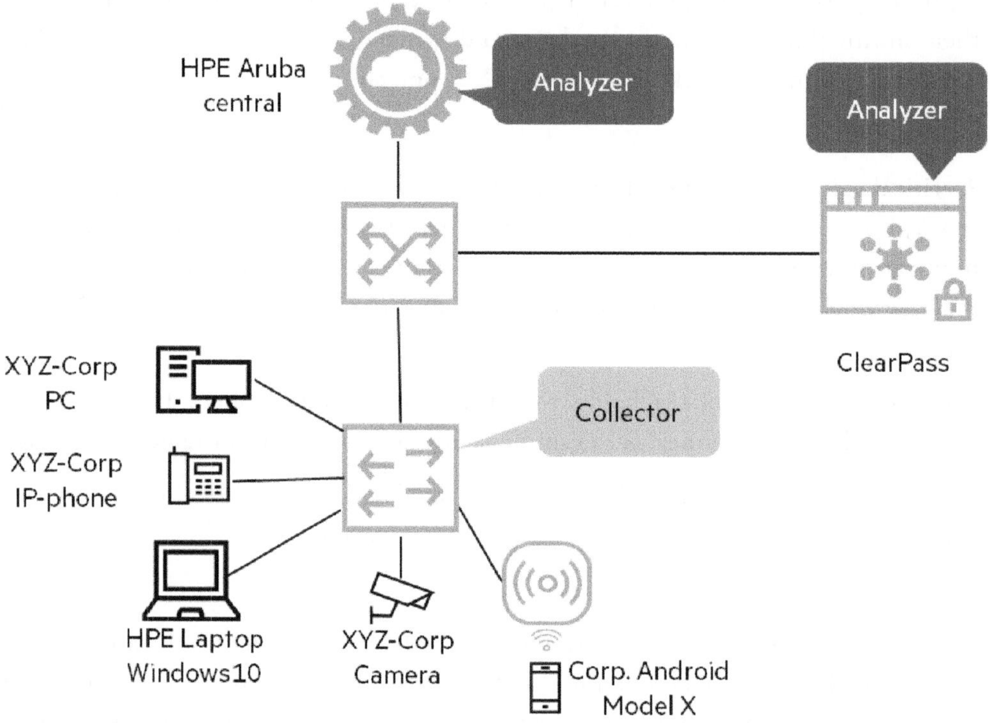

Figure 10-32: DFP collector and analyzers

DFP is achieved by configuring a switch to collect the traffic patterns and send only required information or attributes to an analyzer.

As show in the previous example, this solution consists of two parts:

Collector

Collects the information from the packets that are obtained from the clients. The information collected is based on the certain type of traffic from the clients. AOS-CX switches will act as collectors.

Analyzer

Is an entity that processes the information collected from the collector and fingerprints the device with the details of the device such as type, host name, vendor type, and so forth. An example of an analyzer can be HPE Aruba Networking Central or ClearPass.

DFP using AOS-CX switches

AOS-CX switches will act as collectors and forward the details to the analyzers for the analysis and device fingerprinting. The analyzer helps the network operator to have better visibility into what types of devices are plugging into the access layer. By obtaining as much information as possible from a client, more granular security policies can be created and enforced to maintain client and network stability. This feature is also helpful in detecting and blocking unauthorized or rogue devices attempting to connect to the network. You can apply security policies based on device type, such as restricting access for certain device categories.

AOS-CX switches currently support DHCP, HTTP, LLDP, and CDP protocols to collect fingerprinting.

HPE Aruba Networking Central uses MAC OUI of the client for the device finger printing.

If you are using PEAP authentication, you know who the user is but not what device they are using. ClearPass profiling capabilities use fingerprinting to determine the device type and operating system.

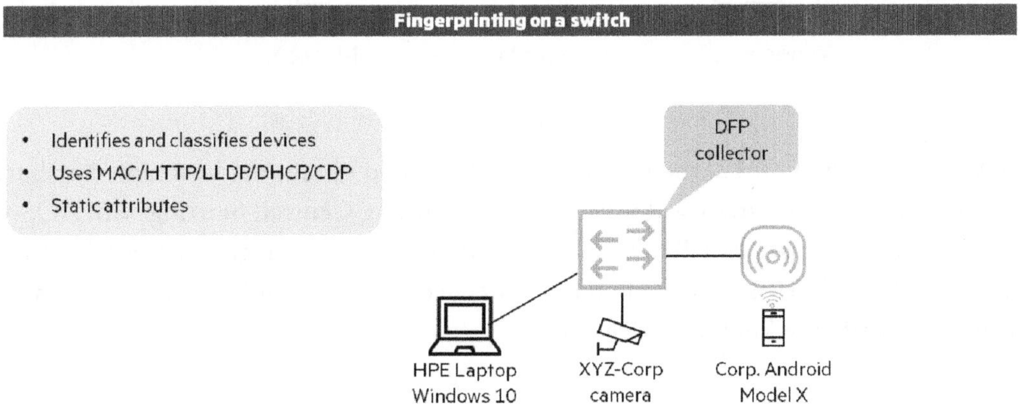

Figure 10-33: DFP with AOS-CX switches

This feature is helpful in detecting and blocking unauthorized or rogue devices attempting to connect to the network. You can apply security policies based on device type, such as restricting access for certain device categories. You can even optimize network performance by understanding the types of devices and their expected network behavior.

If you are using PEAP authentication, you know who the user is but not what device they are using. ClearPass profiling capabilities use fingerprinting to determine the device type and operating system.

DFP with ClearPass

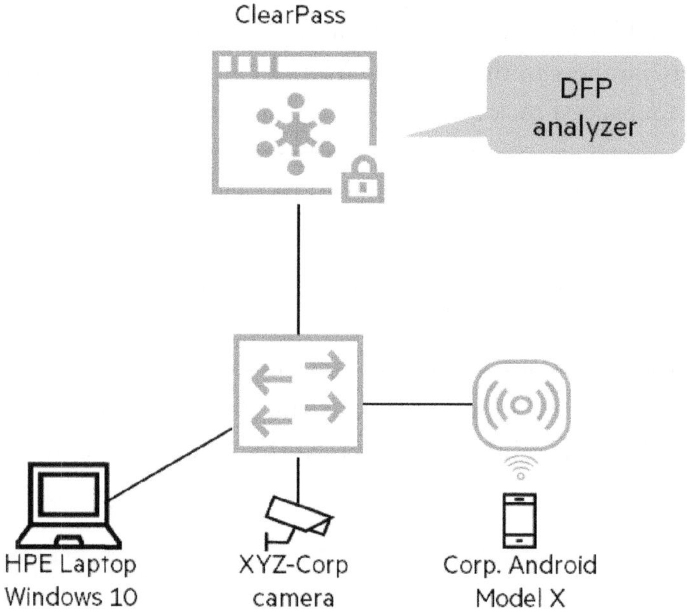

Figure 10-34: ClearPass as an analyzer

Prior to AOS-CX version 10.12, the only analyzer that could obtain the hostname attribute from the switch fingerprint attributes was HPE Aruba Networking Central. Starting with AOS-CX 10.12, ClearPass can be used as a RADIUS accounting server and integrated as a DFP analyzer. When you integrate ClearPass, you can take advantage of its profiling features, which use fingerprinting to identify the kind of device and OS.

How DFP works with ClearPass

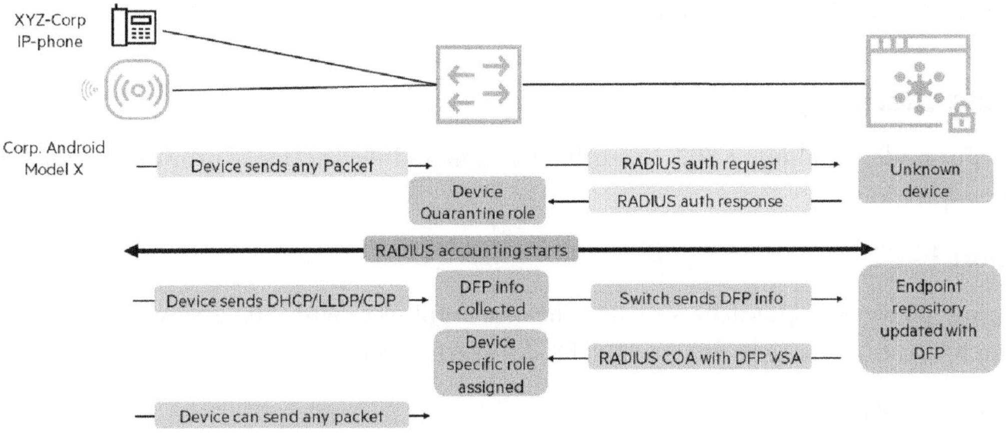

Figure 10-35: How DFP works with ClearPass

1. The device starts sending the packets. The switch sends the RADIUS access request to ClearPass; as device details are unknown to ClearPass, it classifies the device as an unknown device.
2. When ClearPass returns an "unknown-device" status, the switch maps the device to a "quarantine" role.
3. The device sends DHCP/LLDP/CDP/HTTP packets. The switch (collector) captures the DFP information and forwards it to ClearPass (analyzer). ClearPass then analyzes, classifies the device, and sends a device-specific role (VSA) packet to the switch as a RADIUS COA packet.
4. The switch assigns the device specific–role to the device and allows the device to send the traffic as per the policy.

Enable fingerprinting on a switch

Device fingerprinting on AOS-CX switches can be integrated with ClearPass or other network management and security systems to enhance visibility and control. This integration allows for more sophisticated policy enforcement and analytics based on the comprehensive data gathered from device fingerprinting.

Enter global configuration mode and enable device fingerprinting.

```
Switch(config)# device-fingerprinting enable
```

Configure DHCP fingerprinting if DHCP fingerprinting is not enabled by default.

```
Switch(config)# dhcp fingerprinting enable.
```

After enabling device fingerprinting, you can verify the configuration and view identified devices.

```
Switch# show device-fingerprinting
```

This command will display a list of devices that have been fingerprinted, including their MAC addresses, detected device types, and any relevant details.

Learning check

Check your understanding of 802.1X authentication and device finger printing by answering the following questions.

Question 1

A company plans to use ClearPass to authenticate employees with 802.1X. Which role do AOS-CX switch at the access layer play in 802.1X authentication?

- O supplicants
- O authenticators
- O authentication servers
- O authentication aggregators

Question 2

A network administrator is specifying ClearPass as a RADIUS server on an AOS-CX switch. What is should the administrator configure to prevent issues of accepting and using VSAs? (Select two.)

- ☐ radius-server dyn-authorization enable
- ☐ radius-server dyn-authorization port set 1812
- ☐ time window set to 0
- ☐ enable NTP

Question 3

Which EAP method is known for using digital certificates for mutual authentication and is commonly used in switches?

- O EAP-PEAP
- O EAP-TTLS
- O EAP-TLS
- O EAP-MD5

Question 4

What is the primary function of user roles in AOS-CX switches?

- ○ To configure routing protocols
- ○ To provide differentiated network access and permissions for users
- ○ To assign IP addresses to devices
- ○ To manage switch firmware updates

Question 5

Which of the following attributes is commonly used in device fingerprinting on AOS-CX switches? (Select two.)

- ☐ Device MAC address
- ☐ Device hostname
- ☐ Device operating system
- ☐ All of the above

11 MAC authentication

This chapter teaches you how to implement MAC authentication (MAC-auth) technology on AOS-CX switches. MAC authentication is a network access control method used to authenticate devices based on their MAC addresses. It is particularly useful in environments where devices do not support more complex authentication mechanisms, such as 802.1X. AOS-CX switches support MAC authentication to provide a flexible and straightforward way to manage network access for the endpoints.

MAC authentication leverages the unique MAC address of a device as its identifier for network access control. Unlike 802.1X, which requires client-side software and user credentials, MAC authentication relies solely on the MAC address of the device. This method is especially useful for devices like printers, IP phones, and IoT devices that do not support traditional authentication methods.

This chapter also introduces you a new security feature called Media Access Control Security (MACsec). MACsec is a security protocol defined by the IEEE 802.1AE standard. It provides point-to-point security on Ethernet links, ensuring data integrity and confidentiality by encrypting and authenticating Ethernet frames. AOS-CX switches support MACsec to enhance network security, particularly in environments where sensitive data traverses the network.

OBJECTIVES

After completing this chapter, you should be able to:

- ✓ Describe how MAC-auth works.
- ✓ Implement RADIUS-based MAC-auth.
- ✓ Implement dynamic switch port settings based on LLDP or CDP information.
- ✓ Explain how MACsec works.

MAC authentication overview

MAC authentication on AOS-CX switches provides a simple and effective way to manage network access for devices that do not support more complex authentication mechanisms such as 802.1X and captive portal. By leveraging the unique MAC addresses of devices, network administrators can ensure that only authorized devices can access the network, enhancing security and control. This method is particularly useful with a range of devices, including printers, IP cameras, and IoT devices.

In addition to permitting a device network access based on its MAC address, the authentication server can assign the dynamic settings, such as a VLAN ID and QoS priority, to the device's session.

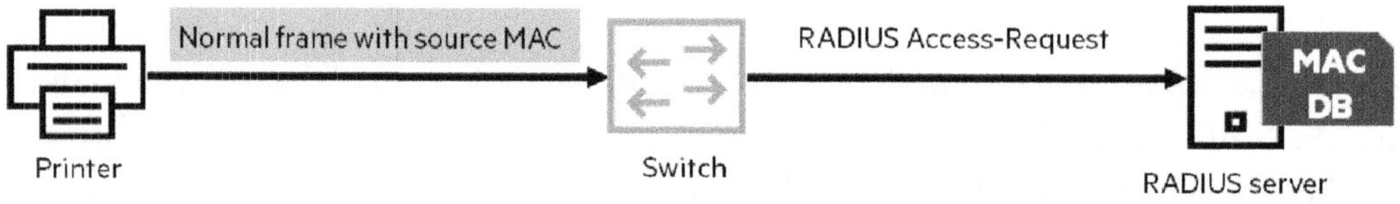

Figure 11-1: MAC authentication example

These devices simply connect and send a normal Ethernet frame, which, of course, includes their own MAC address as the source. The switch harvests this MAC address and creates a RADIUS Access-Request message to the RADIUS server. No client configuration is required.

When you need rigorous security, you should use 802.1X, whenever possible, since it supports username/password and digital certificate authentication at Layer 2. Often, it is trivial for a hacker to discover MAC addresses used in your system, whether by eavesdropping or accessing a legitimate device. The hacker can then easily spoof an authorized MAC address. MAC authentication might also be more difficult to manage than other forms of authentication, requiring directory or domain administrators to create accounts for each MAC address on each device, whereas 802.1X can use the user accounts already in the directory or domain.

MAC authentication relies on a RADIUS server. No need to configure MAC addresses on each switch—a single RADIUS server has a MAC DB for centralized control. You can use up to three RADIUS servers for redundancy, in case access to the primary server fails. It also means the same credentials can be used for authentication, regardless of which switch or switch port is the current access point (AP) into the LAN.

The switch acts as a port access authenticator, using a RADIUS server and the CHAP and PAP protocols. Inbound traffic is processed by the switch alone, until authentication occurs. Some traffic from the switch to an unauthorized client is supported (for example, broadcast or unknown destination packets) before authentication occurs.

RADIUS-based MAC-auth

You can enable RADIUS-based MAC-auth on individual switch ports to enable the following process:

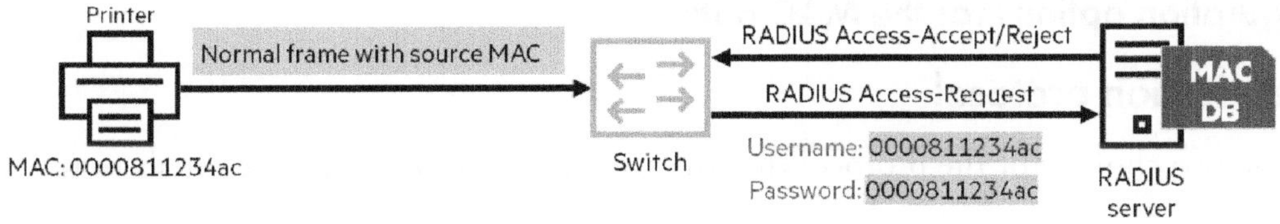

Figure 11-2: RADIUS-based MAC authentication

Step 1

The client is initially unauthenticated. The switch does not transmit traffic on the line; authentication is not triggered until the client begins to send traffic on its own. This traffic is initially dropped while the switch completes the next steps.

Step 2

The switch submits credentials to a RADIUS server on the client's behalf. Often, these credentials consist of the client's MAC address as the username and password. However, the password can be different from the MAC address.

Step 3

The RADIUS server authenticates the MAC address against a user account. Only the switch and the server are involved in the exchange of authentication messages, unlike with other port-based authentication methods. The RADIUS server uses its user database—which should have an account for the client's MAC address—and policies to determine whether to grant access. It communicates its decision to the switch in a RADIUS Accept-Access or Reject-Access message.

Step 4

The switch handles the client according to the RADIUS server's decision.

1. If the server accepted access, the switch allows all traffic to and from the client with that MAC address.
2. If the server rejected access, the client remains in the unauthenticated state, and the switch handles its traffic as indicated in step 1.

The switch continues to examine incoming frames. If a new client connects, the switch will authenticate that client. The switch also attempts to again authenticate the client for which authentication failed. By default, the switch suppresses authentication requests for such clients over the quiet period—by default, 60 seconds. It then sends another authentication request.

Configuration options for the MAC-auth

Authentication protocol

You have two choices for the protocol that the switch uses to send the credentials to the RADIUS server:

- Password Authentication Protocol (PAP)
- Challenge Handshake Authentication Protocol (CHAP)

You learned about how RADIUS works with EAP in 802.1X. AOS-CX switches can use PAP or CHAP for MAC-auth. In this case, the switch acts as the authentication client rather than the client itself, generating the PAP request messages. Note that PAP sends passwords in cleartext, which can pose a security threat.

RADIUS has specific messages for CHAP authentication, and most RADIUS servers support this option. You should be aware that the mechanisms of CHAP require passwords to be stored with reversible encryption, which poses a potential security hazard and is not typical for most directories and domains. This limitation would mean, for example, that the directory, domain, or RADIUS administrator who creates the accounts for the clients would have to take special measures to store the passwords with reversible encryption. (For a Microsoft Windows domain, you can edit the Group Policy Object (GPO) and enable reversal encryption in the computer policies for account settings.)

Credentials

You must decide the precise credentials that switches will send so that the RADIUS, directory, or domain administrator can set up the accounts accordingly.

By default, AOS-CX switches send the MAC address for both the username and password. You can set up a global password, which the switch uses for the password instead of the device's MAC address. The switch itself places the password in the authentication request. No configuration is required on the client. The switch continues to use the client's MAC address for the username, so every client requires a user account on the RADIUS server (or background directory).

The password option is intended to add a bit of security to the solution. However, keep in mind that hackers can still spoof an authorized MAC address, connect to the switch, and gain access to the network because the switch adds the password on its end. Remember: MAC-auth provides basic security; you should use 802.1X when you require more rigorous security.

MAC address format

When you are using MAC addresses for the username, password, or both, you must ensure that the switch sends the credentials with the same format used in the user accounts. Configure the delimiter format as shown in **Table 11-1**. The default is no-delimiter. The following table shows the options and the syntax for configuring a specific format for AOS-CX switches.

Table 11-1: MAC address format

Address format	AOS-CX switch parameter
aabbccddeeff	no-delimiter
aabbcc-ddeeff	single-dash
aa-bb-cc-dd-ee-ff	multi-dash
aa:bb:cc:dd:ee:ff	multi-colon
AABBCCDDEEFF	no-delimiter-uppercase
AABBCC-DDEEFF	single-dash-uppercase
AA-BB-CC-DD-EE-FF	multi-dash-uppercase
AA:BB:CC:DD:EE:FF	multi-colon-uppercase

Accounts in Microsoft Active Directory (AD) domains cannot use colons, so choose a format that does not include colons if you use AD.

MAC-auth with HPE Aruba Networking ClearPass

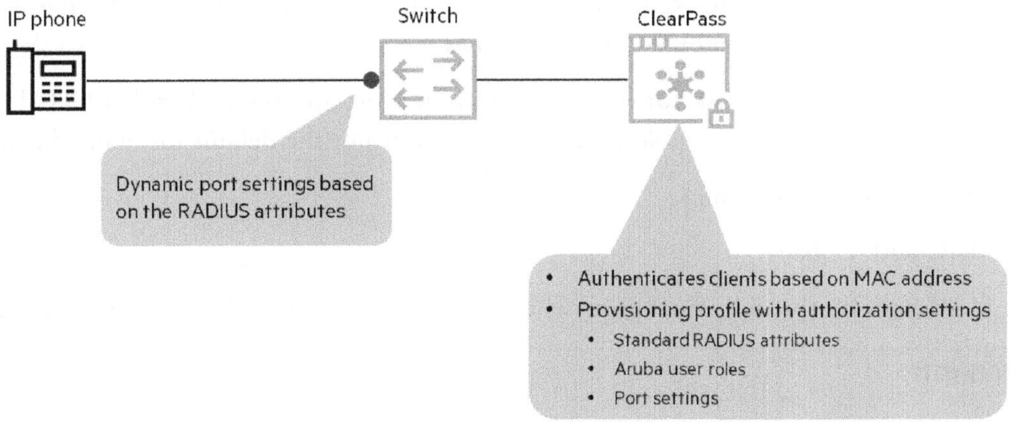

Figure 11-3: Using ClearPass for MAC-auth

Configuring MAC-auth with RADIUS

Specify your RADIUS server. In the following example, you are using only the global group. These commands specify the server and set the proper options for ClearPass. Remember to set the key to match the secret configured for this switch on ClearPass.

```
radius-server host <SERVER-IP> plaintext <KEY>
radius-server dyn-authorization enable
```

MAC-auth settings

```
aaa authentication port-access mac-auth
addr-format <DELIMITER_FORMAT>
auth-method {chap | pap}
radius server-group {radius | <GROUP_NAME>}
enable
```

Enabling MAC-auth on an interface

```
interface <IFACE-ID>
aaa authentication port-access mac-auth
enable
```

Optional settings

```
aaa authentication port-access mac-auth
addr-format <DELIMITER_FORMAT>
password plaintext <PASSWORD>
quiet-period <PERIOD>  reauth  reauth-period <PERIOD>
interface 1/1/1
aaa authentication port-access allow-lldp-bpdu
```

Configure the format for the MAC-auth credentials so that they match the credentials configured on the RADIUS server (or its backend directory). ClearPass often uses the no-delimiter option, which is also the default option on AOS-CX switches.

Choose between PAP and CHAP for the authentication protocol. Remember that CHAP requires passwords to be stored with reversible encryption, which is not the default for many directories.

Specify the RADIUS server group.

Enable MAC-auth globally.

Enable MAC-auth on the individual ports where you want to enforce this type of authentication.

Verification

You can use the following commands to verify the configuration and operation of MAC-auth:

Table 11-2: Verification commands

Command	Description
`show aaa authentication port-access mac-auth interface {all|<IFACE-NAME>} client-status [mac <MAC-ADDRESS>]`	Shows information about MAC authentication clients status. The output can be filtered by interface or MAC address.
`clear mac-auth statistics [interface <IFACE-NAME>]`	Clears the MAC authentication statistics associated with the port and all the authenticator state machines associated to this port. If no interface is specified, the statistics is cleared for all MAC authentication enabled ports.

`show tech mac-auth [local-file]`	Displays the output for the following commands:
	`show aaa authentication port-access mac-auth interface all port-statistics`
	`show aaa authentication port-access mac-auth interface all client-status`
	`show aaa authentication port-access interface all client-status`

`diag-dump mac-auth basic`	Provides a diagnostic dump for the MAC-auth feature.

MAC-auth with multiple clients

Previous scenarios have featured edge ports that support one endpoint at a time, which is very common. However, you might have a Voice over IP (VoIP) deployment, where the VoIP phone connects directly to the switch and a PC connects to the phone. This is smart, because each user only uses one switch port for both devices. You might apply Local MAC authentication (LMA) or RADIUS MAC-auth to the port for IP phones, and 802.1X authentication for the PC, on the same port.

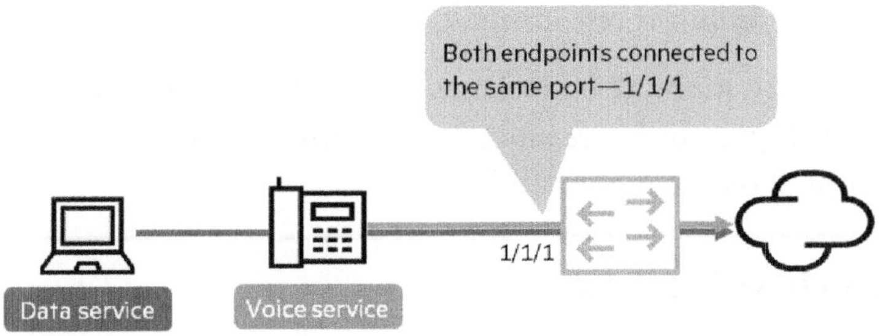

Figure 11-4: Scenario 1: both data and voice clients connected to the same edge port

You must raise the MAC address limit to two, and you typically assign tagged VLANs to IP phones and untagged VLANs to computers. If you want to enforce authentication for the computers, you can assign the untagged VLAN through the RADIUS profile, or you can have computers use the port's static VLAN.

Another scenario is to set up LMA or RADIUS MAC-auth as needed. By default, the switch allows only one MAC address to authenticate on the port at the same time. After a connected client passes authentication, the switch drops traffic received from any other MAC address, even if that address would otherwise pass authentication. You must raise the MAC address limit as needed; you can set the value from 1 (default) to 256.

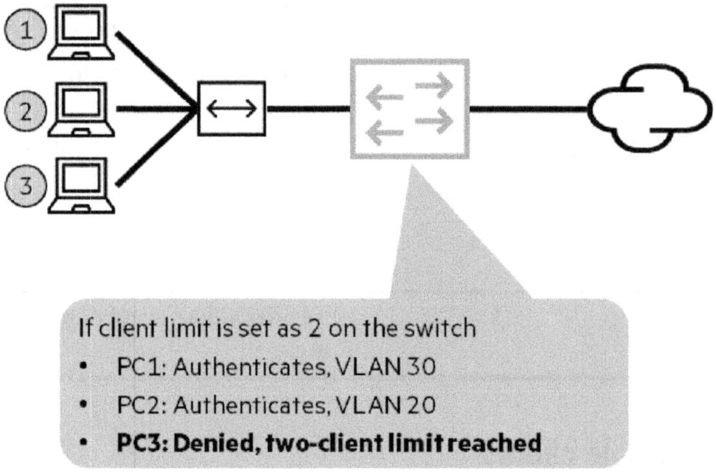

Figure 11-5: Scenario 2: multiple clients are connected to the same port that is enabled with client limit set as 2

802.1X port access and MAC authentication can be configured at the same time on a port. A total of 256 clients can be configured per port and 16,384 clients on the entire switch, irrespective of the authentication method. After the limit of 16,384 clients is reached, no additional authentication clients are allowed on any port for any method. The default is one client.

MAC authentication, MAC lockout, and port security are mutually exclusive on a given port. If you configure any of these authentication methods on a port, you must disable LACP on the port.

Combining MAC-auth and 802.1X

A single port can support multiple authentication methods, enabling it to authenticate different types of devices and users. For example, you have devices that use 802.1X authentication, but also some devices that do not. By default, 802.1X is performed first. If 802.1X times out, the switch assumes that the device does not support it. If MAC-auth is also enabled on the port, then the switch will authenticate the device with MAC-auth.

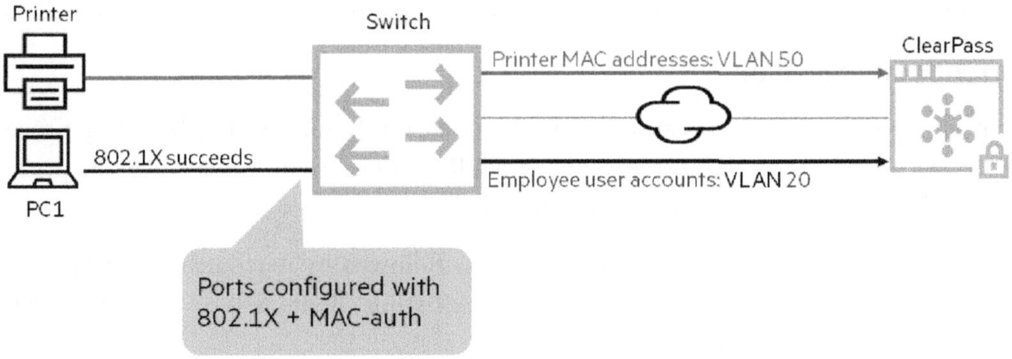

Figure 11-6: Switch port configured with both MAC and 802.1X authentication

You can change the order, per interface, using the `aaa authentication port-access auth-precedence [dot1x mac-auth | mac-auth dot1x]` command.

In this scenario, each port supports a single device, but you are not sure which type of device will connect. You want any port to support any device type—those that support 802.1X and those that only support RADIUS MAC-auth. Supposing you left it at the default, 802.1X is tried first.

When the printer connects, the switch initiates 802.1X authentication, which the printer does not support and so does not respond. 802.1X times out, and the switch tries MAC-auth, which succeeds. Policy defined on ClearPass ensures that printer ports are assigned to VLAN 50. When the PC connects, the switch initiates 802.1X authentication, which the PC supports and so authentication succeeds. ClearPass policy ensures that employees are assigned to VLAN 20.

Guidelines for MAC-auth with 802.1X

Here are some important guidelines when enabling MAC-auth or LMA and 802.1X on the same port.

- Ensure that ports are in user-based mode for 802.1X. Set the client limit, even if that client limit is one.
- Validate proper VLAN assignment per device type/client. The RADIUS server profile should ensure that MAC-auth clients and 802.1X users are on different VLANs.
- MAC-auth is vulnerable to spoofing. Filter traffic so clients can only access required resources.
- Validate RADIUS server policies/accounts. Only provide MAC-auth for devices that require it.

- MAC-auth is the weakest link in the security system, so you often want MAC-auth to be used only if 802.1X fails.
- A client that has successful MAC-auth sometimes might not initiate 802.1X.

Concurrent authentication

This helps in starting all authentication methods concurrently for faster onboarding process. If authentication priority is not configured when enabling concurrent onboarding, the priority will be 802.1x followed by mac-auth and device-profile.

Default priority for concurrent onboarding is 802.1x followed by mac-auth and device-profile. When enabling concurrent onboarding on the port, existing clients will be de-authenticated and freshly onboarded concurrently.

When concurrent onboarding is enabled, then auth-precedence will be ignored. If concurrent onboarding is configured, the client will stay in pre-auth role till it gets succeeded by one authentication method or gets failed by all the authentication methods. When the authentication method with the highest priority fails, the profile of the next successful authentication method is applied. If all methods fail, the reject or critical role is applied based on the 802.1X authentication failure reason and continues to reauthenticate with the 802.1X method. Reauthentication will be triggered for all high priority methods and not just the final successful authentication method.

Enabling concurrent onboarding on a port

```
switch(config)# interface  1/1/1
switch(config-if)# port-access onboarding-method concurrent enable
```

Disabling concurrent onboarding on a port

```
switch(config)# interface 1/1/1
switch(config-if)# port-access onboarding-method concurrent disable
```

Some RADIUS server may block the client when it receives two requests, mac-auth and 802.1X, from the same client at the same time. This is because the RADIUS server allows only one authentication request. In such cases, concurrent onboarding is not feasible. To prevent such scenarios, configure auth-precedence with auth-priority.

MACsec overview

MACsec is a security protocol that provides point-to-point security on Ethernet links, ensuring data integrity and confidentiality by encrypting and authenticating Ethernet frames. HPE Aruba Networking CX Switch Operating System switches support MACsec to enhance network security, particularly in environments where sensitive data traverses the network.

MACsec is intended for wired LANs and enables a bidirectional secure link after an exchange and verification of security keys between two connected devices. It uses MACsec Key Agreement (MKA) protocol and static Connectivity Association Key (CAK). A combination of data integrity checks and encryption is used to safeguard the transmitted data. It protects against range of attacks including denial of service, man-in-the-middle, and eavesdropping.

Some of its salient features are:

- Connectionless data integrity: Unauthorized changes to data cannot be made without being detected. Each MAC frame carries a separate integrity verification code.
- Data origin authenticity: A received MAC frame is guaranteed to have been sent by the authenticated device.
- Confidentiality: The data payload of each MAC frame is encrypted to prevent it from being eavesdropped on by unauthorized parties.
- Replay protection: MAC frames copied from the network by an attacker cannot be resent into the network without being detected.

MACsec frame format

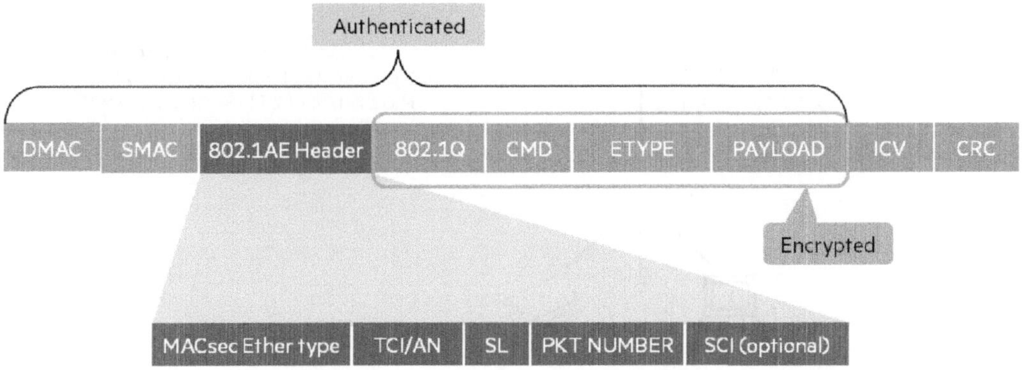

Figure 11-7: MACsec frame format

MACsec adds a 32-byte header (SecTAG) to every MACsec-protected data frame that exits the switch ASIC. Notice that the 802.1Q VLAN tag as well as the payload are encrypted, but the source and destination MAC addresses are in the clear.

Supported modes

MACsec supports two modes:

- Pre-Shared Key (PSK)
- Dynamic EAP MACsec: 802.1X authenticator with 802.1X supplicant switch

PSK-based mode uses a static PSK for the encryption and is more suitable for enabling MACsec between switches, and the dynamic EAP MACsec mode uses dynamic keys derived automatically from an authentication server. This support is typically limited to a small subset of uplink ports.

Host-to-Switch and Switch-to-Switch Dynamic EAP MACsec support are verified with CPPM 6.11 and with CX 6300 Series and CX 8360 Series switches.

802.1X (authentication) with MACsec (encryption)

This combination eliminates the need for static MACsec PSK configuration on devices:

- MKA policy automatically generated on the switches
- Associated with predefined MACsec policy and a configured port access role (local user role or downloadable)

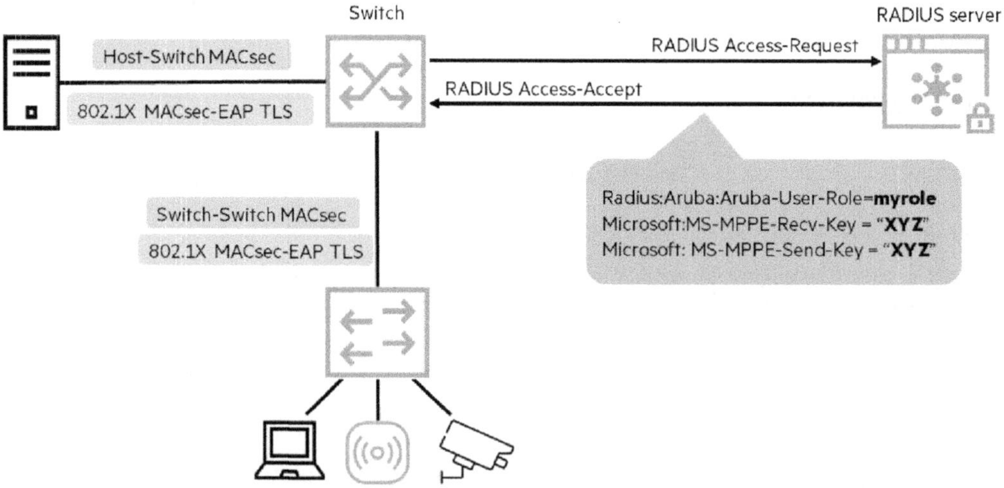

Figure 11-8: 802.1X authentication with MACsec encryption

When a Windows server running with a MACsec-capable 802.1X client isconnected to an authenticator switch, during 802.1X, an authentication request is sent from the authenticator switch to RADIUS. The RADIUS server accepts the request and replies back with the local user role, "myrole," which you can see is mapped to a MACsec policy. The policy is using 256-bit cipher suites and is set to "Must-Secure."

After successful authentication, the Master Secret Key (MSK) and session ID are distributed to the supplicant and the authenticator to derive the CAK and CKN to be used for the MKA policy. This is what the MKA policy would look like on both the Supplicant and Authenticator switches post authentication.

Learning check

Check your understanding of MAC authentication by answering the following questions.

Question 1

A network administrator sets up both 802.1X and RADIUS-based MAC-auth on a port. For which clients does the RADIUS server need accounts with their MAC addresses?

- O All clients.
- O Clients that do not support 802.1X.
- O Clients that support 802.1X.
- O Clients that support CHAP.

Question 2

An administrator enables both 802.1X and MAC-auth on a switch port. By default, the switch implements MAC authentication first. True or false?

- O True
- O False

Question 3

What is the primary purpose of the MACsec feature on AOS-CX switches?

- O To authenticate users based on their usernames and passwords.
- O To provide encryption and integrity for data at Layer 2 of the OSI model.
- O To manage VLAN configurations on the switch.
- O To enable wireless connectivity on the switch.

Question 4

When MACsec is enabled, the 802.1Q tag of the Ethernet frame will be encrypted.

- O True
- O False

12 Dynamic Segmentation

This chapter teaches you about Dynamic Segmentation with AOS-CX switches. Dynamic Segmentation is comprehensive network security and management by enabling fine-grained control over user and device access. This approach leverages the advanced capabilities of AOS-CX switches in conjunction with HPE Aruba Networking Mobility Controllers (MCs) and ClearPass to dynamically apply security policies based on user roles, device types, and contextual factors. Instead of relying on static VLAN assignments, Dynamic Segmentation allows for the creation of User-Based Tunnels (UBT) that segment traffic at the network edge, directing it to centralized controllers where policies can be enforced consistently. This results in enhanced security, as it isolates user traffic and applies specific security measures tailored to each user or device, mitigating the risk of lateral movement within the network.

Implementing Dynamic Segmentation with AOS-CX switches also streamlines network operations and enhances scalability. By automating the assignment of policies and network configurations, IT teams can reduce the complexity and time associated with manual configurations and adjustments. This approach supports a highly dynamic and flexible network environment, accommodating a growing number of devices and users without compromising on security or performance. Additionally, the integration with ClearPass ensures that access policies are consistently applied across wired and wireless networks, providing a unified and secure network infrastructure.

OBJECTIVES

After completing this chapter, you should be able to:

- ✓ Explain the significance of dynamic segmentation.
- ✓ Describe user-based tunneling operations.
- ✓ Configure dynamic segmentation on AOS-CX switches.
- ✓ Describe when and how to configure PAPI enhanced security, high availability, and fallback switching for tunneled node.

Dynamic Segmentation overview

Dynamic Segmentation (User-Based Tunneling (UBT)) in AOS-CX is a feature that enhances network security and management by allowing user traffic to be tunneled directly from access switches to a centralized controller. This approach leverages Generic Routing Encapsulation (GRE) or VXLAN (Virtual Extensible LAN) tunnels to route user traffic through a secure path, ensuring that policies and controls are uniformly enforced regardless of the user's physical location within the network. By centralizing the control of user traffic, UBT simplifies the application of security policies, QoS, and monitoring, making the network more manageable and secure.

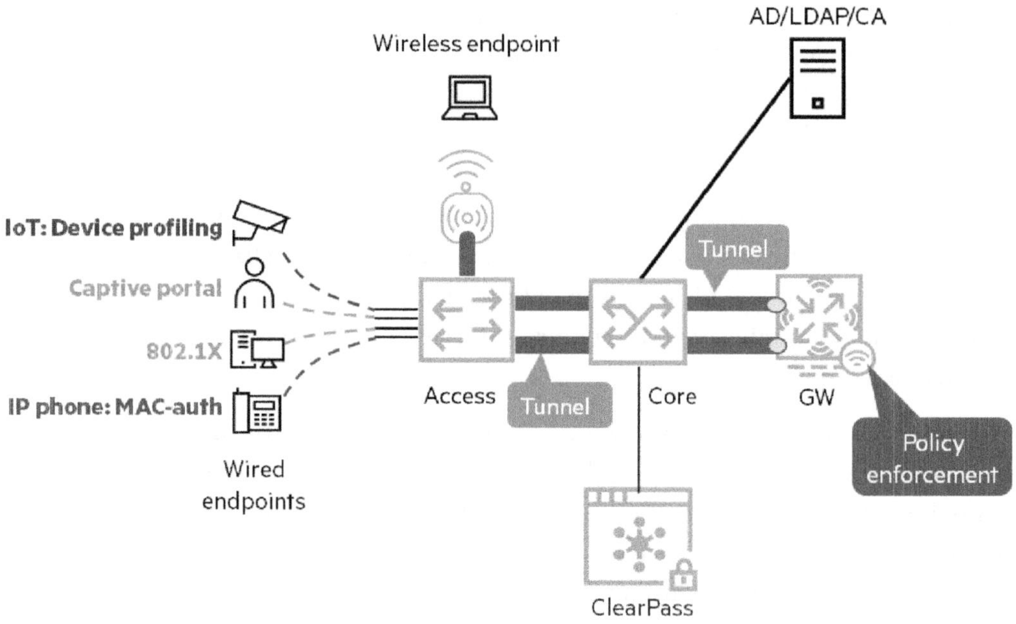

Figure 12-1: Typical Dynamic Segmentation implementation

Dynamic Segmentation allows network administrators to assign specific policies based on user roles, ensuring that each user receives the appropriate level of access and security. For instance, at a university, IP phones, students, and a faculty member might connect to the same access network (wired and wireless), but UBT ensures their traffic is tunneled and treated according to their respective policies (roles). This level of granularity not only enhances security but also optimizes network performance by ensuring that bandwidth and resources are appropriately allocated.

How Dynamic Segmentation addresses security challenges

All modern desktops, laptops, tablets, and smartphones support robust security mechanisms like 802.1X/EAP. But many devices lack these security mechanisms. This includes specialty devices and some devices that rely on Power over Ethernet (PoE), such as security cameras, payment card readers, IoT, and medical devices.

Without Dynamic Segmentation, security policies are often static and applied broadly across the network. This approach lacks the granularity needed to address varying security requirements for different devices and users, increasing the risk of unauthorized access and data breaches.

A switch with Dynamic Segmentation improves your security stance while easing administration. It authenticates these devices using ClearPass and tunnels the traffic to a mobility controller, which centralizes formerly disparate ACLs and firewall rules into a single, intuitive policy system. You can continue to use these devices without compromising network security.

Operational overview of Dynamic Segmentation

ClearPass profiles each device, deems appropriate endpoints as "acceptable," and sends accept messages to the controller. The messages include tunnel information, assigned VLAN, and secondary role. In **Figure 12-2**, a user was profiled into the Finance role, and so their port is configured for VLAN 15 and their traffic is securely tunneled to the controller. The controller applies centralized policy enforcement to limit finance traffic as appropriate. Printers and cameras are similarly treated, as shown in **Figure 12-2**.

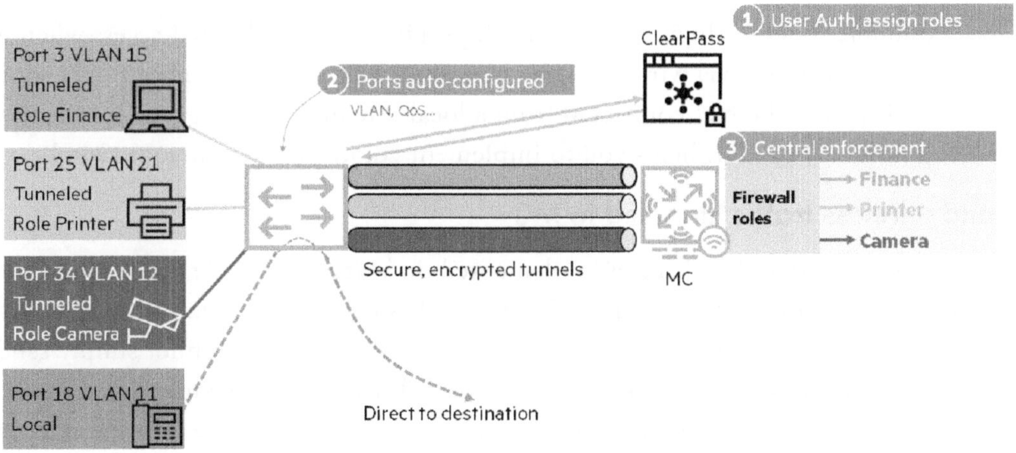

Figure 12-2: Dynamic Segmentation operation

ClearPass can also indicate if device traffic should be locally forwarded by the switch directly toward its ultimate destination, without tunneling to the controller. You sacrifice some of that centralized Policy Enforcement Firewall (PEF) enforcement, but you accommodate devices that are sensitive to delay and jitter. You just do not get all the benefits of this "tunneled node" solution.

Role of ClearPass

ClearPass identifies and classifies devices connecting to the network by analyzing attributes such as device type, operating system, and manufacturer. This device fingerprinting helps in applying specific policies tailored to different device types. It continuously monitors network traffic and updates user and device profiles with contextual information such as location, time of access, and behavior patterns. This profiling enables adaptive policy enforcement that responds to changes in context.

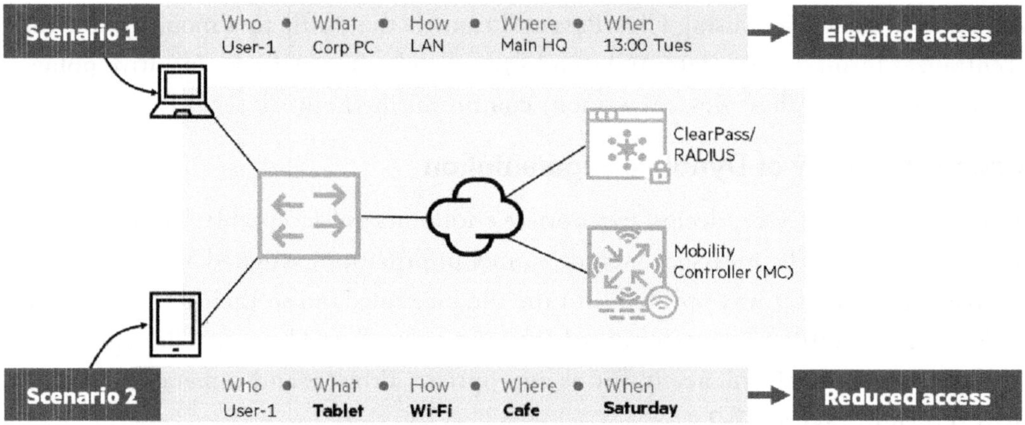

Figure 12-3: ClearPass example

In the above example, User-1's login credentials might give the user elevated access when using their corporate PC connected via wireless at the main building, during business hours. But when the user uses their personal tablet connected from a local café on Saturday, they may have reduced access. Among other things, this helps you to implement colorless ports on your switches.

Wired and wireless traffic tunneled to MC

ClearPass and port-based authentication implemented on switches provide one method for unifying access policies. Both the MCs and AOS-CX switches can enforce 802.1X authentication to users and apply the policies returned by ClearPass. However, ClearPass cannot simply send the same settings to both MCs and switches. For example, ClearPass often simply sends a user role assigned to an MC, and the MC has firewall rules associated with that role. On the other hand, ClearPass would need to send a policy with dynamic ACLs to apply similar controls to an AOS-CX switch.

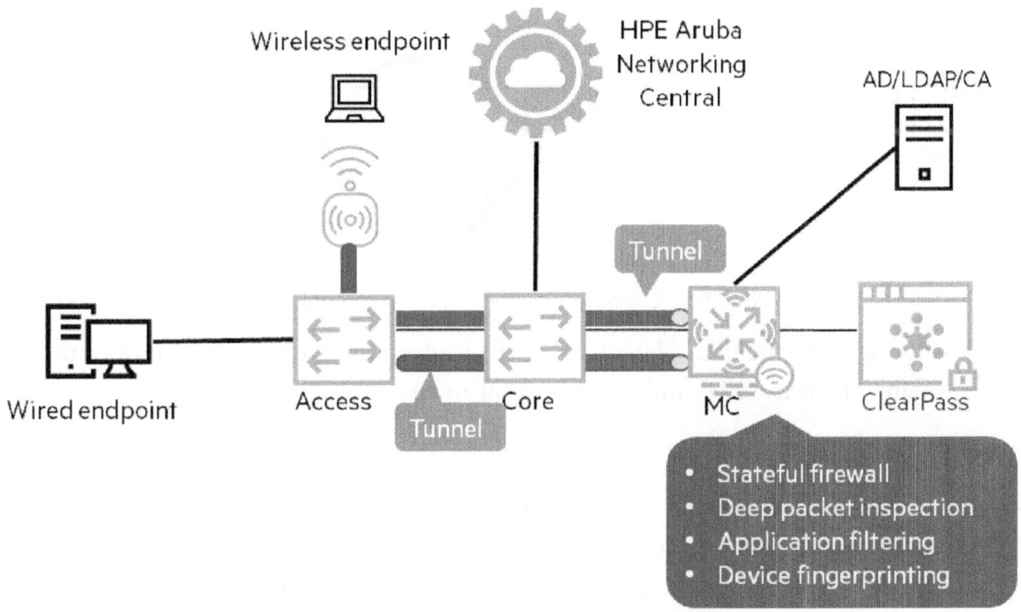

Figure 12-4: How wired and wireless traffic is tunneled to MC

Tunneled node provides tighter and simpler unification of wired and wireless access. Switches simply tunnel traffic to the MC, which handles the traffic much as it would handle traffic tunneled from an AP. An employee can connect through their desktop, log in, be assigned a role, and receive customized access. Later, the employee can connect wirelessly using a laptop in a meeting room, log in, be assigned the same role, and receive the same access.

The bottom line is this: Use the tunneled-node feature whenever you want to apply HPE controller-based security/control mechanisms to both wireless and wired traffic. You get unified access control—the same policies, regardless of whether they have a wired or wireless connection.

Dynamic Segmentation is supported on both HPE Aruba Networking switching products: AOS-CX and AOS-S switches.

Communication between switch and MC

The switch uses two protocols to connect to an MC.

The control plane uses PAPI (UDP port 8211), the same protocol used by AP-to-MC communications. However, where APs use IPSec to protect the PAPI connection between the AP and MC, AOS-CX switches do not support this protection. Instead, you can optionally implement an MD5 HMAC function to protect PAPI between the AOS-CX switches and MCs.

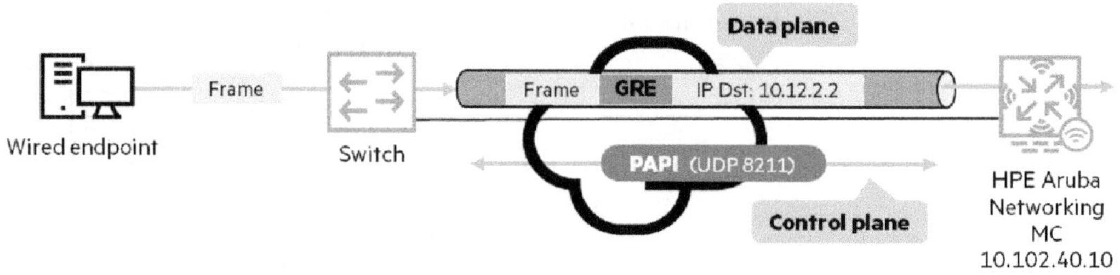

Figure 12-5: Communication between switch and MC

As shown in **Figure 12-5**, the data plane uses GRE. Endpoints send standard frames to the switch, which encapsulates them in GRE and tunnels them to the MC.

Enhanced PAPI security

The switch-MC PAPI control plane traffic creates a potential security risk. Rogue devices could send PAPI messages to APs, MCs, and switches and disrupt functionality. They could also try to insert spoofed PAPI messages on the controller to disrupt communications or obtain information about the network.

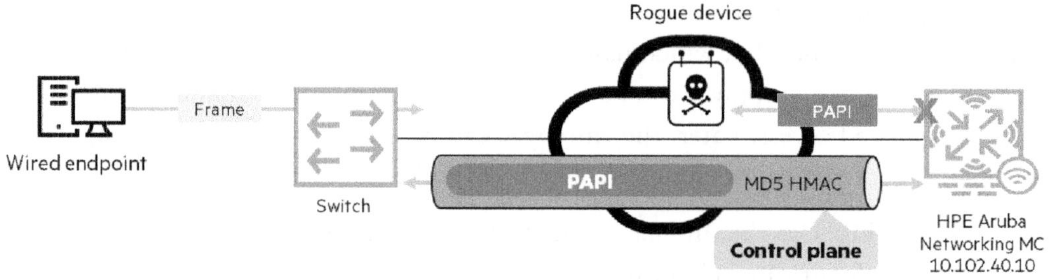

Figure 12-6: PAPI security

PAPI enhanced security minimizes this risk by using an MD5 and user-configurable key to hash the message. If a rogue device tries to spoof or tamper with a message, the hash will not check out because the rogue does not know the correct key to create the hash. A primary reason for this enhanced security is to prevent rogue switches from consuming AP licenses on the controller.

You must configure matching keys on the MC and switch, between 10 and 64 characters.

> Enhanced security mode is disabled by default.

How the MC handles tunneled-node traffic

The MC divides the arriving tunneled traffic into one tunnel per tunneled-node port. It decapsulates the traffic and applies various controls. Learning how to configure the MC to control tunneled-node traffic is beyond the scope of this guide, but you should understand how the configuration affects the MC's decisions.

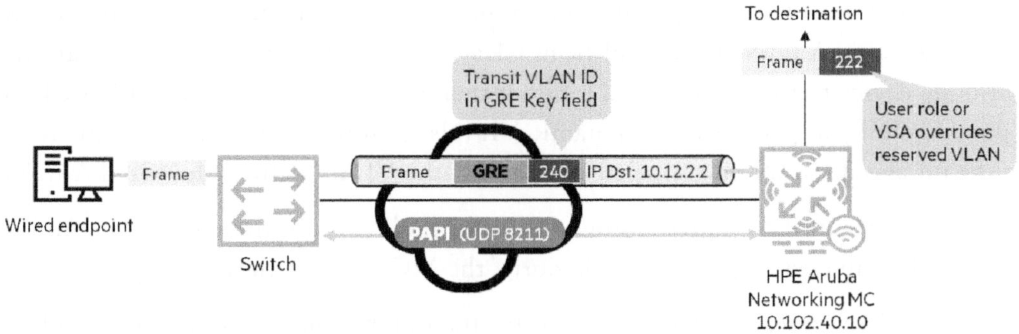

Figure 12-7: How MC handles traffic

The switch communicates the untagged VLAN configured on a tunneled-node port to the MC. The MC must have this VLAN configured on it. It then handles the tunneled traffic based on the settings in the wired AAA profile assigned to that VLAN. For example, the MC might apply MAC-auth or 802.1X, or some combination. After successful authentication, the controller applies a role to the traffic. Based on that role, it controls traffic with firewall policies and other policy actions. Finally, it forwards the packet towards its destination.

You must also set up the MC to forward the traffic in the tunneled-node ports' untagged VLAN at Layer 2 or Layer 3, as desired. In **Figure 12-7**, the MC uses Layer 2 forwarding; it forwards the frame tagged for VLAN 240. If the MC were using Layer 3 forwarding, it would have an IP address on VLAN 240 and route traffic out of this VLAN.

You can configure different VLANs on different tunneled-node ports to inform the controller to assign those nodes to different VLANs. However, the MC can also apply different firewall rules to different devices in the same VLAN, again based on dynamic settings applied during authentication. For example, the MC could enforce MAC-auth and use the devices' MAC addresses to assign different roles to different devices. You should coordinate with MC administrators and make sure to match the switch configuration with their desired configuration.

Keep in mind that the switch is simply using the untagged VLAN on the tunneled node port to communicate the desired default VLAN to the controller. It is not switching the tunneled-nodes' traffic in this VLAN, so you do not need to tag it on the switches' uplinks as you normally would. (Using the fallback switching mode makes an exception, as you will learn later.) Instead, the VLAN exists on the MC at the distribution or core. In keeping with this design, the switch is restricted from implementing many normal features—such as an IP address—on any untagged VLAN on a tunneled-node port. HPE recommends reserving the tunneled-node VLANs for tunneled-node ports only.

You should also understand that the MC can override the default VLAN assignment after it authenticates the device. For example, it might apply 802.1X authentication to tunneled nodes and receive a dynamic VLAN assignment from ClearPass. The MC would then forward the tunneled-node device traffic in this VLAN, rather than the untagged VLAN configured on the switch port. Even if you plan to apply dynamic VLANs for some devices, though, the MC must support the untagged VLAN on tunneled-node ports; the VLAN defines the wired AAA profile.

Finally, note that **Figure 12-7** shows a single tunneled node port on the switch for simplicity. When you apply tunneled-node to multiple ports, the MC sees one tunnel per port.

The term "transit" or "reserved" VLAN is used for the number placed in the GRE header's Key field. This helps the MC determine the profile policies that should be used, by default, if nothing else is received, like the security policies and VLAN that should be applied to the user. In the previous figure, the AAA server did not send a VLAN attribute, so what was used in the VLAN header is the final VLAN for the user on the MC. Note that this VLAN must be pre-configured on the MC. If the AAA server had passed down a specific VLAN attribute, like VLAN 888, then this would have become the user's final VLAN number and what the MC would use to tag the user's traffic before forwarding it onward.

Allowing traffic between the switch and MC

You should check the network infrastructure between the AOS-CX switches that implement tunneled node and the MCs. Make sure that the infrastructure will not interfere with communications in any way.

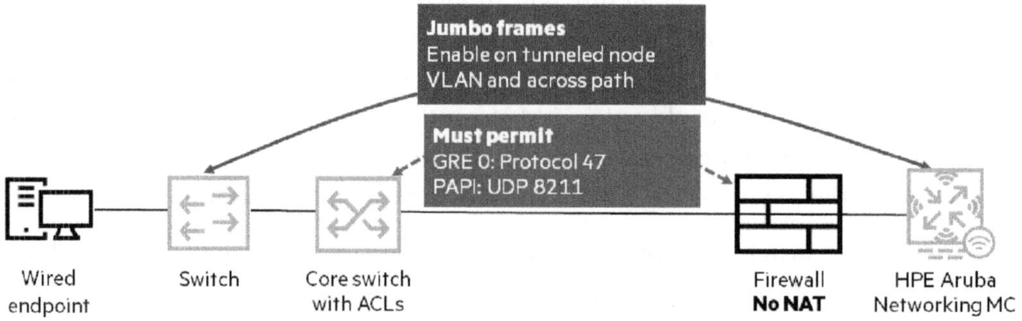

Figure 12-8: Traffic that needs to be allowed between the switch and MC

If any firewalls or network infrastructure devices with ACLs are in the path, they must allow GRE and PAPI traffic. Enable GRE on IP protocol 47 and PAPI on UDP port 8211.

You must also understand that NAT cannot be applied in the path between the switch and the controller. It is best if you can direct the tunneled-node switch to a local MC.

Also keep in mind that encapsulating frames can cause them to exceed the MTU. However, tunneled-node ports do not support the features typically used to prevent MTU violations, including fragmentation, PMTU detection, or reassembly. HPE recommends that you enable jumbo frames across the path from the switch to the MC. On AOS-CX switches, you enable jumbo frames on a VLAN basis. Make sure that every VLAN that carries traffic between the switch and the MC supports jumbo frames.

Make sure that you know the IP address that your AOS-CX switches will use to establish the tunnel with the MC. Typically, the switch acts at Layer 2 and has a single IP address and a default gateway for reaching all other subnets. But if your switch has multiple IP addresses, it uses the IP address on the forwarding VLAN for its route to the MC. Make sure that the MC can also reach this IP address (through its own route or its default gateway).

User-based tunneling

UBT uses the concept of a colorless access port. It does not matter what you connect to the port; roles and policies are assigned per device.

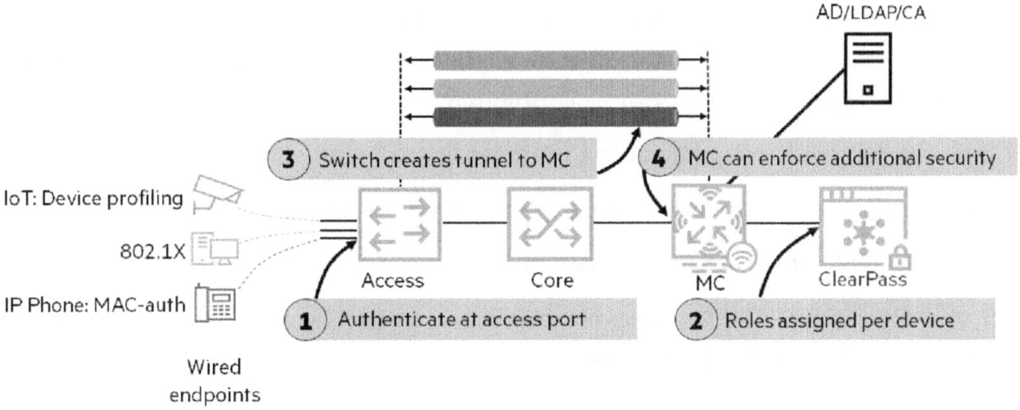

Figure 12-9: Example of UBT

As shown in **Figure 12-9**, a device connects to an access port—maybe an IoT device. Authentication happens at the port level, and ClearPass assigns a role. This role dictates how to color the port with VLAN, QoS, ACL, and so forth. The switch tunnels the end user traffic to the MC, which can enforce additional security. Perhaps later, the IoT device is disconnected, and a corporate user or IP phone connects to that same port. The port is colored, tunneled, and secured as appropriate.

UBT features

UBT offers the following features:

- **Secured and flexible control of access layer:** With ClearPass or switch configuration, only traffic from a specific user/device role is sent to the MC Policies (such as QoS, ACL, rate-limit) can be enforced at tunneled ports or at the controller.
- **Access to controller's applications:** Users can access the controller's applications, such as stateful firewall and AppRF.
- **Policy enforcement** is achieved by local user roles or downloadable user roles.
- **Local user roles** are configured on the switch.
- **Downloadable user roles** are configured on ClearPass and pushed to the switch.
- **High availability and scalability**.
- **Load balance** to multiple controllers for high scalability.
- **Stateful failover** to standby mobility controller.

Port access user roles

Every device that connects to a switch port where port access is enabled is associated with a role. Roles are associated with all clients, both authenticated and unauthenticated, and applied to each user session. By default, roles are enabled on a switch. Following are a few examples of user role names and the access privileges that can be configured:

- Employee—Provide complete access to network resources.
- Contractor—Provide limited access to network resources.
- Guest—Provide only Internet browsing access.

Each user role determines the client network privileges, frequency of re-authentication, applicable bandwidth contracts, and other permissions

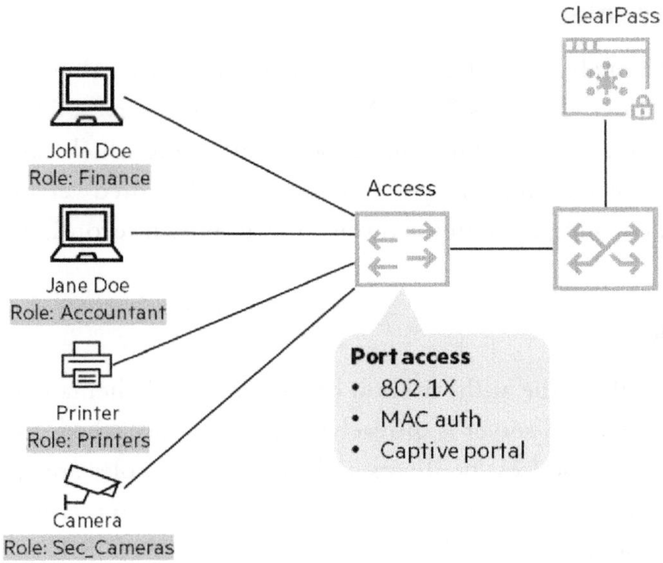

Figure 12-10: User role assignment

When roles are enabled, they are applied to all devices connected to ports where authentication is configured. Special roles, such as, critical, reject, pre-auth, and auth are applied depending on the authentication state of the device.

Roles can be applied in one of the following two ways:

Method 1: Vendor-Specific Attribute (VSA)-Derived Role

Type: RADIUS: Aruba

Name: Aruba-User-Role

ID: 25

Value: <user-role> -

The RADIUS server (ClearPass Policy Manager server) determines how the VSA-Derived Role is applied to the user. The role is sent to the switch through a RADIUS VSA. The VSA derived role will have the same precedence order as the authentication type (802.1X, MAC authentication).

Method 2: User Derived Role (UDR)

The UDR is applied when the roles are enabled. UDR will have the same precedence order as the authentication type (802.1X, MAC authentication).

Downloadable user roles

Downloadable user roles enable AOS-CX switches to download user roles, policy, and class from the HPE Aruba Networking ClearPass Policy Manager Platform server. The download facilitates the configuration of policies and attributes for a specific user role which can then be stored locally on the switch. New users can then be assigned the same locally stored version of the user role in Policy Manager server, enabling the administrator to save time in reconfiguring each user individually.

Cached-critical role

The cached-critical role allows the authorization of authenticated clients with the previously applied roles when the RADIUS server is unreachable. When the cached-critical user role feature support is enabled, the MAC address of clients and their applied roles are cached in the system during the client log-off or re-authentication. When the RADIUS server is unreachable, the cached-critical role is applied as a special role to the client. Once the RADIUS server is reachable, cache details are cleared from the switch. The cached-critical role can be enabled at the global or per-interface level.

The cached-critical role can be applied only to the authentication-enabled ports. A maximum of 1024 client entries can be cached in the switch. You can configure a timeout period during which the client details remain cached. On a timeout, the cached entry is removed from the switch within the buffer time of 30 minutes. When the number of clients exceeds the maximum limit of 1024, the first in, first out (fifo) replace mode can be used to replace the oldest cached entry in the switch with a new entry. When the maximum limit of 1024 is reached, the following event log is generated: Reached the maximum cached-clients limit of 1024 on the switch for Limited Auth-Survivability.

Special roles

Special roles are always local roles that are created to support instances when the deployments are not yet complete. They are also helpful when any network issues occur during authentication of devices. Special roles are always purpose-based, that is, they are applied only in instances such as the RADIUS server is not reachable for authentication or when a single role has to be applied across all switches. The following special roles are available:

- Critical role
- Reject role
- Pre-authentication role
- Auth-role
- Fallback role

Critical role

The critical role is applied to devices when the RADIUS server is unreachable during the first authentication process or during reauthentication. This role helps ensure that the devices have limited access to the network even though the authentication is not completed. Once the RADIUS server is available for authentication, the devices are authenticated and the ultimate role is applied.

Reject role

The reject role is applied when the RADIUS server rejects a device during authentication. The reject role gives restricted access to the device compared to a full access role.

Pre-authentication role

The pre-authentication (pre-auth) role allows a device, such as an IP phone, to have network access before the device is authenticated. The pre-auth role is triggered when a MAC-based client is connected to a switch before being authenticated by the RADIUS server. Devices must be assigned a VLAN to provide network connectivity. Two new VLANs are created for pre-auth role functionality, one for voice traffic and one for data traffic. Pre-auth role VLANs can be configured on the switch individually or within a user-role. Devices that can be connected to the switch without authentication are divided into two categories:

- Devices that send voice traffic.
- Devices that send data traffic

Auth-role

After the RADIUS server authenticates a device, if no role is configured on the server, it sends an empty access accept packet to the switch. The switch translates this empty packet to assign an auth-role to the device. The switch then checks if an auth-role is configured on that port and assigns this role to the device.

Fallback role

The fallback role is applied to onboarding devices when there is no derived role available for the devices. Following are the conditions for the fallback role to be applied on onboarding devices:

- The device profile local MAC match feature with block-until-profile-applied mode is configured.
- Device profile along with AAA is configured but no match was found for the device profile client.
- AAA method with no reject or critical role is configured, and the connection to RADIUS server failed.

- 802.1X authentication is enabled on the port, but the supplicant of the device timed out to respond to the authentication request.

One advantage of using user roles together with user or device authentication is the option to ensure seamless access levels for users anywhere in the network, including supporting user Mobility while keeping the security and access levels.

User roles - parameters

A user role consists of the following optional parameters:

- Inactivity-timeout - The inactivity timeout period in seconds with a range of 300 to 4294967295 for the authenticated client for an implicit logoff.
- Reauth-period - Sets the reauthentication period in seconds or 0 to disable.
- VLAN access - Sets the untagged VLAN ID.
- VLAN trunk - Sets the tagged VLAN ID.
- Auth-mode - Configures the authentication mode for the clients that are associated with the current role. Available modes are: client-mode, device-mode.
- MTU - Configures the MTU support for the client.
- VLAN trunk allowed - Specifies the list of tagged VLANs configured for the interface.
- Trust-mode - Configures the QoS trust mode for the client.
- Private-VLAN - Configures PVLAN port type for a user role. The following are the attributes: o promiscuous Configures the port type as promiscuous. o secondary Configures the port type as secondary.

UBT user roles

Local User Role (LUR): You define user roles locally on the switch. First, a device connects and authenticates. Then, the RADIUS server tells the switch which of its LURs to apply by sending a User-Role name VSA. You can use ClearPass or a a third-party AAA server with LUR.

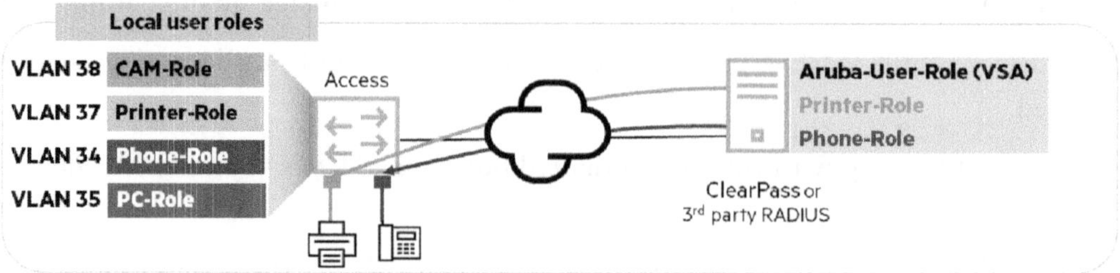

Figure 12-11: Local user roles

Downloadable User Role (DUR): You do not define roles on the switch. Instead, you define them centrally on a ClearPass server—no third-party RADIUS support. When a device connects and authenticates successfully, the switch downloads the User-Role to the switch and the switch applies it to the device.

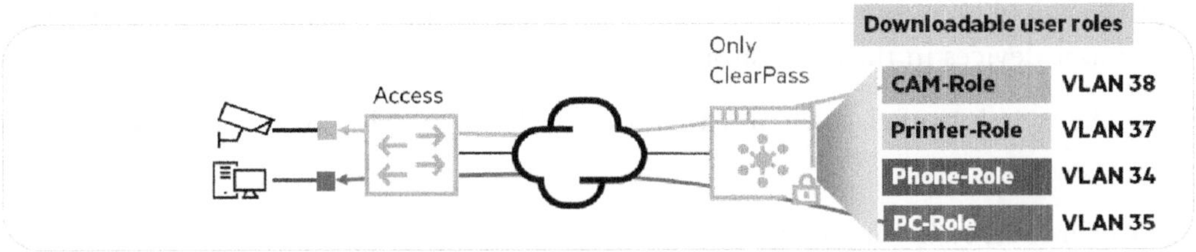

Figure 12-12: Downloadable user roles

On the older AOS-S switches, the term "secondary role" was used instead of gateway role. With AOS-CX switches, the appropriate term used to describe the role the switch passes to the controller is the gateway role.

With AOS-CX release 10.4, HPE does not support role download for a gateway (controller) or secondary role to the mobility controller. Either the role must preexist on the MC, or the user goes through a second authentication and the role can then be downloaded to the MC.

Local switched traffic versus UBT traffic

Local switching is typically used when delay-sensitive traffic is involved between access-layer devices, like voice or video communications (VoIP phones, for example), or a third-party firewall already exists in the network and the company wants to continue using the policy function of that firewall.

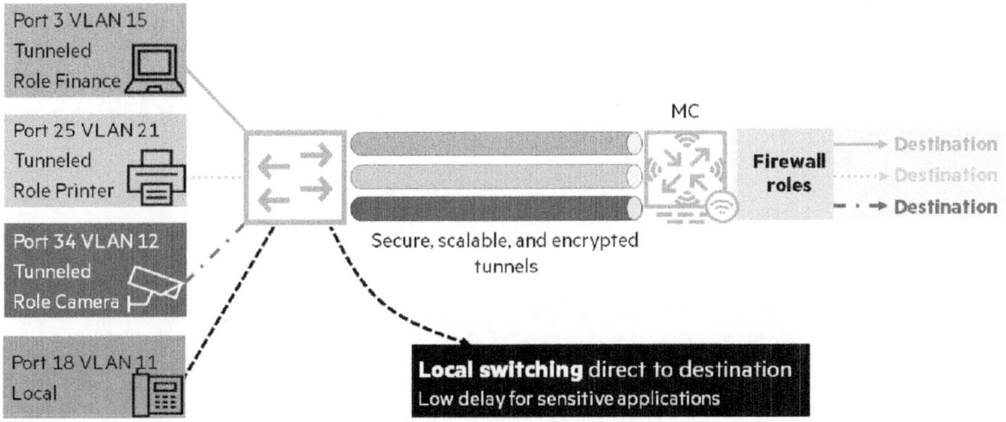

Figure 12-13: VoIP traffic switches locally

However, local switching lacks the centralized policy and enhanced security of UBT. Traffic filtering is limited to ACLs, with no stateful or application inspection features. In many cases, companies use only VLANs for security at the access layer, where devices in a VLAN can freely interact with each other. Their traffic is only processed by the firewall if it leaves the subnet. Allowing this type of promiscuous behavior can lead to many security issues—user devices can attack other user devices in the same subnet.

With UBT, the switch tunnels authenticated user traffic to an MC, to be processed by security policies.

Advantages over local switching include:

- **Centralized security policies for both wired and wireless traffic.** Users have a consistent experience, whether they connect via wired Ethernet or Wi-Fi.
- **Endpoint traffic must be processed by the MC** before it can be forwarded to any other device. This allows the MC to apply stateful and application inspection policies, as well as other security features, to better secure the traffic, greatly limiting user-to-user threats in the same VLAN.
- **Easy scalability is provided by the MCs**, where the 7280 MC supports a theoretical stateful firewall throughput of 100 Gbps.
- **Easier to implement security policies because only one product is primarily used for the user traffic MCs**. The problem with a company that has MCs for wireless and a third-party firewall is that it becomes very difficult to ensure that policies are consistent between the third-party firewall and the MCs.

Traffic switching process

The following flowchart helps you understand how a switch can easily decide that the client traffic is tunneled to the MC or is locally switched.

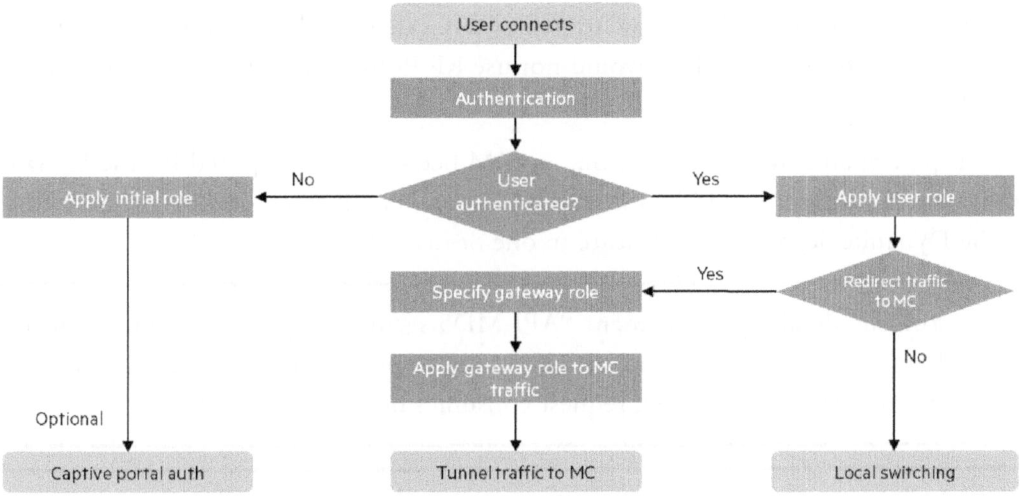

Figure 12-14: Traffic switching process

License requirements for UBT

Initially, HPE did not require any licensing to implement Dynamic Segmentation. Starting with AOS-8.4 on the MCs, the switch is treated as an AP and will consume licenses as an AP. Note that if you set up a virtual switch (VSF) that is comprised of multiple physical AOS-CX switches, only the virtual switch will consume licenses.

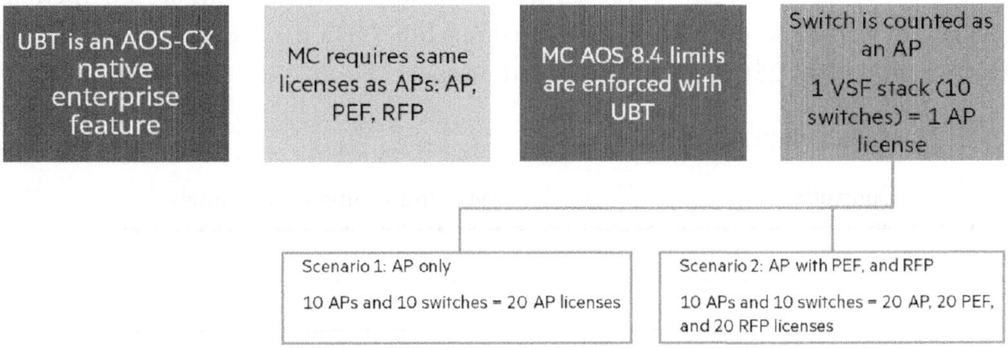

Figure 12-15: Licenses required for UBT

A standalone MC (or centralized MM-based solution) needs one AP license for each switch that is running Dynamic Segmentation. If the solution is using firewall functionality, it also needs one PEF license for each switch implementing Dynamic Segmentation.

If the MC is using the RF Protect features, it should have one RFP license for each Dynamic Segmentation switch as well. The RFP license is not strictly required. However, licensing works such that when RF Protect is enabled, the MC should have the same number of AP, PEF, and RFP licenses; if it has fewer RFP licenses, it disables the excess AP and PEF licenses until it has the same count. The company might choose to deploy MCs dedicated to supporting the wired tunneled nodes. In this case, the MC would not use RF Protect, and the RFP licenses would not be required.

If the MC is part of an MM-based solution, an MM license is also required for the Dynamic Segmentation switch, just as it would be for an AP. To conserve licenses, you may choose to consolidate the Dynamic Segmentation feature to one or just a few switches.

You should always implement PAPI MD5 security to protect communications between the controller and switch, as well as protecting against malicious misuses of licenses, since each switch request consumes licenses on the MC.

Licenses are consumed per switch not per port. A switch stack is considered a single switch for licensing purposes.

Dynamic Segmentation scalability on MC and switch

The following table shows the number of supported tunnels for both the MCs and the AOS-CX switches. Typically, the number of tunnels required to implement Dynamic Segmentation is not a concern if you are implementing one tunnel per port on a switch. However, if you are implementing multiple tunnels on a port because there are multiple downstream devices (perhaps another switch or a blade server with many virtual machines in a data center), then you must ensure that no more than 256 tunnels are created per port.

Table 12-1: Tunnel scalability on MC

Controller	Maximum supported tunnels
7280	32768
7240/7240XM	32768
7220	16384
7210	8192
7205	4096
7030	1024
7024	512
7010	512
7008	256
7005	256

Table 12-2: Tunnel scalability on AOS-CX

Switch series	Maximum supported tunnels
CX 6300 Series—Per switch/stack	1000
CX 6300 Series—Per port	256
CX 6400 Series—Per switch/per system	1000
CX 6400 Series—Per port	256

As mentioned earlier, architects should consider the Dynamic Segmentation feature as they plan. In addition to the licensing requirements, they should keep in mind tunneling requirements. The MC treats each user as a separate tunnel. Architects should also remember that a single tunneled-node port could connect to multiple devices (if, for example, the port connects to an unmanaged switch), which could raise the user count. The MC must support these additional tunnels.

Finally, architects should consider the effects of the wired traffic on the firewall throughput and whether the controller can support the bandwidth; if not, additional controllers will need to be purchased.

Of course, you must also implement the policies for wired traffic control. However, configuring the MC to control and forward the traffic is beyond the scope of this guide and should be handled by an experienced HPE Aruba Networking controller administrator.

Also, make sure that your MCs are running up-to-date software. Minimally, AOS-8.4 is required on your MCs to support user-based mode on the AOS-CX switches.

Configuring UBT

Configuring UBT in AOS-CX switches involves setting up tunnel interfaces, creating LURs if required, integrating with ClearPass for role-based access control, and defining policies that dictate how traffic is managed based on user roles and device types. This setup not only enhances security by isolating user traffic but also simplifies network management by automating policy application and ensuring consistent enforcement across the network.

Configuring UBT with LURs

Configuring LURs for UBT in AOS-CX switches involves creating user roles directly on the switch, defining policies for these roles, and then applying these roles to user sessions.

To configure LURs, you must first configure 802.1X, MAC-auth, or captive portal on your AOS-CX switch, along with connectivity to an AAA server like ClearPass.

Configuring the zone

To set up a UBT LUR, you first need to define a zone:

First, you create the zone and define the VRF instance that will be used. Next, you define the primary, and optionally, backup IP address of the controllers. Unlike the APs, the switches currently do not support Control Plane Security (CPSec). Instead, you can optionally protect the PAPI control messages using MD5 signatures. The key that you specify must match what the MCs have configured. Last, you need to enable the zone. Optionally, you can define which IP address the switch should use when tunneling the GRE packets (by default, this is based on the local routing table).

```
switch(config)# ubt zone <ZONE-NAME> vrf <VRF-NAME>
switch(config-ubt-zone)# primary-controller ip <IP-ADDRESS>
switch(config-ubt-zone)# backup-controller ip <IP-ADDRESS>
switch(config-ubt-zone)# papi-security-key <KEY>
switch(config-ubt-zone)# enable
switch(config)# ip source-interface ubt {interface <IFNAME> | <IPV4-ADDR>} [vrf
<VRF-NAME>]
```

> In AOS-CX 10.13, eight zones are supported per switch/stack.

To verify your configuration, there are various `show` commands that you can use:

```
switch# show ubt

Zone Name                    : zone1
UBT Mode                     : local-vlan
Primary Controller           : 10.116.51.10
Backup Controller            : 10.116.51.11
SAC HeartBeat Interval       : 1
UAC KeepAlive Interval       : 60
VLAN Identifier              : 4094
VRF Name                     : default
Wake-on-LAN Enabled-VLANS    : -NAAdmin
State                        : ENABLED
PAPI Security Key            : AQBapdxySvGPvdTl ... bL4FE=
Operational State            : Up
Zone Name                    : zone2
UBT Mode                     : local-vlan
Primary Controller           : 1.1.5.10
Backup Controller            : 1.1.5.11
SAC HeartBeat Interval       : 1
UAC KeepAlive Interval       : 60
VLAN Identifier              : 4094
```

```
VRF Name                     : blue
Wake-on-Lan Enabled-VLANS : -NAAdmin
State                        : ENABLED
PAPI Security Key            : TRQapdxySvGPvdTlkYn1 ... zP4FE=
Operational State            : Up

switch# show ubt information

==========================================================
Zone zone1:
==========================================================
SAC Information :

  Active        : 192.168.10.8
  Standby       : -NA

Controller is in Standalone mode

Wake-on-LAN VLAN boostrap information:
    Active-SAC registered vlans    :
    Active-SAC failed vlans        : 200
    Active-SAC failure reason      : Failed in controller
    Standby-SAC registered vlans   :
    Standby-SAC failed vlans       :
    Standby-SAC failure reason     :

==========================================================
Zone zone2:
==========================================================
SAC Information :

  Active        : 20.1.1.2
  Standby       : 20.1.1.3

Node List Information :

  Cluster Name          : cluster2

  Cluster Alias Name    :

  Node List      :
  ----------------
    20.1.1.2
```

```
    20.1.1.3
    20.1.1.4

Bucket Map Information :

Bucket Map Active   : [0...255]

Bucket ID  A-UAC        S-UAC       Connectivity
-----------------------------------------------
0          20.1.1.2     20.1.1.3    L2
1          20.1.1.3     20.1.1.4    L2
2          20.1.1.4     20.1.1.2    L2
Wake-on-LAN VLAN boostrap information:
    Active-SAC registered vlans  : 10-20,30,40
    Active-SAC failed vlans      :
    Active-SAC failure reason    :
    Standby-SAC registered vlans : 10-20,30,40
    Standby-SAC failed vlans     :
    Standby-SAC failure reason   :
```

Define transit VLAN

To define the transit VLAN, use the following configuration:

```
Switch(config)# vlan <VLAN-ID>
Switch(config-vlan)# exit
Switch(config)# ubt-client-vlan <VLAN-ID>
```

Remember that the transit VLAN is what the switch will use when tunneling the user traffic to an MC (if the switch is not performing traffic switching).

Creating local user roles

```
switch# configure terminal
switch(config)# user-role <ROLE-NAME>
switch(config-user-role)# rule 10 action permit any
switch(config-user-role)# exit
```

Configuring a gateway role

Next, you need to define the role names locally on the switch:

```
switch(config)# port-access role <ROLE-NAME>
switch(config-pa-role)# gateway-zone zone <ZONE-NAME> gateway-role <CONTROLLER-
ROLE-NAME>
```

The role name in the `port-access role` command must exactly match the Aruba-User-Role VSA that the AAA server, like ClearPass, will pass back to the switch. In the role configuration, you must define the zone that this is used for and then the role name that is passed to the MC. Note that the switch role name and the MC role name could be the same, or different, names. The controller role name is also case-sensitive. See the *Security Guide* for other configuration options for the role. For example, you could define a policy to define a preauthentication VLAN, ACL, QoS, and other parameters that should be used before the user is authenticated.

Enable user authentication

After you have configured the role information, you then need to set up user authentication, like MAC or 802.1X (captive portal is also supported). Here is an example with 802.1X:

```
Switch(config)# aaa authentication port-access dot1x authenticator auth-method
eap-radius
Switch(config)# aaa authentication port-access dot1x authenticator radius server-
group radius
Switch(config)# aaa authentication port-access dot1x authenticator enable
Switch(config)# interface <interface-id>
Switch(config-if)# aaa authentication port-access dot1x authenticator
Switch(config-if-dot1x-auth)# enable
Switch(config-if-dot1x-auth)# exit
```

You can define a preauthentication role on the interface—this is applied to the user's traffic prior to authentication.

```
Switch(config-if)# aaa authentication port-access preauth-role <ROLE-
NAME>
```

Also, if you need to exempt LLDP, CDP, and/or BPDU traffic, use the following two commands:

```
switch(config-if)# aaa authentication port-access allow-lldp-bpdu
switch(config-if)# aaa authentication port-access allow-cdp-bpdu
```

This configuration might be necessary for devices like a phone or camera, where LLDP (or Cisco's CDP) must be exchanged for the device to operate correctly prior to authentication occurring on the port.

DUR setup

Recall that you define DURs, which are downloaded to the switch after successful authentication, on the ClearPass server. Because this happens via a REST API using an SSL connection, you must install an HTTPS certificate on the edge switch. Also, a username/password must be configured to authenticate the switch with ClearPass. Here is how it works:

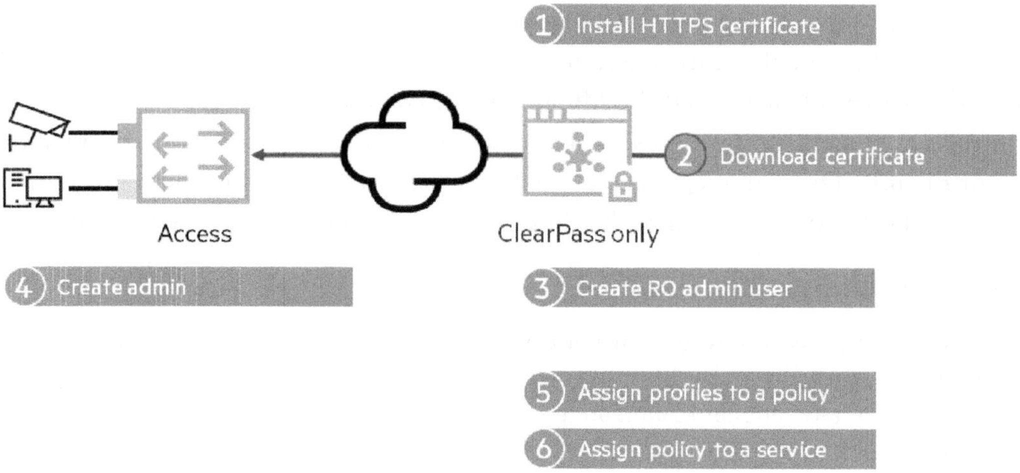

Figure 12-16: DUR setup

1. First, ensure that ClearPass has obtained a valid HTTPS certificate.
2. Automatically download the certificate from ClearPass (requires ClearPass 6.7.8+ or later).
3. On ClearPass, create a read-only admin user. The switch uses this to authenticate and download role information.
4. On the switch, create a user that matches the one created on ClearPass. Enable downloadable user roles and configure the downloadable user roles profiles.
5. Assign the profiles to a policy and the policy to a service on the ClearPass server.

You can do this with a cluster of controllers, or with a stand-alone controller.

Because the login process from the AOS-CX switch uses HTTPS, the switch must validate the ClearPass server certificate. On the older AOS-S switches, they would automatically download and install the ClearPass certificate. Unfortunately, in AOS-CX 10.4, this process is not automatic. You must access the ClearPass server, download its SSL certificate, and install it on the AOS-CX switch.

UBT/local switching and user roles are two different, but related, concepts. They are both used to implement Dynamic Segmentation. With AOS-CX switches, some devices could be locally switched with a predefined local role, perhaps for VoIP phones. For other devices, their role information could be downloaded from ClearPass and the traffic tunneled to an MC. It is flexible in how you want to set it up. However, HPE recommends that if you are using ClearPass and have either

AOS-CX or the older AOS-S switches, that you implement DUR because of the scalability and ease of centralized policy management that it provides.

Configuration tasks

You must complete the following actions to set up the DUR with ClearPass:

1. Enable MAC/802.1X authentication on the switch.
2. Configure the ClearPass server to send Aruba-User-Role (VSA) to the switch.
3. Configure the zone on the switch:
    a. Define the controller.
    b. Enable MD5 security for PAPI.
4. Define the transit VLAN on the switch and the MC.
5. Configure ClearPass with user roles for the DUR.
6. Define the ClearPass RADIUS server and key to enable the secure communication between the switch and ClearPass.
7. Download, install, and validate the ClearPass server certificate on the switch.
8. Define REST API access credentials: Create an admin account on ClearPass with the appropriate privilege level and specify the administrator credentials on the switch.

As a good practice, set up DNS on the switch and use the ClearPass server host-name FQDN on the server SSL certificate.

As a good practice, manually download/import the ClearPass certificate on the switch and configure ClearPass and the switch to use the same NTP source.

Configuring ClearPass

The process for DUR is very much the same as LUR. Instead of having the user role configured on the edge switch, however, the user role is configured on ClearPass and sent to the edge switch after successful authentication. The method for sending the user role is through REST API using an SSL connection. This means that an HTTPS certificate has to be installed on the edge switch. In addition, a username/password has to be configured to authenticate the switch with ClearPass.

Make sure that the following tasks are successfully completed on ClearPass:

1. First, ensure that ClearPass has obtained a valid HTTPS certificate.
2. Automatically download the certificate from ClearPass (6.7.8 or a later version is required).

3. On ClearPass, create a read-only admin user that will be used by the switch for authentication in order to download the actual role information.
4. On the switch, create the user entry for the read-only user, enable downloadable user roles, and configure the downloadable user roles profiles.
5. Assign the profiles to a policy and the policy to a service on the ClearPass server.

You need to configure a role enforcement policy, enforcement profile, and an authentication service by using the following steps:

Step 1

Ensure the profile matches local Aruba-User-Role (switch) to allow user access.

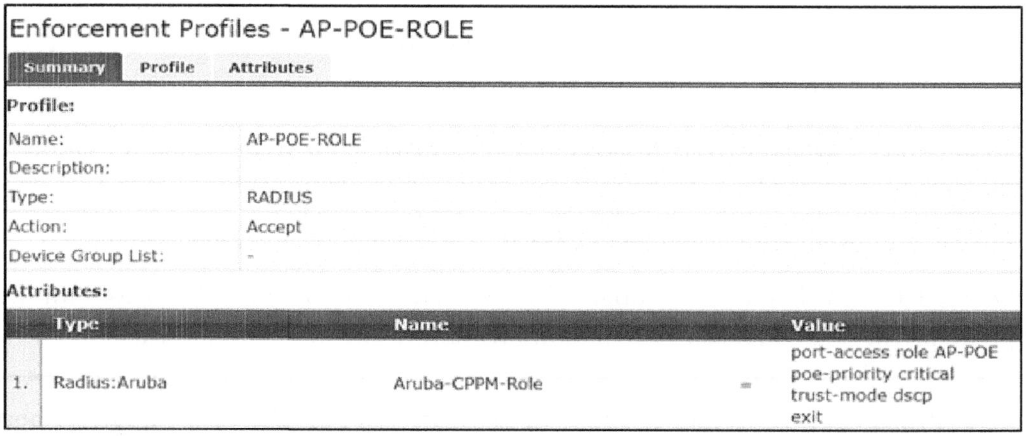

Figure 12-17: Configuring an enforcement profile

Step 2

Ensure enforcement policies contain rules that authorize by attributes (device type, CP, user role), then apply the desired profile.

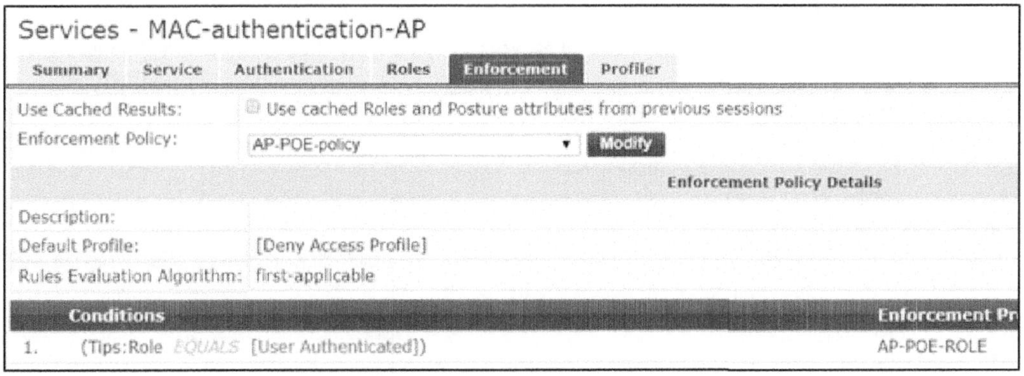

Figure 12-18: Enforcement profile attributes

Step 3

Configure an authentication service. Authentication methods and sources are defined within the service

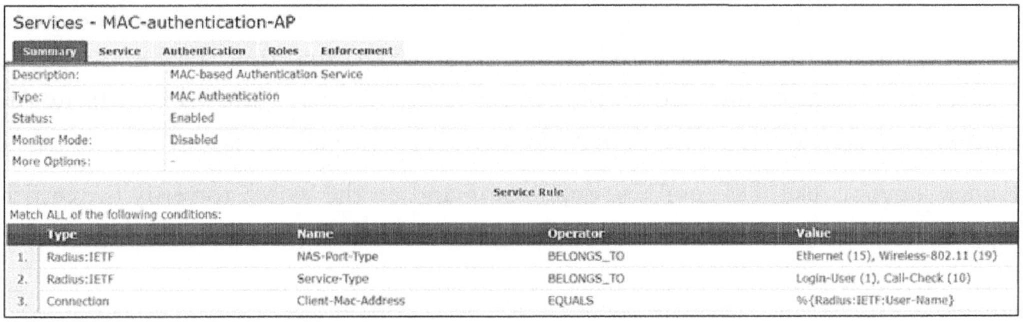

Figure 12-19: Configuring an authentication service

LUR should be used if you have a generic AAA server product. Either LUR or DUR can be used if you have ClearPass.

The previous steps show the authentication service/enforcement policy and profile configuration on ClearPass. See the ClearPass documentation for more information on configure services, enforcement policies, and enforcement profiles.

Server certificate on ClearPass

Figure 12-20 shows a sample server certificate installed on ClearPass. Make sure the RADIUS server host name configured on the switch matches the Common Name (CN) field of the ClearPass HTTPS server certificate. If an FQDN is used, you will need to specify a DNS server that the switch should use to resolve the server's name to an IP address.

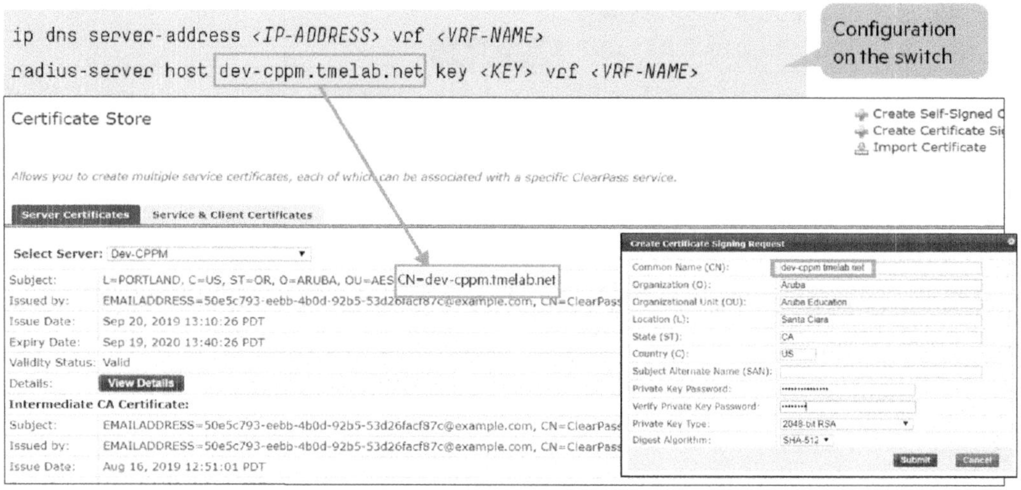

Figure 12-20: Sample server certificate on ClearPass

Validating the server certificate on the switch

ClearPass will send a user role to the edge switch after successful authentication. The method for sending the user role is through the REST API using an SSL connection. This means that a valid HTTPS certificate has to be installed on the edge switch, one from a trusted Certificate Authority (CA) of both the switch and the ClearPass server.

Use the `show crypto pki ta-profile cppm` command on the switch to verify the certificate details.

```
T1-ACC-1# sh crypto pki ta-profile cppm
 TA Profile Name            : cppm
 Revocation check           : disabled
   OCSP Primary url         : Not configured
   OCSP secondary url     : Not configured
   OCSP Enforcement-level: strict
   OCSP Disable Nonce     : false
   OCSP VRF                   : mgmt
  TA certificate :installed and valid
    Certificate :
      Data:
        version: 3 (0x2)
        Serial Number: 1 (0x1)
        Signature Algorithm: sha512withRSAEncryption
          Issuer: C=US, ST=California, L=Sunnyvale, O=Aruba Networks,
                26facf87c@example.com
          Validity
            Not Before: Aug 16 19:51:01 2019 GMT
            Not After : Aug 16 20:21:01 2029 GMT
          Subject: C=us, ST=california, L=Sunnyvale, o=Aruba Networks,
            26facf87c@example.com
          <output deleted for clarity>
```

Figure 12-21: The certificate installed on the switch

You will also need to configure NTP to ensure the certificate for the switch is valid based on the current date and time and the time range defined on the certificate. If the switch's time falls outside the range of the time on the certificate, the certificate is automatically invalidated.

Creating an admin user role on ClearPass

In addition, a username and password have to be configured to authenticate the switch with the ClearPass server. This is defined on the ClearPass server but referenced on the switch. The privilege level can be "Aruba User Role Download" or "Read-Only Administrator."

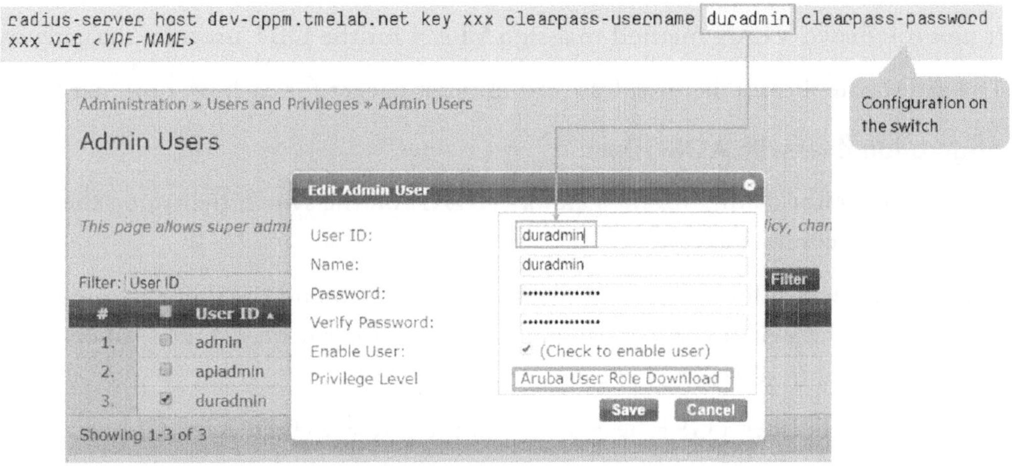

Figure 12-22: Creating a user role on ClearPass

UBT with AOS-10

AOS-10 supports UBT 1.0 and 2.0 versions. UBT 1.0 version provides VLAN configuration on the switch and UBT 2.0 provides VLAN config on gateway.

UBT 1.0 is supported from AOS-CX 10.06 and later versions, and UBT 2.0 is supported from AOS-CX 10.04.0020 and later versions.

UBT 1.0

- It creates multicast tunnel between the switch and the gateway (UDG).
- The UBT user role does not have any VLAN configured on the gateway.
- It supports both tagged and untagged UBT users.
- The switch forwards the multicast and broadcast traffic to the UBT clients of the same VLAN.
- The unicast traffic from the gateway to the switch is sent through the gateway (UDG) tunnel.
- The unicast, broadcast, or multicast traffic from the switch to the controller is always sent through the gateway (UDG) tunnel.

UBT 2.0

- It does not support UBT user VLAN configuration on the switch.
- It supports only untagged UBT users.
- It uses role-based VLAN method to assign VLAN for the UBT users on the gateway.
- The gateway replicates the broadcast or multicast packet for each user on that VLAN.

UBT configuration flow with AOS-10.x

1. The administrator configures the gateway (DDG) on the switch by issuing the `tunneled-node-server` command.
2. Once the gateway information is available on the switch and the mode is configured as "role-based", the switch performs a handshake with the gateway (DDG) to determine its reachability and to discover the version information.
3. The switch establishes a GRE heartbeat with the gateway (DDG) by sending a bootstrap message.
4. After it is acknowledged, the switch updates its local data structures with cluster information including Standby Device Designated Gateway (S-DDG), node list, and bucket map.
5. After the bucket map list is downloaded to the switch, a GRE heartbeat is established between the switch and the S-DDG.
6. The switch identifies the UBT user based on the user role, and checks the bucket map to obtain the UDG information of the client.
7. The switch establishes UDG GRE tunnel, and the UDG creates the user entry and assigns the user-role and VLAN sent by the switch as part of the user bootstrap.

UBT with MC cluster

When the switch uses a role-based tunneled node, it can tunnel traffic to clustered MCs. Typically, the switches will be connecting to a cluster of MCs. Clustering enables load balancing and failover capabilities over other redundancy solutions.

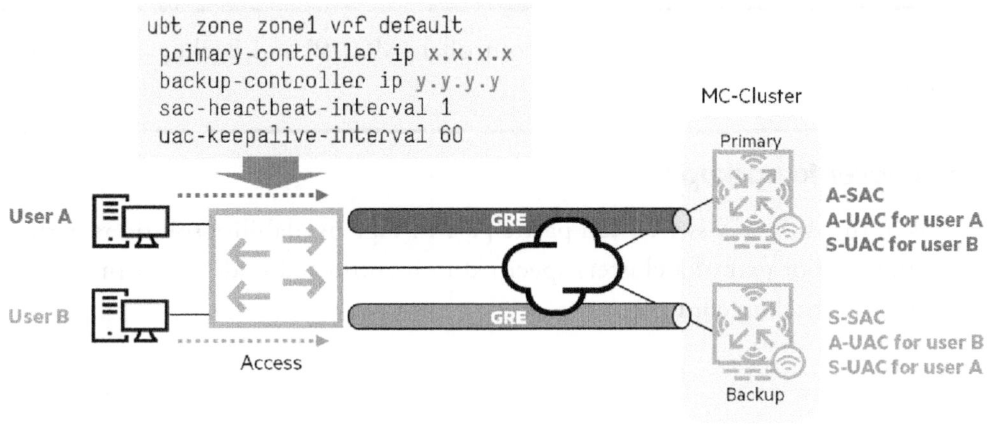

Figure 12-23: UBT with MC cluster

With this method, again, you only need to set a single tunnel node server (controller) IP address on the AOS-CX switch. Define at least one IP address of a cluster member and, optionally, a secondary address as a backup. When the switch contacts a member of the cluster, it receives assignments with the actual IP addresses of two MCs. One is the switch's active Switch Anchor Controller (A-SAC), and the other is the switch's standby SAC (S-SAC). The switch establishes PAPI tunnels with both controllers.

In addition, the switch receives a bucket map. When the switch starts tunneling a new device's traffic, the switch hashes information about the session. It looks up the hash in the bucket map to determine the correct active User Anchor Controller (A-UAC) and standby UAC (S-UAC) for that device. The switch tunnels the device's traffic to the A-UAC, but the state for the session is synchronized between the A-UAC and the S-UAC, enabling more seamless failover. (The degree of synchronized information depends on the device's needs.) In addition, because the cluster dictates where the switch tunnels each user's traffic, load balancing is improved.

If the switch's A-SAC or any of its A-UACs fail, the cluster can detect that failure very quickly and signal to the switch to failover to the appropriate standby tunnel, rather than the switch having to wait for its heartbeats to time out.

In **Figure 12-23**, notice that users A and B's active and standby tunnels are not on the same controller. User A's active tunnel is on the left controller, while user B's active tunnel is on the right controller.

As you have seen, the AOS-CX switch can participate in high availability solutions with a single controller IP address specified in its tunneled node server profile. For additional redundancy, the primary and secondary addresses can be controllers in two different clusters. The zone profile can also include a backup controller, though, specified with the configuration shown in **Figure 12-23**.

UBT zone gateway IP = Primary controller (MC) IP and Backup controller (MC) IP (optional).

UBT tunnel failover to backup MC

As you have seen, the AOS-CX switch can participate in high availability solutions with a single controller IP address (not part of a cluster) specified in its tunneled node server profile. This profile can also include a backup controller.

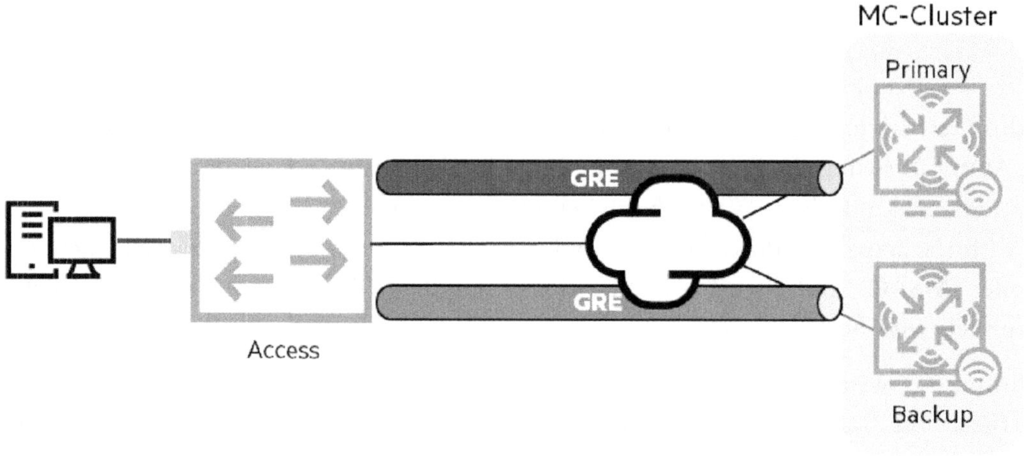

Figure 12-24: Tunnel failover to the backup controller (MC)

As discussed earlier, the switch and its MC—or any of its SACs and UACs when clustering is used—send each other keepalives over the GRE tunnel every second to assess the tunnel's health. If the switch fails to hear keepalives over the keepalive timeout period, it attempts to re-establish communications with the primary controller. In this scenario, though, the company has a standalone MC at a different location. After failing to connect to the primary MC (or cluster), the switch establishes a tunnel with the backup controller at the different location.

After switching over to the backup controller, the switch continues to check whether the primary controller is available every 30 seconds. If it re-establishes communication with the primary controller, it moves its tunnel back to that controller.

Again, MC administrators might choose an active-standby or active-active deployment. Communicate with them to determine whether all switches will use the same primary controller IP and backup controller IP (active-standby), or whether you should alternate these roles on different switches (active-active). Often, the MCs are deployed in different locations, and each switch uses the local MC as its primary MC and the remote MC as its backup.

Troubleshooting

Understanding the underlying architecture of AOS-CX and the specific UBT implementation details is essential for resolving these issues efficiently and ensuring seamless network operation. Troubleshooting UBT issues can be complex, given the intricate nature of network configurations and policies. Common challenges include authentication failures, tunnel establishment problems, and communication with the ClearPass server. Effective troubleshooting requires a systematic approach verifying basic connectivity with the server, checking configuration settings, and utilizing diagnostic tools. Some of the solutions for the known issues are presented here.

Authentication failed or server not reachable

You can use the `show aaa authentication port-access interface all client-status` command on the switch to diagnose the issue.

A sample output of this command is shown in **Figure 12-25**.

```
Access-1# show aaa authentication port-access interface all client-status
Port Access Client Status Details

Client 00:50:56:b1:a5:1c, icx-avp1
============================
  Session Details
  ---------------
   Port         : 1/1/3
   Session Time : 1871s

  Authentication Details
  ----------------------
   Status          : Authentication Failed, Server-Timeout
   Auth Precedence : dot1x - Unauthenticated, mac-auth - Not attempted

  Authorization Details
  -----------------
   Role   :
   Status :
```

Role missing

Figure 12-25: Authentication failed or server not reachable

The `show aaa authentication port-access interface all client-status` command allows you to see the status of clients on ports, including their authentication status and assigned user role in any. This example shows an issue contacting the ClearPass server. Mostly commonly, certificate errors and the ClearPass username and password being misconfigured cause this issue.

Authenticated, but no downloadable user role

You can use the same command to diagnose this issue.

```
Access-1# show aaa authentication port-access interface all client-status
Port Access Client Status Details

Client 00:50:56:b1:a5:1c, icx-avp1
============================
  Session Details
  ---------------
   Port         : 1/1/3
   Session Time : 9s

  Authentication Details
  ----------------------
   Status          : dot1x Authenticated
   Auth Precedence : dot1x - Authenticated, mac-auth - Not attempted

  Authorization Details
  ----------------------
   Role   :
   Status : Not Ready
```

Role missing

Figure 12-26: Authenticated, but no downloadable user role

This example shows a status of "Not Ready." In this example, ClearPass passed down the user role called "Authenticated," but this does not exist in a `port-access role` command.

Authenticated, role downloaded. Success!

The following example (**Figure 12-27**) shows a status of "Applied," indicating that everything is working normally and that the authenticated user role is applied to the client connected to port 1/1/3.

```
Access-1# show aaa authentication port-access interface all client-status
Port Access Client Status Details

Client 00:50:56:b1:a5:1c, icx-avp1
============================
  Session Details
  ---------------
   Port         : 1/1/3
   Session Time : 9s

  Authentication Details
  ----------------------
   Status          : dot1x Authenticated
   Auth Precedence : dot1x - Authenticated, mac-auth - Not attempted

  Authorization Details
  ----------------------
   Role   : employee-tunnel
   Status : Applied
```

Status is "Applied"
and a role is mapped

Figure 12-27: Authenticated, role downloaded. Success!

Applying the profile makes the port function like a controlled HPE Aruba Networking AP. The switch no longer forwards traffic that arrives on the ports locally, as it normally would. Instead, it uses the GRE tunnel to send all the traffic to the IP address that you configured in the tunnel server profile. The tunnel acts at Layer 2. It encapsulates not just the IP payload but also the MAC header, and it forwards all traffic, including broadcasts and multicasts.

The main difference between the tunneled node and AP-based tunnels is that tunneled-node encapsulates 802.3 instead of 802.11 frames. Note also that often wireless traffic is encrypted; however, wired traffic is not encrypted, and GRE does not provide encryption either. However, the wired network infrastructure is typically trusted, so encryption is not required.

The switch continues to function as normal for all traffic that arrives on the non-tunneled-node ports.

Tunneled-node ports are intended for wired endpoints. You should avoid plugging controlled HPE Aruba Networking APs into wired tunneled ports because these APs are already fulfilling similar functions on their own. You would be causing the network to encapsulate wireless clients' traffic twice. Lab tests show that the tunneled node will still operate properly, but the double encapsulation can create performance and resource issues.

Validating the UBT details from MC

The following example shows how to validate connectivity from the MC with the `show user` command. From left to right, you see the end user's IP and MAC address, name and role, time of connectivity, and authentication.

```
(MC1) [mynode] #show user
This operation can take a while depending on number of users. Please be patient ....

Users IP       MAC                Name       Role           Age(d:h:m)   Auth
-----          ----------         --------   ------         ----         ----------
10.3.11.204    00:50:56:b1:a5:1c  icx-avp1   authenticated  00:00:05     Tunneled-User-802.1X

AP name    Roaming   Essid/Bssid/Phy                       Profile                 Forward mode  User Type
-------    --------  ---------------                       -------                 -----------   ---------
10.3.41.4  Tunneled  tunnel 10/10:4f:58:f5:bf:40/1/1/3     default-tunneled-user   tunnel         TUNNELED USER

User Entries: 1/1
 Curr/Cum Alloc:1/35 Free:0/34 Dyn:1 AllocErr:0 FreeErr:0
```

Figure 12-28: Output of the `show user` command

The next columns start with the source IP address of the access switch, under the heading "AP name." You see that this is a tunneled connection, along with tunnel information—switch MAC address and port number. Then you see the assigned profile, forwarding mode, and the user type: TUNNELED USER.

Identifying the issues with "show port-access role" command

The `show port-access role` command can be used to verify the connectivity to the ClearPass server when using DUR. The example in **Figure 12-31** shows a successful connection and interoperation with the ClearPass server: the status is Completed. **Figure 12-29** shows a parsing_failed status, typically indicative of either a DNS or network connectivity issue. The other example in **Figure 12-30** shows a certificate error, which could be caused by a number of things:

- It could be a time issue (use NTP to ensure the correct time).
- It could be that the CN name of the ClearPass certificate is not defined on the switch correctly (or at all).
- The certificate might have expired.

```
T1-ACC-1# sh port-access role
Role information:

Name : AP_POE_ROLE-3009-5
Type : clearpass
status: failed,parsing_failed
-----------------------------------
Reauthentication Period
Authentication Mode
Session Timeout
```

Figure 12-29: Identifying DNS issues

```
T1-ACC-1# sh port-access role
Role information:

Name : AP_POE_ROLE-3009-5
Type : clearpass
status: failed,certificate_invalid
----------------------------------------
Reauthentication Period
Authentication Mode
Session Timeout
```

Figure 12-30: Identifying certificate issues

```
T1-ACC-1# sh port-access role
Role Information:
Name   : AP_POE_ROLE-3009-5
Type   : clearpass
status: completed,(null)
----------------------------------------------
Reauthentication Period                    :
Authentication Mode                        :
session Timeout                            :
Client inactivity Timeout                  :
Description                                :
Tunneled Node Server zone                  :
Tunneled Node server secondary Role        :
Access vlan                                :
Native vlan                                :
Allowed Trunk vlans                        :
MTU                                        :
QOS Trust Mode                             : dscp
PoE Priority                               : critical
Captive Portal Profile                     :
Policy                                     :
```

Figure 12-31: Output without any issues

The other commands that can be run on the MC are:

- Use the `show tunneled-node-mgr tunneled-nodes` command on the MC to verify the connectivity from the controller side.
- Use the `show tunneled-node-mgr-tunneled-users` command on the MC to see the users that are being tunneled from the switch.

- Use the `show tunneled-node-mgr trace-buf` command on the MC to see the log messages related to tunneled node on the MCs.
- Use the `show datapath tunnel table` command on the MC to see the tunneled connections in the MC's state table.

Learning check

Check your understanding of Dynamic Segmentation by answering the following questions.

Question 1

What is the primary advantage of using User-Based Tunneling (UBT) with AOS-CX switches?

- Simplified switch hardware setup
- Decentralized network management
- Enhanced security through centralized policy enforcement
- Increased physical port count on switches

Question 2

When deploying Dynamic Segmentation v2.0, how many AP licenses are needed for one VSF stack formed by four CX 6300 Series switches?

- 4
- 5
- 1
- 2

Question 3

Which component is essential for implementing User-Based Tunneling (UBT) with AOS-CX switches to ensure specific user traffic management?

- HPE Aruba Networking (Legacy) Management Software
- HPE Aruba Networking Mobility Controller
- HPE Aruba Networking ClearPass
- HPE Aruba Networking Instant Access Point

Question 4

In the context of AOS-CX switches, what is the primary function of User-Based Tunneling (UBT) when combined with downloadable user roles (DUR)?

- ○ To download user-specific policies from ClearPass.
- ○ To use the user-specific policies from the switch itself (local).
- ○ To enforce user-specific policies directly from the Mobility Controller.
- ○ To enable direct communication between switches without a controller.

Question 5

Which of the following commands is useful to verify communication with ClearPass?

- ○ `show aaa authentication port-access interface all client-status`
- ○ `show port-access role`
- ○ `show ubt tunnels`
- ○ `show aaa authentication port-access all cppm-status`

13 REST API

Switches running the AOS-CX software are fully programmable with a REST (REpresentational State Transfer) API, allowing easy integration with other devices—on premises and in the cloud. This programmability, combined with the HPE Aruba Networking Network Analytics Engine, accelerates network administrator understanding of and response to network issues.

This chapter explains how to use the REST API on AOS-CX switches.

OBJECTIVES

After completing this chapter, you should be able to:

- ✓ Describe the AOS-CX REST APIs and their concepts.
- ✓ Examine the REST API (swagger) interface.
- ✓ Understand the REST status codes.

REST API introduction

AOS-CX switches are unique in the level of scripting and automation that they support. You can unlock this power in the OS using the REST API.

In this chapter, you will learn what a REST API is and how it works. You will explore REST resources and the different methods REST uses to gather information about those resources or change them. You will also learn how to enable the REST API on AOS-CX switches and log in and out of the interface from a client. Finally, you will learn how to interpret response codes.

You can use various tools and scripting languages to make REST requests. You will learn to use the API reference interface to practice requests and find the correct formatting.

What is REST API?

An Application Programming Interface (API) defines how users and other applications interact with an application's components. An API can use common web-based interactions or communication protocols. An API is a logical interface; it allows other programs to communicate with the AOS-CX switch to collect information about the switch and change settings. The REST API is the point of contact for scripting AOS-CX switches. It creates a structured system for reporting data about the switch status and statistics and defining switch settings.

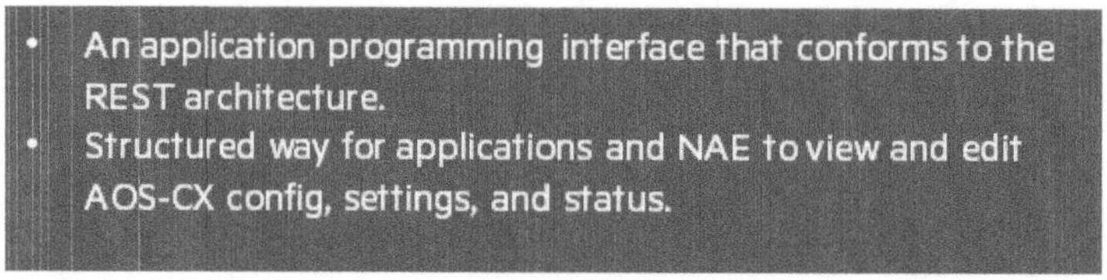

Figure 13-1: REST API definition

Scripting commands to a CLI requires complicated filters for sorting through content returned by show and other commands. However, because the REST API uses structured data, programs can search through and interpret the data more easily and submit settings to reconfigure the switch. When you understand how to use the API, you will be able to create scripts to automate network configuration. You will also be able to automate monitoring and troubleshooting, especially when combined with the AOS-CX HPE Aruba Networking Network Analytics Engine (NAE).

REST refers to the type of API. REST uses familiar HTTP/HTTPS as the communication protocol. The client (the program the manager uses to deploy scripts) contacts the server (the AOS-CX switch) and issues basic HTTP commands to create, read, update, and delete resources on the switch. The requests and responses contain information formatted in a particular media type. AOS-CX switches use JavaScript Object Notation (JSON), a standard media type with which developers should be familiar.

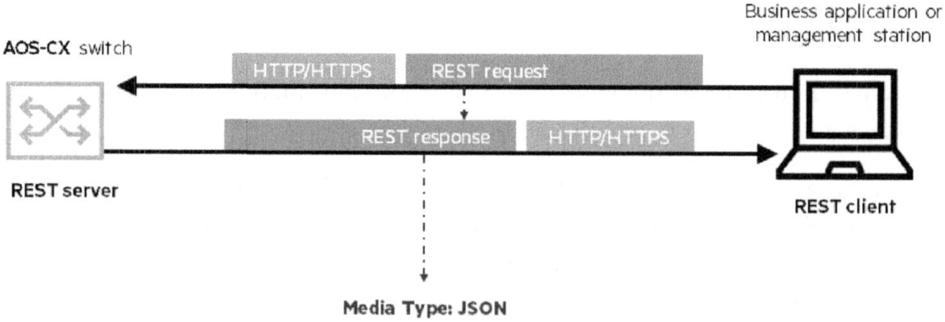

Figure 13-2: REST requests and responses

REST basic concepts

There are four core components in HTTP. When a browser requests a webpage from a web server, the browser sends a few pieces of information to tell the server exactly what it wants from the server.

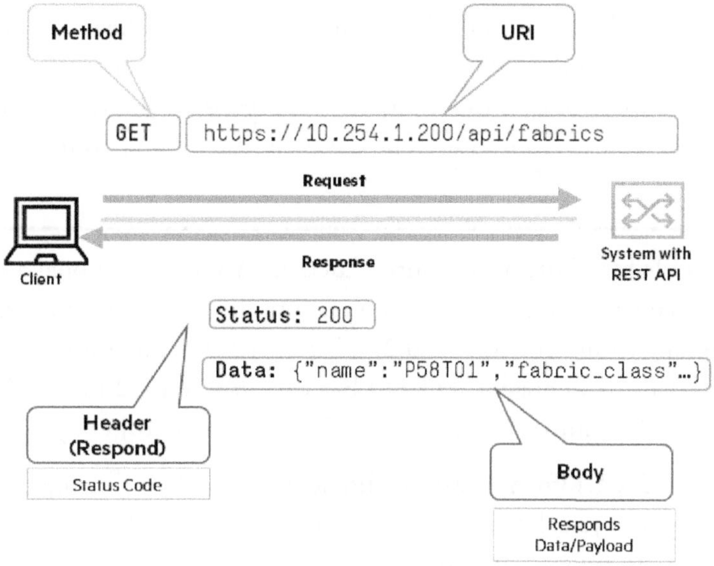

Figure 13-3: The core components in HTTP

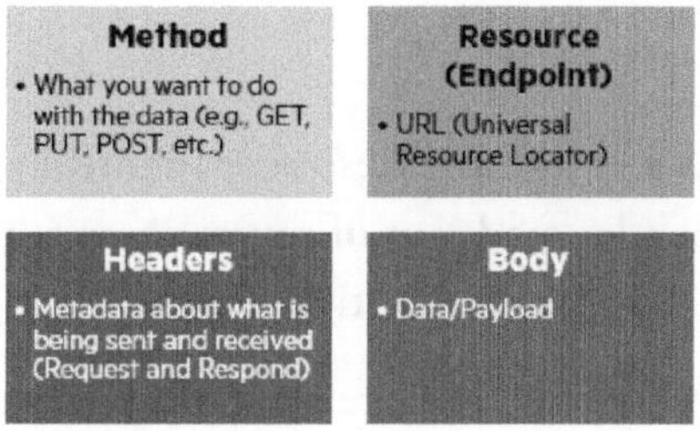

Figure 13-4: The core components in HTTP summary

URL: A resource that resides on the server. This could be a file, image, database record, and so forth.

Headers: The request header lists key/value pairs containing metadata about what is being sent and received. Think of this as a list of details or instructions for the server to provide additional information to help the server deal with the request.

Method: Tells the server what you want to do with the resource.

- GET the resource
- PUT to update resource
- POST creates a new resource

Body: Used in some requests to provide any other information that the server might need. The POST and PUT methods are typically used to send a new or updated resource to the server.

Recap: the URL identifies the location of the resource—one of the headers specifies HOW the server should represent the resource when it is returned in the response. The method tells the server what operation to carry out on the resource—in **Figure 13-3**, GET the list of fabrics on HPE Aruba Networking Fabric Composer. The data/payload (body) contains the list of all configured fabrics.

URL stands for Uniform Resource Locator. A URL is a global address of documents and protocols to retrieve resources on a computer network, for example, https://-domainname.com/ or https://10.2.3.4/ server ID or the host. URLs are most frequently used for webpages (HTTP) but can also be used for database access using JDBC, email (mailto), file transfer (FTP), and many other applications.

URI stands for Uniform Resource Identifier. A URI is a string containing characters identifying a physical or logical resource, for example, fabrics, integrations, or VMs. A URI follows syntax rules to ensure uniformity. Moreover, it also maintains extensibility via a hierarchical naming scheme.

URI

A switch resource is indicated by its Uniform Resource Identifier (URI). A URI is the location of a specific web resource. A URI can be made up of several components, including the host name or IP address, port number, the path, and an optional query string.

Parts of a URI : The two main parts of a URI are the path and the (optional) query component.

- **URI path:** The path is the part of the URI starting with the server URL and ending with the resource ID. In URIs that have a query component, the path is everything before the question mark (?). The path has a hierarchy. In a path, the forward slash (/) indicates the hierarchical relationship between resources. Because the forward slash has a special meaning, the forward slash characters that are part of the URI path must be percent-encoded with the code %2F, which represents the forward slash. For example, the following URI represents the resource utilization for the management module in slot 1/5: *https://192.0.2.5/rest/v10.04/system/subsystems/management_module,1%2F5?attributes=resource_ utilization*
- **Query component:** In many cases, the unique identification of a resource requires a URI that contains both a path and a query component. The query component is sometimes called the query string. For example, CPU utilization is a resource represented by the above URI: *https://192.0.2.5/rest/v10.04/system/subsystems/management_module,1%2F5?attributes=resource_ utilization* In a URI, the question mark (?) indicates the

beginning of the query component. The query component contains nonhierarchical data, and the format of the query string depends on the implementation of the REST API. The query component often contains "=" pairs separated by the ampersand (&) character. Multiple attribute values are supported and are separated by commas. For example: *https://192.0.2.5/rest/v10.04/system/vlans?depth=2&attributes=id,name,type*

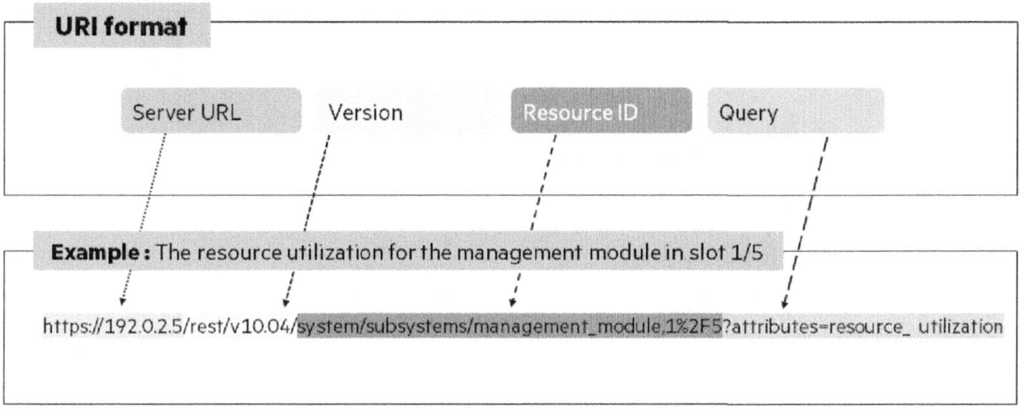

Figure 13-5: URI format

Resources

In a REST API, the primary representation of data is called a resource. A resource is a representation of an entity in the system as a URI. The entities can include hardware objects, statistical information, configuration information, and status information. The URI might or might not include a query component. Resources are nouns—anything that can be named can be a resource.

Collections: A collection is a directory of resources managed by the server. Typically, a resource collection contains multiple resource instances, and the collection name is in the plural form. For example,

- `/system/vlans`
- `/system/vrfs/default/routes`
- `/system/users`
- `/fullconfigs`

A GET request to a collection returns the set of JSON objects representing the members of the collection.

Subcollections: A single resource instance can also contain subcollections of resources. In the example below, **vlans** is a subcollection of the **system** resource: **/system/vlans** , routes is a subcollection of the **default** VRF resource instance.

Singletons There are some resources that can only have one instance. These resources are called singletons, and the resource collection name is in the singular form.

For example:

- `/system`
- `/system/vsx`
- `/firmware`

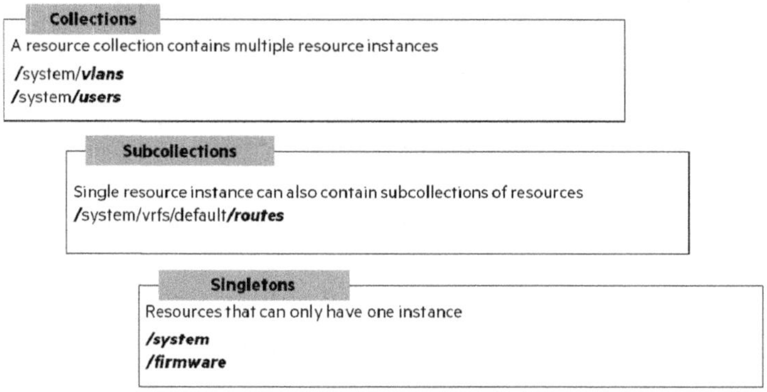

Figure 13-6: Resources

Hierarchy

The REST API defines resources, each of which is a URI that a client can access. For example, a switch with the IP address 10.1.1.1 has the URI 10.1.1.1/rest/v1/system/bridge/vlans, which lists VLANs on the switch. A REST client can contact this URL to obtain information about this list or add to it.

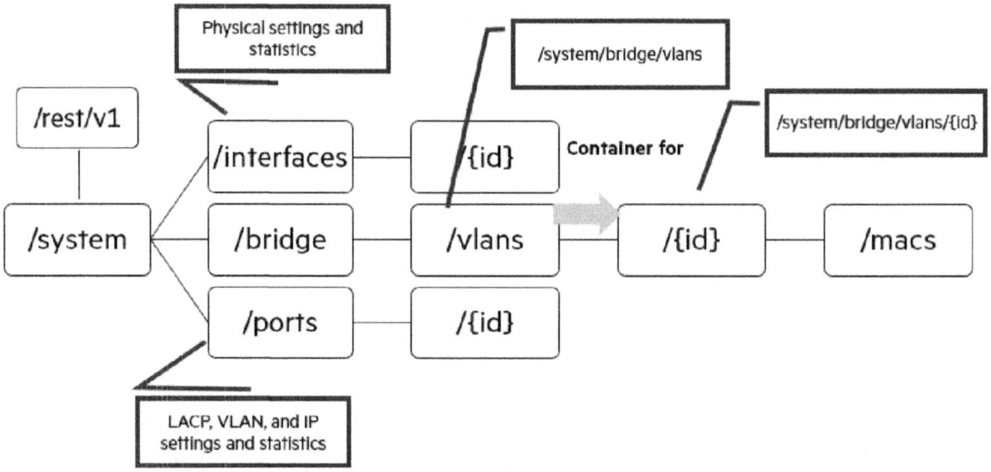

Figure 13-7: Hierarchical structure of URIs

As you see in **Figure 13-7**, the URIs for resources follow a hierarchical structure. This means that a resource can also be a container for other resources. For example, the /system/bridge/vlans resource contains multiple VLAN resources, each identified with an ID. The same concept holds true for the system/interfaces resource, a container for multiple interface IDs, and several other resources on the switch.

For the most part, you can navigate through the REST interface and find the intuitively named resources. However, a bit more explanation about where settings for VLANs and interfaces are stored could be helpful. The /rest/v1/systems/interfaces resource stores physical interfaces, VLAN interfaces, and loopback interfaces. The /rest/v1/system/ports resource also stores physical interfaces, VLAN interfaces, loopback interfaces, and link aggregations.

The "interfaces" resource stores physical and lower-level information. Settings configured through the interfaces resource include admin status and UDLD. Statistics include the ports resource, which stores VLAN and IP/IPv6 protocol information. Settings configured through the ports resource include LACP, IP, OSPF, PIM, IGMP, and VLAN settings. The /rest/v1/system/bridge/vlans resource stores Layer 2 settings for VLANs such as name, ID, and IGMP snooping.

REST API methods

Every REST request a client sends is associated with a REST method, which is an HTTP command. A GET command reads an existing resource. A POST command creates a resource. A DELETE command removes a resource, and a PUT command replaces attributes for an existing resource.

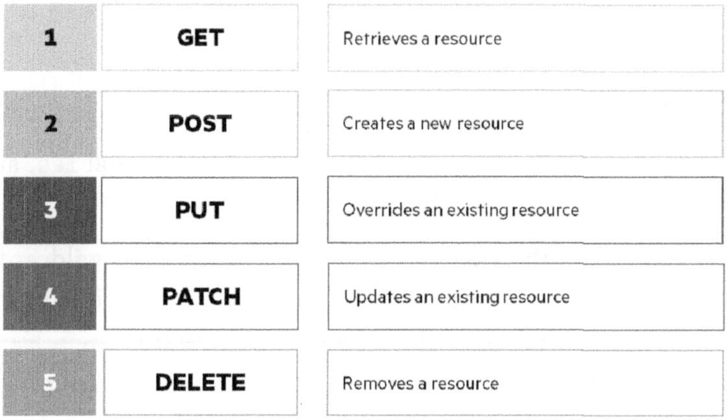

Figure 13-8: REST methods

You can use different HTTP methods for read, create, update, or delete operations. Within AOS-CX switches, there are five methods used with the REST API interface:

HTTP methods

- GET – Retrieve object instance properties
- POST – Create a new object instance
- PUT – Modify object instance properties
- PATCH – Modify a group of object properties
- DELETE – Remove an object instance

Using GET method

The GET method is a read method that gets the resource specified by the URI. Using GET on a resource collection results in a list of URIs. Each URI in the list corresponds to a specific resource in the collection.

Example:

```
curl GET -k -b /tmp/auth_cookie "https://10.17.0.1/rest/v1/system/bridge/vlans"
[ "/rest/v1/system/bridge/vlans/1"]
```

The above example shows a GET request to the vlans collection. The response body is a list in JSON format of the configured VLANs. Each VLAN resource is listed in URI format.

Using GET on a resource collection results in a list of URIs. Each URI in the list corresponds to a specific resource in the collection.

GET on a specific resource returns the attributes of that resource. The following example shows a GET request on VLAN 10. The URI for VLAN 10 specifies resource ID 10 in the vlans collection.

Example:

```
curl GET -k -b /tmp/auth_cookie
"https://10.17.0.1/rest/v1/system/bridge/vlans/10"
{ "id": 10, "name": "vlan10", "type": "static", "flood_enabled_subsystems": [
"/rest/v1/system/subsystems/system/base" ] }
```

When you use the GET method, the URI can contain the asterisk (*) wildcard character instead of a component in URI path. You can use wildcard characters in multiple places in the path.

> You cannot use a wildcard character as part of the query string.

Example: The following URI specifies all routes regardless of VRF:

"https://192.0.2.5/rest/v1/system/vrfs/*/routes"

The GET method supports the following parameters in the query string of the URI:

- Attributes
- Count
- Depth
- Filter
- Selector

A parameter is specified as a "key=value" pair. When permitted, multiple values are separated by the comma (,) character.

Using GET method

- GET is a read method that gets the resource specified by the URI
- GET on a resource collection returns a list of resource URIs
- GET on a resource returns the attributes of that resource
- GET supports wildcard characters in the path

- Attributes
- Count
- Depth
- Filter
- Selector

Figure 13-9: Using GET method summary

Using PUT method

The PUT method updates an instance of a resource by replacing the existing resource with the resource provided in the request body.

- Not all resources support the PUT method. For information about the methods supported for a resource, see the HPE Aruba Networking CX Switch Operating System REST API Reference. The REST API must be in read-write mode to see all the PUT methods supported.
- The URI must specify a specific resource, not a collection.
- The URI must specify a resource that currently exists.
- The JSON model you use for a resource PUT operation is different from the JSON model used for the GET operation. For example, immutable attributes are omitted from the PUT JSON model of a resource. See the AOS-CX REST API Reference for the JSON model used the PUT operation of a resource.

- For almost all resources, the PUT method is implemented as a strict replace operation. All changeable configuration attributes are replaced and must be provided in the JSON data in the request body.
- When you are using the PUT method, The JSON data must not include any immutable attributes. The JSON data in the request body must include the mutable or changeable configuration attributes only. Any mutable attributes not supplied in the request body are set to their default values.

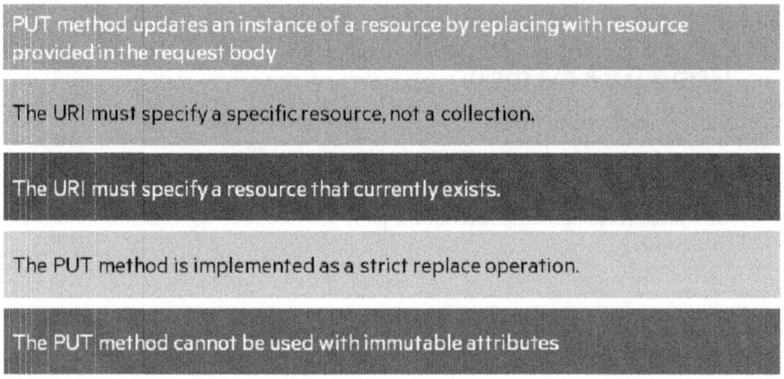

Figure 13-10: Using PUT method summary

Using other write methods

PUT method:

The PUT method updates an instance of a resource by replacing the existing resource with the resource provided in the request body. Note that the REST API must be in read-write mode to see all the PUT methods supported. The URI must specify a specific resource, not a collection, and that should currently exist. The PUT operation is a replace operation not an update operation because the resource instance in the request body replaces every changeable configuration attribute of the existing resource therefore Partial updates are not supported.

The JSON model for a resource for a PUT operation is different from the JSON model for the GET operation. For example, immutable attributes are omitted from the PUT JSON model of a resource. For almost all resources, the PUT method is implemented as a strict replace operation. All changeable configuration attributes are replaced and must be provided in the JSON data in the request body.

Configuration attributes that are set at the time a resource is created and that cannot be changed afterward are called **immutable** attributes. The JSON data must not include any immutable attributes.

DELETE method:

The DELETE method deletes an instance of a resource. Like the PUT method, not all resources support the DELETE method, and the REST API must be in read-write mode to see all the DELETE methods supported.

The URI must specify a specific resource instance. The URI must not specify a collection. Child subcollections and resources are deleted when you delete the parent resource. DELETE requests do not contain a request body, and the do not return a response body.

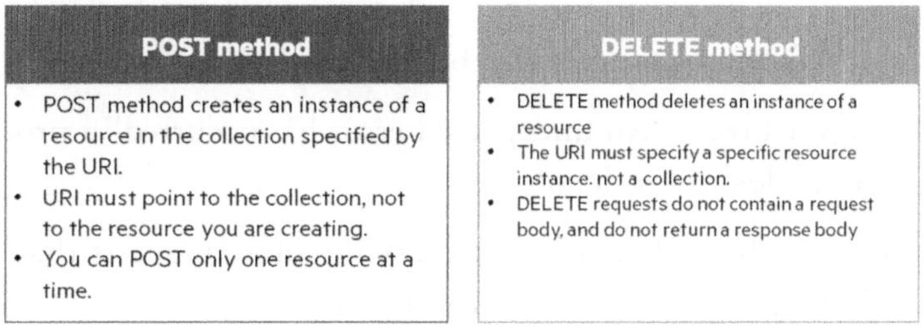

Figure 13-11: Summary of other using other write methods

REST API concepts – HTTP response

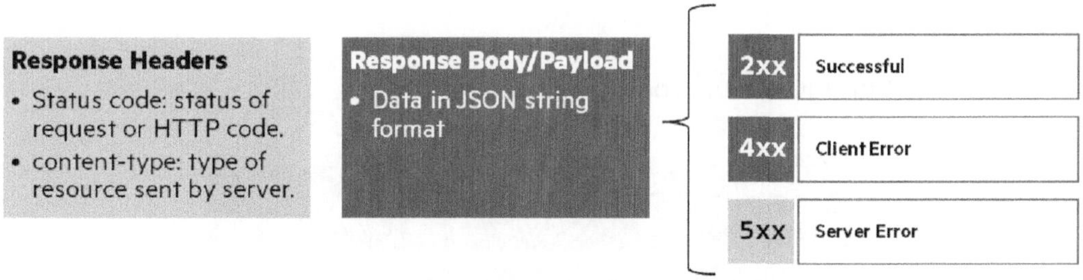

Figure 13-12: HTTP response

Responses of REST calls will have a status code and a body/payload. An example of the possible response codes:

Success status codes:

- 200 – OK – Returned from GET and PUT operations and nonconfiguration API calls such as Login or Logout when the request is successfully completed.
- 201 – Created – Returned from POST operations when a new resource was successfully created.
- 204 – No Content – Returned from an operation when the request was successfully processed and there is no content to return.

Error status codes:

- 400 – Bad Request – The request was invalid or cannot be served.
- 401 – Unauthorized – The request requires a user authentication.
- 403 – Forbidden – The server understood the request but refused it, or access is not allowed.
- 404 – Not Found – There is no resource behind the URI.
- 500 – Internal Server Error – An unexpected error has occurred in processing the request. View the logs on the device for details.
- 503 – Service Unavailable – The device is receiving more requests than it can process and is defensively rejecting requests to protect resources.

Enabling the REST interface on an AOS-CX switch

Now that you understand how REST works, you will learn how to enable the REST API on an AOS-CX switch.

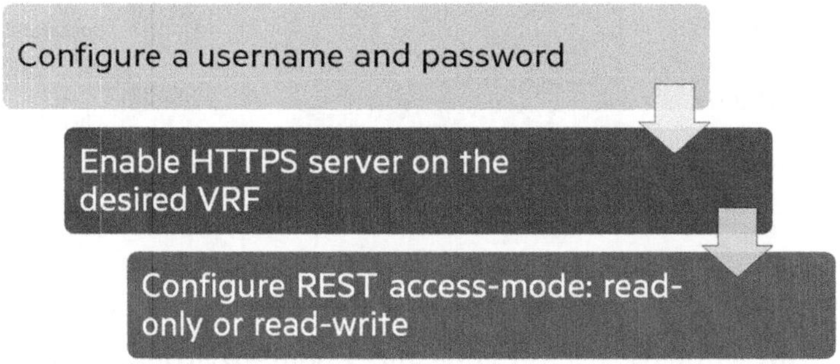

Figure 13-13: Steps to enable REST on an AOS-CX switch

First, make sure the switch has at least one user account configured with a password. The switch has an admin account by default but no password on that account. You can set up multiple users if you want and assign those users to the administrators.

Optionally, certificate-based authentication can be used. The trust anchor (TA) profile is needed for the certificate-based authentication method to validate the client certificate. Also, a RADIUS server must be configured on the switch. For more information on configuring certificates and managing certificates, see the *Security Guide* for your switch and software version.

Next, enable the HTTPS server on one or more VRFs. All non-OOBM interfaces belong to the "default" VRF by default, so use that name if you have not set up a VRF. Keep in mind that the out-of-band management (OOBM) port automatically belongs to the "mgmt" VRF, so enable the HTTP server on that VRF if you want to access the REST API through the OOBM port. Enabling the HTTPS server allows access to the web browser interface and the REST interface.

You can then set the mode for the REST API. Read-only access restricts clients to GET requests for the most part. They can send POST requests to do things like log into the interface. Read-write access allows clients to send any type of request—as long as they are authorized.

Logging in and out of the interface

To become authorized, the client must authenticate to the interface by sending a POST request to the rest/v1/login resource. The URL must include a query string with a username and password that match a user configured on the AOS-CX switch.

The switch sends back a cookie, which must be included in all subsequent requests. The types of requests authorized depend on the user's role; operators can only use the GET method while managers can use all methods, assuming that the HTTP server's REST access mode permits them. An exception is that operators can also POST to log out of the interface.

User role	Privileges
operator	GET requests
manager	All requests

After sending all the desired requests, the script should end by logging out of the interface, sending a request to the logout resource. The AOS-CX switch can only support up to five sessions at once. Logging out ensures that sessions are appropriately tied up.

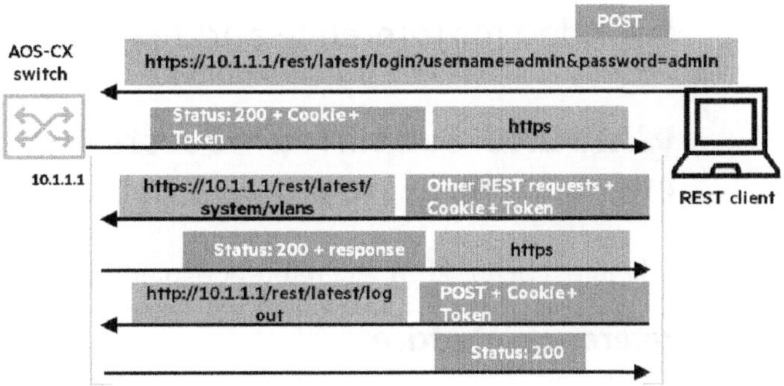

Figure 13-14: Logging in and out of the interface

Sending requests to the REST API

You will now learn how to use the REST API interface of AOS-CX switches.

AOS-CX REST API reference interface

AOS-CX switches support a reference interface for the REST API. HPE used Swagger to create this interface from the API, so it is sometimes referred to as the Swagger interface.

The primary purpose of the reference interface is to document the API for developers rather than for actually managing and monitoring the switch. The reference interface lets you explore all of the resources in the API and find the correct URI for each. It also helps you to construct valid requests and shows you the `curl` command for issuing that request. cURL is a command line utility for sending and receiving HTTP/HTTPS requests. You can try out various formats for request bodies and validate that they work. You can then use this knowledge when scripting an outside program.

However, be aware that the actions you take in the interface are real. For example, a PUT request to /system/interfaces/ID will change an interface's settings. Therefore, you should be careful not to submit actual PUT, POST, and DELETE requests on a live system.

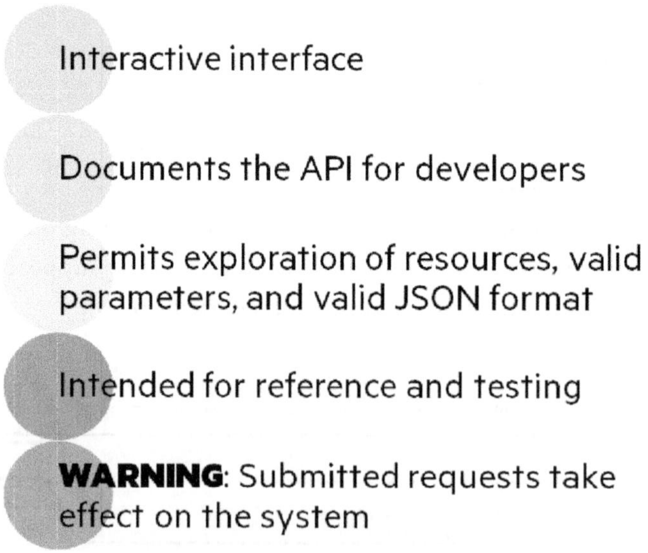

Figure 13-15: REST API reference interface

Accessing the REST API reference interface

You can reach the interface by browsing to the AOS-CX switch's IP address or hostname /api, or you can log into the Web UI and select the System icon and then select API.

Figure 13-16: Accessing the REST API

The reference interface then launches. You can scroll down the page to view all the resource categories on the switch.

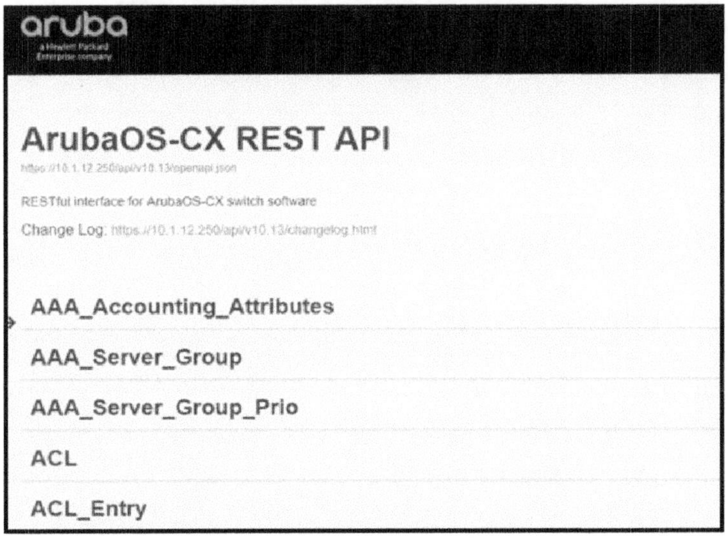

Figure 13-17: Reference interface

Exploring the REST API reference interface

When you select a resource, it expands so you can see the URI for the resource and all the REST methods, or operations, supported for that particular resource. You can select a method to expand and see details about submitting the request correctly.

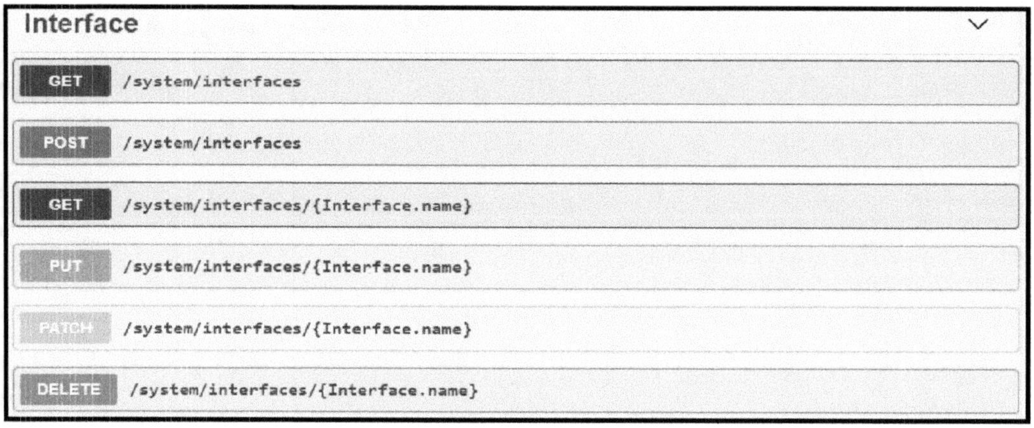

Figure 13-18: REST methods interface

The Response Class shows the response you should expect for a successful request. The parameters list shows all the parameters for refining a request. These include attributes, depth, count, selector, and filters for a GET request.

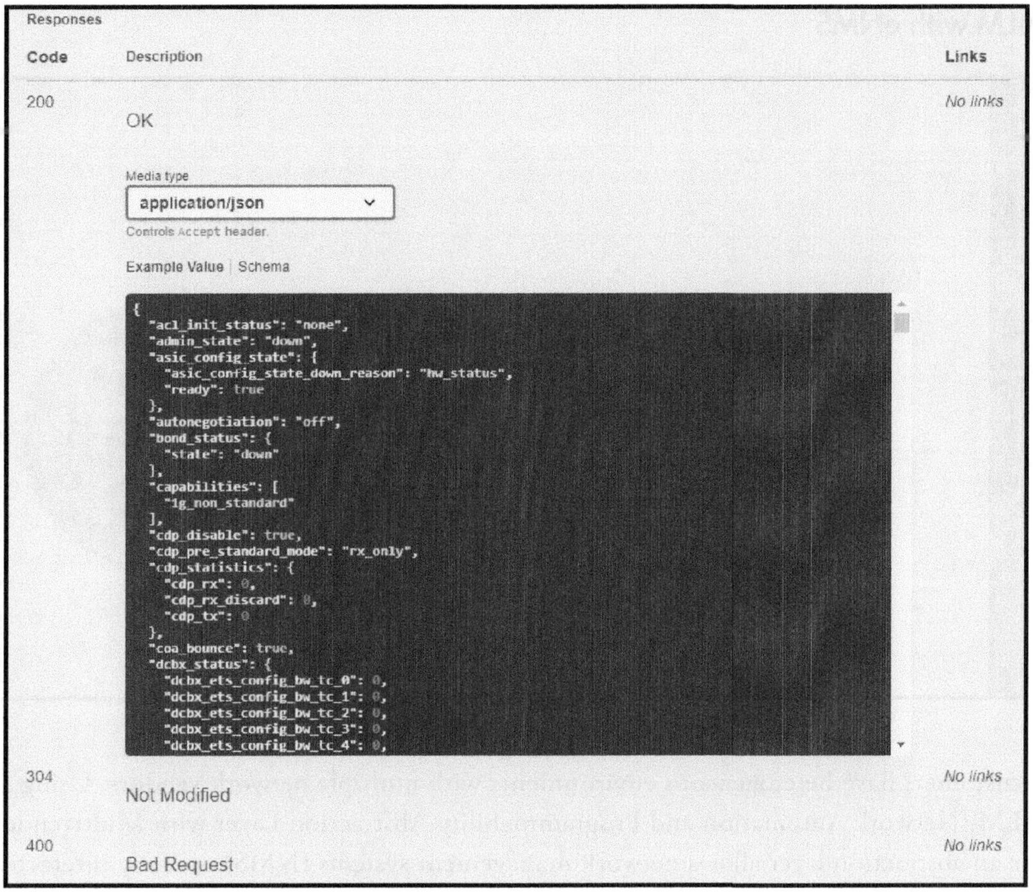

Figure 13-19: REST responses

Use cases and resources

In this section, you will review some important use cases of APIs in data centers and a set of useful resources.

REST API use cases

Data centers are a dynamic and unique environment. APIs can be used to integrate different platforms and applications, bringing agility and opening a universe of possibilities.

Two examples of possible use cases of APIs in the data center environment are:

NAPALM with eNMS

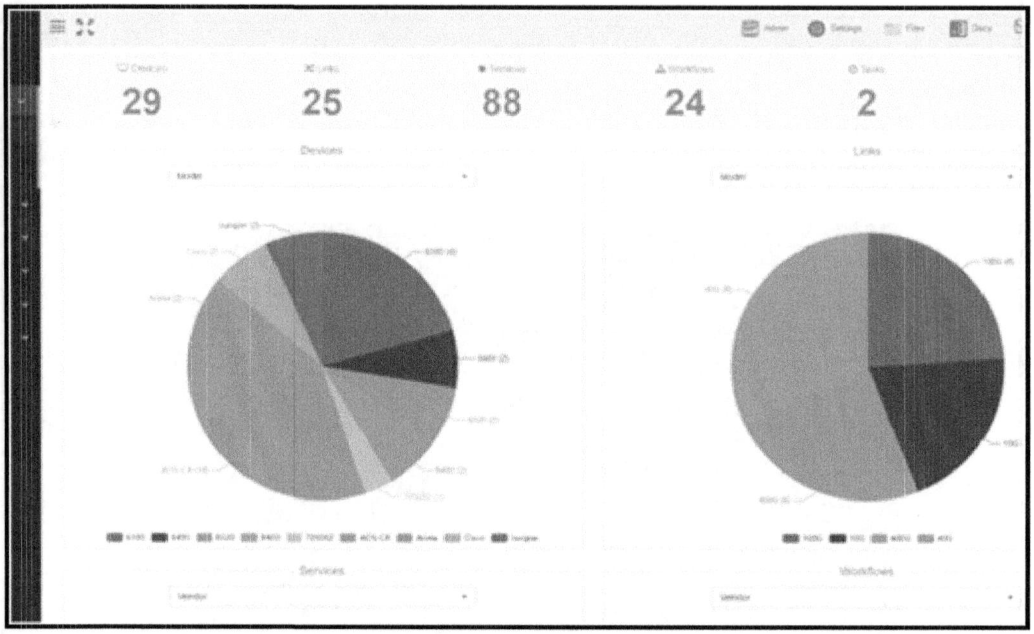

Figure 13-20: NAPALM with eNMS

Many customers have heterogeneous environments with multiple network vendors. Using NAPALM (Network Automation and Programmability Abstraction Layer with Multivendor support) as an abstraction layer allows network management systems (NMS) to easily interact with several switch platforms, regardless of the vendor. eNMS is a vendor-agnostic NMS designed around this idea, using NAPALM to help build workflow-based network automation solutions.

StackStorm with Ansible and Syslog

StackStorm differs from other automation tools in the sense that it was not designed for endpoint configuration or communication; however, in conjunction with tools like Ansible, StackStorm becomes a catalyst for creating efficient and intelligent workflows. For example, if an interface link comes up or down on an AOS-CX switch, an Ansible playbook can be executed reacting to the network change.

HPE Aruba Networking Developer Hub and HPE Aruba Networking Code Exchange

HPE Aruba Networking Developer Hub is where to find help and resources to automate networks and quickly build powerful applications by integrating with HPE Aruba Networking products.

- Supercharge your network operations with HPE Aruba Networking Developer Hub.
- Learn how to integrate with HPE Aruba Networking products.

- Master the game with the complete API reference.
- Collaborate and contribute to the developer community.

Figure 13-21: HPE Aruba Networking Developer Hub

To access the HPE Aruba Networking Developer Hub portal, navigate to: https://devhub.arubanetworks.com.

HPE Aruba Networking Code Exchange

HPE Aruba Networking Code Exchange will help you build automation on the intelligent edge.

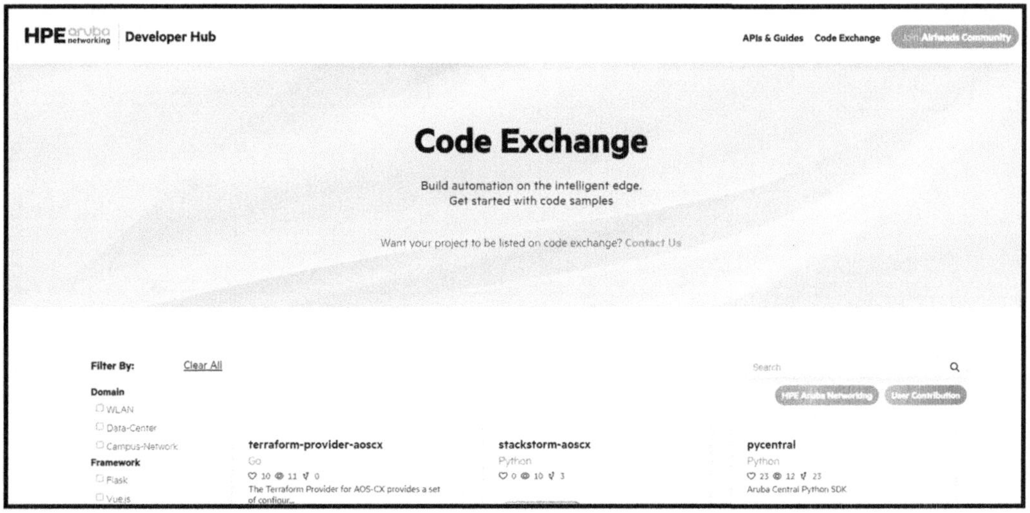

Figure 13-22: HPE Aruba Networking Code Exchange

To access the HPE Aruba Networking Code Exchange portal, navigate to: https://devhub.arubanetworks.com/code-exchange.

Learning check

Check your understanding of the REST API by answering the following questions.

Question 1

Which REST method should you use to create a new resource?

- O GET
- O PUT
- O DELETE
- O POST

Question 2

The AOS-CX Swagger API interface is intended for testing. It will not affect or change any parameter in the switch.

- O True
- O False

14 Quality of Service

In this chapter, you will learn how to implement Quality of Service (QoS) mechanisms that improve the experience of users running a variety of network applications from traditional ones to voice and video applications. The chapter teaches you how to implement the many sophisticated QoS features on AOS-CX switches. You begin by learning about QoS objectives and how AOS-CX switches use a variety of mechanisms to classify and service traffic. Then you dive into each of those mechanisms in more detail—trust settings, queuing, scheduling, rate limiting, and device profiles.

OBJECTIVES

After completing this chapter, you should be able to:

- ✓ Determine how an AOS-CX switch will prioritize traffic based on incoming priority value and QoS trust settings.
- ✓ Implement rate limiting.
- ✓ Use classifier-based policies to assign traffic a new priority based on customer requirements.
- ✓ Customize CoS maps, DSCP maps, and queue profiles.
- ✓ Understand how scheduling mechanisms differ and select an appropriate scheduling profile.

QoS overview

QoS enables networks to provide better or "special" service to a set of users/applications, perhaps to the detriment of other users/applications. You make the best possible use of available bandwidth.

Different traffic types compete for the same network resources, such as bandwidth and fast processing. Without QoS, packets are served as they arrive, typically First In, First Out (FIFO). Your network management traffic may be more important to you than some users watching videos during their lunch break, but there is no differentiation between this traffic.

On the highway, an ambulance gets higher priority than other vehicles. Likewise, QoS is not being fair when dividing the network resources, but rather selecting certain traffic for improved service, perhaps at the expense of other traffic. QoS does not make the road wider; it just decides who goes first and, consequently, who must wait.

Understand that QoS is not a standalone service or product, but rather a concept that supports a framework of protocols, techniques, and network attributes that span across the infrastructure.

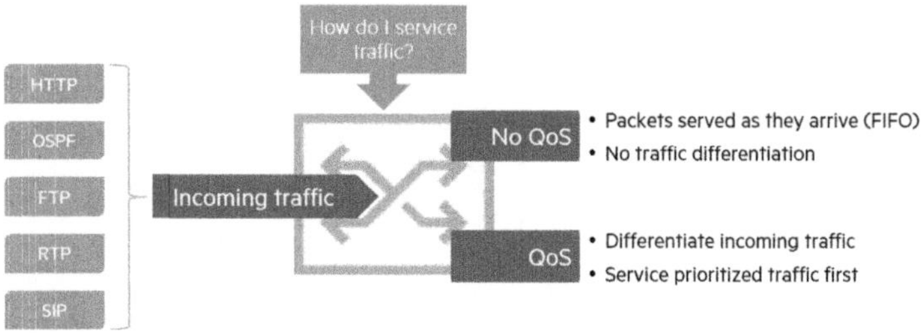

Figure 14-1: QoS overview

Dealing with congestion

Even in a well-designed network, certain links may become oversubscribed; traffic arrives faster than it can be serviced. On a moment-to-moment basis, fluctuating traffic patterns and bursts of traffic can cause congestion. Lower-bandwidth links act as a bottleneck, and if traffic continues to arrive faster than the port can forward it, the port's buffers fill and it drops traffic.

A properly designed network architecture can help to eliminate common sources of congestion. For example, a three-tier Access-Distribution-Core design can sometimes be flattened into a two-tier Access-Core design. This can reduce latency by eliminating the Distribution layer. You should properly provision up-links based on user and device needs. You should not design a network where several 48 Gigabit access links must all be serviced by two 10 Gigabit Access-to-Core links, which must then be serviced by two 1 Gigabit Core-to-WAN Edge links. As shown in the previous figure, this likely creates near-constant congestion.

Still, even a well-designed network will experience occasional over subscription and momentary bursts of congestion. Business-critical traffic might be dropped, and delay-sensitive applications might lose functionality. You should implement QoS mechanisms to protect sensitive traffic, ensuring a good user experience.

QoS mechanisms help to enforce business policies that prioritize certain applications or network segments. You can customize how different traffic types are serviced. You accommodate unique application characteristics and their importance within the organization. QoS ensures uniform and efficient traffic handling, keeping the most important traffic moving at an acceptable speed, regardless of current bandwidth usage. It also enables you to control the priority settings of ingress traffic for each network device.

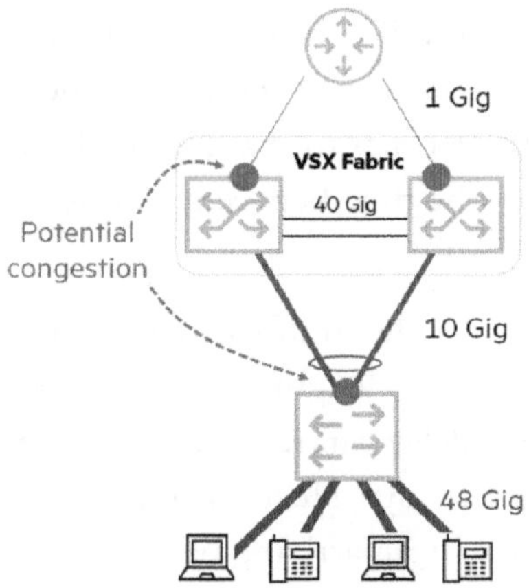

Figure 14-2: Dealing with congestion

TCP and UDP

In general, "best-effort" delivery is satisfactory for typical data traffic. Applications that use TCP for transport are less affected by packet loss because TCP provides acknowledgement and flow-control mechanisms. TCP detects and retransmits dropped packets. Each segment of a TCP/IP message includes sequence numbers that inform receivers of the proper order and number of packets in the transmission. The receiver in a TCP transmission acknowledges packets by returning messages that include the highest sequence number it has received. TCP applications include a "window" parameter that determines how many bytes can be outstanding and unacknowledged by the receiver at any given moment. This parameter limits the number of packets that can be acknowledged by a single control packet. Delay in delivery of packets or excessive packet loss prompts the sender and receiver to establish a smaller window size. Furthermore, TCP applications back off when network congestion is detected, enabling applications to share bandwidth.

UDP, on the other hand, does not provide acknowledgments or flow control. This reduced overhead can benefit delay-sensitive applications. Also, some applications include their own error-checking procedures and do not need TCP to do it for them. However, in most cases, UDP applications are designed to minimize the need for retransmission and use UDP as a transport mechanism precisely because of its lower overhead and network efficiency. UDP also does not provide any back-off procedures.

Real-time traffic

Voice traffic is sensitive to delay, sometimes called latency, which is the amount of time that passes between the sending of a transmission and its arrival at the receiving station.

The next figure shows small packets traveling from sender to receiver. In both examples, the packets are sent at a consistent rate. In the latency example, each device in the network cloud requires time to process each packet. Add these times to that required for packets to traverse each link in the cloud, and you get total latency: the time between a packet sent from Host 1 to arrival at Host 2.

In the case of unidirectional traffic (as video traffic often is), this predictable, measurable time period is usually tolerable, regardless of its length. However, because multidirectional voice traffic is nearly always interactive, excessive delay can result in conversation collisions. Suppose that Host 1 begins transmitting in the belief that Host 2 is not transmitting. However, because Host 2 is still transmitting, bits from Host 2 begin arriving before Host 1 concludes its transmission. The conversations overlap.

Voice traffic is also sensitive to jitter, which is the variation in intervals between the arrivals of packets or, in other words, the difference between when a packet is expected and when it actually arrives. Video is also sensitive to jitter.

In the following Jitter example, Host 1 sends data at a constant rate to Host 2. However, devices within the cloud that represents the network experience varying levels of congestion that result in some packets being delayed more than others. The outcome is a variation in the interval between packet arrival, which results in a choppy voice or video stream. Jitter can have a significant effect on real-time applications, even if they require very small data streams. For instance, an IP telephone sends very small packets every 20–30 milliseconds (ms). For the receiver to properly interpret the stream, the packets must arrive at the same rate. If the interval between packets grows to 50 ms, for instance, the transmission will be unsatisfactory for many users.

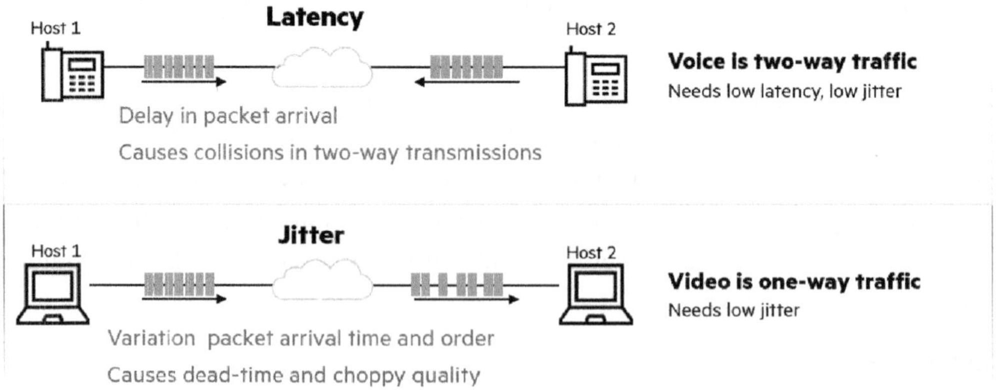

Figure 14-3: Real-time traffic example

Characterizing QoS needs

All traffic does not require the same handling to receive an appropriate level of service. You should understand common applications and their characteristics, as relates to QoS. This is summarized as follows.

TCP/file transfer

Many TCP applications can tolerate some service degradation due to TCP's retransmission and congestion–handling mechanisms. File transfer applications have higher bandwidth requirements but are not sensitive to delay and jitter— the user cares only about the result. You may be familiar with the File Transfer Protocol (FTP), which relies on TCP, and the Trivial File Transfer Protocol (TFTP), which uses UDP. Even so, TFTP has the same high bandwidth, low sensitivity as FTP. This traffic rarely requires special QoS treatment.

UDP/Voice/Video

Time-sensitive traffic like voice and video requires prioritization over other traffic so that it is transmitted immediately. This prioritization ensures that traffic transmitted at a constant rate by an endpoint such as an IP phone continues to be transmitted at a constant rate throughout the network, minimizing jitter.

This traffic is nearly always a candidate for QoS prioritization. But remember, less important UDP traffic must be prevented from monopolizing a link because the application will not back off on its own. You might want to enforce rate limits that ensure certain classes of traffic use only their fair share of bandwidth.

You know that UDP does not respond well to congestion, but often it does not matter. Suppose you transmit a video at 30 frames per second (fps) and one packet, containing one small part of one frame, is dropped. It is unlikely anyone will notice—no need to retransmit video or voice traffic, just move on without it. However, if too many packets are lost at once, the user experience suffers. Users have a better experience when the network infrastructure responds to congestion gracefully by dropping less important or sensitive traffic.

Delay over media

Delay is the interval between the time that a message leaves the sender's mouth and the time that it reaches the receiver's ear. Some delay is inevitable, especially in transmissions over significant distances. While the speed of light in a vacuum is 186,000 miles per second (267,000 km per second), signals move through fiber or copper cable at approximately 125,000 miles per second (200,000 km per second). Consequently, a fiber network stretching halfway around the Earth introduces a delay of approximately 70 milliseconds under the best of circumstances, with additional delay introduced by the physical transport medium. During periods of congestion, the delay can be much longer.

The effects of delay upon application performance depend heavily upon the expectations of users. The International Telecommunications Union (ITU) recommends no more than 150 ms end-to-end delay for voice traffic. Greater delay can cause conversation collisions when each party begins to speak because the line appears quiet. Most users will find this situation to be unacceptable.

QoS use case

The following figure shows times when you should implement QoS. Understand that use cases 1-4 are all a part of typical scenarios. Remember, well-designed networks can experience temporary bursts of congestion, during which there will be insufficient uplink capacity. During these times, some edge traffic may require special handling; you want to give special traffic some type of priority service.

Understand that if your network experiences near-constant or very frequent congestion, QoS will not help you. You need faster links and/or more capable devices. QoS is for well-designed networks that experience relatively short, temporary bursts of congestion.

It is wise to think about QoS for any network deployment. Many network engineers think of this as a typical part of network design. During the almost inevitable short bursts of temporary oversubscription, which traffic in your network is business-critical? Which traffic requires high bandwidth, low packet drop, and/or is sensitive to delay and jitter? When you can answer these questions, you are ready to then implement QoS, often by configuring network devices to classify traffic and apply QoS policies.

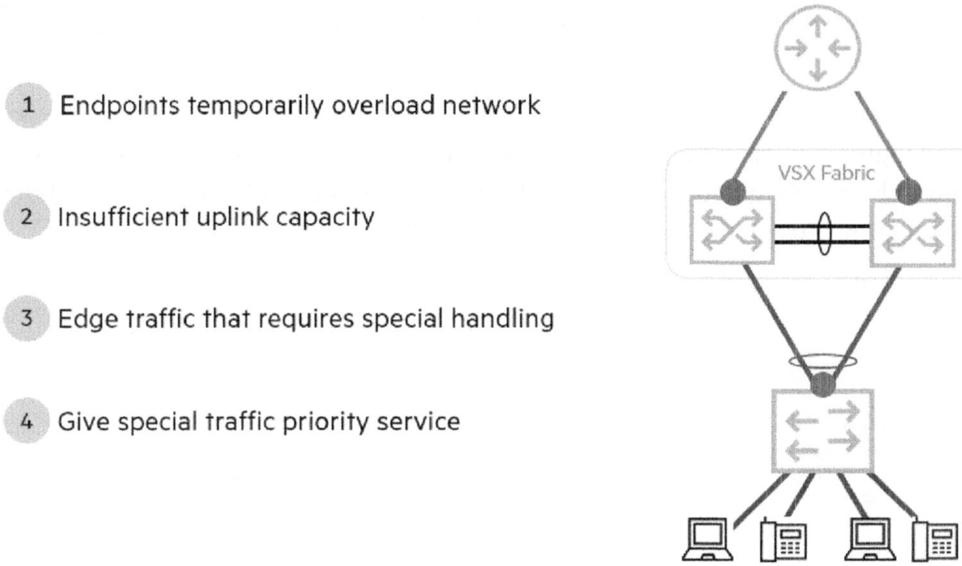

Figure 14-4: QoS use cases scenarios

Classifying traffic and applying policies

Now you understand what QoS is, you should understand how it works —with a high-level, slightly simplified QoS scenario. Packets ingress at some switch or router interface —some video packets, some FTP packets, and some voice packets. In this section, we are going to starting witha queuing operation overview to develop a deeper knowledge on classifying traffic and applying policies.

First the switch must classify packets. It can assess source/destination MAC and IP addresses, the ingress port, the application type, and more, depending on the scenario. Perhaps all packets arriving on port 1/1/1 are low priority, or all VoIP packets are high priority.

The purpose of classification is marking. Suppose you arrive at a music performance. An attendant classifies you based on your identification and your entry ticket. Then you are "marked" with a wrist band. Certain folks are on the performer's guest list and are given a gold "all access" band. Others paid a premium for their ticket and are given a silver band – they get a special VIP line to quickly get food and beverages. Others paid a low price, are given a blue band, and must stand in a long line to get warm lemonade.

In the figure below, some packets were identified as video. These packets are marked upon ingress by placing a priority value of 5 in a special frame or packet header. Other packets classified as FTP and are marked with Priority=1. The VoIP traffic is marked with Priority=6. Notice that the classification process can also use marking. Perhaps an upstream device has already marked packets, or a VoIP phone has pre-marked its packets.

Then packets are queued up for service. Video packets are placed in the medium queue, FTP in the low queue, and VoIP in the High Queue.

Now the Scheduler takes over, often based on a configured Service Policy. Perhaps the policy states that the high queue shall be serviced exclusively, until it is empty. The router services other queues if and only if the high queue is empty. It goes back and forth between the medium and low queues, spending 70% more time servicing the medium queue.

Let's look at classification and marking in more detail.

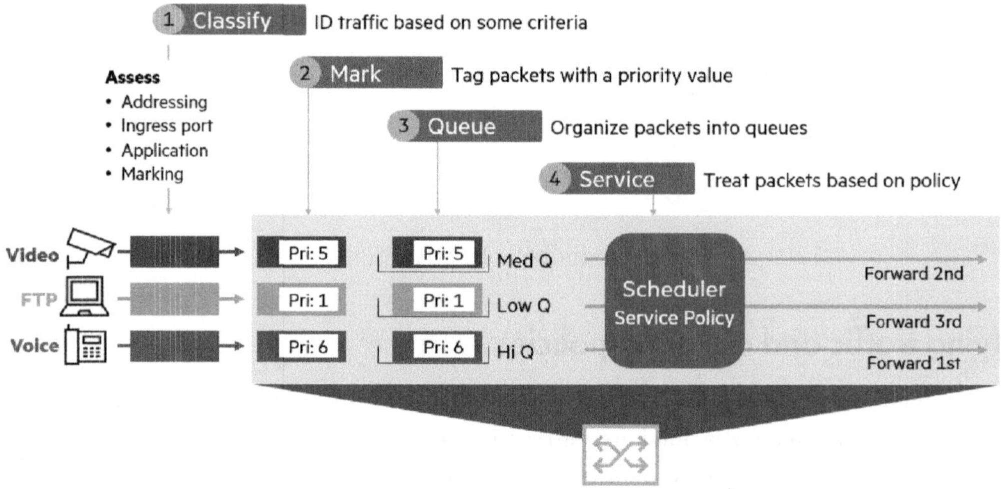

Figure 14-5: Classifying traffic and applying policies overview

L2-L3 marking

To communicate the desired priority for traffic between network infrastructure devices, the devices use QoS marking. These mechanisms define specific values, which request specific priorities for traffic marked with that value. AOS-CX switches can leverage these values to determine how to classify and service traffic. The values reside within frames or packet headers and so are passed on with the traffic from device to device.

At Layer 2, Ethernet frames can have 802.1Q information, used for VLAN tagging. One of the fields inside the 802.1Q fields is used to do QoS tagging, as defined in the 802.1p standard. This is a 3-bit field, and so supports 8 priority values – 0 – 7. These fields are called the Class of Service (CoS) bits. If you use 802.1p user priority field that is a part of the 802.1Q standard, you are said to using CoS marking.

At Layer 3, the IP header has a 8-bit Type of Service (ToS) field. Initially, 5 bits were unused, and so 3 bits were used for ToS or "IP Precedence". Then more of these 8 bits were defined in a standard. If using this standard, you are said to be using Differentiated Services Code Point (DSCP) marking.

Best effort service

This is the simplest service type. All traffic is treated equally in a first-come, first-served manner. If the traffic load is low in relation to the capacity of the network links, then there is no need for the administrative complexity and costs of maintaining a more complex end-to-end policy. If you intend to use this method, you must ensure that your network is over-designed – a so-called over-provisioned network where all link speeds are much higher than peak loads on the network.

802.1p or CoS

CoS uses a three-bit field in the 802.1Q VLAN tag to indicate priority. It defines eight values between 0 and 7. Note that because the VLAN tag carries the priority value, native VLAN traffic cannot contain a CoS value. Although this example below shows an Ethernet frame that encapsulates an IP packet, an Ethernet frame with an 802.1p field can encapsulate any Layer 3 protocol an Ethernet supports.

ToS and IP precedence

ToS is rarely used today, but it is helpful to understand what ToS is if you hear it referenced.

The figure below shows that every IP packet includes an eight-bit field called the Type of Service (ToS) field. As originally defined, the ToS field contained two subfields—a three-bit IP precedence field used for relative prioritization and a four-bit subfield for the specific ToS desired by an individual packet. The remaining bit, bit 7, is called the must be zero (MBZ) bit and is unused.

The three IP precedence bits in the most significant position were to be used to provide eight levels of precedence or priority. The levels are labeled 0 to 7, with 7 being the highest priority and 0 the lowest. The values 6 and 7 are reserved for "Internet" and "network" uses, respectively. This enables routing updates and other critical traffic to have a higher priority than user-generated traffic. The highest user-defined value is 5, which represents "critical" priority.

The four ToS bits were designed to allow applications to instruct routers to choose routing paths with one of the following four characteristics: minimized delay, maximized throughput, maximized reliability, and minimized cost. Only one of the four bits could be "turned on," that is, set to 1.

DiffServ / DSCP

Differentiated Services (DiffServ), defined in RFC 2474, supersedes a legacy IP QoS protocol called a type of Service (ToS). In DiffServ, the first six bits of the eight bit ToS field define 64 DiffServ "code points" (DSCPs). Instead of only defining priority relative to each other, the 64

DSCPs are also intended to define distinct forwarding behaviors, or Per Hop Behaviors (PHBs).

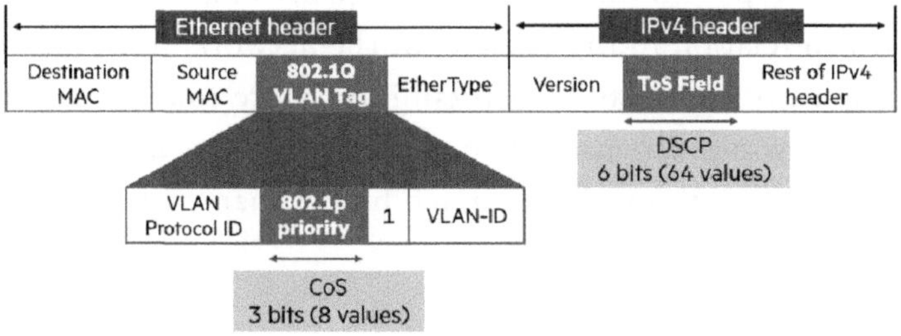

Figure 14-6: L2-L3 marking illustration

Layer 2 CoS classification

CoS values request up to eight different priority levels for different types of traffic. Use the 3-bit Priority Code Point (PCP) 802.1p field within the 16-bit 802.1Q Ethernet VLAN tag to mark CoS. The standard recommendation differs based on whether the traffic type is 802.1Q or 802.1D.

For 802.1Q traffic, the you should use the most up-to-date recommendations in the IEEE 802.1Q-2014 standard. This standard recommends that you reserve the two highest values (7 and 6) for network control and internetwork control traffic. Network control traffic includes frames for Layer 2 control protocols such as spanning tree BPDUs while internetwork control includes Layer 3 control frames such as OSPF hellos and LSUs.

AOS-CX switches automatically mark their control traffic for high priority. Voice, value 5, and video, value 4, are given the next highest priority. These recommendations are based on their low latency requirements. Mission critical applications receive the next highest priority, value 3. The standard recommends assigning Value 2 to traffic that requires excellent effort. It is up to you to define what is considered critical or excellent effort applications. Next the switch gives its best effort to traffic with the default priority, which is 0. Finally, 802.1p value 1 is assigned to background traffic, which is less important or can tolerate delay. CoS 1 is deliberately set as the lowest CoS. This enables a traffic service level below the default (best effort) traffic level to be specified.

802.1D recommendations are based on the original standard that defines CoS. Some network infrastructures still use these recommendations, so you should be aware of them. In this scheme, network control traffic has just one value (7), which shifts voice and video traffic up one priority level. Controlled load (rather than critical) has the next highest level and then excellent effort. But again, companies can choose themselves what applications fit in these categories. One of the most crucial differences in the 802.1D standard is that it assigns value 2 to "spare" traffic and is intended for traffic with a priority less than best effort. It is crucial that you understand whether other devices in the network consider value 2 as higher than 0 or less than 0.

802.1p	Traffic type (802.1Q)	Traffic type (802.1d)
7	Network control	Network control
6	Internetwork control	Voice (<10ms latency)
5	Voice (<10ms latency)	Video (<100ms latency
4	Video (<100ms latency	Controlled load
3	Critical applications	Excellent effort
2	Excellent effort	Spare (less than 0)
0	Best effort	Best effort
1	Background	Background

Figure 14-7: Layer 2 CoS classification

L3 classification: usage of ToS bits

You know that every IP header includes an 8-bit ToS field. Originally, only the high-order 3 bits were used for IP Precedence. With three bits you have 8 possible values: decimal 0 – 7.

Much later, this 8-bit ToS field was redefined in a set of RFC standards to use 6 bits. Of those 6 bits, for nearly all common implementations, the lowest order bit is not used.

> Newer specifications can make specialized use of formerly unused bits, but they are not relevant to this discussion of QoS.

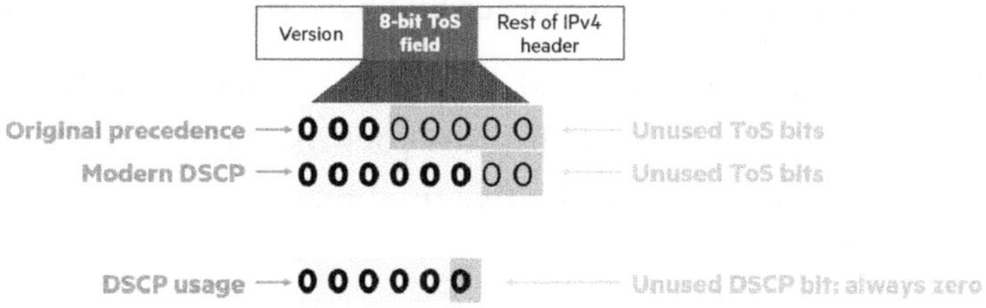

Figure 14-8: L3 classification: usage of ToS bits

L3 classification: interpret DSCP as a decimal number

Consider the decimal number 9,325. You know that the 5 is in the one's column, the 2 is in the 10's column, 3 is in the hundred's column, and 9 is in the thousand's column.

With the 6-bit DSCP binary number, the columns are 1, 2, 4, 8, 16, and 32 —as shown in the figure below. So, the binary number 001010 = 10 in decimal —there is a 1 in the eight's column, and a 1 in the 2's column, and 8 + 2 = 10.

Similarly, 001100 in binary = 12 in decimal, and 001110 = 14 in decimal.

So, when communicating with others you can say, "I have created a policy to mark certain packets with a DSCP decimal value of 10". (or 12, 14, etc.) But there is a second way to interpret these bits.

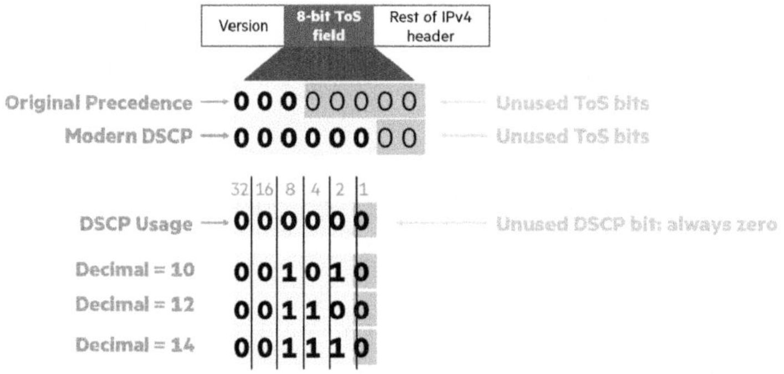

Figure 14-9: L3 classification: interpret DSCP as a decimal number

L3 classification: interpret DSCP as a class

Instead of perceiving the 6-bit number as a single value, think of the 6-bit field as two separate binary numbers. The three high-order bits represent the DSCP Class, and next two low-order bits represent a Drop Probability (DP), to further distinguish each class. That lowest order bit is simply ignored, as if it did not exist. Since they are two separate binary numbers, the high order bits have columns 1, 2, and 4, while the low-order bits have the 2 and 1 columns.

Now convert from binary to decimal as normal. Any packets marked with 001 in the high-order bits are in Assured Forwarding (AF) Class 1. If those packets also have 01 in the low-order bits, the packets are said to be in class AF 11. If the five used bits are set to 001 10, AF 12 is indicated. If the bits are 001 11, that is AF 13.

Packets with high-order bits 010 are in Assured Forwarding Class 2. Depending on the DP bits, they could be AF 21, AF22, or AF23.

Look at how this plays out.

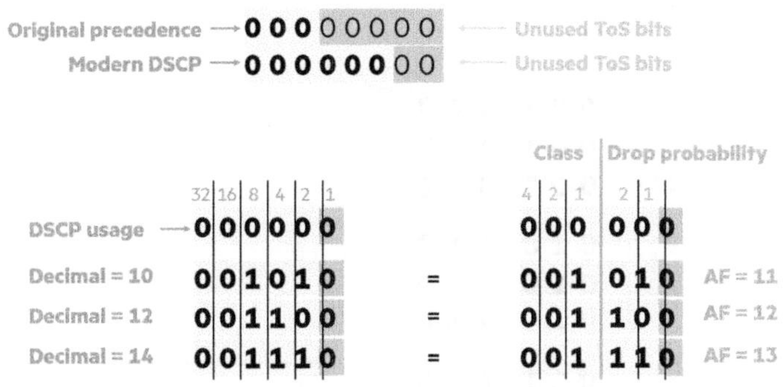

Figure 14-10: L3 classification: interpret DSCP as a class

L3 classification: decimal DSCP vs AF class

Here are the four DSCP Assured Forwarding (AF) Classes, each with 3 Drop Probability values, as defined in RFC 2597. Each value is also interpreted as its equivalent decimal number. Spend some time looking at this. Patterns will become obvious to you, and it will become more intuitive.

Also, there is yet another way to interpret these bits, but its easy —just for backward compatibility with IP Precedence— the original, now largely unused practice of using only the three high order bits of the IP header's ToS field. Just focus on the highest order three bits in the figure, with columns 1, 2, and 4 —just like for the DSCP class, but without any lower order drop probability bits. You can simply count from 0 (binary 000) to 7 (binary 111). 000 is called DSCP "Class Selector (CS) 0", 001 is "CS1", and so on up to 111— "CS7". DSCP and IP precedence can exist in parallel because of this interpretation.

If some applications only look for the three high order bits, they will recognize the IP Precedence value. An application or device that uses the DiffServ interpretation will interpret those same three bits from its own perspective. Applications and devices will arrive at the same relative priority, although the actual values are different.

You may be thinking, "Why are there two ways to talk about the same values that do the same thing?". Perhaps to accommodate differing preferences of different folks. The reason why is not for us to dwell upon. What is important is that we understand how these values are to be used in the real world.

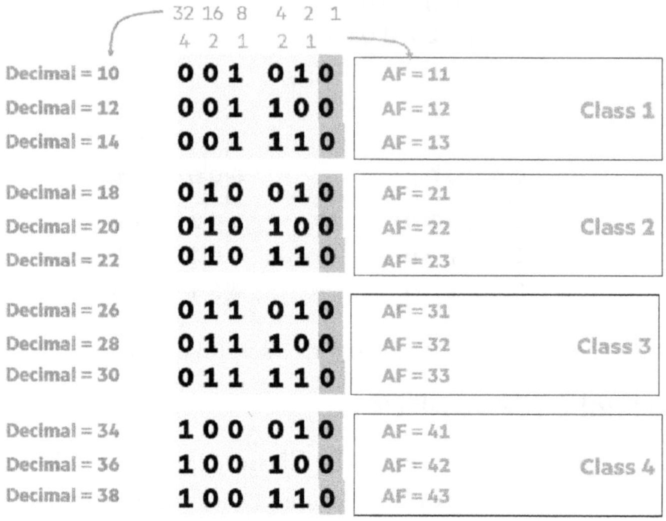

Figure 14-11: L3 classification: decimal DSCP vs AF class

Using L2 and L3 classification

The second table on the right side below recommends the type of applications that you might assign to each AF class. For example, you could assign multimedia conferencing applications to AF class 4. These applications are interactive, so they need very low latency. You could assign multimedia streaming, which can use buffering and has a bit less strict latency requirements to AF class 3. You can assign AF class 2 to applications that require low latency, but less than multimedia. For example, interactive network management applications such as SSH could use this class. Finally, AF class 1 is for bulk data that can tolerate delays—it has priority less than 0.

If needed you can use the CS values for legacy IP precedence values. Also, RFC 3246 defines Expedited Forwarding (EF) class 46, to be used for applications most sensitive to jitter and delay. HPE often recommends this DSCP for voice over IP (VoIP) traffic.

VLAN Protocol ID	802.1p priority	1	VLAN-ID

802.1p	Traffic type (802.1Q)
7	Network control
6	Internetwork control
5	Voice (<10ms latency)
4	Video (<100ms latency
3	Critical applications
2	Excellent effort
0	Best effort
1	Background

Version	ToS field	Rest of IPv4 header

DSCP	Name	Examples
56	CS6	Network control
46	EF	Telephony (VoIP)
40	CS5	Signaling
34,36,38	AF41 / AF42/ AF43	Multimedia conferencing
32	CS4	Real-time interactive
26,28,30	AF31 / AF32 / AF33	Multimedia streaming
24	CS3	Broadcast video
18,20,22	AF21 / AF22 / AF23	Low-latency data
16	CS2	OAM
00	CS0 / BE / DF	Best effort
10,12,14	AF11 / AF12 / AF13	Bulk data
08	CS1	Low-priority data

Figure 14-12: Using L2 and L3 classification

QoS basics

This **Figure 14-13** below illustrates the basic steps that traffic experiences when you implement a QoS solution.

1. A device sends a frame, with or without classification markings.
2. The switch receives the frame and places it in the ingress queue.
3. The switch determines if the ingress packet was pre-marked by the host application. And if so, does the switch trust these markings. If there are not trusted markings, the switch can mark or re-mark the frame based on some policy you define. If ingress marking is trusted, the switch can use those markings for any policy you may have defined. If you expect both L3 DSCP and L2 802.1p markings, you should define a policy to control which marking to trust. You might also define an ingress rate limiting policy - applied certain ports.
4. The switch then uses the classification information and Virtual Output Queuing (VOQ) to determine when to move the frame to the egress queue.
5. The switch then implements the egress queuing policy, based on the traffic classification, to determine when the frame is sent out the port, along with applying any defined outbound rate limiter.

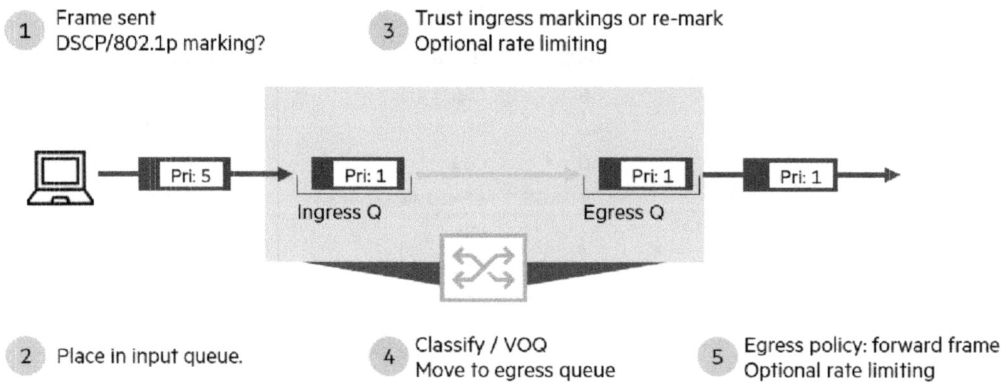

Figure 14-13: Basic steps that traffic experiences when you implement a QoS solution.

Ingress rate limiting

Port rate limiting helps control undesirable traffic. The goal is to allow only enough Broadcast, Unknown unicast, and Multicast (BUM) traffic for the network to function properly, while preventing flooding and traffic storms.

A certain amount of BUM traffic is required for normal network operation. Broadcast packets may include ARP and DHCP traffic, for instance. Video streams and certain types of network protocol packets are multicasts. Unknown-unicast packets may be intended for devices whose addresses have temporarily aged out of network-forwarding caches. Configuring rate limits can help provide the balance between necessary traffic and flooded traffic.

Rate limiting is a traffic policer QoS tool. It ensures that traffic conforms to a defined rate: a bandwidth limit. An important characteristic of rate limiting is that it does not add delay to traffic to be forwarded—that which is under the threshold and is accepted by the switch. You configure it per physical or LAG interface.

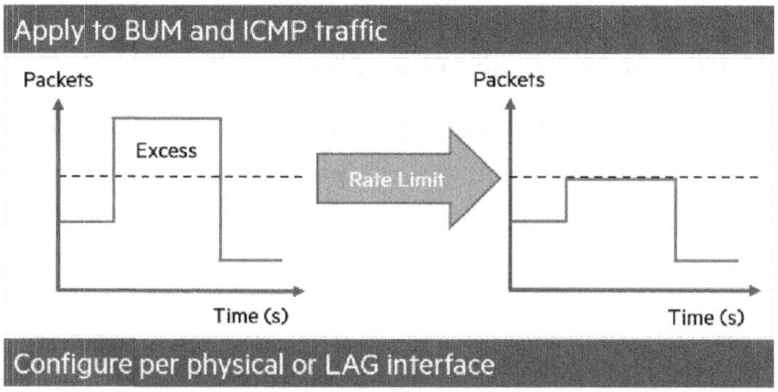

Figure 14-14: Ingress rate limiting

Configure rate limiting

You can implement the rate limiting QoS feature on all AOS-CX platforms. However, supported parameters are platform dependent:

Broadcast: Supported on all AOS-CX platforms, limit only the broadcast traffic

Multicast: Supported on all AOS-CX platforms, limit only the multicast traffic.

Unknown Unicast: Supported only on CX 8400 Series, CX 8325 Series, and CX 8320 Series switch platforms, limit only the unknown unicast traffic.

ICMP: Supported only on CX 6400 Series and CX 6300 Series switch platforms, the ICMP rate limit can be configured to apply to IPv4, IPv6, or all IP traffic. Only one ICMP rate limit can be configured at a time. Applying a new ICMP rate limit replaces any previous ICMP rate limit.

When multicast and broadcast rate limits are applied at the same time on the same interface, broadcast packets are limited to the lower of the two rate values.

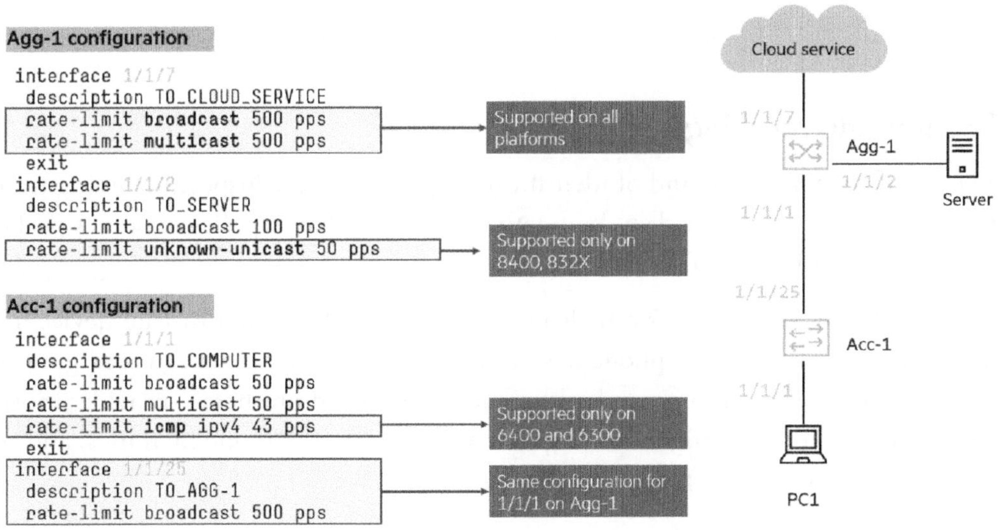

Figure 14-15: Configure rate limiting summary

Verify rate limiting

The `show interface <INTERFACE-ID> qos` and `show interface <INTERFACE-ID>` commands are useful to verify the rate limiting feature implemented on the interface.

```
AGG-1# show interface 1/1/7 qos
Interface 1/1/7 is up
 Admin state is up
 qos trust none (global)
 qos queue-profile factory-default (global)
 qos schedule-profile factory-default (global)
 rate-limit unknown-unicast 500 pps (500 actual)
 rate-limit multicast 500 pps (500 actual)

AGG-1# show interface 1/1/7
Interface 1/1/7 is up
 Admin state is up
 Link transitions: 1
 Description:
 Hardware: Ethernet, MAC Address: 90:20:c2:c0:8e:fd
 MTU 1500
 Type SFP+DAC1
 Full-duplex
 qos trust none
 rate-limit unknown-unicast 500 pps (500 actual)
 rate-limit multicast 500 pps (500 actual)
 Speed 10000 Mb/s
 Auto-negotiation is off
  <<Omitted output>>
```

Figure 14-16: Verify rate limiting commands

Classification and marking

Recall that classifiers perform a kind of identification process. They inspect incoming traffic, identify it, and then assign it to a class. You define a classifier-based policy, analyzes packet parameters to apply a specific class of service.

Notice that the classification process can also use marking. Perhaps an upstream device has already marked packets, or a VoIP phone has pre-marked its packets. You can configure certain switch ports to trust these markings, and take some action based on them—provide priority service levels perhaps. You can configure other ports to ignore ingress frame markings, and re-mark them according to a local policy that you define.

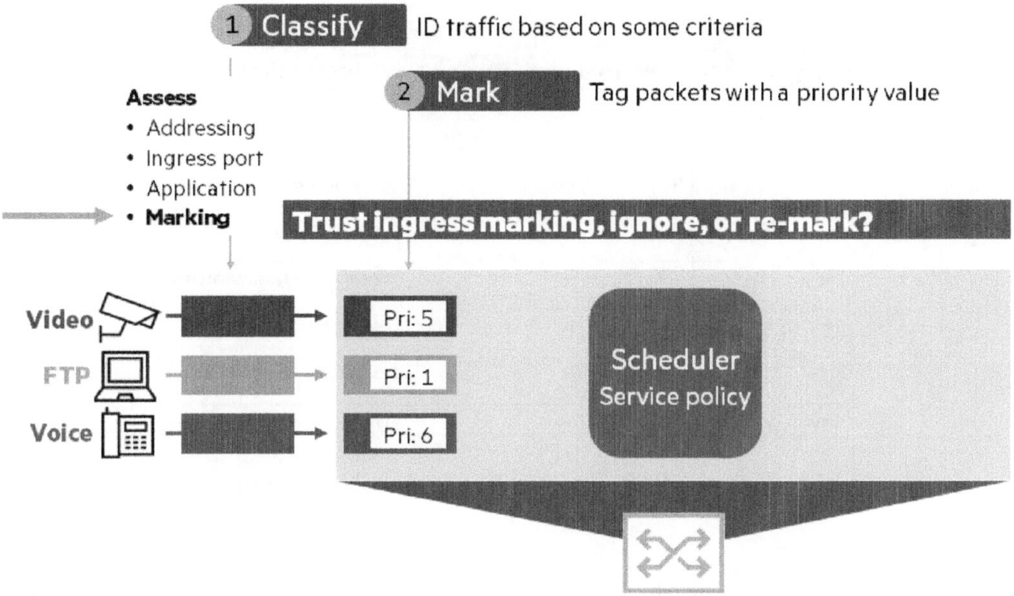

Figure 14-17: Classification and marking illustration

Prioritization stage

The objective of the prioritization stage is to determine the Internal Local priority value and the Color (drop eligibility) for a specific traffic. The AOS-CX switch uses these parameters to specify the behavior, or the class of service used.

The decision to assign a Local Priority and Color value could be based on ingress packet L3 DSCP or L2 CoS marks. Each switch maintains a mapping table that helps with the decision. You can display the default mapping table with the `show qos dscp-map` default command or `show qos cos-map` default command. AOS-CX supports the creation of custom mapping. The other possibility is to use a classifier-based policy where the interesting traffic is selected based on specific characteristics. In this case the marking may or may not be present.

Trust is an important setting for QoS in AOS-CX. When it is set on the incoming interface the switch can use the incoming marks to determine the Local Priority. However, if the trust is not set (or set to none) then incoming marking cannot be used and all traffic is placed in the Default Best Effort behavior, represented by the Local Priority 1. Notice that classify policy-based assignment is not dependent on the trust status and overrides default and custom mapping. The following table summaries this discussion.

Trust (Ingress Interface)	Custom mapping	Classifier-based policy	Classifier-based policy
Yes	No	No	Default mapping
Yes	Yes	No	Custom mapping
Yes	Yes	Yes	Classifier-based policy mapping
None	No	No	Default (LP = 1 - Best Effort)
None	Yes	No	Default (LP = 1 - Best Effort)
None	Yes	Yes	Classifier-based policy mapping

Figure 14-18: Prioritization stage table summary

Overview

When traffic arrives at the switch, the switch can trust existing markings or not. The term trust refers to whether the packet markings are valid input for the classifier, which decides each packet's service class.

You know that trust configuration plays an important role in assigning Local Priority and Color values. In AOS-CX, you can configure trust at the global or Interface level. When the trust level is set at both levels, the interface level configuration takes precedence.

AOS-CX supports three possible configurations: Trust CoS, Trust DSCP and Trust none. The first two options act as the name suggests —the switch accepts the incoming marking and uses it to determine class of service. In AOS-CX 10.5 an interface can only support one trust setting at the time, meaning that you can set DSCP or CoS but not both at the same time, on the same interface.

It is not always recommended to trust ingress markings, especially if traffic is received from non-trusted devices. In these situations, you can protect the network by ignoring the incoming marking setting the trust setting to none.

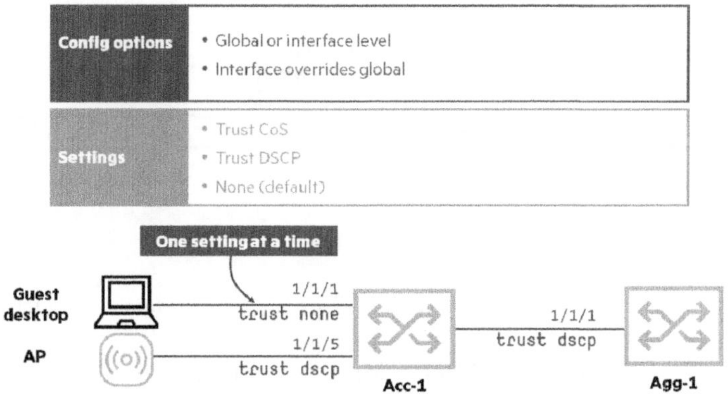

Figure 14-19: Overview summary

Global and interface trust

If a port has no setting, the port takes its setting from the global setting. Otherwise, the port setting, including "none," overrides the global setting. In this way, you could globally not trust incoming marks, but trust DSCP on certain ports. You could globally trust DSCP, but use CoS instead on certain ports, and so on.

By default, the global setting is "none," and all ports have no trust setting. This means that all ports use the global setting of "none" —the switch ignores any markings in ingress frames and assigns all traffic the LP and color associated with CoS 0 in the CoS map. By default, this is LP=0 and color=green. The switch preserves any incoming DSCP when it forwards the traffic, as well as an incoming CoS value if the forwarded traffic is tagged. However, these values have no effect on how the switch queues the traffic.

RFC 2697 defines the use of colors with DSCP codes. A Single Rate Three Color Marker (srTCM) meters an IP packet stream and marks its packets either green, yellow, or red. Marking is based on a Committed Information Rate (CIR) and two associated burst sizes, a Committed Burst Size (CBS) and an Excess Burst Size (EBS). A packet is marked green if it doesn't exceed the CBS, yellow if it does exceed the CBS, but not the EBS, and red otherwise. The srTCM is useful, for example, for ingress policing of a service, where only the length, not the peak rate, of the burst determines service eligibility. See RFC 2697 for more information: https://www.ietf.org/rfc/rfc2697.txt. The use of colors is optional.

It is important to clarify that a port not having a trust configuration is different from the port having a configuration of **none**. If a port has no setting, the port takes its setting from the global setting. Otherwise, the port setting, including "none," overrides the global setting.

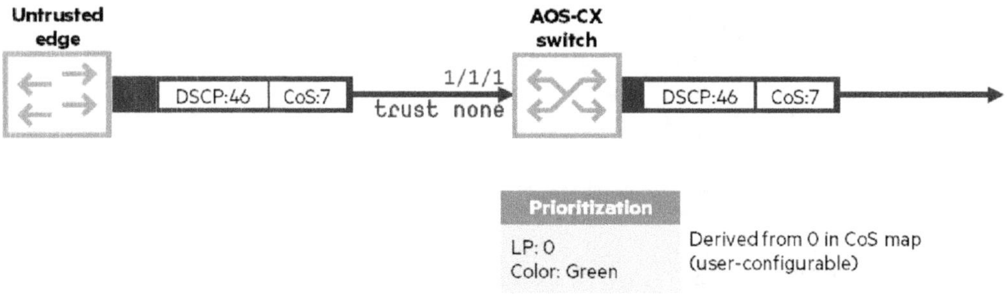

Figure 14-20: Global and interface trust

Trust set to DSCP

Many networks classify traffic at the edge, so it can be useful for the AOS-CX switch to use the already applied classifications to determine the correct priority for the traffic. If other network infrastructure devices are using DSCP to classify traffic, set the global or port trust mode to dscp. When a port has a dscp trust setting, whether that is set on the port or taken from the global setting, the switch looks at the incoming DSCP in traffic's IP headers and uses its DSCP map to assign the traffic an LP and color.

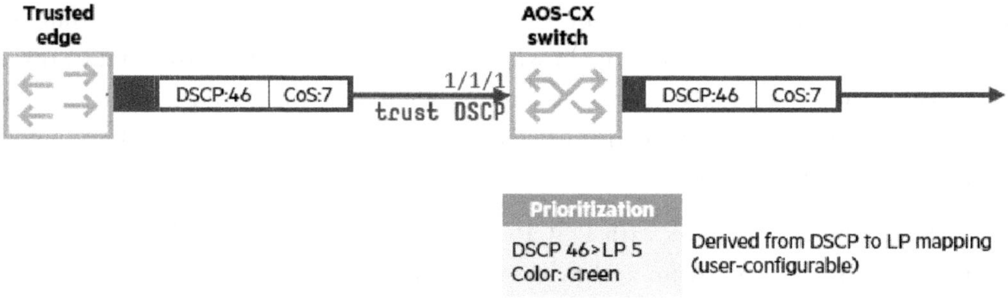

Figure 14-21: Trust set to DSCP

Default DSCP map

Here is the default DSCP map, which follows the standards and works in many circumstances.

In the following example, the map has an entry for DSCP 46, which assigns the incoming traffic LP 5 and color green. However, you can customize this map, setting your own local priority, color, and name for each DSCP. Your goal in customizing the map should be to match the system the devices use from where the switch is receiving the incoming traffic. For example, you may want to change DSCP 46, used for VoIP traffic, to LP 7 if the company wants to give this traffic highest priority.

Note that, if the incoming traffic has a CoS value as well as a DSCP, the switch preserves this value in the outgoing frame, if tagged. However, the CoS value and map have no effect on the traffic's LP and color. Those derive from the DSCP only.

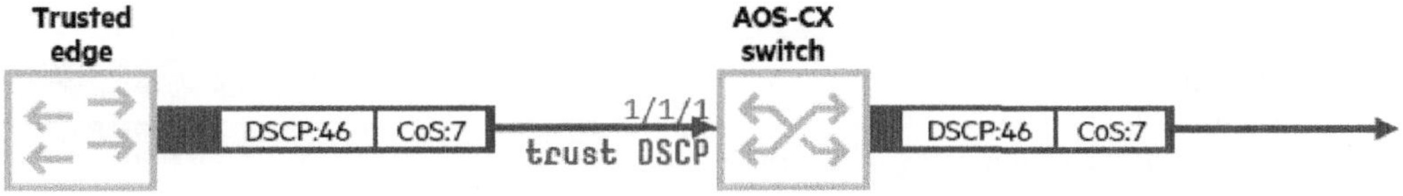

Default DSCP map

DSCP	LP	Color name
0	1	green CS0
1-7	1	green
8	0	green CS1
	<<omitted output>>	
46	5	green EF
47	5	green
48	6	green CS6
49-55	6	green
57	7	green CS7
58-63	7	green

Prioritization

DSCP 46>LP 5
Color: Green

Derived from DSCP to LP mapping (user-configurable)

Figure 14-22: Default DSCP map

Configure prioritization

You can configure the prioritization stage as shown in the following figure.

For the custom map has two options. Use the **cos-map** option to define the local priority assigned to incoming packets for a specific 802.1 VLAN priority code point (CoS) value. The CoS map values are used to mark incoming packets when QoS trust mode is set to **cos**.

Use the **dscp-map** option to defines the local priority assigned to incoming packets for a specific IP differentiated services code point (DSCP) value. The DSCP map values are used to prioritize incoming packets when QoS trust mode is set to **dscp**.

Trust global configuration

```
qos trust <cos | dscp | none>
```

Trust interface configuration

```
interface <INTERFACE-ID>
 qos trust <cos | dscp | none>
```

Custom map

```
qos cos-map <CODE-POINT> local-priority <PRIORITY-NUMBER> [color <COLOR>] [name
<DESCRIPTION>]
qos dscp-map <CODE-POINT> local-priority <PRIORITY-NUMBER> [color <COLOR>] [cos <PCP-VALUE>]
[name <DESCRIPTION>]
```

Figure 14-23: Configure prioritization

Classifier-based configuration

The following example shows how you can use a classifier-based policy configuration to assign the Local Priority.

```
ACC-1(config)# qos trust none
ACC-1(config)# class ip VOICE
ACC-1(config-clas-ip)# 10 match any any any count
ACC-1(config)# policy POLICY-VOICE
ACC-1(config-policy)# 10 class ip VOICE action local-priority 5
ACC-1(config)# interface 1/1/1
ACC-1(config-if)# apply policy POLICY-VOICE in
```

Global trust none
Select interesting traffic
Set the local priority
Apply the policy

Figure 14-24: Classifier-based configuration

Validate classifier-based configuration

The following **Figure 14-25** shows how to validate the classifier-based policy configuration from the previous example.

Use the `show policy interface <interface-id> in` command to verify if the policy was correctly set on the interface. The command `show policy hitcounts <Policy-name>` displays the number of hits counts that the policy received.

```
ACC-1# show policy interface 1/1/1 in
Direction
           Name
           Additional Policy Parameters
  Sequence Comment
           Class Type
                    action
-------------------------------------------
Inbound
           POLICY-VOICE
           10
           VOICE  ipv4
                  local-priority 5
```

Policy applied

```
ACC-1# show policy hitcounts POLICY-VOICE
Statistics for Policy POLICY-VOICE:
Interface 1/1/1* (in):
           Hit Count  Configuration
10 class ip VOICE action local-priority 5
                402  10 match any any any count
```

402 hits

Figure 14-25: Validate classifier-based configuration

Trust set to CoS

Instead of using DSCP to assign the LP and color, you can use the CoS value by configuring ports to trust to CoS. You can set the trust globally and leave ports without individual trust settings, or you can set the trust setting on individual ports. Keep in mind that the CoS is carried in the VLAN tag, so you should not use this setting for ports that receive native VLAN traffic or operate in route-only mode. All traffic that arrives without a VLAN tag is considered as having CoS value 0. In general, you might use the CoS trust setting if port(s) receive non-IP traffic or if the rest of the network uses CoS in preference to DSCP.

In the following **Figure 14-26** of the default CoS map with the LP and color values associated with each CoS value.

This table below provides the traffic LP 7 and color green. AOS-CX switches use the 802.1Q-2014 recommendations in which CoS 1 has the lowest priority, LP 0 and CoS 0 has the next lowest, LP 1. The rest of the CoS values are mapped to an LP with the same value. Again, you can adjust the map to assign each CoS value the LP and color that you desire.

When a port trusts CoS, whether based on the global or port setting, incoming DSCP has no effect on the traffic's LP and color. However, the switch does preserve this DSCP in forwarding the traffic.

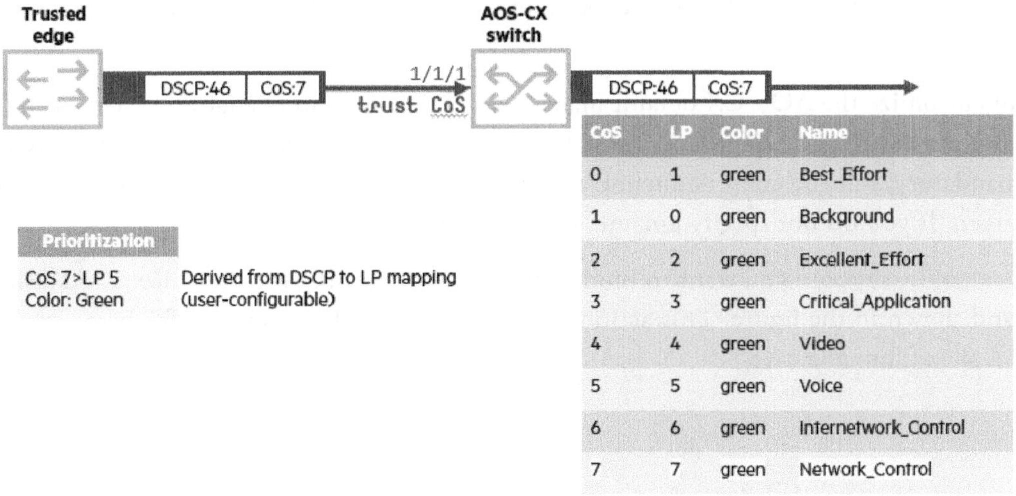

CoS	LP	Color	Name
0	1	green	Best_Effort
1	0	green	Background
2	2	green	Excellent_Effort
3	3	green	Critical_Application
4	4	green	Video
5	5	green	Voice
6	6	green	Internetwork_Control
7	7	green	Network_Control

Figure 14-26: Trust set to CoS example

Using interface DSCP to remark QoS values

Rather than use the incoming priority to assign the LP, AOS-CX switches can also classify traffic on their own. You should use this approach when connected devices are not capable of or not trusted to assign the correct values. The simplest approach is to assign a DSCP value to the Ethernet interface or LAG. When you use this approach, the trust setting on the interface must be none.

Then the port overwrites any incoming DSCP with the specified DSCP value on all traffic that arrives on the port. The switch uses that new value and the DSCP map to assign the traffic its LP and color.

This does not affect the CoS, also does not affect the prioritization. In the example below, traffic is untagged (native VLAN) and has no CoS.

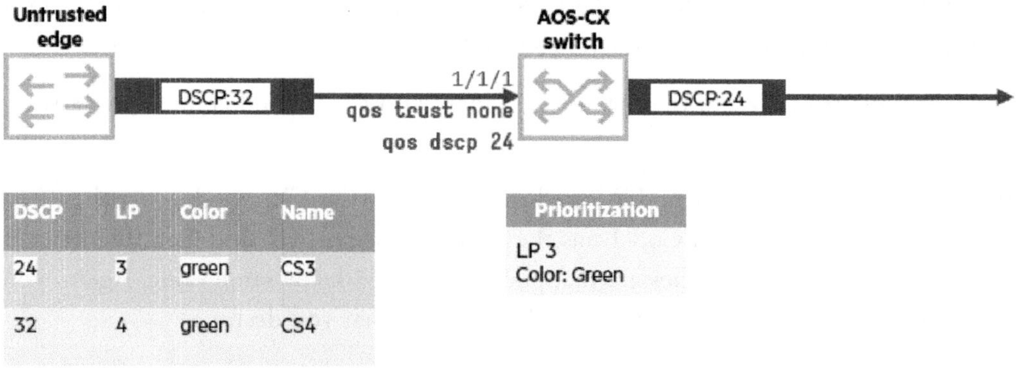

Figure 14-27: Using interface DSCP to remark QoS values

Modify default CoS mapping

Use the `show qos cos-map default` command to verify the default mapping for incoming CoS marks.

You can customize the AOS-CX default mappings. The command changes one row in the map, specifying the desired LP, color, and name to associate with a specific CoS value. Specifying the LP is mandatory, but the color and name are optional. If you do not specify a color, the color is set to green. If you do not specify a name, the row in the map has no name.

In this example, suppose you want to modify the default mapping for the voice marks using the command shown in the figure. This action overrides the default mapping, which you can verify with the `show qos cos-map`.

```
AGG-1# show qos cos-map default
code_point local_priority color   name
---------- -------------- ------- ----
0          1              green   Best_Effort
1          0              green   Background
2          2              green   Excellent_Effort
3          3              green   Critical_Applications
4          4              green   Video
5          5              green   Voice
6          6              green   Internetwork_Control
7          7              green   Network_Control
```

Default CoS mapping

Modify CoS mapping

```
qos cos-map 5 local-priority 6 color yellow name NEW-VOICE-COS
```

```
AGG-1# show qos cos-map
code_point local_priority color   name
---------- -------------- ------- ----
0          1              green   Best_Effort
1          0              green   Background
2          2              green   Excellent_Effort
3          3              green   Critical_Applications
4          4              green   Video
5          6              yellow  NEW-VOICE-COS
6          6              green   Internetwork_Control
7          7              green   Network_Control
```

New LP/color mapping overrides defaults

Figure 14-28: Modify default CoS mapping

Modify default DHCP mapping

Use the `show qos dscp-map default` command to verify the default mapping for incoming DSCP marks.

Customize these defaults with the command shown in the figure. The command changes one row in the map, specifying the desired LP, color, and name to associate with a specific DSCP value. Specifying the LP is mandatory, but the color and name are optional. If you do not specify a color, the color is set to green. If you do not specify a name, the row in the map has no name.

This action overrides the default mapping, which you can verify with the `show qos dscp-map` command.

```
AGG-1# show qos dscp-map default
DSCP     code_point local_priority color    name
-------- ---------- -------------- -------- ----
000000   0          1              green    CS0
000001   1          1              green
000010   2          1              green
<<Omitted Output>>
101101   45         5              green
101110   46         5              green    EF
101111   47         5              green
<<Omitted Output>>
```

Default DSCP mapping

Modify DHCP mapping

```
qos dscp-map 46 local-priority 6 color yellow name NEW-VOICE-DSCP
```

```
AGG-1# show qos dscp-map
DSCP     code_point local_priority color    name
-------- ---------- -------------- -------- ----
000000   0          1              green    CS0
000001   1          1              green
000010   2          1              green
<<Omitted Output>>
101101   45         5              green
101110   46         4              yellow   NEW-VOICE-DSCP
101111   47         5              green
<<Omitted Output>>
```

New LP/color mapping overrides defaults

Figure 14-29: Modify default DHCP mapping

Queue packets

We are going to discuss packet queuing.

FIFO queuing

As a switch prepares to transmit traffic, it places the traffic in an egress queue for the egress port. Then the device must schedule packets for service, or transmit them out the egress port. If the queue works in a First In First Out (FIFO) manner, packets are sent in the order received, as shown in the following figure. Using a single queue offers no advantage to sensitive traffic.

Perhaps the green VoIP packets are stuck waiting behind large FTP packets. The FTP traffic would not be affected by an extra 100 ms of delay, but the voice quality suffers because of this delay.

The main parameter that defines the queue is its length, which is how many packets it can store. In AOS-CX, each interface has an egress buffer up to 1 MB.

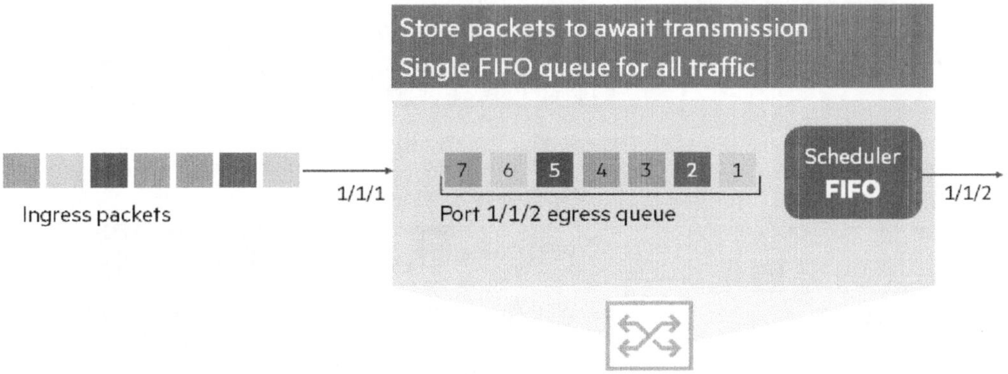

Figure 14-30: FIFO queuing

Priority queuing

To prioritize traffic, the switch can use a queuing mechanism that sorts various packets into multiple queues. AOS-CX switches support up to eight egress queues per port.

The following figure shows a simplified scenario with four queues.

- The green VoIP packets are placed in one queue.
- Various other traffic is placed in other queues.

Now that traffic is organized into queues, the port's scheduling mechanism selects the next packet for forwarding. Scheduling works in tandem with queuing, ensuring that packets in high-priority queues are transmitted more promptly than packets in other queues.

AOS-CX switches support Strict Priority (SP) queuing and Weighted Fair Queuing (WFQ) or Deficit Weighted Round Robin (DWRR) queuing. Basically, the SP queue is serviced until it is empty. Then it services the other queues using some variation of a round-robin, ensuring that all traffic types get their fair share of service.

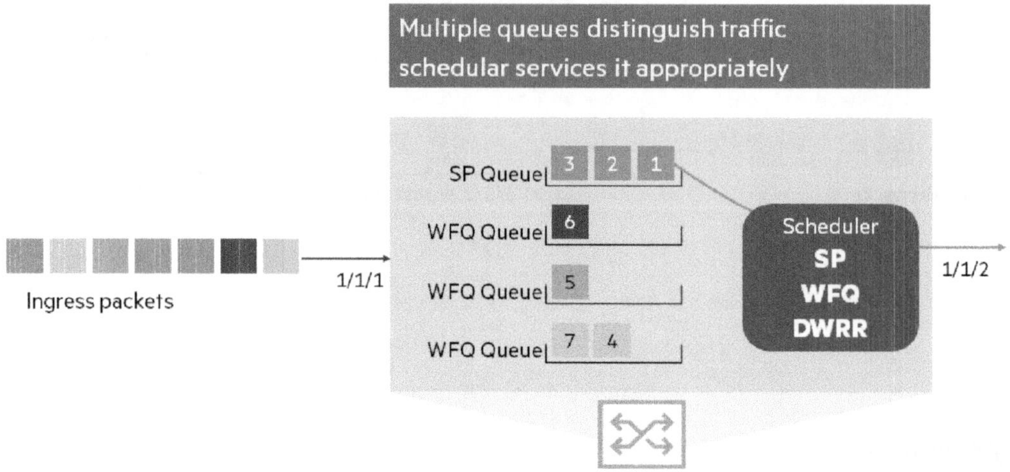

Figure 14-31: Priority queuing

Class-based QoS

With QoS, every port has eight egress queues, numbered 0 to 7. You can define class-based policies and apply them to appropriate interfaces.

Based on this policy, the switch classifies each ingress packet. This classification could be based on ingress interface, IP addresses, MAC addresses, preexisting ingress marking, and more. That priority then determines the correct queue for the traffic.

The queue number is not relevant to how queues are serviced. Scheduling determines how the switch selects which queue forwards traffic at any given moment in a port's service cycle.

Perhaps the scheduler services queue 7 first, then queue 5, then the multicast traffic in queue 1, and then the HTTP traffic in queue 3.

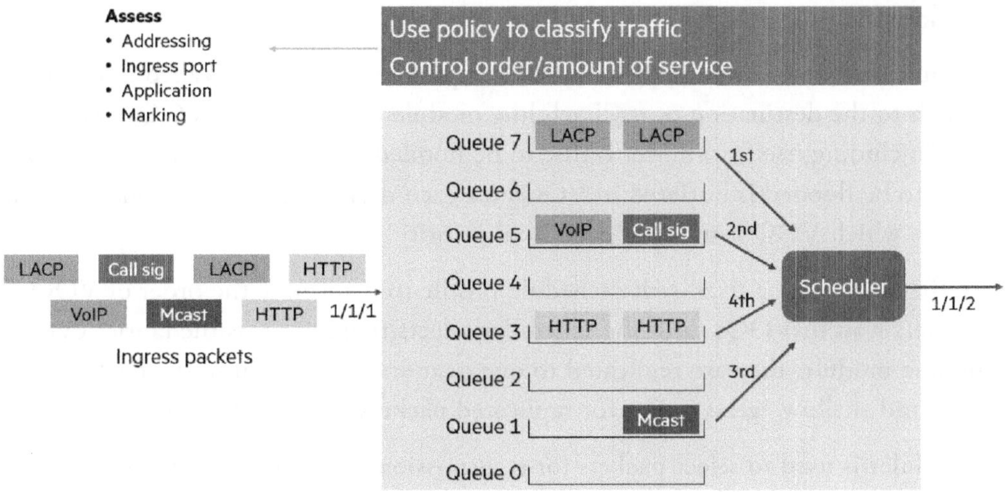

Figure 14-32: Class-based QoS

VOQ: unicast

AOS-CX switches use queue and schedule profiles for making these decisions. The queue profile is global, while you can apply schedule profiles globally or to individual ports.

The switch uses a Virtual Output Queue (VOQ) architecture, where most packet buffering occurs on the ingress line module. Unicast traffic destined for one port uses different buffering and scheduling than BUM traffic destined to be flooded out multiple ports. The relative priority and the amount of packet transmission for these two types of traffic differ.

For unicast traffic, each line module contains eight VOQs for every destination port in the chassis (one per local priority). The queue profile determines which VOQ is used to buffer a local priority. The schedule profile determines the order of VOQ servicing. Unicast packets wait in VOQs until the scheduler selects them to cross the fabric. Each destination port has a shallow egress transmit queue that buffers unicast packets.

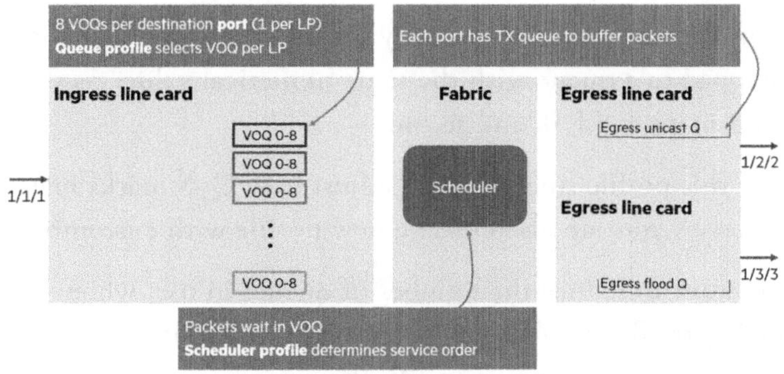

Figure 14-33: VOQ: unicast

VOQ: multicast

Broadcast, multicast, and unknown-unicast (BUM) packets, collectively called flooded traffic, use a separate path to the destination ports. Each line module contains eight VOQs per destination line module (including itself) to buffer traffic to be flooded (one VOQ per local priority). A copy of the packet to be flooded is buffered in VOQs for each destination line module. The queue profile determines which VOQ is used for each local priority.

Flooded traffic VOQs use a fixed strict schedule profile to determine the order of VOQ servicing. Flooded packets wait in VOQs until the scheduler selects them to cross the fabric. On the destination line module, they are replicated to one or more destination ports. Each destination port has a second shallow egress queue for replicated packets buffered for transmission.

A WFQ scheduler is used to select packets for transmission from the unicast and replicated traffic egress queues. When selecting packets between two non-empty queues, WFQ uses a fixed data weight of 4 for unicast traffic, and a weight of 1 for replicated traffic. As long as both queues are non-empty, replicated (flooded) packets comprise approximately 20% of the transmitted traffic, independent of the unicast scheduled percentages.

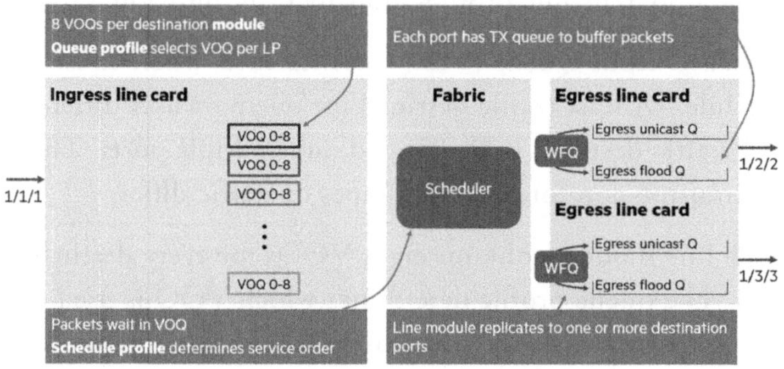

Figure 14-34: VOQ: multicast

Queue profile

The queue profile translates the LP to the priority queue from the traffic that is forwarded. By default, each LP is mapped to a queue with the same numerical value. As shown in the following figure, queue number 0 maps to LP 0, and so on.

You will typically keep this profile since you can adjust how QoS marks map to LP in the CoS and DSCP maps. However, you can also create a new profile with customized mappings.

For example, you may want to reduce the number of queues in use, which can simplify when your network is not classifying traffic granularly and some LPs are unused. You can assign more than one LP to the same queue to ensure that even LPs that you do not think you will use map to a queue.

Default profile works in most situations

Applied globally

```
AGG-1# show qos queue-profile factory-default
queue_num local_priorities name
--------- ---------------- ----
0         0                Scavenger_and_backup_data
1         1
2         2
3         3
4         4
5         5
6         6
7         7
```

Might change to reduce number of queues in use

Assign multiple LPs to same queue

Figure 14-35: Queue profile

Schedule profiles

Now that traffic is in the correct queue, the port schedules it for forwarding. A port's schedule profile determines how the port determines the queue service order. You can apply a schedule profile to each individual port or LAG. LAG port members inherit the LAG schedule profile; you cannot apply a profile to LAG port members directly. If you do not apply a schedule profile to a port or LAG, it takes its schedule profile from the global profile.

The schedule profile must include an entry for every queue that you define in the global queue profile. The factory default profile has eight queues, numbered 0–7. For each queue, you specify a scheduling algorithm and settings associated with that algorithm. All the AOS-CX switch models support strict priority for the algorithm. In addition, the CX 8325 Series, CX 8320 Series, CX 6400 Series, and CX 6300 Series support deficit weighted round robin, while the CX 8400 Series supports weighted fair queuing.

Every queue in the profile must use the same algorithm with one exception: the highest–priority queue can use SP while the other queues use DWRR or WFQ. This is shown in the following figure, a typical scenario. Some jitter/delay sensitive traffic like VoIP uses SP queuing, and the rest of the queues use either WFQ or DWRR. Basically, the SP queue is serviced until it is empty. Then it uses WFQ or DWRR to service other queues, ensuring that all traffic types get their fair share of service.

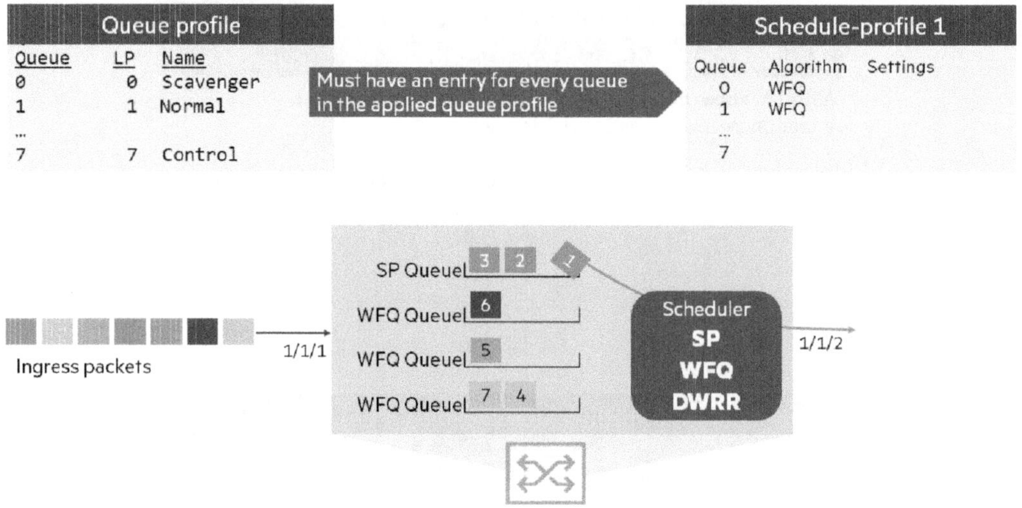

Figure 14-36: Schedule profiles

Transmission scheduling

Packets are queued, and so now they are scheduled to be transmitted.

A schedule profile determines the order in which queues are selected for transmission, and the amount of service available for each queue. A schedule profile must be configured on every interface at all times. A schedule profile can be applied globally to all interfaces, or only to specific interfaces. In this guide, we will discuss the following scheduling profiles:

- **Deficit weighted round robin**
- **Weighted fair queuing**
- **Strict priority**
- **Combining SP with DWRR or WFQ**

Deficit weighted round robin

Deficit algorithms can handle packets of variable size without knowing the mean size. A maximum packet size number is subtracted from the packet length, and packets that exceed that number are held back until the next scheduling turn.

DWRR serves packets at the head of every non-empty queue whose deficit counter is greater than the size of the packet at the head of the queue. If the deficit counter is lower, the queue is skipped and its credit value, called a *quantum*, is increased. The increased value is used to calculate the deficit counter the next time around, when the scheduler examines the queue for service. If the queue is served, the credit is decremented by the size of packet being served.

There are four key elements that affect how DWRR schedules queue service:

- **Weight:** reflects a proportion of the bandwidth on the outgoing interface.
- **Quantum:** translates the weight value into bytes. With each scheduling turn, a quantum value is added in proportion to the weight. Thus, the quantum is the throttle mechanism that allows the scheduling to be aware of the bandwidth.
- The **value of credits** can be positive or negative. Positive credits accrue when, at the end of a scheduler turn, there are leftover bytes that were not used in the queue's scheduling turn. This value is deferred to the queue's next scheduling turn. Negative credits accrue when the queue has transmitted more than its bandwidth value in a scheduling turn, and thus the queue is in debt when the next scheduling turn comes around.
- The **deficit counter**, which provides bandwidth fairness, is the sum of the quantum and the credits. The scheduler removes packets until the deficit counter reaches a value of zero or until the size of the packets is larger than the remaining deficits. But if the queue does not have enough deficits to schedule a packet, the value of the deficit is retained until the next scheduling round, and the scheduler moves to the next queue.

Benefits

The DWRR algorithms limit the shortcomings with WRR because they are a more modern form of WFQ that incorporates scheduling awareness of bits and packets.

Limitation

Services that have very strict demand on delay and jitter can be affected by other queues by the scheduling order. DWRR has no way to prioritize its scheduling.

Weighted fair queuing

Currently, only the CX 8400 Series switches support WFQ; the other AOS-CX switches support DWRR.

WFQ is commonly referred to as "bit-by-bit round robin" because it implements a queuing and scheduling mechanism in which queue servicing is based on bits instead of packets.

In WFQ, each queue or flow is allocated a weight that is a proportion of the interface rate or the shaping rate. WFQ is aware of packet sizes and can thus support variable-sized packets. The benefit is that sessions with big packets do not get more scheduling time than sessions with smaller packets because effectively, the focus of WFQ is bits and not packets. So, there is no unfairness in the scheduling for sessions with smaller packet sizes. With WFQ, each queue is scheduled based on a computation performed on the bits of each packet at the head of the queue.

WFQ brings queue weight into the calculation of the virtual finish time. You can approximate how this process works by thinking of each queue as having its own bandwidth, which is its share of the port bandwidth relative to its weight. Based on this weighted bandwidth, queues with higher weight will have relatively sooner finish times for their packets than queues with lower weights.

In the following example, all queues have the same weight, and thus the number of bytes scheduled in each scheduling turn is the same for all queues and reflects the weight value. Queue 7 removes three packets with a total value of 900 bytes, queue 6 removes one packet with 900 bytes, and queue 5 removes two packets with 900 bytes. The weight factor is effectively an allowance for how many resources can be assigned and used.

> Queuing does not fragment frames. If a queue could transmit 450 more bytes, but there is a 900-byte frame in the queue, it is not transmitted.

DWRR and WFQ

DWRR and WFQ both attempt to avoid the shortcomings of SP and ensure that even low–priority queues receive a level of service. With both methods, the scheduling algorithm works at the bit level, although in different ways. They aim to give fair treatment to every queue regardless of its average packet size. This principle is important for reducing jitter for small packets and can be crucial for applications such as VoIP.

In both algorithms, each queue receives a predictable share of the bandwidth based on the queue's relative priority, or weight. You can calculate the queue's bandwidth by taking the queue weight and dividing it by the total weight of all non-empty queues, and then multiplying that by the port bandwidth.

Using weights that add up to 100 makes it easy to estimate the bandwidth: the weight converts to a percentage of the bandwidth. Pay attention to the "non-empty" part of the formula.

For example, a port has four queues with weights 40, 30, 20, and 10. If all queues are contending for bandwidth, the queue with weight 40 is guaranteed up to 40% of the bandwidth. If a queue empties and does not need its full share of the bandwidth, that bandwidth is available for other queues, shared out according to their weight.

Both DWRR and WFQ can impose a bit higher latency for high priority queues as compared to SP.

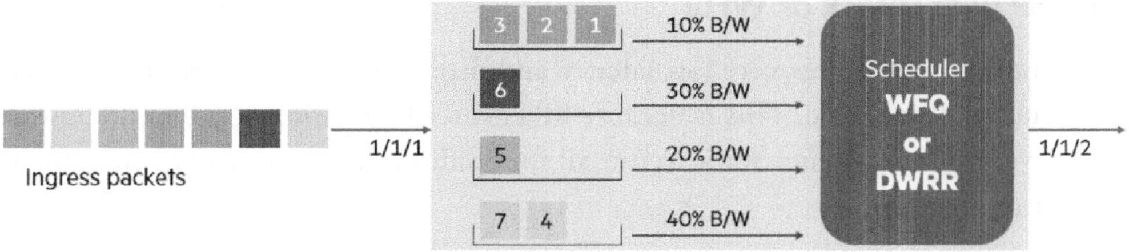

Figure 14-37: DWRR and WFQ

Strict priority

SP scheduling always forwards the highest priority traffic first. That is, the port forwards all packets in queue 7. Packets in queue 6 are serviced if and only if queue 7 is empty. Queue 5 is serviced if and only if packets in queues 7 and 6 are empty, and so on. If packets in queue 3 are being serviced and a packet arrives in queue 7, the packet in the middle of being serviced is finished up, and then queue 7 is serviced.

Under SP, the port will never forward a packet in a lower priority queue if packets show up in a higher priority queue. If queue 7 receives more packets before queue 4 has a chance to be served, the lower priority traffic is starved out entirely. Therefore, SP works best when the network assigns only a few applications, which take a relatively low amount of bandwidth, to the higher priority queues—network control traffic and VoIP. The switch then forwards this traffic immediately when it arrives but still has bandwidth left over for other traffic in the best effort queue.

If you want to use SP alone, you could apply egress queue shaping to the high priority queues to prevent starvation of low priority queues. Egress queue shaping allows you to apply a maximum bandwidth to a priority queue, as well as a burst size. The port buffers excess traffic up to the burst size and sends the buffered traffic at the max rate, smoothing out bursts while also preventing the high priority queue from exceeding its maximum rate and starving out lower priority queues.

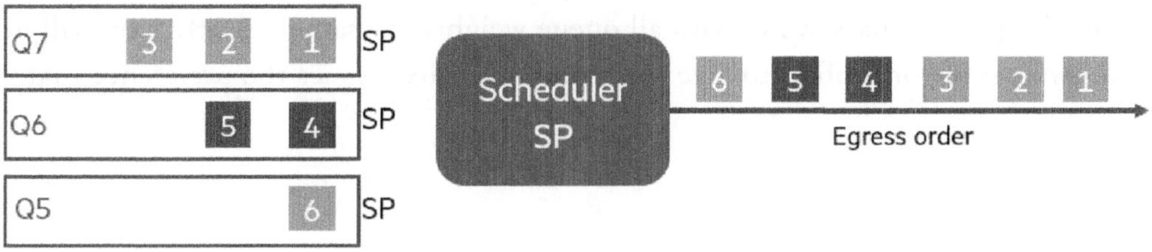

Figure 14-38: Strict priority

Combining SP with DWRR or WFQ

For network traffic that requires very low latency and jitter, such as VoIP traffic, you may want to use an SP queue for that traffic. This is because SP ensures high priority traffic the lowest latency and jitter. However, you may not want to use SP for medium high–priority queues due to the risk of starving out other traffic.

AOS-CX switches give you the best of both worlds. You can create a schedule profile that combines either DWRR or WFQ, depending on the switch model, with SP. In a hybrid profile, only the highest priority queue can use SP. This is queue 7 if the profile uses all queues or the highest queue used if the profile does not. You should assign traffic like VoIP to this queue; network control traffic is also automatically assigned to it. The switch forwards traffic from this queue if the queue has traffic. When the queue is empty, the port uses the remaining bandwidth to serve other queues. It allocates this remaining bandwidth using DWRR or WFQ. Because VoIP and network control traffic typically have low bandwidth requirements, the risks of starving the other traffic is minimal.

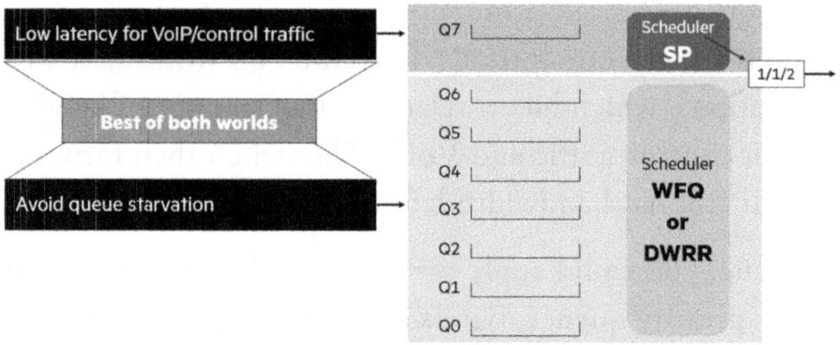

Figure 14-39: Combining SP with DWRR or WFQ

Profile configuration

The CX 832x Series, CX 6400 Series, and CX 6300 Series switches default scheduling profile uses DWRR for all queues with all DWRR queue weights the same. The CX 8400 Series switches default scheduling profile uses WFQ with all queue weights the same. Clearly, you will need to create your own profile or profiles to give the higher–priority queues the preference that you desire.

Factory default scheduler	CX 8320/6400/6300: All queues DWRR, weight 1 CX 8400: All queues WFQ, weight 1

Step	Command: config context	Notes
Create a non-default schedule profile	`qos schedule-profile` *<NAME>*	Enters schedule profile context.

Step	Command: schedule profile context	Notes
Assign a queue to use SP.	`strict queue` *<QUEUE>* [max-bandwidth *<KBPS>* [burst *<KBPS>*]]	All queues in profile must use same algorithm. Exception: SP for highest priority queue, WFQ/DWRR for the rest. Use max-bandwidth and burst options with SP for egress queue shaping.
Assign a queue to use DWRR or WFQ	[`dwrr` | `wfq`] `queue` *<QUEUE>* `weight` *<WEIGHT>*	

Figure 14-40: Profile configuration

Profile activation

You will apply the global schedule profile when you apply the queue profile. You need to apply the queue and schedule profile together. The queue and schedule profile must use the same queues.

If you want, you can keep the factory-default queue profile and apply a new schedule profile with this command: `apply qos queue-profile factory-default schedule-profile` *<PROFILE-NAME>*.

If you want to override the global schedule profile on a particular physical interface or LAG, access its context and enter the `apply qos schedule-profile` command. Remember that the global schedule profile and individual interface schedule profiles must use the same queues that you used in the queue profile. For the factory-default queue profile, there are eight queues from 0 to 7.

Step	Command: global config context	Default	Notes
Applies the queue profile to all interfaces and LAGs.	`apply qos queue-profile` *<QUEUE-NAME>* `schedule-profile` *<SCHEDULE-NAME>*	Factory-default profile	Schedule profile must be applied at the same time. The queue and schedule profile must use the same queues.

Step	Command: interface context	Notes
Override the global schedule profile on this interface.	`apply qos schedule-profile` *<SCHEDULE-NAME>*	The per-interface schedule profile must also use the same queues as the global queue profile.

Figure 14-41: Profile activation

Verification

Use the `show interface` command to display the number of packets and bytes queued for transmission.

Here is a description of each of the column outputs for the queues:

- **Tx Bytes:** Total bytes transmitted. The byte count may include packet headers and internal metadata that are removed before the packet is transmitted. Packet headers added when the packet is transmitted may not be included.
- **Tx Packets:** Total packets transmitted.
- **Tx Errors:** The amount of traffic dropped on an egress interface before it was sent. When traffic cannot be forwarded out an egress interface, it backs up on ingress. The more servicing assigned to a queue by a schedule profile, the less likely traffic destined for that queue will back up and be dropped. Tx Errors shows the sum of packets that were dropped across all line modules (due to insufficient capacity) by the ingress VOQs destined for the egress port. As the counts are read separately from each line module, the sum is not an instantaneous snapshot.
- **Tx Byte Depth:** Largest byte depth (or high watermark) found on any ingress line module VOQ destined for the egress port.

```
switch# show interface 1/1/5 queues
Interface 1/1/5 is up
 Admin state is up
             Tx Bytes           Tx Packets         Tx Errors          Tx Byte Depth
 Q0          157113373520       1890863919                 0                   1362
 Q1          233312143017       2808451320                18                  65550
 Q2          156814056423       1887257650                 0                   1392
 Q3          157441358980       1894815504                 0                   1374
 Q4          157700809294       1897941370                 0                   1362
 Q5          157872849381       1900014146                 0                   1392
 Q6          183486049854       2208268429                 0                   4398
 Q7          231607534141       2787913734                 0                  65544
```

Figure 14-42: Verification

Example

Here is an example configuration for a custom queue and schedule profile that works with that queue profile. Notice that only four queues are being used (0, 2, 4, and 6). Based on the schedule profile, DWRR is being used and the queue and schedule profile are applied globally.

```
qos queue-profile 4queues
        map queue 0 local-priority 1
        map queue 0 local-priority 2
        map queue 2 local-priority 0
        map queue 2 local-priority 3
        map queue 4 local-priority 4
        map queue 4 local-priority 5
```

```
        map queue 6 local-priority 6
        map queue 6 local-priority 7
qos schedule-profile 4qDwrrStrict
        dwrr queue 0 weight 10
        dwrr queue 2 weight 35
        dwrr queue 4 weight 55
        strict queue 6
apply qos queue-profile 4queues schedule-profile 4qDwrrStrict
```

LLDP-MED and device profiles

You have learned how to classify, prioritize, and apply policies to traffic. You will now learn how to support VoIP phones using LLDP-MED.

LLDP and LLDP-MED

The Link Layer Discovery Protocol (LLDP) is an open-standard link layer protocol used by network devices for advertising their identity, capabilities, and neighbors on a local area network based on IEEE 802 technology, primarily Ethernet. It can be used to verify Layer 2 connectivity between devices as well as learning basic information about directly connected devices.

The Media Endpoint Discovery (MED) is an enhancement of LLDP, known as LLDP-MED. It is commonly used to provision devices—like determine the correct power settings for a connected VoIP phone or AP— or automatically assign the correct VLAN and QoS settings for a connected device. It can enable plug-and-play networking. Features include the following:

- Set LAN policies based on the connected device: VLAN, Layer 2 802.1p CoS values, Layer 3 DSCP settings, and more.
- Device location discovery allows the creation of location databases: VoIP and enhanced 911 services.
- Extended and automated power management of Power-over-Ethernet (PoE) end devices.
- Inventory management: Track your network devices, and determine device-specific information—manufacturer, software and hardware versions, serial or asset number.
- Support for the fast-start capability.

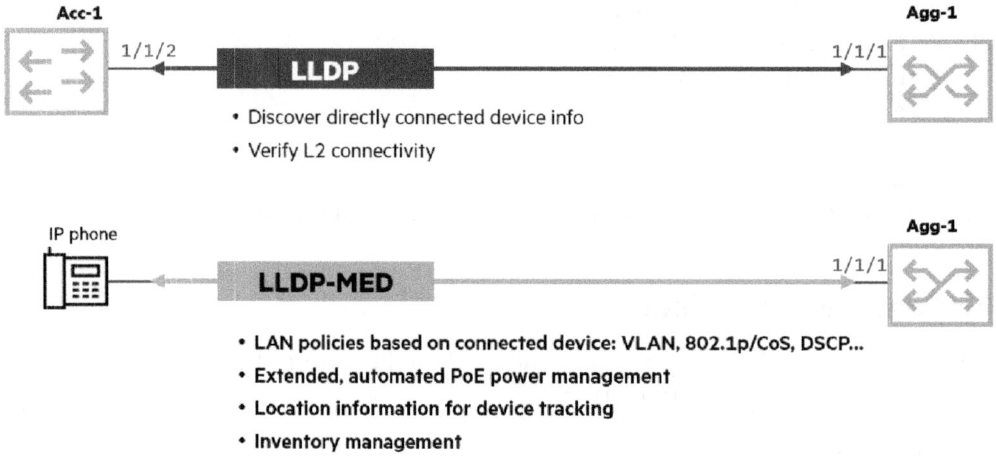

Figure 14-43: LLDP and LLDP-MED

LLDP-MED use case

You know to place VoIP phones in their own VLAN and configure a high priority for that VLAN—typically, a tagged VLAN. You connect a PC via Ethernet cable to the phone. The PC sends its frames on a native, untagged VLAN. But setting up the switch port to receive the tagged traffic is not enough. The phone must also know what VLAN ID to use for its packets. You can manually configure that on each phone, which is tedious.

With LLDP-MED, the switch automatically sends the correct VLAN ID to the IP phone, along with 802.1p and DSCP settings. You get automatic, plug-and-play provisioning for LLDP-MED–capable VoIP phones. The switch can also communicate other information to the phone, including its physical location. This information can help to support Emergency Call Services (ECS).

When phones receive PoE from the switch, LLDP-MED can help the switch allocate and deliver exactly the power that the phone needs.

LLDP-MED can also help in tracking and troubleshooting VoIP phones through a central management platform. SNMP servers can read detailed VoIP endpoint data inventory, PoE status, and information to help troubleshoot the IP telephony network and call quality issues.

Review interoperation with authentication

You manually configure switch ports with the tagged VLAN for VoIP, or rely on dynamic assignment using 8021.X and RADIUS attributes.

You manually configure each IP phone with a matching tagged VLAN, or rely on LLDP-MED to automatically configure it for you.

You can use 802.1X with LLDP-MED; simply set up 802.1X on the port. Make sure that IP phones support 802.1X and that the RADIUS server has policies and accounts for authenticating them. If you want the server to send the tagged VLAN assignment dynamically, make sure that it has the proper policies for doing so.

For this scenario, you probably want the computer and IP phone to authenticate separately so that an unauthorized user cannot piggyback on the IP phone's session. Make sure to set the 802.1X client limit to two so that the port operates in user mode and authenticates each device separately.

If you want to enforce 802.1X but the IP phones do not support it, combine 802.1X for users and MAC-auth for IP phones.

Figure 14-44: LLDP-MED use case

LLDP operation

The IEEE 802.1AB Link LLDP provides a standards-based method for network devices to discover each other and exchange information about their capabilities. Devices send LLDP data packets out all interfaces with outbound LLDP enabled. Devices accept LLDP advertisements from L2-connected neighbors on ports with inbound LLDP enabled.

Inbound packets from neighbor devices are stored in a special LLDP Management Information Base (MIB). This information can then be queried by other devices through SNMP. Network management tools can use LLDP information to create accurate physical network topologies. LLDP helps to determine which devices are neighbors and through which interfaces they connect. LLDP operates at Layer 2 and requires an LLDP agent to be active on each interface that sends and receives LLDP advertisements.

LLDP advertisements can contain a variable number of Type Length Value (TLV) information elements. Each TLV describes a single attribute of a device, such as system capabilities, management IP address, device ID, port ID.

Packet boundaries

LLDP packets travel only to directly connected peers. These packets are never forwarded to any other devices, regardless of whether they are LLDP-enabled.

An intervening hub or repeater forwards the LLDP packets it receives in the same manner as any other multicast packets it receives. Therefore, two LLDP switches joined by a hub or repeater handle LLDP traffic in the same way that they would if directly connected.

Any intervening 802.1D device or Layer 3 device that is either LLDP-unaware or has disabled LLDP operation drops the packet.

LLDP-MED

LLDP-MED in the switches uses the standard LLDP commands described earlier in this section, with some extensions, and also introduces new commands unique to LLDP-MED operation.

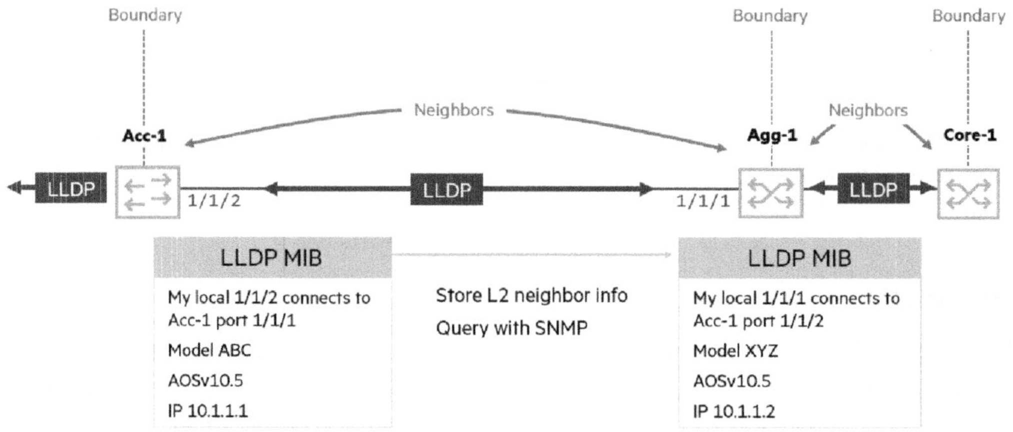

Figure 14-45: LLDP operation

LLDP configuration

LLDP is enabled by default, both globally and on interfaces. Configure LLDP if needed, using the syntax shown in the figure. Validate as shown in the figure.

A portion of the output is not shown. At the bottom of the `show LLDP` configuration, you would see the LLDP, advertised TLVs. But you can also use the `show lldp tlv` command, as shown in the figure. It is good that the switch advertises information about itself, but it is not currently advertising TLVs for LLDP-MED

You should look for the following TLVs:

- **MED Capability:** The capabilities TLV enables the switch to detect LLDP-MED–capable endpoints, their class, and their capabilities.
- **MED Network Policy:** The switch communicates the proper VLAN ID and QoS settings in this TLV. You will learn how to configure those on the next page.

- **MED Power (PoE):** The switch advertises its capabilities as a power sourcing equipment (PSE). The VoIP phone can also use this TLV to request a specific amount of power after it first powers up or dynamically throughout the connection. Note that PoE+ defines its own LLDP TLV for negotiating the power (a dot3 TLV called `poeplus_config`), and the phone might use this TLV instead. In either case, the switch will allocate the requested amount of power and reserve it for this port even if the phone is not drawing the full power at the moment.

If you need to enable any of these TLVs, use these commands:

```
Switch(config)# interface <INTERFACE-ID>
Switch(config-if)# lldp med capability
Switch(config-if)# lldp med network-policy
Switch(config-if)# lldp med poe [priority-override]
```

The `capability` TLV specifies advertisement of supported LLDP MED TLVs. The capability TLV is always sent with other MED TLVs; therefore it cannot be disabled when other MED TLVs are enabled. It is enabled by default.

The `network-policy` TLV lets endpoints and network devices advertise their VLAN IDs and IEEE 802.1p (PCP and DSCP) values for voice applications. This TLV is only sent when a voice VLAN policy is present. It is enabled by default.

The `priority-override` option overrides user-configured port priority for Power over Ethernet. When both lldp dot3 poe and lldp med poe are enabled, the lldp dot3 poe3 setting takes precedence.

Use the `lldp select-tlv` command to change what TLVs the switch will send and receive. See the *Switching Fundamentals Guide* for more information on this command. The show `lldp tlv` command shows the LLDP TLVs that are configured for send and receive:

```
Switch# show lldp tlv

TLVs Advertised
===============

Management Address
Port Description
Port VLAN-ID
System Capabilities
System Description
System Name
OUI
```

MED location TLV

It is a best practice to populate the location ID TLV for each switch port that might connect to a VoIP phone with the correct location information. For example, the command for specifying the civic address of a media endpoint such as a phone is:

```
Switch(config)# interface <INTERFACE-ID>
Switch(config-if)# lldp med-location civic-addr <COUNTRY-CODE> <CA-TYPE> <CA-VALUE> [<CA-TYPE> <CA-VALUE>]
```

A CA type specifies the type of location being indicated, such as 3 for a city. You can specify many CA types and values in order to program the city, street, address, and building, for example. For a list of CA types, see your switch *Fundamentals Guide.*

You can alternatively specify an Emergency Line Identification Number (ELIN), which is used in North America, if such numbers have been assigned to your phones:

```
Switch(config)# interface <INTERFACE-ID>
Switch(config-if)# lldp med-location <OCTET>
```

Voice VLANs

To create a voice VLAN, configure the voice command in the VLAN context, like this:

```
Switch(config)# vlan <VLAN-ID>
Switch(config-vlan)# voice
```

LLDP-MED extensions support voice VLANs.

```
lldp
interface 1/1/1
  lldp receive
  lldp transmit
```

```
Switch# show lldp configuration 1/1/1
LLDP Global Configuration
=========================
LLDP Enabled : Yes
LLDP Transmit Interval : 8
LLDP Hold Time Multiplier : 4
LLDP Transmit Delay Interval : 2
LLDP Reinit Time Interval : 2
LLDP Port Configuration
=======================
PORT TX-ENABLED RX-ENABLED
------------------------------------
1/1/1 Yes Yes
```

```
switch# show lldp tlv

TLVs Advertised
===============
Management Address
Port Description
Port VLAN-ID
System Capabilities
System Description
System Name
```

Figure 14-46: LLDP configuration

LLDP-MED configuration

You should look for the following TLVs related to LLDP-MED:

- **MED Capability:** Enables the switch to detect LLDP-MED–capable endpoints, their class, and their capabilities.
- **MED Network Policy:** The switch sends VLAN IDs and QoS settings in this TLV.
- **MED Power (PoE):** The switch advertises its capabilities as a PSE. The VoIP phone can also use this TLV to request a specific amount of power after it first powers up or dynamically throughout the connection. Note that PoE+ defines its own LLDP TLV for negotiating the power (a dot3 TLV called poeplus_config), and the phone might use this TLV instead. In either case, the switch will allocate the requested amount of power and reserve it for this port, even if the phone is not drawing the full power at the moment.

To enable these TLVs, use the configuration shown in the figure.

- The **capability** TLV advertises supported LLDP-MED TLVs. The capability TLV is always sent with other MED TLVs; therefore, it cannot be disabled when other MED TLVs are enabled. It is enabled by default.
- The **network-policy** TLV lets endpoints and network devices advertise their VLAN IDs and IEEE 802.1p (PCP and DSCP) values for voice applications. This TLV is only sent when a voice VLAN policy is present. It is enabled by default.
- The **priority-override** option overrides user-configured port priority for Power over Ethernet. When both lldp dot3 poe and lldp med poe are enabled, the lldp dot3 poe3 setting takes precedence.

Use the `lldp select-tlv` command to change what TLVs the switch will send and receive. See the *Fundamentals Guide* for more information on this command.

MED location TLV

It is a best practice to populate the location ID TLV for each switch port that might connect to a VoIP phone with the correct location information. For example, the command for specifying the civic address of a media endpoint such as a phone is shown in the following figure.

A CA type specifies the type of location being indicated, such as 3 for a city. You can specify many CA types and values in order to program the city, street, address, and building, for example. For a list of CA types, see your switch *Fundamentals Guide.*

You can alternatively specify an Emergency Line Identification Number (ELIN), which is used in North America, if such numbers have been assigned to your phones, as shown in the following figure.

Voice VLANs

To create a voice VLAN, configure the voice command in the VLAN context, as shown in the following figure. LLDP-MED extensions support voice VLANs.

Advertise MED TLVs

```
interface 1/1/1
  lldp receive
  lldp transmit
  lldp med capability
  lldp med network-policy
  lldp med poe [priority-override]
```

Advertise MED location TLV

```
interface 1/1/1
 lldp med-location civic-addr <country-code> <CA
type><CA value>[<CA type><CA value>]
```

Specify an ELIN

```
interface 1/1/1
 lldp med-location <octet>
```

Create a Voice VLAN

```
vlan 123
  voice
```

Figure 14-47: LLDP-MED configuration

Device profiles

AOS-CX supports device profiles to make it even simpler to deploy HPE Aruba Networking Instant Access Points (IAPs) and other devices. Use it when you are not sure what switch port the device might connect to. Typically, you have a standard configuration that applies to all AP-connected ports. This would include the native, untagged VLAN where the IAPs have their IP addresses. It also includes tagged VLAN assignments for static and dynamic VLANs assigned to WLANs for IAPs, the PoE settings such as a critical PoE priority, and so on.

A device profile lets you define these settings once on a switch, before connecting the APs. The switch knows to apply settings when it detects a target device connection. You can match on a variety of criteria, but the most common is the LLDP device type field. You associate the device profile with IAP's system description, for example, in its LLDP message. Installers can then connect IAP's to any switch port without worry. The switch detects the IAP type in an AP's LLDP messages and automatically applies the settings in the device profile (that is the user role associated with the device profile).

The switch provides support for LLDP and CDP to enable automatic discovery and configuration of other devices on the network.

When the LLDP information on the interface ages out, the profile/role is revoked from the interface. Only devices directly connected to the switch are detected and processed for device profile application.

You can apply the following interface settings when a target device connects:

- **Native (untagged) VLANs:** The VLAN on which untagged frames (normal Ethernet frames) are classified. The process for applying the setting depends on how the VLAN interface is configured. See VLAN interfaces.
- **Tagged VLANs:** The list of allowed tagged VLANs on the interface.
- **Policies:** The policy associated with a role. This mode is set by the policy command.
- **PoE priority:** The PoE priority applied to the interface. Supported values are high, low, and critical.
- **QoS Trust:** QoS Trust mode applied to the interface. Supported values are dscp, tos, and none. This mode is set by the qos trust command.
- **Jumbo frames:** Enable or disable jumbo frame support for the interface. This mode is set by the mtu command.

The following figure lists the settings defined in a device profile, which is associated with the system description that contains "Aruba IAP." Use the `show lldp neighbor-info` command of an associated device connected to the switch to determine what you should actually match on and include in your device profile matching parameters.

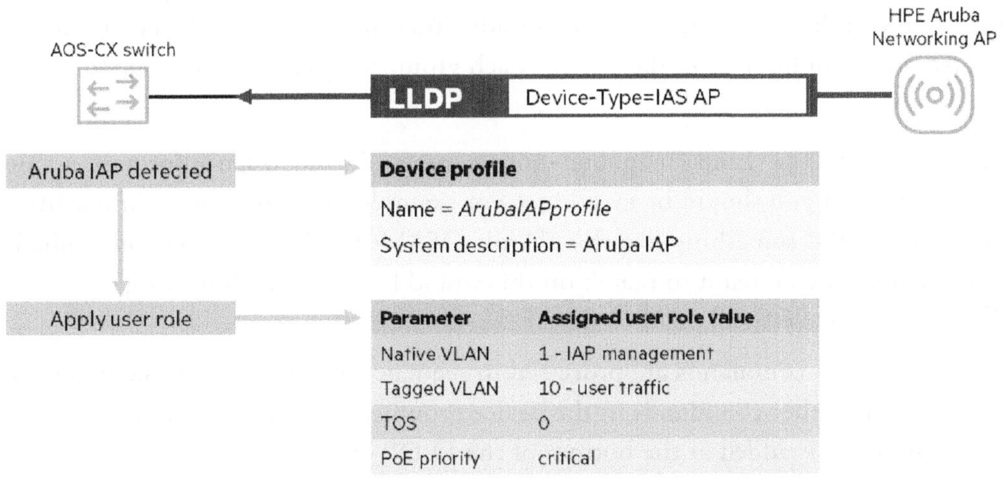

Figure 14-48: Device profiles

Device profiles with LLDP-MED

Suppose that an LLDP-MED–capable phone boots up and connects to an AOS-CX LLDP-MED–capable switch port. First, the devices do a handshake to negotiate PoE, if required. The devices also exchange LLDP messages. The switch's LLDP message includes many TLVs, as you just learned.

If you are implementing authentication on the port, like 802.1X, MAC authentication, or captive portal, remember to allow LLDP:

```
switch(config)# interface <INTERFACE-ID>
switch(config-if)# aaa authentication port-access allow-lldp-bpdu
```

Creating an LLDP group

The LLDP group defines the devices to which you want to assign dynamic settings. To create an LLDP group, perform the following:

```
switch(config)# port-access lldp-group <LLDP-GROUP-NAME>
switch(config-lldp-group)# match <PARAMETER> <PARAMETER-VALUE>
switch(config-lldp-group)# ignore <PARAMETER> <PARAMETER-VALUE>
```

The full `match` command syntax is as follows:

```
switch(config-lldp-group)# match sys-desc <DESCRIPTION>
switch(config-lldp-group)# match sysname <SYS-NAME>
switch(config-lldp-group)# match vendo-oui <OUI-MAC-ADDRESS>
switch(config-lldp-group)# match type <OUI-SUB-TYPE-KEY>
switch(config-lldp-group)# match value <OUI-SUB-TYPE-NUMBER>
```

This creates an LLDP group or modifies an existing LLDP group. An LLDP group is used to classify connected devices based on the LLDP TLVs advertised by the device. A maximum of 32 LLDP groups can be configured on the switch. Each group accepts 64 `match and/or ignore` commands.

You can easily use the `show lldp neighbor-info` command to see what information a device is sending, and thus what you should be matching. For example, to match on an Aruba IAP, the system description contains something like this: "ArubaOS (MODEL: 325), Version Aruba IAP". So a corresponding match command to match on this would look like the following:

```
switch(config-lldp-group)#  match sys-desc "Aruba IAP"
```

The `match` and `ignore` commands are processed in sequence. Once a device matches on a `match` or `ignore` command, other commands in the device group are ignored. If you omit the sequence number, the command is added at the bottom of the LLDP group.

CDP groups are also supported. See the *Fundamentals Guide* for configuring and using CDP groups.

Local User Roles (LURs)

The role defines the policies to use. To create a local role, perform the following:

```
switch(config)# port-access role <ROLE-NAME>
switch(config-pa-role)# vlan access <VLAN-ID>
switch(config-pa-role)# vlan trunk native <VLAN-ID>
switch(config-pa-role)# vlan trunk allow <VLAN-ID-RANGE>
switch(config-pa-role)# trust-mode [cos | dscp | none]
switch(config-pa-role)# associate policy <POLICY-NAME>
```

Creating a device profile

Next, you need to create a device profile and associate the LLDP group to it.

```
switch(config)# port-access device-profile <DEVICE-PROFILE-NAME>
switch(config-device-profile)# associate lldp-group <NAME>
switch(config-device-profile)# associate role <NAME>
switch(config-device-profile)# enable
```

The role must be a local user role (downloadable user roles are not supported for this feature). Use the `show port-access device-profile` command to verify your configuration. The device profile is disabled by default and must be enabled with the `enable` command. You can also disable it with the `disable` command.

As you can see, you can specify VLAN and QoS polices. To define a specific policy, you need to create class and policy configurations. A class specifies what to match on, and the policy map specifies the actions for the class map. The policy map is then referenced in the role with the `associate policy` command.

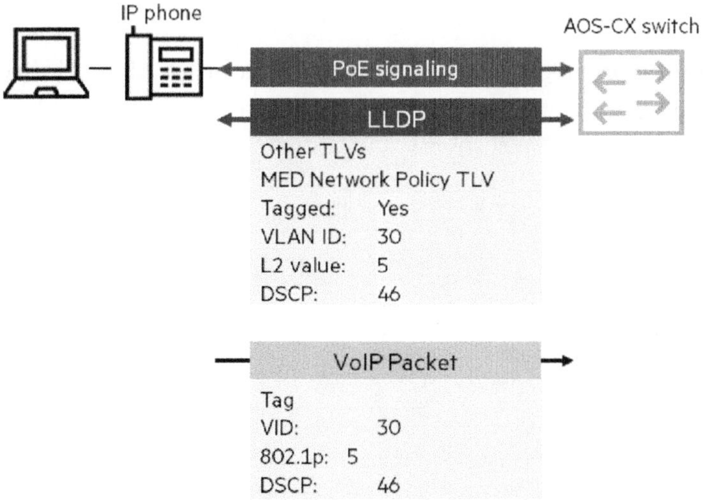

Figure 14-49: Device profiles with LLDP-MED

Learning check

Check your understanding of Quality of Service by answering the following questions.

Question 1

Which classification assigns traffic a QoS mark in the VLAN tag?

- O DiffServ
- O ToS
- O IP precedence
- O CoS or 802.1p

Question 2

The global QoS trust setting is defined as DSCP, and a port has no trust setting configured on it. How does the port derive the LP and color for incoming traffic?

- O It looks at the incoming DSCP and maps it to an LP and color with the DSCP map.
- O It ignores the incoming DSCP and uses LP 0 and color green.
- O It ignores the incoming DSCP and uses the incoming CoS and CoS map to derive an LP and color.
- O It looks at the incoming DSCP, maps it to a CoS value with the DSCP map, and then uses the CoS map to assign an LP and color.

Question 3

Which type of scheduling provides very low latency and jitter for high priority queues but can starve lower priority queues?

- O DWRR
- O WRR
- O WFQ
- O SP

15 NAE

This chapter teaches you all about the HPE Aruba Networking Network Analytics Engine (NAE). By the time you have completed this chapter, you will be able to explain how NAE helps you to monitor and troubleshoot, to analyze NAE data in the Web UI, to install scripts from the HPE Aruba Networking Solution Exchange (ASE), and to create agents based on the scripts.

OBJECTIVES

After completing this chapter, you should be able to:

- ✓ Explain how the AOS-CX NAE helps you to monitor and troubleshoot.
- ✓ Analyze NAE data in the Web UI.
- ✓ Install scripts from ASE and create agents based on the scripts.
- ✓ Implement sFlow and port mirroring.

Overview

NAE in AOS-CX uses Python-based agents, which were developed by HPE and the HPE community, to help companies better monitor and troubleshoot their network environment through an automated process. Agents monitor the state, settings, and statistics around specific system components within the current state database. For example, an agent can monitor the number of bytes transmitted on an interface or the switch's operating temperature. The agent then writes the ongoing results to the time series database for ongoing analysis.

Administrators can interact with the NAE through the Web UI's Analytics dashboard. They can also automate the NAE through the REST API using tools such as Postman, Ansible, etc., and create or obtain Python scripts for new agents. Admins can then load the agents on the switch using the Web UI or the REST API.

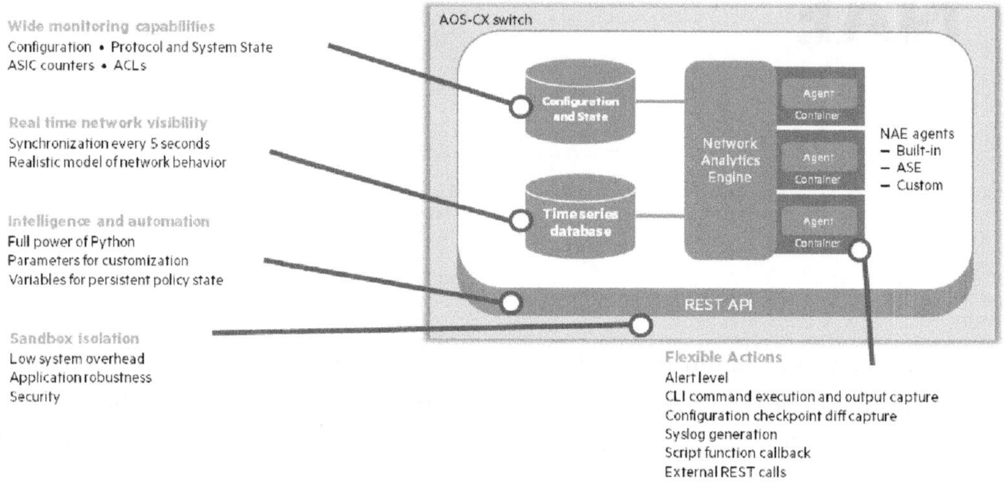

Figure 15-1: NAE monitoring and troubleshooting characteristics

Identify issues before they impact users

Using NAE, the intelligent AOS-CX switch provides a foundation for security, DevOps/operation automation, supportability, capacity planning, and monitoring/root-cause analysis solutions.

Agents create baselines for specific system components and monitor excessive deviations from those baselines and other system aspects. When predefined conditions occur, including excessive deviation from a baseline, NAE can take automatic actions to help alert administrators or even mitigate the problem.

NAE can also feed information to external management solutions using REST calls. These solutions include HPE Aruba Networking Central, HPE Aruba Networking ClearPass, and third-party ticketing IT services solutions such as ServiceNow and TOPdesk.

Examples of possible use cases for NAE are:

- **Application Health:** Visibility into the health/performance of a specific application (TCP/UDP ports), helping figure out if the problem of the slowness lies on the network, the application, or the link to that server
- **Dynamic Segmentation Health:** Monitoring overlay traffic
- **Client Services Health:** For example, RADIUS authentication, DHCP, and DNS services
- **L3 Network Health:** Static routes and OSPF
- **L1/L2 Network Health:** Interfaces, LAGs, STP, and LLDP
- **Device Health:** Hardware: fans, power supply, CPU, memory

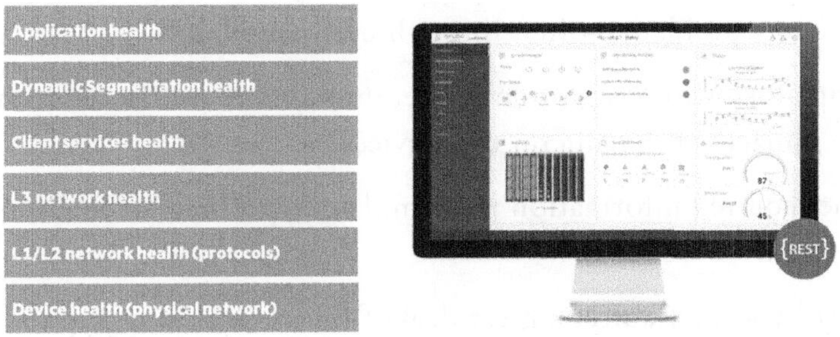

Figure 15-2: NAE use cases examples

Root cause analytics

IT staff often spend 70% or more of their time troubleshooting. Staff members see symptoms such as poor VoIP quality or slow internet access, and they have to take this symptom and connect it to a bad cable or misconfigured QoS configuration. Searching for the root cause of issues manually can be time-consuming and frustrating.

NAE uses rule-based, real-time monitoring to track issues as they occur and send alerts. It helps to automate diagnostics and data collection, correlating all data with time using the time series database so that staff can quickly find the information needed to arrive at the root cause of issues.

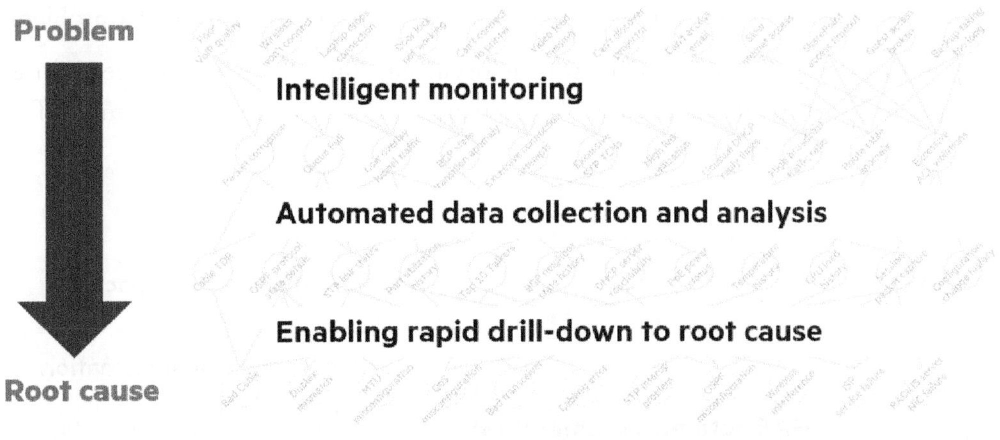

Figure 15-3: Root cause analytics characteristics

Simplified resolution process

A typical troubleshooting and resolution process includes many steps, which can take a long time to complete:

- Become aware of a problem—often through user complaints.
- Convert complaints, which are often vague, into a more precise assessment of the actual issue and the impact it is having on the services.
- Gather all the required information to formulate hypotheses as to a root cause, which is a laborious process.
- Make and test hypothesis, after a great deal of initial work slogging through configs and log files.
- Deploy a change as needed.
- Verify that change had the intended effect.

NAE simplifies this process in several ways because NAE agents:

- Continuously monitor different aspects of the system and detect issues before they cause problems for the users.
- Collect all relevant information when an issue is occurring, including changes on the system, output from CLI show commands, log files, configuration outputs, and tests such as traceroute, ping, and packet captures.
- Trigger an alert so admins know more quickly that they need to address a problem.
- Pinpoint the exact point at which an issue occurred, thus alleviating the need for administrators to sift through mountains of information and making it much simpler for them to find the cause and fix it.

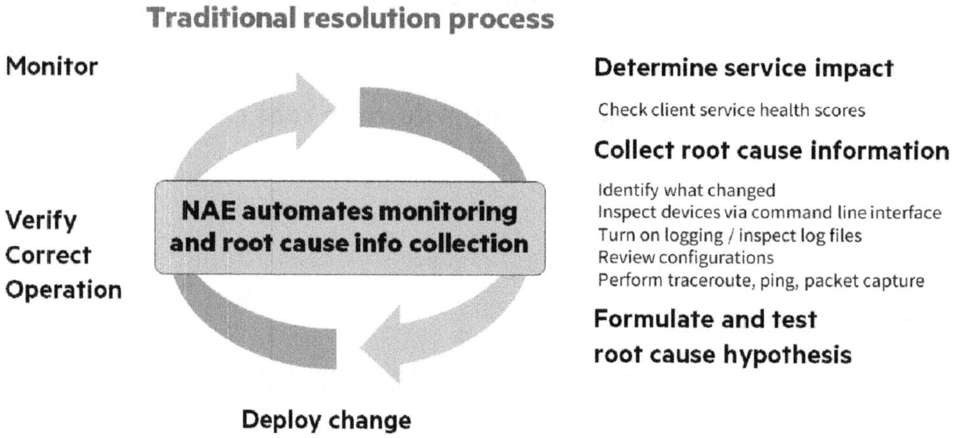

Figure 15-4: Traditional resolution process

Network agility

To provide network assurance, administrators need visibility, analytics, and remediation. HPE addressed this need with a first-of-its-kind Network Analytics Engine. When a problem arises, a network operator typically starts manual troubleshooting by looking at the network's stats at that moment. It takes time, many CLI-based commands, and could even require running probes and other traffic generators to re-create the problem. Localizing the problem could take hours, and you may never reproduce it.

What if you had intelligent scripting agents already in place to monitor symptoms related to poor application performance?

This native capability comes with a built-in time series database with simple software scripting to monitor for any set of parameters, with automated actions ready to go deeper and remediate. So, when a problem occurs, you are ahead of the game because you already have follow-up actions in place, historic visibility to show you what happened, and integration into your workflow systems for reporting.

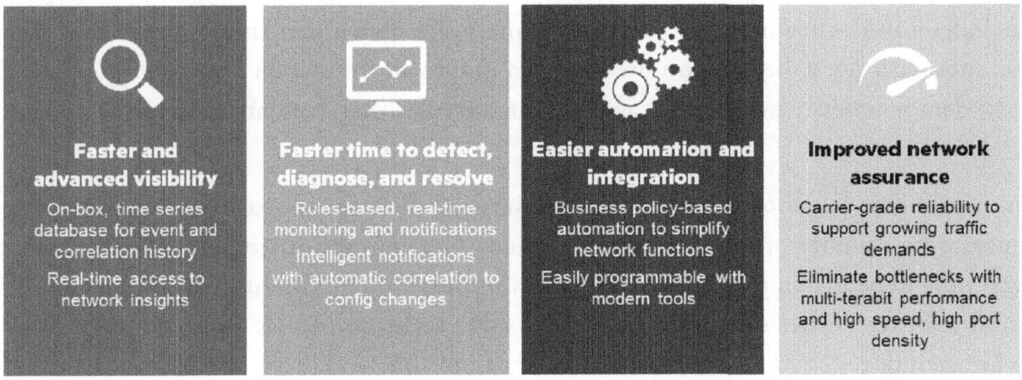

Figure 15-5: Network agility summary

NAE and the CSDB

Developers can use the REST URI to access a specific configuration, statistic, or status result for any aspect of the switch. The following image shows several example URIs.

Note that switches that are members of a stack are treated as a single switch, and that %2F is equal to a space.

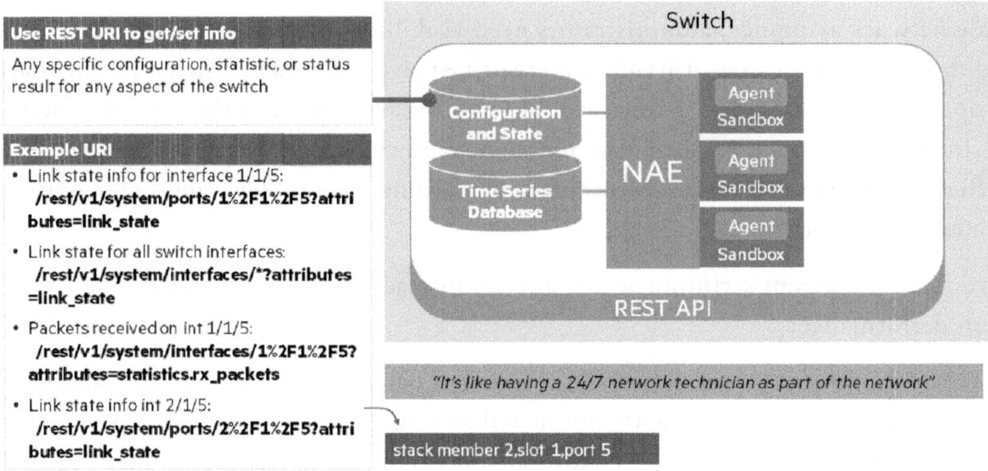

Figure 15-6: REST URI examples

NAE and the TSDB

NAE includes a built-in time series database. Time series data about the resources monitored by agents is automatically collected and presented in graphs in the switch's Web UI. The database makes the data seamlessly available to agents that use rules that evaluate network conditions over time.

Old time series data is removed automatically, either as the storage space on the switch is used or as the maximum number of days of data is reached. The amount of storage space consumed at any given time depends on the number of switch resources being monitored at that time. Each monitored resource creates one time series. Each time series consumes approximately 240 KB of storage for each day.

When creating a script, software developers do not interact with this database directly. Use of the database is automatically handled by the Monitor and Rule functions.

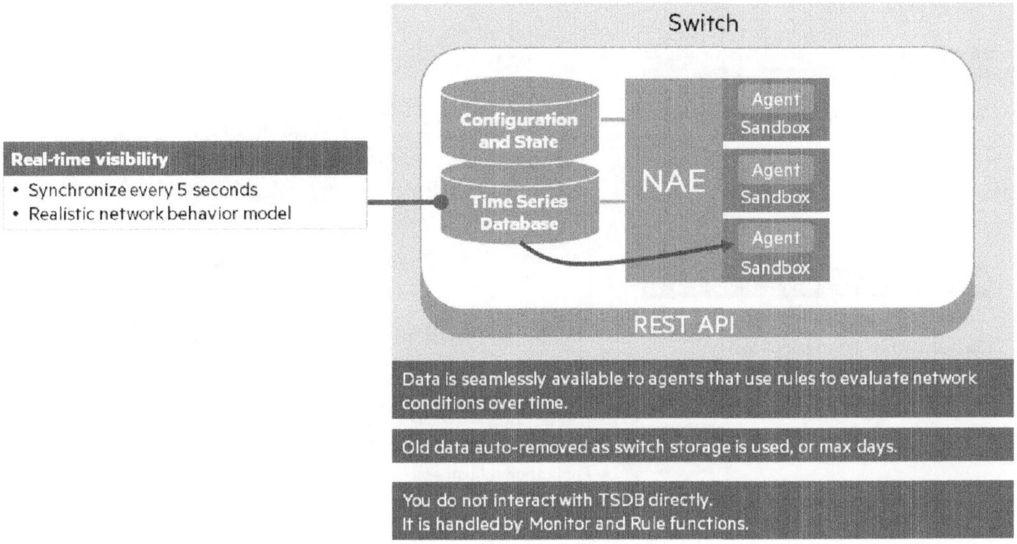

Figure 15-7: Built-in time series database

NAE, agents, and sandboxes

When an agent performs an action, the action is performed in a "sandbox" that is created when the action starts and removed when the action completes. The sandbox is in the default VRF, so it does not have access to the management network.

A sandbox is an isolated, tightly controlled environment in which programs can be run. Sandboxes restrict what a program can do, giving it the appropriate permissions and computing resources without allowing it access to the entire computing environment.

This design has the following benefits:

- Agents coexist and are prevented from using an excessive amount of CPU resources.
- Agents can benefit from the high-availability features of AOS-CX. During a switch failover event, the daemon that handles the sandbox can recover its state information and continue operations as before.
- Agents are prevented from accessing sensitive information—such as certificate files—in the switch operating system.

Soon you will learn about the three sources for the scripts that create NAE agents: built-in system-created scripts, scripts installed from the ASE portal, and custom scripts created by you.

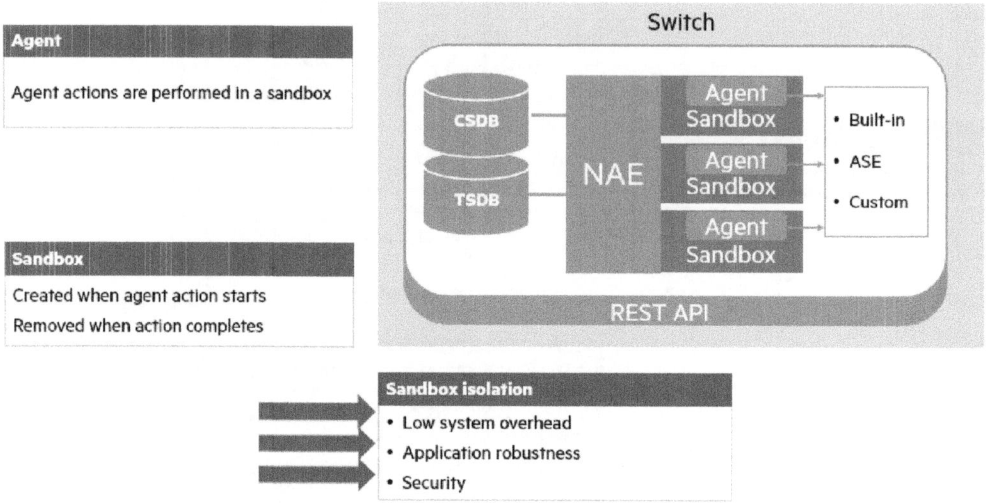

Figure 15-8: NAE, agents, and sandboxes

NAE agents

An NAE agent is the fundamental component of the AOS-CX NAE. Every agent includes one or more monitors, which track data about a specific attribute on the switch in the current state database. An attribute is any state or statistic defined within the AOS-CX REST API. For example, it could be the number of transmitted bytes on a particular interface, or it could be CPU utilization on a management module. This gives you great flexibility in terms of collecting the data that is relevant for your environment.

The agent collects the data every five seconds and writes the data to the time series database on the switch's hard disk. This creates an ongoing record of the resource's state. You can see a graph of the monitored resource state over time in the AOS-CX Web UI.

If programmed with the appropriate rules, the NAE agent can act when a particular condition is met. For example, if the agent is monitoring CPU utilization, it can set an alert when the utilization rises above a specific percentage.

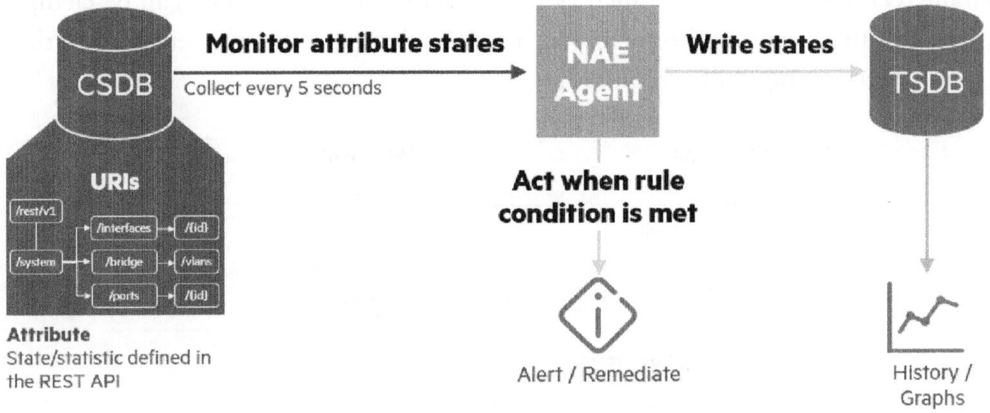

Figure 15-9: NAE agent's function

Relationship between NAE scripts and agents

Each NAE agent is defined by a Python script that is loaded on the AOS-CX switch. The script defines the agent, including which attributes it monitors and rules for agent actions. The script might also define parameters—user configurable settings. For example, instead of specifying that an agent monitors transmitted bytes per second (Bps) on a specific interface, the script might specify that the agent monitors transmitted Bps on a variable interface parameter. Then users can choose the target interface when they create agents. Parameters can also define values for rule conditions, and much more; it depends on how you write the script.

> An HPE Aruba Networking Network Analytics Engine agent is a specifically configured executable instance of an NAE script on a switch. When the agent is enabled, it performs the tasks defined by the script. Agents have administrator rights.

An NAE script has no effect until it has an associated agent, which instantiates the script and starts monitoring the attribute and taking actions. To create an agent, you define parameter values. You can create different agents from the same script using different parameters, as shown in the following figure. Perhaps the script parameter indicates the interface to monitor, and you create multiple agents, each monitoring a different interface.

Other examples of tasks a script can define include the following:

- Monitor average CPU usage and send a system log message when the CPU usage is greater than 70% for five minutes.
- Monitor the connection state of a particular BGP neighbor and execute a CLI command when the state transitions from UP to DOWN.

Some parameters that are integers—such as a CPU utilization threshold—can be changed for a given agent after that agent has been created. You can change such parameters easily through the Web UI; no programming skill is required.

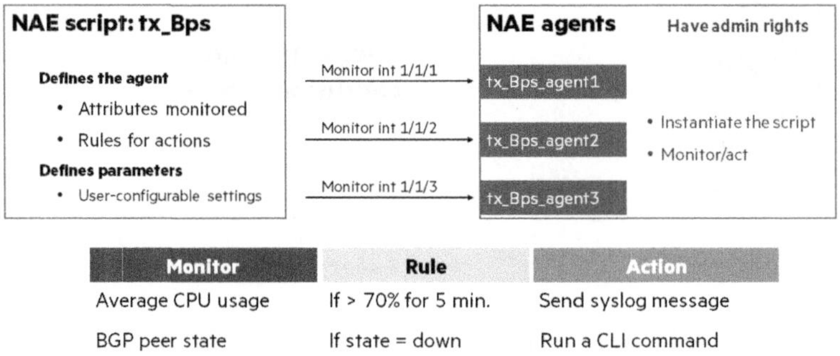

Figure 15-10: Relationship between NAE scripts and agents

Sources for scripts

There are three sources for an agent script:

System created, built-in scripts

Built-in scripts and agents are installed on the switch before the switch is shipped from the factory. All scripts and agents include origin information—"system" for the built-in ones, marked System Created in the Web UI.

Built-in scripts and agents cannot be deleted. You can enable, disable, and change the configuration of built-in agents:

- Built-in scripts are not displayed.
- Built-in agents are displayed only if one of more parameters has been changed and saved.
- If you create an additional agent from a built-in script, that agent is considered a user-created agent, which can be deleted.

The current software release includes a single built-in script and agent that monitors several system resources:

- Built-in script: system_resource_monitor
- Built-in agent: system_resource_monitor.default

HPE certified scripts

You can also obtain agent scripts from community sources. Scripts listed as HPE certified are written and validated by HPE to work as advertised. Community members can also create and upload scripts—use at your own risk. Always evaluate a script before loading it on the switch to ensure that the actions executed are as expected. You can disable and remove these agents after loading them as desired. Before doing a software upgrade, you may need to remove an old version of the agent. After the upgrade, upload a new one.

Create your own scripts

Finally, you can create agents on your own scripts with Python script. These agents are managed like the ones that you install from a community source. You can disable, remove, and update them separately from the software.

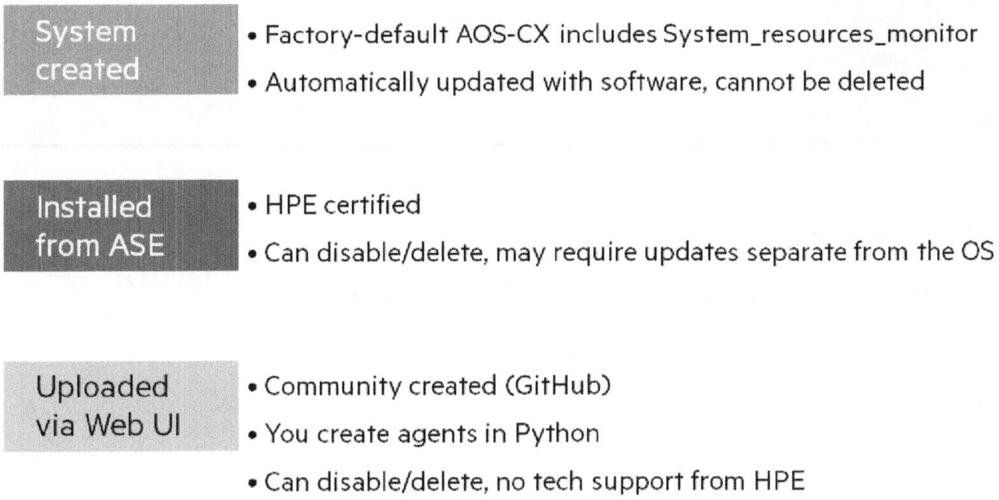

Figure 15-11: Sources for scripts

System-created script

Use the AOS-CX Web UI to access information for NAE agents, scripts, and alerts. This includes time series data graphs and other information generated by the enabled agents.

Select **Analytics** in the UI navigation pane to see the Analytics dashboard. You can display time series graph panels for up to nine agents on the dashboard. However, many more agents can be enabled on a switch.

From the Analytics dashboard, you can open Analytics detail pages. Analytics detail pages allow you to enable, disable, create, delete, or edit agents; upload scripts; and view detailed data about monitored resources. Administrator rights are required for actions that modify an agent.

As shown in the following figure, the Script Management page shows all scripts loaded on the switch and the agents associated with them. Look in the System Created column to find the system-created script, as highlighted in the figure.

AOS-CX supports a single system-created script, which has an associated agent by default. This agent monitors CPU and memory utilization over short, medium, or long time periods. It triggers alerts and log messages when utilization passes a particular threshold. The alert severity depends on whether the high usage continues over a short, medium, or long period.

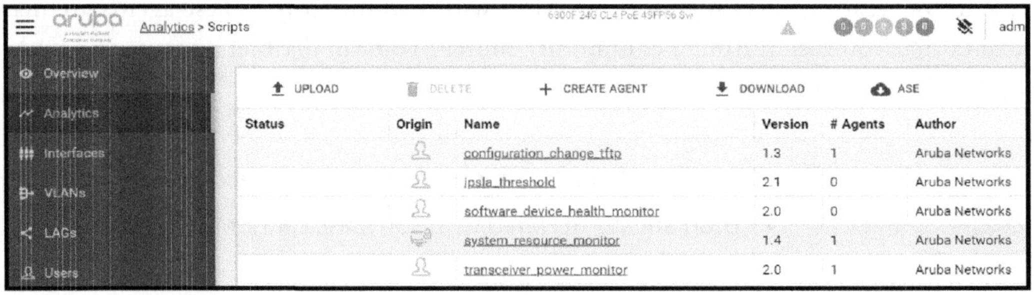

Figure 15-12: Analytics dashboard

HPE certified scripts

You can obtain HPE certified scripts from three NAE communities:

- HPE Aruba Networking Solution Exchange
- GitHub
- Airheads

Key aspects of the three NAE communities are:

HPE Aruba Networking Solution Exchange

The primary source for additional NAE scripts is the HPE Aruba Networking Solution Exchange (ASE), at http://ase.arubanetworks.com. You can browse agents and search for particular agent types. To view or download the agent script, you must log in with your HPE SSO system user account. You can also access ASE directly through the AOS-CX switch UI, assuming good internet connectivity—no log in required.

GitHub

GitHub is a wide-ranging developer community with custom scripts for many different applications. HPE will post all HPE-certified NAE scripts here, as well as to ASE. The GitHub community can enhance these scripts and create new scripts for sharing within the community.

Airheads

The Airheads community is a place for members or participants to search for information, read and post about topics of interest, and learn from each other. Unregistered guests can browse or search the community for information. Registered members can post messages or comments, track discussions, and get email notifications on posting activity and other community actions.

The Airheads community is the glue that holds all NAE solution components together. Members can share their expertise and tips and point each other toward relevant agents. The Developer Community forum under the Technology category has discussion forums about automation and NAE questions and support.

HPE is committed to expanding the NAE value with extensive research and development (R&D) investment in building NAE scripts and supporting the community. The Airheads Developer Community group is specific to APIs, programming, and automation. This is the recommended place to post NAE-related questions. Go to https://community.arubanetworks.com/t5/Developer-Community/bd-p/DeveloperCommunity. See the Airheads community at https://community.arubanetworks.com/home.

Creating and editing agents for existing scripts

When you install a new script, it must have an associated agent. To create an agent based on a script, access the **Agent Management** page—just select **Agents** in the Analytics dashboard. Then click **Create**. You can also select **Create Agent** in the **Script Management** page.

A window appears where you can choose the script on which to base the agent. Define the agent's name, customize the parameters, and save.

You can also select an agent in the Agent Management page and select **Edit** to adjust the parameters on an existing agent.

> After updating a script, disabling and re-enabling the script is recommended.

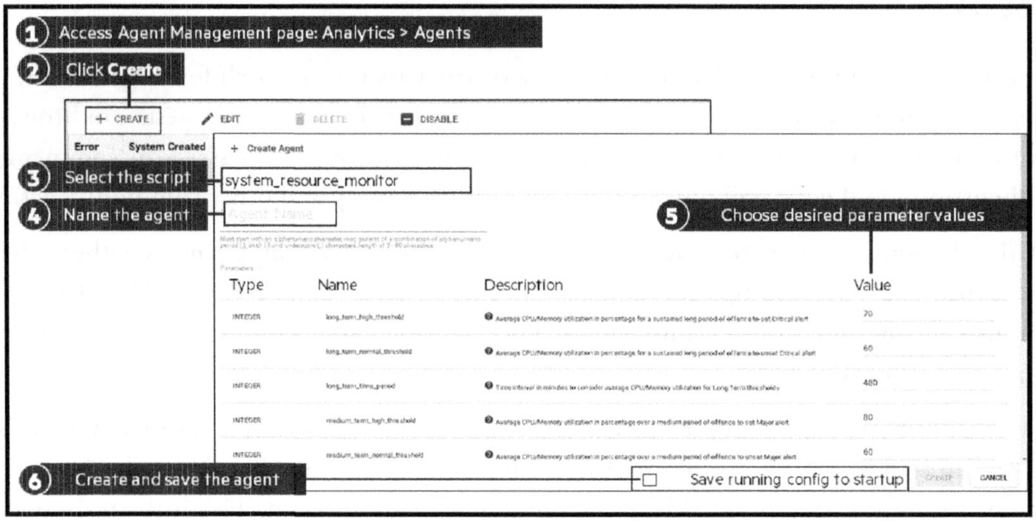

Figure 15-13: Steps to create agents for existing scripts

Agent actions

Agents can execute CLI commands or shell commands. Be very careful using shell commands in an agent! Mistakes can damage switch functionality and render it unusable. Agents can make REST API calls to themselves or remote devices and applications. This allows for interaction with all HPE applications and devices, which have their own REST interfaces. NAE can use REST calls to integrate with third party applications, including ticket generation systems like ServiceNow and TOPdesk.

An action can also send log messages to a system log or remote Syslog server, indicating information about the condition that has occurred. Finally, an action can generate a custom report with multiple pieces of relevant information about the condition.

For more complex scenarios, a rule can combine multiple actions. Note that AOS-CX uses a safe sandbox environment to execute actions, which prevents agents from using excessive CPU or obtaining access to sensitive files such as certificates. The sandbox has access to the default VRF, not the management one. If actions need access to the internet to execute properly, make sure that the default VRF has such access.

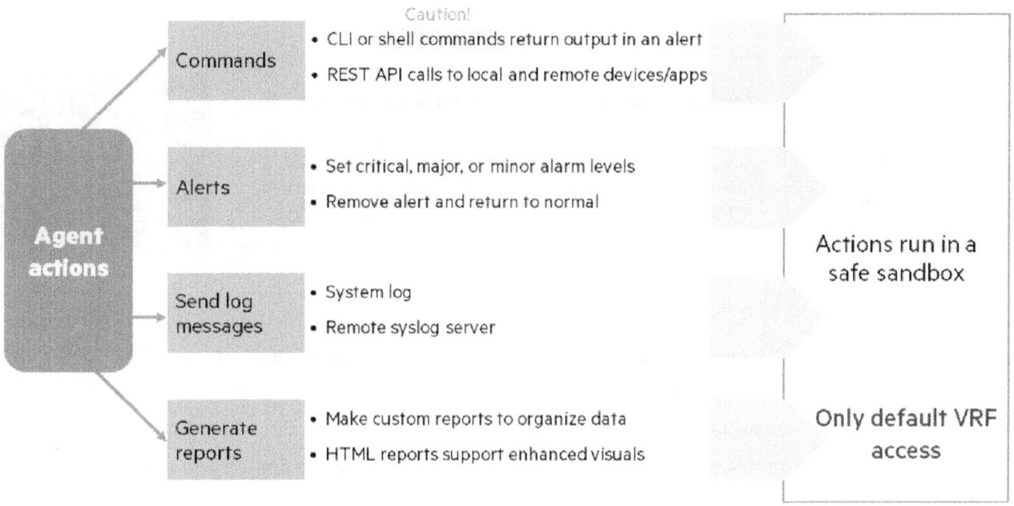

Figure 15-14: Action options summary

How are actions triggered?

- **Clear condition not met:** Sometimes you only want an agent to track information in the time series database and produce a graph. But you may also want the agent to act based on information obtained by the monitor. You define these actions with one or more rules. A rule must have a condition, which defines a particular state for the monitored value. At any given moment, the state is either true or false. Generally, the state being true indicates a potential issue that you should know about. Suppose that an agent monitors transmitted Bps on an Ethernet interface. A rule condition might be that the monitored value is greater than 500 million Bps.

  The agent checks the state against the condition every five seconds. The rule's action executes when the condition changes from false to true. At that point, the condition becomes inactive, and the agent does not re-execute actions if the condition remains true. This prevents constant alerts for the same ongoing issue. If the condition becomes false again (for example, the Tx Bps drops lower), the condition becomes active again, and the agent starts monitoring it once again.

  In some cases, a condition might toggle frequently between true and false. For example, the condition is more than 500 million Bps transmitted on an interface. The interface might transmit 600 million Bps and change the condition from false to true. Then it transmits 490 million Bps and transitions from true to false. Then it transmits 510 million Bps and triggers the action again.

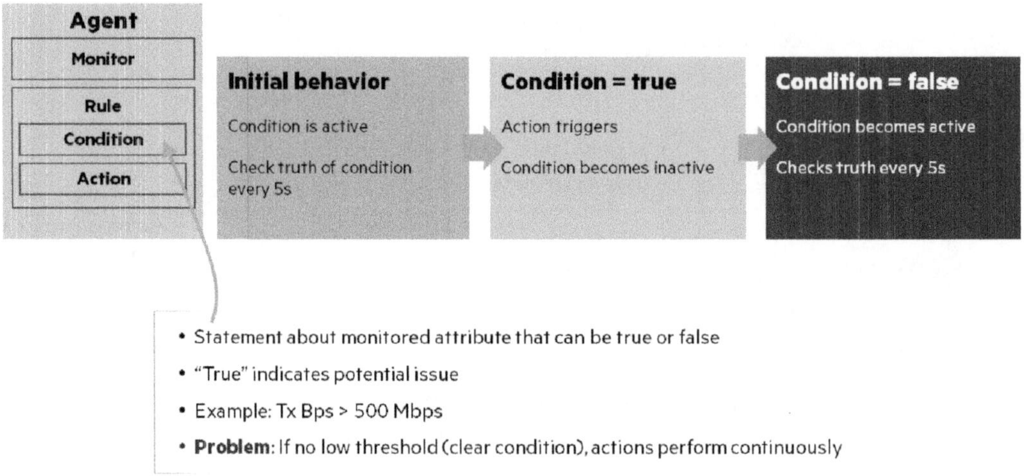

Figure 15-15: How actions are triggered: clear condition not met

- **Clear condition met:** to avoid that "continuous action" problem, you can define a clear condition, which must be met to free the main condition. The "clear" condition is a low threshold that the monitored attribute must fall below before the action is triggered again. The clear condition becomes active when the condition first transitions from false to true, and the agent begins checking the state of the clear condition every five seconds.

  Even if the condition becomes false again, the clear condition remains active and the condition inactive, as long as the clear condition is also false. Only when the clear condition becomes true does the condition transition to active. For example, the clear condition could be the transmission of less than 300 million Bps. Now the rate must fall below that (transition to true) before the original condition can become active again. You can optionally define a clear action, which also executes when the clear condition changes from false to true.

  The following example shows just one rule, but agents can customize their behavior to nuanced conditions by using multiple rules, each with a different condition and action. For example, a condition in one rule might indicates a less severe problem, and a condition in another rule might indicate a more severe problem.

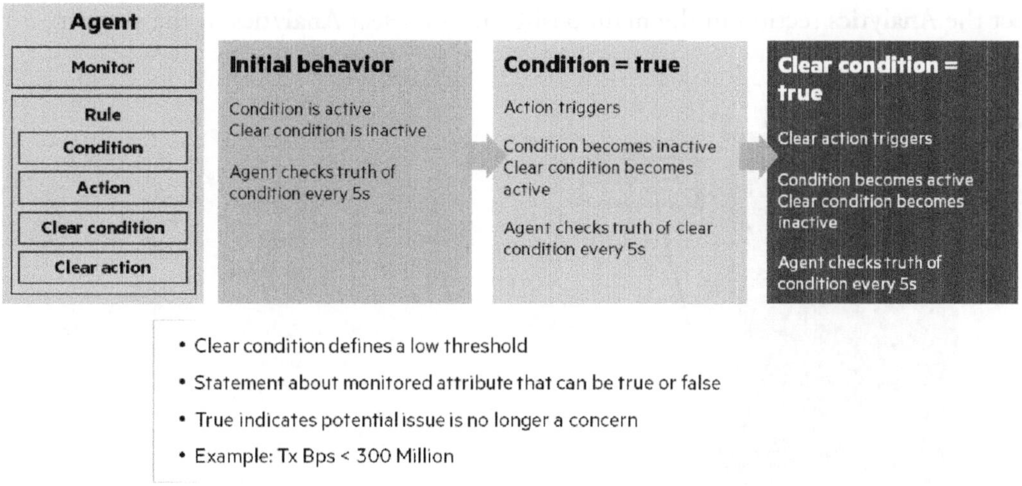

Figure 15-16: How actions are triggered: clear condition met

Feature: Analytics data collector

You can set up agents with scripts that monitor traffic with specific filtering criteria, as shown in the following figure.

Here are some examples:

- A specific private server
- Subnets of web addresses, such as social media traffic
- Application traffic usage of cloud services, such as Office 365

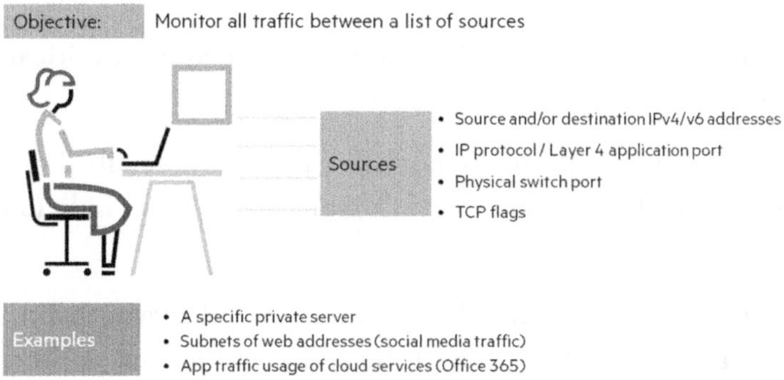

Figure 15-17: Analytics data collector

Viewing NAE information

When you enter the Web UI, you see a main dashboard. The Analytics section in this dashboard refers to the NAE functions. It shows how many critical, major, or minor alerts are active. If you want more detailed information, you should view the Analytics dashboard. To get there, either select the **Analytics** section in the main dashboard or select **Analytics** in the sidebar.

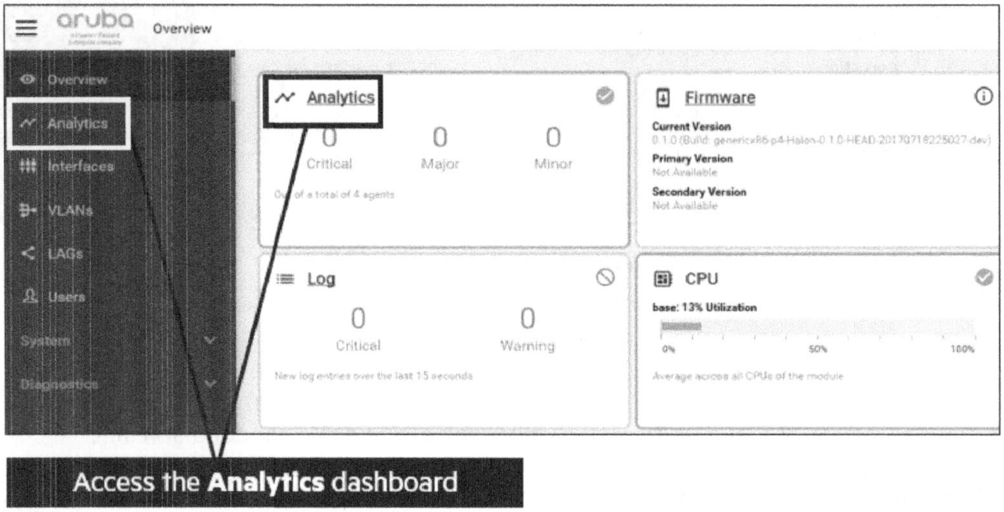

Analytics dashboard

On the Analytics dashboard, the Agents section lists all of the agents on NAE and indicates whether they are operating normally.

The dashboard also shows one graph for each of the NAE agents that has been added to it.

You can click the plus icons in the Agents section to add the graph for another agent.

The Scripts section shows the Python scripts associated with those agents.

Look to the Alerts section to see more details on any minor, major, or critical alerts that the NAE agents have generated.

You can select the name of these sections to move to a page with details. For example, click the **Agents** link to see the Agent Management page. Or you can select an individual agent to move to that agent's detail page.

You can also change the layout for the dashboard. Click the **Edit layout** icon at the top of the bar and select **Unlock Page Layout**. Each section in the dashboard is then outlined with dashed lines. You can click a section and move it around. When you are done editing, you can click the icon again to lock the page layout.

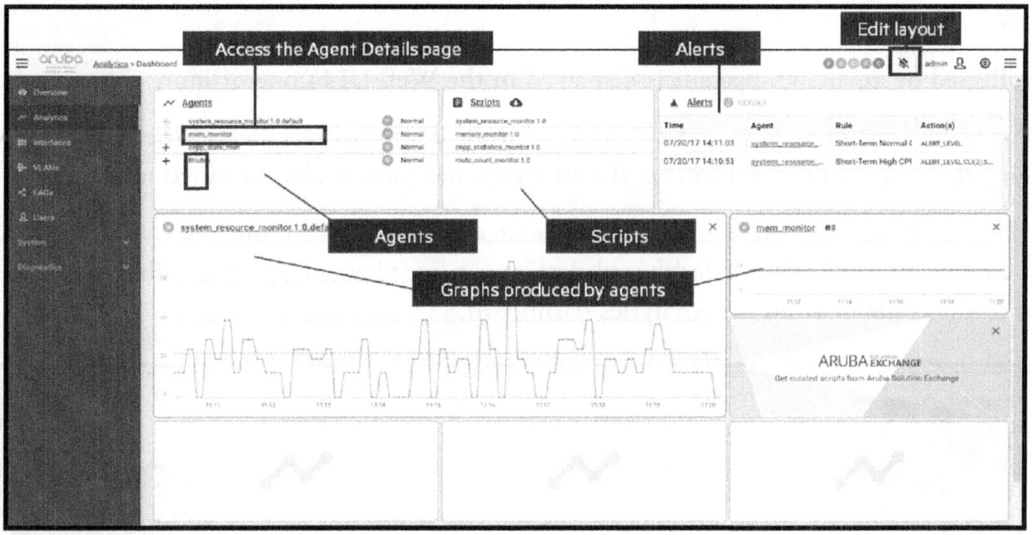

Figure 15-18: Analytics dashboard

NAE agent details

The Agent Details page provides more detailed information about the agent, including its version and status. If the agent has multiple monitors, this page shows all of the graphs associated with those monitors. The Parameters section shows agent settings that you can configure. You can also click icons to edit the agent or look at its script.

Check the Alerts section to see information about the alerts this agent generates so that you can take the proper steps to resolve them. You can click a particular alert and then select Details to see a window with alert details.

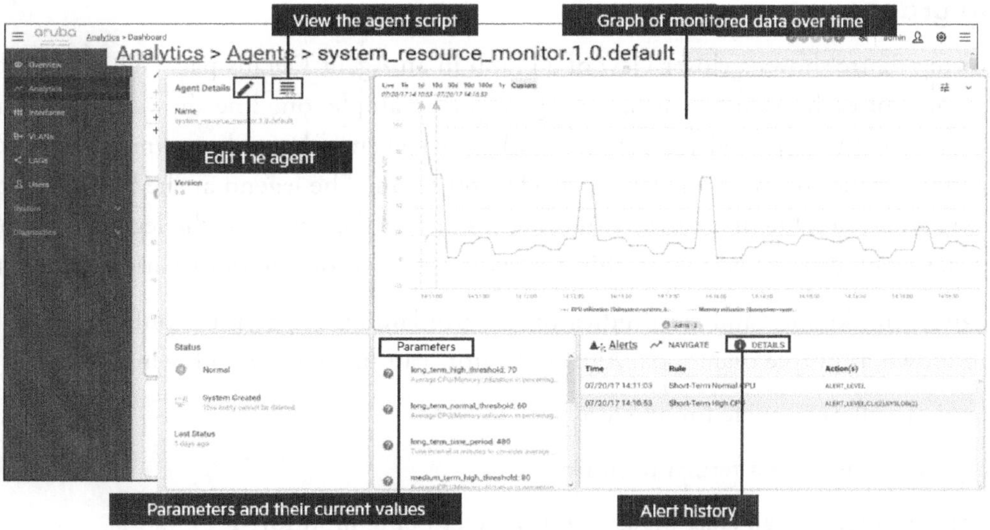

Figure 15-19: Agent details

Graphs

Data collected by an analytics agent is displayed in the Web UI in one or more time series graphs.

An agent has at least one graph and can have multiple graphs as specified in the script. However, only one graph represents the agent on the Analytics dashboard, also specified in the script.

If the dashboard does not include a graph for an agent, you can add one from the Analytics dashboard. Graphs on the Analytics dashboard represent a live view only. The graph customization toolbar is not available from the Analytics dashboard.

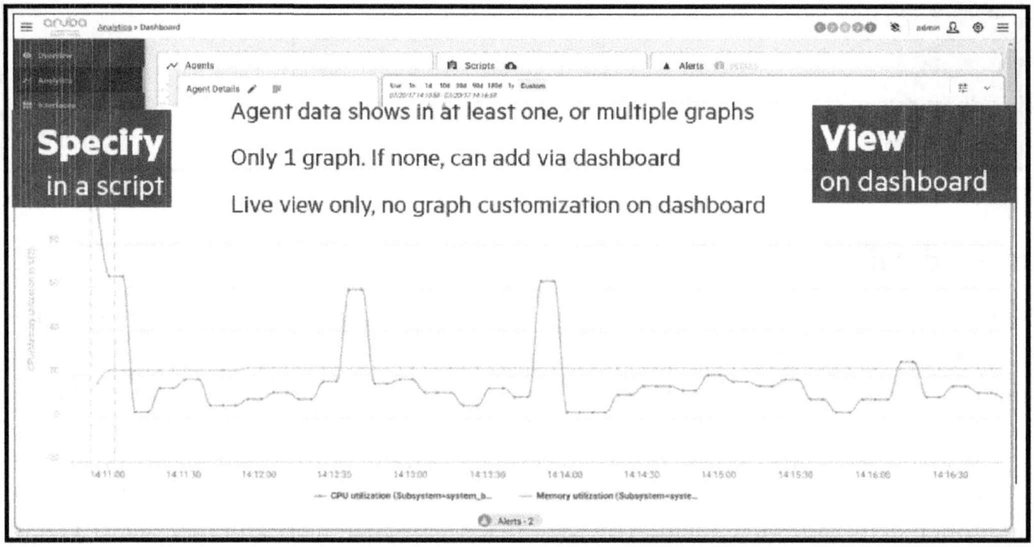

Figure 15-20: Graphs in NAE

Viewing graphs

The following figure shows a closer view of a graph in an Agent Details page. This graph has multiple lines, one for each monitor within the agent. For example, one line shows the CPU usage averaged over short time periods, one shows CPU usage averaged over long time periods, one shows memory averaged over short time periods, and so on. The legend at the bottom of the graph shows what each line indicates. You can click any of the names in the legend to remove that line from the graph or to re-add it. In this way, you can focus on the information that you want.

Configuration checkpoints and alert indicators are overlaid on the graph. Configuration checkpoints are shown as purple diamonds. Alert indicators can include the following:

- A red or yellow triangle for an alert
- A green triangle for a return to normal
- A blue triangle for an alert on several resources being monitored

You can also change the graph to show different time periods, such as the last hour or day up to the last year. Or you can specify a custom time period.

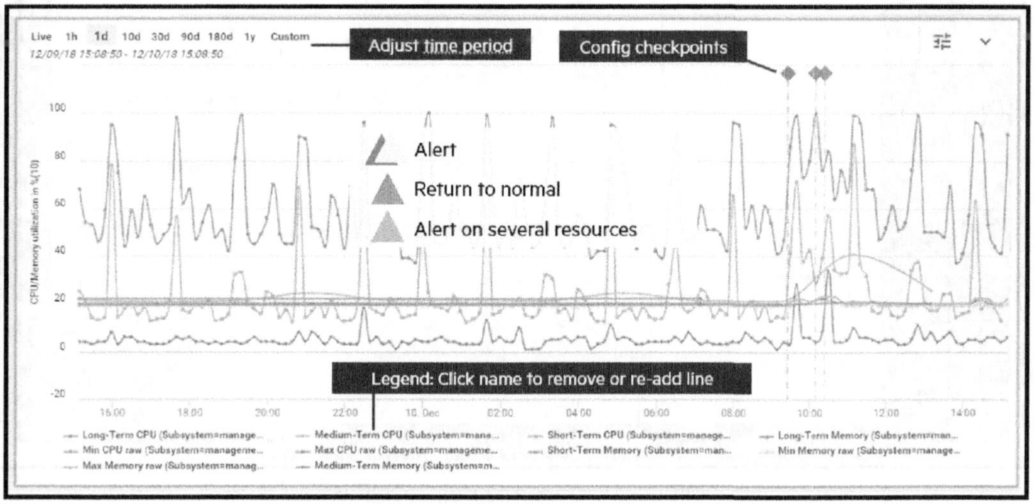

Figure 15-21: Example graph explanation

NAE alert details

The Alert Details window shows when and why the alert occurred, the alert level, and a history of actions taken for the condition that triggered this alert. In the following example, these actions include syslog messages and CLI commands. You can also collect any information that those actions generated.

From the Alert Details window, select **Output** to see the output from the CLI commands. |f the alert generated a custom report, you can click and view that report.

You can also use the REST API to obtain information about NAE and NAE agents. Refer to the documentation and reference interface for more details.

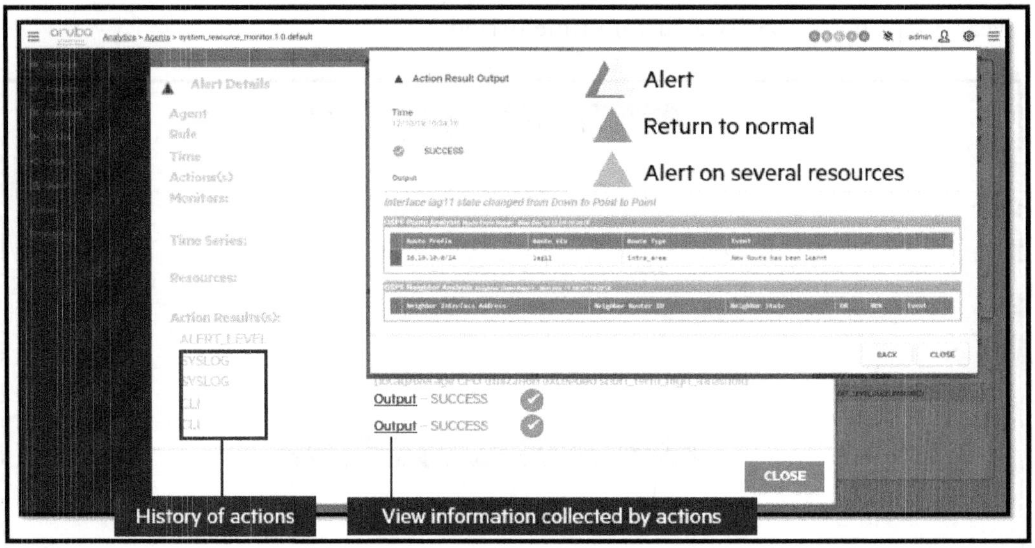

Figure 15-22: NAE alert details

CLI verification

You can see which scripts and agents are available from the CLI with the `show nae-script` and `show nae-agent` commands.

```
6300# show nae-script
-----------------------------------------------------------------------
Script Name                          Version   Origin    Status
-----------------------------------------------------------------------
configuration_change_tftp            1.3       user      VALIDATED
ipsla_threshold                      2.1       user      VALIDATED
software_device_health_monitor       2.0       user      VALIDATED
system_resource_monitor              1.4       system    VALIDATED
transceiver_power_monitor            2.0       user      VALIDATED

6300# show nae-agent
-------------------------------------------------------------------------------------------
----------------------------------
Script Name                      Version  Origin   Disabled  Status   Time Series Count
Count  Rules   Error
-------------------------------------------------------------------------------------------
----------------------------------
Configuration_Change_To_TF       1.3      user     false     NORMAL   1
 0            3       NONE
system_resource_monitor.default  1.4      system   false     NORMAL   10
    0             6       NONE
Transceiver_Power_Monitor        2.0      user     false     NORMAL   0
    0             0       NONE
```

Troubleshooting tips

In this section, we will cover the following things to look for when troubleshooting issues encountered when viewing information in NAE:

- **Switch limits**
- **Software updates**
- **Time sync**
- **CPU/memory usage**

Switch limits

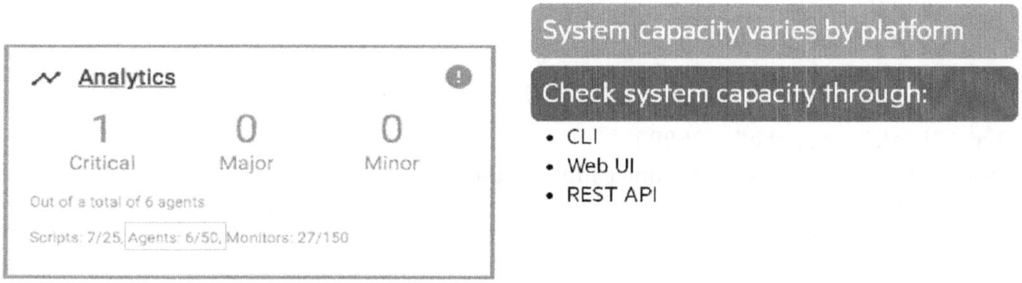

Figure 15-23: Switch limits

Each AOS-CX switch platform has its own limitations regarding the maximum number of supported scripts, agents, and monitors. When those limits are reached, the platform will deny the execution of new items.

To use the CLI, enter the `show capacities-status nae` command. For example:

```
switch# show capacities-status nae
System Capacities Status: Filter NAE
Capacities Status Name                                              Value  Maximum
-----------------------------------------------------------------------------------
Number of configured NAE agents currently active in the system       1       25
Number of configured NAE monitors currently active in the system     7       50
Number of configured NAE scripts currently active in the system      1       12
```

For detailed information about the `show capacities-status` command, see the *Command-Line Interface Guide* for your switch.

- To use the Web UI on the overview page, look at the analytics panel to see the total number of scripts, agents, and monitors compared to the total number supported on the switch. For example, agents: 6/50 indicates that there are total of six enabled and disabled agents out of a maximum of 50 agents supported on this switch.

- To use the REST API or get information about maximum number of scripts, agents, and monitors, send a GET request to the /system resource, specifying the attributes=capacities query parameter. For example:

GET /rest/v1/system/?attributes=capacities

The response body contains capacity information about multiple switch features. The information about the HPE Aruba Networking Network Analytics Engine starts with the string: nae_ . For example:

```
{
"capacities": {
...
... "nae_agents": 50,
"nae_monitors": 150,
"nae_notification_handlers": 8,
"nae_notification_queue_size": 25000,
"nae_notifications_rate": 25000,
"nae_notifications_rate_per_monoitor": 500,
"nae_scripts": 25,
"nae_tsdb_disk_quota": 9,
...
...
}
}
```

- To get information about the current number of scripts, agents, and monitors, send a GET request to the /system resource, specifying the attributes=capacities_status query parameter. For example:

GET /rest/v1/system/?attributes=capacities_status

The response body contains information about multiple switch features. The information about the HPE Aruba Networking Network Analytics Engine starts with the string: nae_ . For example:

```
{
"capacities_status": {
...
...
"nae_agents": 6,
"nae_monitors": 27,
"nae_scripts": 7,
...
...
}
}
```

Software updates

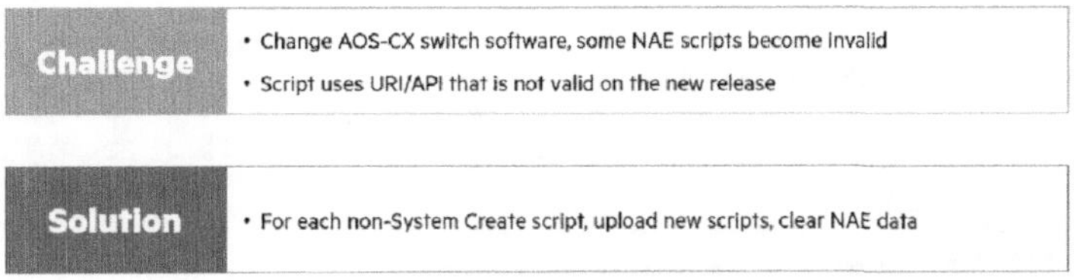

Figure 15-24: Software updates

After you update or downgrade the AOS-CX switch software, some NAE scripts might become invalid. This is likely because the script uses a URI or API function that is not valid on that software release. The existing scripts might generate errors, and the NAE data might not be valid until you upload the new scripts and clear the NAE data.

1. For each script that is not marked System Created in the Web UI, locate the script version that supports the currently running software release.
   - For HPE-certified NAE scripts, the HPE Aruba Networking Solution Exchange (ASE) includes tags that indicate the minimum and maximum supported software release.
   - When you select the ASE download button in the Web UI, the Web UI displays only the HPE-certified NAE scripts that are supported on the software release running on the switch.
   - For scripts that are not HPE-certified NAE scripts downloaded from the ASE, see the information provided by the script author about which script version is supported.
2. Follow the instructions for updating a script. The steps to update a script depend on the switch software release version that was running when the script was installed.
3. On the Analytics dashboard, locate and close any time series graph panels for the default agent from the previous software release.

The default agent is created from the built-in script. The name of the agent is based on the name of the built-in script. Built-in scripts and agents have an origin of "system" and are marked System Created in the Web UI. Built-in scripts and their agents are updated automatically. However, sometimes the time series graph panel for the previous default agent is not closed during the update. After the software update, instead of graphed data, such panels show an error message beginning with "Agent data not found."

Time sync

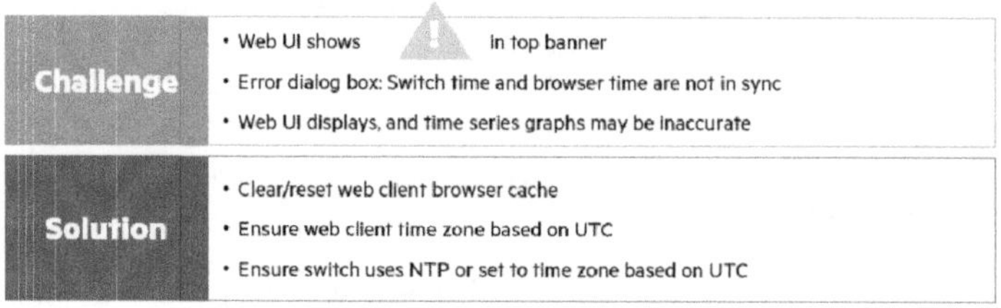

Figure 15-25: Time sync

When the switch time and browser time are not in sync, the Web UI displays a yellow caution triangle in the top banner and an error dialog box with the following title: "Switch time and browser time are not in sync."

The content of the dialog box indicates how many seconds the switch time is ahead of or behind the browser time, and states that the information displayed in the Web UI might not be accurate. In addition, time series graphs on the Analytics page might be missing expected data, might have data shown with inaccurate times, or might display incorrect data.

Things you can do to fix this issue include:

- Try clearing or resetting the web client browser cache.
- Ensure that the web client from which you are viewing the Web UI is set to a time zone based on Coordinated Universal Time (UTC). For example, if your workstation is set to Eastern Standard Time (EST) and you want to use Pacific Standard Time (PST), change the time by setting the time zone instead of by manually resetting the time.
- Ensure that the switch is set to use NTP or to a time zone based on UTC. NTP synchronizes the time of day among a set of distributed time servers and clients to correlate events when receiving system logs and other time-specific events from multiple network devices. All NTP communications use UTC. To show the NTP status, use the `show ntp status` command. After you configure the switch, clear the NAE data by entering the `clear nae-data` command from the manager context.
- If the switch is set to use NTP and there has been a significant clock change, clear the NAE data by using the `clear nae-data` command.

See the *Network Analytics Engine Guide* for more details on using and troubleshooting the NAE feature.

CPU/memory usage

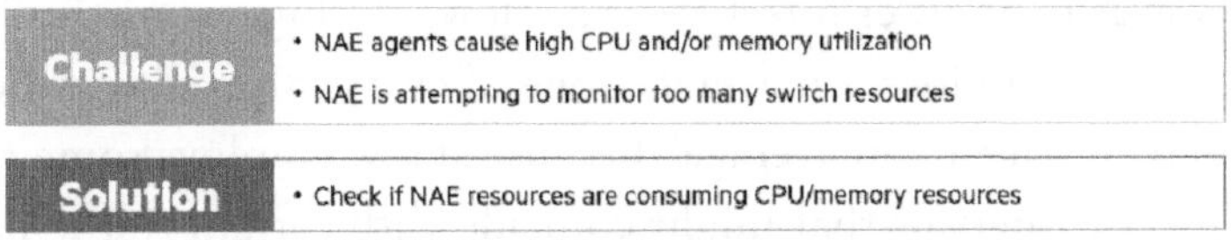

Figure 15-26: CPU/memory usage

Suppose that a switch with NAE agents installed is experiencing high CPU usage, high memory usage, or both, and overall performance is affected. The cause is that the NAE is attempting to monitor too many switch resources. For example, a script uses wildcard characters in a URI to monitor all interfaces, ACLs, or VLANs, and the switch has hundreds or thousands of those items.

To validate this theory, verify that the resources associated with the NAE are consuming the memory and CPU resources by entering the following commands in the switch CLI:

```
AGG-1# show system resource-utilization daemon hpe-tsdbd
Process                 CPU Usage      Memory Usage    Open FD's
-----------------------------------------------------------------
hpe-tsdbd                  0              0              10
T2-AGG-1# show system resource-utilization daemon hpe-policyd
Process                 CPU Usage      Memory Usage    Open FD's
-----------------------------------------------------------------
hpe-policyd                0              0              8
T2-AGG-1# show system resource-utilization daemon prometheus
Process                 CPU Usage      Memory Usage    Open FD's
-----------------------------------------------------------------
prometheus                 5              1              31
```

Basic traffic monitoring

You will now learn about a few more ways to monitor traffic on AOS-CX switches.

sFlow

AOS-CX switches support sFlow, short for "sampled Flow." It is an industry-standard technology that uses statistical sampling to enable network monitoring. sFlow relies on two components:

- sFlow agents
- sFlow collector (also sometimes called a receiver)

sFlow agents run on network devices—such as switches, routers, and access points—gathering information about the packets transmitted on these devices and sending that information to the sFlow collector for analysis. Specifically, agents can send traffic samples and polling counters.

For traffic samples, the sFlow standard defines all the packets that a network device receives on one interface and forwards to another interface as a "flow." Using a statistically accurate algorithm, the sFlow agent examines, on average, one of every *n*th packet, where *n* is the number of packets. The samples are taken "on average" because sFlow employs some randomness to avoid sampling packets at precise intervals that might coincide with certain traffic patterns.

An sFlow agent packages sampled information into small datagrams, which include Layer 2 through 7 information, as shown in the following figure.

An sFlow agent can also use counter polling to gather traffic statistics. It polls data sources at specified intervals, adds this information to the sFlow datagrams, and sends them to a specified sFlow collector. On the switches, the data sources are ports (Ethernet interfaces).

The agents can include information from several packets into a single datagram. Because traffic samples are compact and agents sample only a small percentage of traffic, sFlow does not require a large amount of network bandwidth or processing power on network devices.

To forward the datagram to the sFlow collector, the sFlow agent appends a UDP header and an IP header to the datagram and encapsulates the packet in an Ethernet frame.

AOS-CX switches implement sFlow in hardware, as is typical, for quick, efficient operations.

The sFlow collector analyzes sFlow datagrams from each agent on wireless and wired networks. The collector then creates a statistical model of network traffic that can be used for network traffic management, baselining, security auditing, troubleshooting, and more.

Find more information about sFlow at: https://sflow.org/.

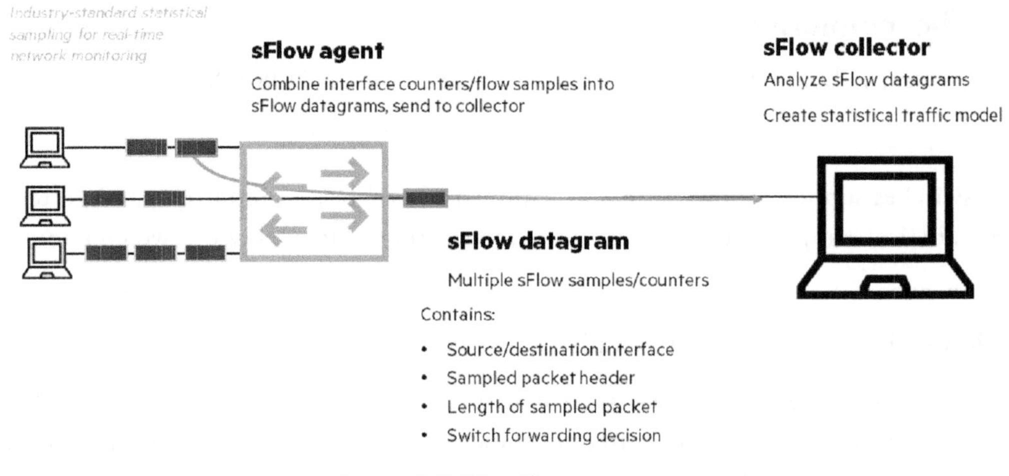

Figure 15-27: sFlow summary

sFlow configuration

The following figure shows how to configure sFlow on AOS-CX switches.

The optional sflow <LOCAL_IP_ADDRESS> command lets you define the source address on the switch to be used when sending sFlow data to a receiver/collector.

You must configure the IP address of the sFlow collector. As shown below, you can enter an optional UDP port number and/or VRF name. If you do not include the port in the command, the AOS-CX switch will use the default UDP destination port, 6343.

```
Switch(config)# sflow collector <IP_ADDRESS> [udp <PORT_NUMBER>] [vrf <VRF_NAME>]
```

You can also specify the polling and packet sampling rate, as shown in the figure. The default sampling rate is 1 in every 4096 packets for a flow. A warning message is displayed when the sampling rate is set to less than 4096 and proceeds only after user confirmation.

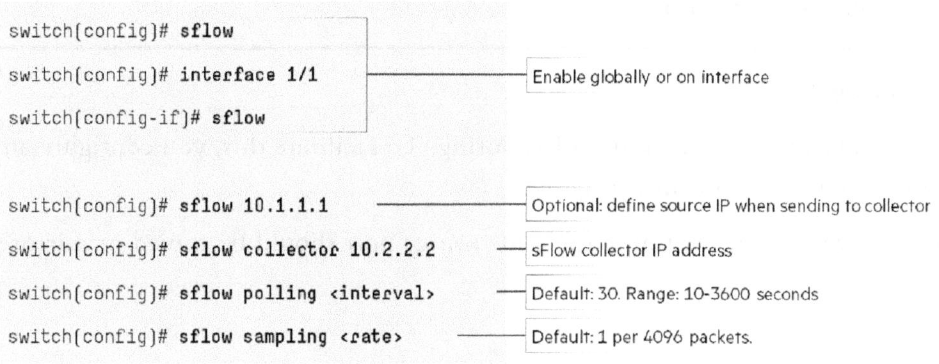

Figure 15-28: sFlow configuration

sFlow validation

As shown in the text below, use the show sflow command to verify your configuration.

```
switch# show sflow
sFlow Global              Configuration
-----------------------------------------
sFlow          enabled
Collector IP/Port/Vrf     10.0.0.2/6343/default
10.0.0.3/6400/default
Agent Address    10.0.0.1
Sampling Rate    1024
Polling Interval          30
Header Size               128
Max Datagram Size         1400

sFlow Status
-----------------------------------------
Running - Yes
```

```
sFlow Statistics
-----------------------------------------
Number of Samples          200
- Agent address is not configured.
```

sFlow and NAE are two different tools that you can use on your AOS-CX switch. sFlow can be used to share information easily with third-party monitoring products, like H3C IMC's Network Traffic Analyzer (NTA) module, which supports protocols like sFlow (open standard), NetFlow (Cisco-proprietary), and NetStream (H3C proprietary). NAE information is RESTful based and thus can be shared with a multitude of systems that support a RESTful API infrastructure. You can use either one of these, or both of these, depending on the type of external solution you need to share information with.

Mirroring overview

Packet captures can facilitate network troubleshooting. To facilitate this, you configure an appropriate switch or switches for port mirroring.

You are telling the switch that the traffic for some source port should be copied or mirrored to some destination.

This could be:

- Local storage on the switch itself.
- A local port on the switch, to which you have attached some capture device—perhaps a Linux or Windows host running the popular Wireshark utility or similar. You can then use the packet capture utilities on your capture device for traffic analysis.
- Tunneled to some remote switch, where a capture device is attached.

This collection of interfaces and settings is called a mirror session, which can have multiple sources but only a single destination. Be careful! Multiple sources can overwhelm a single destination. For example, if you configure multiple 10G source ports to a single 10G destination port, the destination may become saturated, causing packet loss.

Mirroring is sometimes referred to as SPAN, for Switch Port ANalyzer. AOS-CX v10.02 and higher support the ERSPAN feature, in which you send mirrored traffic over a tunnel to a remote switch.

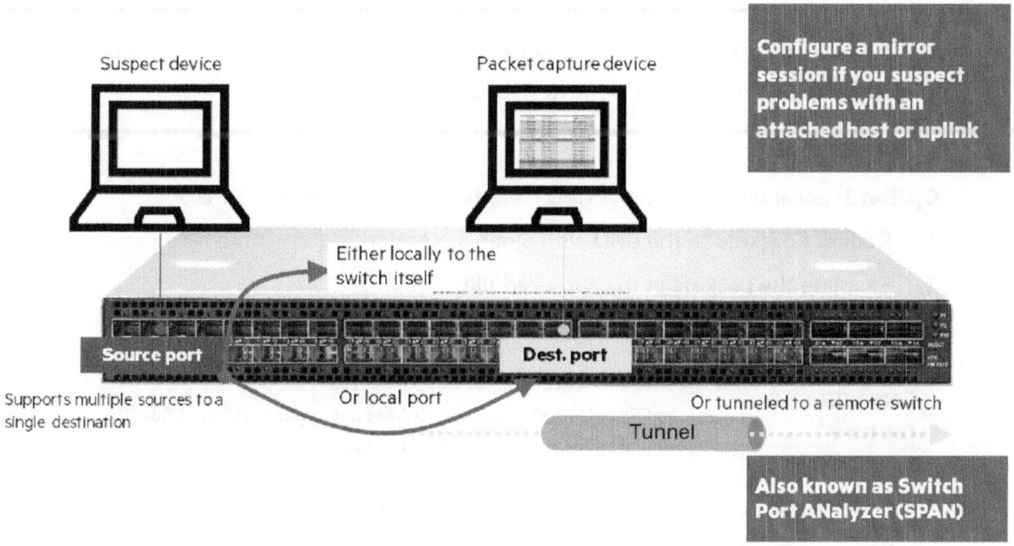

Figure 15-29: Mirroring overview

Local packet capture option 1

There are two options for looking at packets locally on AOS-CX switches via the CLI: local mirroring and tcpdump.

The option 1 configuration is shown in **Figure 15-30**.

The source interface can be a physical port or a LAG. You can configure from one to four mirror sessions to use the CPU (the switch itself). Use the `show mirror` command to verify your configuration.

By default, you don't see any information, even though the traffic being captured is "seen" by the CPU. To start examining the traffic, use the `diagnostic utilities tshark [file]` command, as shown in the figure. Press **Ctrl-C** to stop viewing the capture.

If you specify the file parameter, the switch captures packets from a mirror-to-CPU session and saves the most recent 32MB to a PCAP file, which can then be copied and analyzed. A common option to analyze packet captures is the free WireShark utility. When capturing a mirror-to-CPU session to a file, packets will not be dumped to the console. Otherwise, you are looking at a summary of the captured packets.

If you chose the file option, the file is stored in flash. To delete it, use the `diagnostic utilities tshark delete` command.

To copy the file from the switch to a remote computer, use the `copy tshark-pcap <REMOTE-URL> [vrf <VRF-NAME>]` command.

You can use TFTP or SFTP for the copy.

AOS-CX only supports one packet capture file. If you start a new diagnostic session, the old file is overwritten by a new one.

Option 1: Local mirroring to the CPU (referred to as "destination CPU")

1. Redirect packets to the CPU with a mirror session
2. Examine the packets in the captured file
3. Optionally copy the file to a remote desktop

Configure capture	switch(config)# **mirror session <session-ID>** switch(config-mirror)# **source interface <port-ID-or-LAG-ID>** switch(config-mirror)# **destination cpu**
View capture	switch# **diagnostic utilities tshark [file]** Inspecting traffic mirrored to the CPU until Ctrl-C is entered **^C** Ending traffic inspection.

Figure 15-30: Local packet capture option 1

Local packet capture option 2

Another option is to look at a summary of the packets, in real time, using Linux's tcpdump utility. Note that you do not have access to the Linux CLI, but the AOS-CX switch utilizes this utility from the AOS-CX CLI. This is accomplished using the `diag utilities tcpdump` command. Here's an example:

```
switch# diag utilities tcpdump ethernet-type 35020 count 1 verbosity level4 vrf
default
1  10:50:07.281354 LLDP, length 109
Chassis ID TLV (1), length 7
Subtype MAC address (4): 90:20:c2:bc:97:00 (oui Unknown)
0x0000:  0490 20c2 bc97 00
Port ID TLV (2), length 7
Subtype Interface Name (5): 1/1/46
0x0000:  0531 2f31 2f34 36
<output deleted>
Management Address TLV (8), length 12
Management Address length 5, AFI IPv4 (1): 10.251.12.3
Unknown Interface Numbering (1): 0
0x0000:  0501 0afb 0c03 0100 0000 0000
Port Description TLV (4), length 6: 1/1/46
0x0000:  312f 312f 3436
Organization specific TLV (127), length 6: OUI Ethernet bridged (0x0080c2)
Port VLAN Id Subtype (1)
port vlan id (PVID): 1
```

```
0x0000:  0080 c201 0001
End TLV (0), length 0
1 packet captured
4 packets received by filter
0 packets dropped by kernel
```

WARNING: Any packet capturing, without the proper filtering, and displaying it to the CLI of the switch can be very CPU-intensive and affect the operation of the switch. Care should be taken when using the `diag utilities tcpdump` command.

Local traffic mirroring

Traffic mirroring allows a switch to copy frames that arrive or are sent from a particular port or VLAN and forward them to another port. You can use port mirroring to monitor network traffic to detect threats, troubleshoot problems, or manage the network. For example, you might want to send certain traffic to a security appliance such as an Intrusion Detection System (IDS) device, which can examine the traffic and detect possible threats. Or you may want to send traffic to a network protocol analyzer, which you can use to examine the traffic and troubleshoot a network problem.

The simplest form of traffic mirroring is local mirroring, in which the switch mirroring the packets sends the copies out a local port.

The mirroring feature (when mirroring received traffic) and the sFlow sampling feature both require the receive (rx) capability of a given port. If both features are configured and enabled to use the receive capability on the same port, only one of the features can perform its task. This interaction does not affect transmit (tx) mirroring because sFlow does not use the transmit (tx) capability of a port.

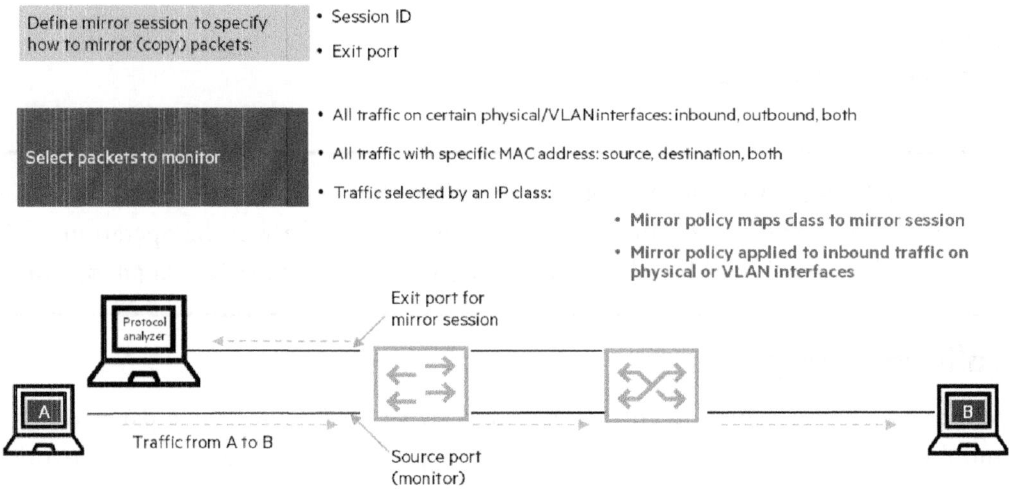

Figure 15-31: Local traffic mirroring summary

Local traffic mirroring configuration

Figure 15-32 shows how to configure local traffic mirroring on AOS-CX switches. You first enter the mirror command, followed by the session number, and then specify the exit port in the interface context. The source and exit port can be a single physical interface or a link aggregation for local mirroring.

You can configure one or more mirror sessions to use the same exit port. Or you can specify different exit ports in different sessions to send mirrored traffic to different destination devices or on different links to the same destination device. In the latter case, you'd want to make sure to set up different sources for each mirroring session. You can create up to four mirror sessions (1-4).

Use the `show mirror` command to verify your configuration.

```
switch(config)# mirror session <SESSION_ID>
switch(config-mirror)# source interface <PORT_OR_LAG_ID> [rx | tx | both}
switch(config-mirror)# destination <PORT_OR_LAG_ID>
```

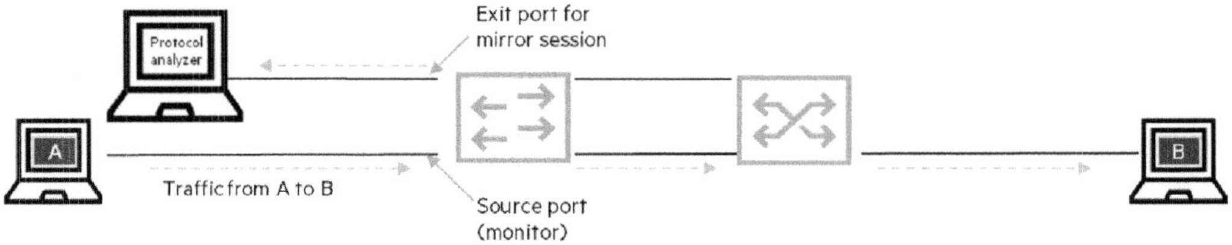

Figure 15-32: Local traffic mirroring configuration

Remote traffic mirroring: ERSPAN

ERSPAN is an acronym that stands for Encapsulated Remote Switched Port Analyzer. ERSPAN mirrors traffic on one or more source ports and forwards the mirrored traffic to a destination on a remote device. The traffic is encapsulated in Generic Routing Encapsulation (GRE) and is, therefore, routable across a Layer 3 network between the source switch and the destination device, like a packet sniffer (for example: Wireshark).

AOS-CX switches also support remote port mirroring, in which the switch forwards the mirrored packets to a destination device. The switch achieves this by encapsulating the mirrored packets in a GRE header that uses the remote switch's IP address as the destination. The remote, or destination, switch is configured to decapsulate traffic from this GRE tunnel and to forward the traffic out an exit port (see the previous section on setting up local mirroring).

This command specifies the tunnel to where all mirrored traffic for the session will be transmitted. Only one tunnel destination is allowed per session.

```
Switch(config)# mirror session <SESSION_ID> {rx | tx | both}
Switch(config-mirror)# source interface <PORT_OR_LAG_ID>
Switch(config-mirror)# destination tunnel <DESTINATION_IP_ADDRESS> source
<SOURCE_IP_ADDRESS> [dscp <DSCP_CODE>] [vrf <VRF_NAME>]
```

You may configure multiple mirror sessions with the same source/destination IP address pair; however, only one of those sessions sharing the same source/destination IP address pair can be enabled at a given time.

ERSPAN is not supported leaving the switch by the OOB port. If VRF management is configured for an ERSPAN session, the session will be in "mirror_err_tunnel_oob_port_not_supported" operation status.

ERSPAN is not supported leaving the switch encapsulated within another tunnel (for example, GRE IPv4). When the path to the destination IP address will leave via a tunnel, the session will be in "tunnel_route_resolution_not_populated" operation status.

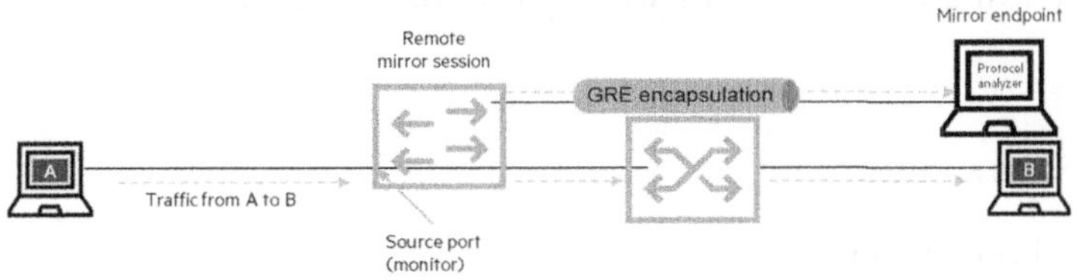

Figure 15-33: ERSPAN representation

AOS-S switches support mirroring to other AOS-S switches. AOS-CX switches, however, do not support this feature. Instead, the remote mirroring must be to a device that supports it, like Wireshark.

ERSPAN output

Here's an example where a remote mirroring session was set up on an AOS-CX switch to forward packets to Wireshark on a remote desktop. As you can see from the highlighted part of the graphic, the remote desktop is receiving the encapsulated (tunneled) traffic denoted by the ERSPAN packets.

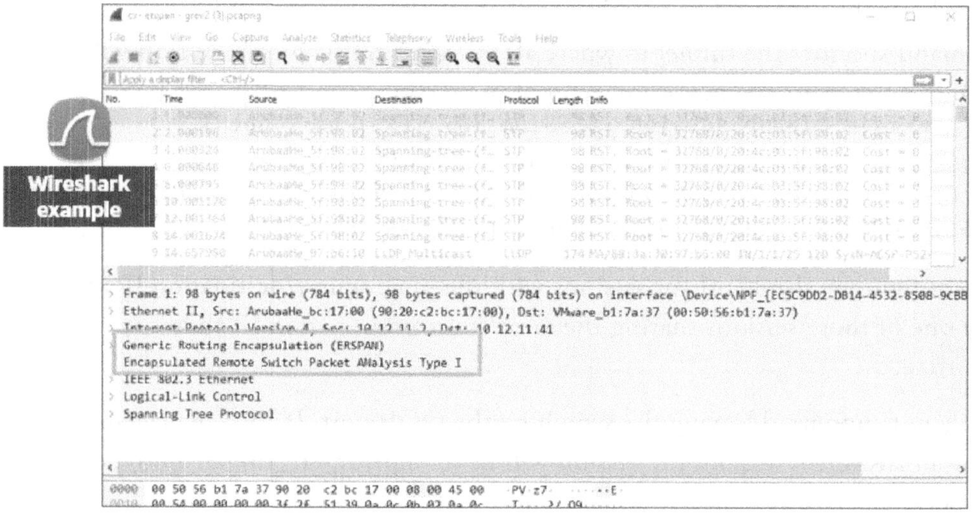

Figure 15-34: Wireshark example

Learning check

Check your understanding of NAE by answering the following questions.

Question 1

What are HPE-certified scripts?

- ○ System created scripts only
- ○ Scripts authored by HPE and posted to ASE and GitHub
- ○ Scripts customized by partners and customers
- ○ Scripts created by third-parties but installed by HPE in the factory software

Question 2

You are not seeing expected NAE data in your web browser. What should you examine?

- O The time on the switch and your desktop.
- O NAE has been enabled from the CLI.
- O The script is digitally signed.
- O The script is written in SQL.

16 Troubleshooting

Network troubleshooting is a critical process of an organization's IT infrastructure. In this chapter, you will explore some common troubleshooting principles and methodologies.

This chapter aims to familiarize you with troubleshooting methodologies. You will examine the principles and some key components of troubleshooting, followed by a commonly used problem-solving methodology. You will also review a list of tools available to perform network troubleshooting.

OBJECTIVES

After completing this chapter, you should be able to:

- ✓ Describe the principles of troubleshooting.
- ✓ Examine the key components of effective troubleshooting.
- ✓ Analyze a typical problem-solving methodology.

Troubleshooting overview

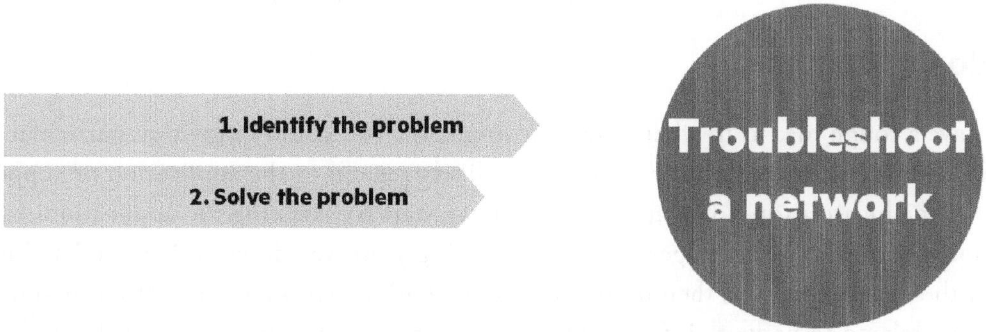

Figure 16-1: Simplified steps of troubleshooting

Network troubleshooting is an important part of any organization's IT infrastructure processes. As businesses grow, it is critical to ensure that the network is up and running to support the business operations at all times. Any network downtime means customers cannot access the services or employees cannot access the applications. This will eventually lead to loss of profit. Therefore, troubleshooting a network is one of the most important skills for a network engineer.

Troubleshooting a network means identifying the problem and its root cause through rigorous processes and then solving the problem using verified methods. How this differs from the general issue resolution process is that it involves a set of organized steps performed in order to identify the problem, rather than randomly trying some steps until the network starts functioning. The identification process is designed such that it can target a network component or a function. This process can then be documented and reused in future.

Troubleshooting principles

Before reviewing an effective troubleshooting methodology, there are several troubleshooting principles to be aware of. Some of them are illustrated in **Figure 16-2**. You can choose among them and use them in conjunction with one another, depending on the scenario.

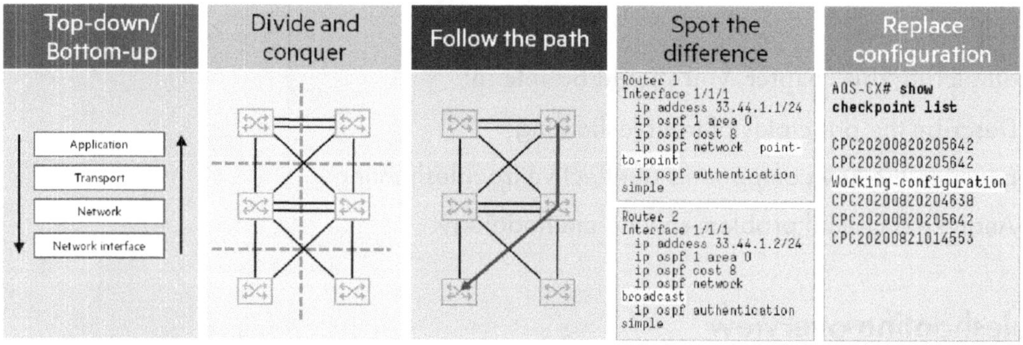

Figure 16-2: Troubleshooting principles

Top-down/Bottom-up

If you have a rough idea as to where the issue might be, you could target that particular layer of the TCP/IP suite. If you think the issue is most likely related to the application or supporting software, you might use a top-down approach. You can start by assessing the application status, configuration, logs and error messages, gradually working your way down to issues related to TCP or UDP at the Transport layer, then on to routing/network issues, saving the physical and data link layer issues related to network interfaces for last. Alternatively, you could start the diagnosis by checking the network interfaces if you think the issue is with the interface configurations and then move up layer by layer.

Divide and conquer

The "divide and conquer" approach is very often quite effective, especially if implemented properly. Take a simple example of a chassis-based switch that will not boot up properly. You might suspect that one of the line cards is faulty. One approach is to remove one line card, then power it on and check for symptoms, then replace this and remove another line card, and so on. You might swap out or eliminate seven lines cards, only to find out that it wasn't related to line cards at all. Perhaps a better approach is to divide the system into two big pieces—"line cards" and "everything else." Remove all the line cards and see if the problem persists. If it does, you have determined that it is not a line card that is faulty in just one step instead of seven steps. If the problem is eliminated, you now know that it is a line card issue. In that case, replace half of the line cards to see if the problem persists. If it does, you know that it is one of those line cards. A similar approach can be used on entire networking infrastructures, like the one shown in **Figure 16-2**.

Follow the path

The "follow the path" method allows for analyzing the network communication in a stepwise manner. You can start from the source and use a set of commands and tools to view the path taken by a packet. Any packet drop or unexpected behavior of the packet can lead to the cause of the issue. To determine the flow of the packet, you can use `show` commands. For example, use commands related to LLDP at Layer 1 and Layer 2, spanning tree at Layer 2, and routing at Layer 3. The typical `ping` and `traceroute` commands are also of great value here. Be cautious because many firewalls drop ping/traceroute traffic, so a ping failure may not necessarily mean that the path is bad. Moreover, packets may travel through several internal processes within that firewall or mobility controller. This is where solid "theory of operation" knowledge becomes vital—you are better equipped to follow the data/packet flow and identify the misconfigured subsystem.

Spot the difference

This approach involves comparing the faulty component with a working component. **Figure 16-2** shows a simple example of comparing two router configurations. It can be concluded that the difference in network type results in the inability to form an OSPF adjacency. You can also use this approach to compare one system that is working as expected with another (identical or very similar) malfunctioning system. The difference between them will often lead you toward a solution.

Replace configuration

According to the "replace configuration" approach, instead of determining what is wrong with a router, simply replace the router and restore the last-known working configuration. Or perhaps replace the configuration on the faulty router with the last-known working configuration checkpoint. This gets the customer up more quickly than troubleshooting some internal router issue during a customer outage. The customer does not care about the technical details of their outages as much as eliminating the outage. It is important to get the network up first and then figure out the exact nature of the problem.

Components of effective troubleshooting

The basics of troubleshooting any kind of networking issue might be summarized as "keep eliminating obvious causes until the real cause presents itself." But understanding what this means requires a systematic approach and real discipline when attempting to identify causes from symptoms and apply the right fixes or work-arounds. Troubleshooting is a skill that all networking professionals learn by trial and error. But skipping some of the more painful or obvious errors can make your learning somewhat less trying than it might otherwise be. The most important characteristic to cultivate when solving problems is calmness. If you can keep a clear head when things fail or start degrading seriously, you will be more capable of assessing your situation and better equipped to solve whatever problems you discover.

The components of effective troubleshooting are mainly the methodology being used, the skill set as well as the set of available tools, and the familiarity with the networking environment and the issues that arise in them.

Methodology

Developing problem-solving techniques is often an on-the-job process. But few can expect much along the lines of formal network troubleshooting training in our job positions for a number of reasons. These reasons may include:

- The fast pace of the day-to-day job tasks and challenges yields little time to pursue formal training on troubleshooting aspects such as learning to use technical tools like a protocol analyzer.
- Few business environments provide the luxury of a "test lab" and the time to hone your skills where a progression of test problems can be examined, worked through, and resolutions tried out.

In the absence of a more ideal situation, a problem-solving methodology can increase the effectiveness of support staff by standardizing the approach used to some extent. With a fairly modest amount of discipline, network technicians can improve their problem resolution efficiency in terms of the effort needed and the number of other people that must be directly involved.

Skill set

There are a variety of skills that can enhance a network technician's success in problem-solving. Some of these skills are purely technical in nature. For instance, it is important to understand the fundamentals of how network devices operate and how they are managed.

Other skills are much less technical, but still very important. As part of the problem-investigation process, a network technician may need to talk with various levels of staff. Having sufficient interpersonal skills can expedite the problem-solving process.

It is also good to have proficiency in the network troubleshooting tools and procedures developed by the IT support groups—software as well as hardware. Some examples of these are reading device logs and device statistics or using tools such as a protocol analyzer.

Experience

A problem-solving methodology that is refined over time can be very beneficial to network technicians. Being methodical and learning from the macro and micro levels of mistakes can help network technicians improve problem-recognition capabilities and yield a more efficient application of a problem-resolution process.

Need for a methodical approach

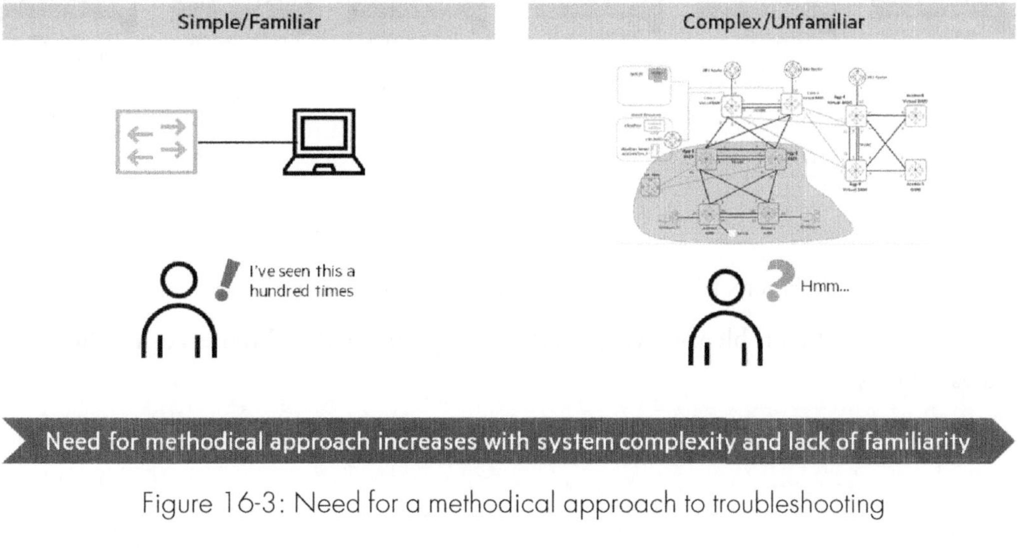

Figure 16-3: Need for a methodical approach to troubleshooting

If you are troubleshooting very simple problems, or those with which you have great familiarity, an elaborate troubleshooting methodology is less practical. If you have been working in the same environment for a long time, you become familiar with most of the issues that arise, and so you may rarely use a more elaborate, systematic approach. This can lead you to believe that such an approach is not necessary or forget how to apply such an approach.

As the complexity of the system increases, it becomes vital to adopt a more scientific approach to troubleshooting. It seems counterintuitive at first, but the experienced engineers know that a slow, methodical approach often provides quicker resolution than a hurried, random method of part-swapping and configuration changes.

Even if you have high familiarity and a simple system, it is often best to use an organized methodology. It enables you to maintain readiness and build good habits for that time when you have a more challenging scenario.

Problem-solving methodology

A problem-solving methodology is a process for managing problem resolution. Although there is no one specific model that can be used in all situations, a general framework can provide guidelines and help ensure efficiency in the efforts made to solve a problem. Applying a methodology can improve the probability of a successful resolution.

Figure 16-4 illustrates the framework for a general problem-solving methodology that has many applications in today's contemporary network environments.

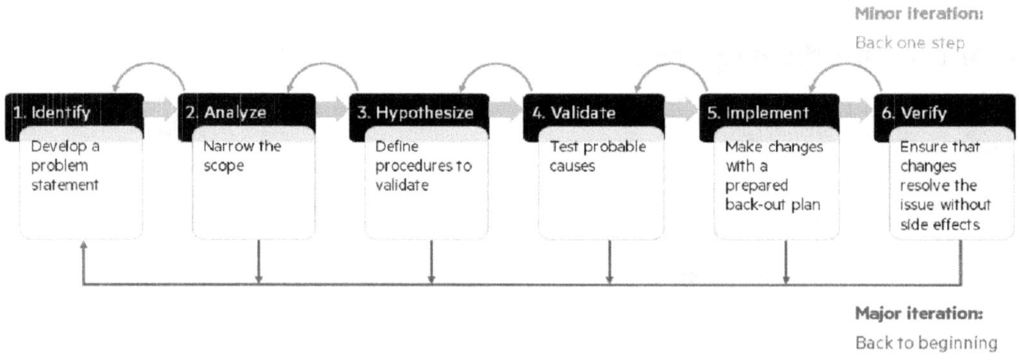

Figure 16-4: A typical problem-solving methodology

There are six steps to the problem-solving methodology outlined in **Figure 16-4**. They are described as follows.

1. **Identification—Understand and document the problem:** Sometimes it is possible to lose sight of what the potential problem is while searching for a cause when we don't consider multiple perspectives. Identifying the problem and documenting it will help keep the problem-solving efforts aligned. This step also helps define success criteria for final verification steps, from the user's point of view. You may have resolved the main issue, but there may still be some lingering aspects that affect the user—"Yes, the network is up, but I still cannot use my application."
2. **Analysis—Evaluate the situation:** One of the initial steps after identifying the problem is to learn more about the problem by investigating, using problem resolution tools, product documentation, and user input.
3. **Hypothesis—Develop possible resolutions:** The next step after analyzing the problem is to formulate a probable resolution based on the analysis and document it. This documentation may be fairly informal, but it is important to be able to explain it in writing. Doing so can reveal a hypothesis that is unclear and for which a possible resolution may not be plausible. A new hypothesis can be formulated in such a case.
4. **Validation—Run a validation process to prove or disprove the hypothesis:** This may not be particularly feasible, for example, if you have no test lab equipment to try out your hypothesis. At the very least, performing a walk-through of the hypothesis in an articulate manner with other team members may help.
5. **Implementation—Develop an implementation plan:** It is important to have an implementation plan along with a back-out plan, and then implement the resolution. For example, have a backup configuration and software image readily available to be used in case the new configuration changes disrupt the network operation.
6. **Verification—Verify the success or failure of the implementation:** Verification can be based on the identification document produced in the first stage. If the verification fails, you can implement the back-out plan.

These steps must be executed in order, starting with identification, and can be iterated as described below:

- **Minor iteration:** The rules of the methodology state that if a step fails, you must go to the preceding step. This is called a minor iteration.
- **Major iteration:** A critical issue or identification of new possibilities may require you to return to the top-level step. This is called a major iteration.

Network troubleshooting tools

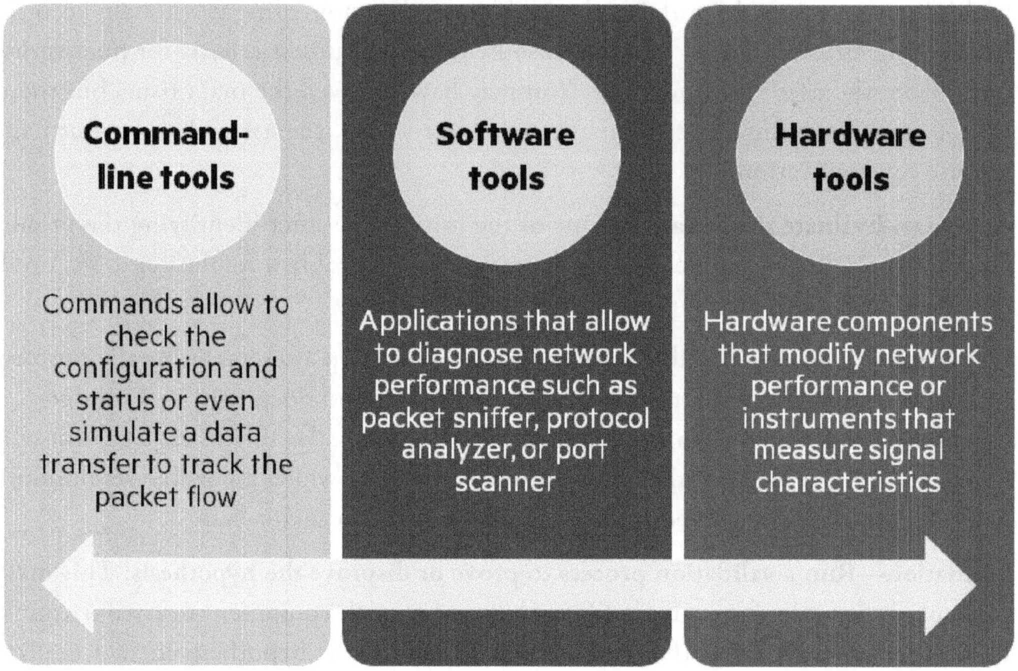

Figure 16-5: Network troubleshooting tools

There are a wide range of tools available that help network technicians diagnose and rectify network issues. They could be software aids, such as troubleshooting commands or applications, or hardware diagnostic tools.

Command-line tools

To start off, commands such as `ipconfig`, `ifconfig`, `netstat`, or `show` commands can be used to display and verify the current configuration or status of the network. Any minor misconfiguration can be easily detected through this. Thereafter, the packet flow can be tested using commands such as `ping` or `traceroute`.

Software tools

In addition to the command-line tools, there are many software applications available for diagnosing network issues. These applications target a specific component or function of the network—applications such as a packet sniffer, protocol analyzer, or port scanner.

Hardware tools

Certain network errors can be caused due to hardware defects. These can be rectified using hardware tools like wire crimpers or loopback adapters. Moreover, there are some electrical measuring instruments that allow you to measure different aspects of a signal. These include tools such as a spectrum analyzer, multimeter, or cable tester.

Learning check

Check your understanding of this material by answering the following questions.

Question 1

A faulty network appears to be unable to launch a specific application. Which troubleshooting principle is it more appropriate to use in this case?

- ○ Divide and conquer
- ○ Spot the difference
- ○ Bottom-up
- ○ Top-down

Question 2

Which tasks are involved in the identification step of the problem-solving methodology discussed in this chapter? (Select two.)

- ☐ Investigating the situation using troubleshooting tools
- ☐ Understanding the problem
- ☐ Developing a possible resolution
- ☐ Documenting the problem
- ☐ Documenting the possible resolution

17 HPE7-A08 practice exam

This chapter provides sample test questions based on the style and difficulty you would find on the real HPE7-A08 exam for the HPE Aruba Networking Certified Professional - Switching certification.

OBJECTIVES

The purpose of this practice exam is to:

- ✓ Reinforce the material you have learned.
- ✓ Become familiar with the type and style of questions found on the HPE Aruba Networking Certified Professional - Switching exam.

Exam overview

The HPE Aruba Networking Certified Professional - Switching exam tests your understanding of wired technologies across multiple sites, including branch, edge, and core environments. It also verifies your advanced skills to implement and operate enterprise-level HPE Aruba Networking switching solutions.

Ideal candidate for this exam

HPE customers, partners, and employees with at least two years of experience working as a team member that supports and maintains multiple campus topologies, edge branches, and data center networks.

Example job roles include network engineer, senior network engineer, network consultant, and tier 2/3 support.

Exam contents

This exam has 75 questions. Here are the types of questions to expect:

- Multiple choice (multiple responses)
- Multiple choice (single response)
- Drag-and-drop

Advice to help you take this exam

- Complete the training and review all materials and documents before you take the exam.
- Exam items are based on expected knowledge acquired from job experience, an expected level of industry-standard knowledge, or other prerequisites (events, supplemental materials, and so forth).
- Successful completion of the course or study materials alone does not ensure you will pass the exam.

Read the entire question and consider all options before you answer. If the question includes an exhibit, study the exhibit and read the question again. Select the answer that fully responds to the question. If the question asks for more than one answer, select all correct answers. There is no partial credit.

Exam objectives and exam information

This section covers the objectives for the HPE7-A08 exam, as well as information about the exam.

HPE7-A08 exam objectives

Table 17-1 breaks down the objectives covered on the exam.

Table 17-1: HPE7-A08 exam objectives

Percentage of exam	Sections/Objectives
18%	1. Create the implementation plan for the wired network solution. a. Describe the HPE switching portfolio and how each product can be used to solve various customer needs. (8%) b. Given a design and/or customer requirements scenario, determine an appropriate implementation plan. (10%)

Percentage of exam	Sections/Objectives
33%	2. Install and configure the wired network solution. a. Provision HPE Aruba Networking switches in HPE Aruba Networking Central. (3%) b. Configure HPE Aruba Networking switches directly via CLI (3%) c. Given an implementation plan, explain how to configure the switches physically. (3%) d. Given the implementation plan, explain how to configure Layer 2 technologies. (5%) e. Given an implementation plan, explain how to configure and validate Layer 3 interfaces, services, routing protocols, and overlays. (6%) f. Explain multicast features and configuration concepts. (2%) g. Explain HPE Aruba Networking switch security features and configuration concepts. (4%) h. Explain QoS HPE Aruba Networking switch features and configuration concepts. (3%) i. Explain HPE Aruba Networking solutions integration and configuration concepts. (4%)
27%	3. Troubleshoot the wired network solution. a. Given a scenario, identify a network failure (IP mismatch, VLAN mismatch, hardware configuration or failure, port configuration). (9%) b. Determine remediation plan. (9%) c. Given a scenario, determine the cause of the performance problem (QoS issue, configuration issue hardware and software, end node). (9%)
22%	4. Manage, maintain, optimize, and monitor the wired network solution. a. Given a scenario, determine a strategy to implement configuration management (maintenance, auditing, backup, archiving). (8%) b. Interpret data that represents the operational state of a network and determine the appropriate action. (11%) c. Audit and configure HPE Aruba Networking switches programmatically. (3%)

Exam information

Table 17-2 summarizes the information to register for the exam. You need an HPE Learner ID and a Pearson Vue login and password. No reference material is allowed at the testing site. This exam may also contain beta test items for experimental purposes. During the exam, you can make comments about the exam items. HPE welcomes these comments as part of our continuous improvement process. To register for the exam, visit https://www.pearsonvue.com/hpe.

Table 17-2: HPE7-A08 exam information

Exam ID	HPE7-A08
Exam type	Proctored
Exam duration	2 hours
Exam length	75 questions
Passing score	68%
Delivery languages	English
Supporting resources	HPE Aruba Networking Certified Professional - Switching, Rev. 24.31 course, available in physical, virtual, and eLearning formats.

More and current information concerning the HPE7-A08 exam can be found here: https://certification-learning.hpe.com/tr/datacard/exam/HPE7-A08.

Practice exam questions and answers

The following questions will help you to measure your understanding of topics that may be covered in the exam. Read all the choices carefully since there may be more than one correct answer. Choose all correct answers for each question.

Questions

1. Which statement correctly describes the role of IGMP in multicast group membership?

   A. IGMP is a router-to-router protocol

   B. Hosts use IGMP to join multicast groups

   C. IGMP uses IP protocol number 5

   D. Only the querier forwards multicast traffic

2. A company has an existing wireless solution deployed with HPE Aruba Networking APs and mobility controllers running 8.x code. The company wants the same security and firewall policies for both wired and wireless traffic. Which solution should the company implement?

   A. 802.1X authentication

   B. Downloadable user roles using ClearPass

   C. IPSec

   D. User based tunneling

3. Which two factors can influence the assignment of Local Priority and Color values in the prioritization stage of AOS-CX? (Select two.)

   A. Ingress packet L3 DSCP marks

   B. Egress interface settings

   C. Egress packet L2 CoS marks

   D. Classifier-based policy

4. Why might network administrators seek alternatives to the old Simple Network Management Protocol (SNMP) and CLI in modern campus network designs? (Select two.)

   A. SNMP's poll-driven approach is outdated

   B. CLI is excellent for machine-to-machine interaction

   C. CLI files are highly secure

   D. SNMP and CLI hinder scalability

5. Which action needs to be configured at the end of an Access Control List (ACL) to forward all the traffic that does not match with any of the rules in the ACL?

   A. The implicit deny behavior of ACLs will be sufficient to achieve this.

   B. Explicitly add a rule to deny all the non-matching traffic.

   C. Create a `permit ip any any` rule.

   D. Create a `deny ip any any rule` with `log` action.

6. What are two benefits of using NAE agents for network troubleshooting? (Select two.)

   A. Manual data collection

   B. Continuous real-time monitoring

C. Automatic alerts

D. Reduced network performance

7. Based on the bellow OSPF configuration, how many routes will routing switches see in Area 1?

   There are two areas. There is the only one ABR for Area 1. The switch has the appropriate adjacencies to route to switches in Areas 0 and 1.

   Configuration on the ABR:

```
router ospf 1
area 0
area 1 stub no-summary
```

   Routes:

   - Area 0: 5 routes (LSA Type 1 and 2)
   - Area 1: 10 routes (LSA Type 1 and 2)
   - External routes: 2 (LSA Type 5)

   A. 15

   B. 6

   C. 11

   D. 12

8. After performing the following configuration on an AOS-CX switch, the administrator notices that the authentication process fails, and switch ports always remain in the EAP start state. What should the administrator do to fix this problem?

   Configuration:

```
Switch-1(config)# aaa authentication port-access dot1x authenticator
radius server-group cppm
Switch-1(config)#aaa authentication port-access dot1x authenticator
enable
Switch-1(config)#radius-server host cppm key plaintext aruba123 vrf mgmt
Switch-1(config)#interface 1/1/6 - 1/1/24
Switch-1(config-if)# aaa authentication port-access dot1x authenticator
Switch-1(config-if-dot1x-auth)#enable
```

   A. Set the ports to client-mode.

   B. Set the ports to device-mode.

C. Configure server group cppm.

D. Enable change of authorization (CoA).

9. Which of the following are true about VSX Multi-Chassis Link Aggregation in AOS-CX switches? (Select two.)

A. VSX MC-LAG allows for active-active forwarding across two switches.

B. VSX MC-LAG is limited to only Layer 2 traffic.

C. VSX MC-LAG can reduce downtime during software upgrades

D. VSX MC-LAG requires a dedicated management switch to operate.

10. Which command is used to configure an interface to use a specific VRF other than the default VRF?

A. `vrf attach <VRF-NAME>`

B. `vrf allowed <VRF-NAME>`

C. `vrf access <VRF-NAME>`

D. `vrf enable <VRF-NAME>`

11. What does the GET method return when used on a specific resource?

A. A list of URIs

B. The attributes of the resource

C. The resource's JSON model

D. The resource's metadata

12. An ecommerce-based company wants to implement BGP for their network. The network has over 25 exit points across 20 different BGP routers. The administrator does not want to implement a fully-meshed iBGP peering between all BGP routers. Which feature should the administrator implement to reduce the number of peers the administrator needs to define?

A. Next-hop-self

B. BFD

C. Peer-Groups

D. Route reflectors

13. What are the key parameters of the Unidirectional Link Detection (UDLD) protocol configuration?

A. mode, interval, and retries
B. aggressive, normal, and Aruba-OS
C. vlan, port, and port-type
D. mode, role, and priority

14. What information can you obtain using the `show ip igmp` command?
    A. Only IGMP group memberships
    B. VLAN configuration details
    C. Querier state and IP address
    D. MAC address table

15. Which of the following statements is true about PIM-Dense Mode?
    A. PIM-DM assumes nobody wants to receive the stream.
    B. PIM-DM is best suited for low-bandwidth environments.
    C. PIM-DM floods multicasts to all neighbours.
    D. PIM-DM requires complex configuration.

16. What is required for an NAE script to start monitoring attributes and taking actions?
    A. An administrator login.
    B. An enabled NAE agent.
    C. CLI command execution.
    D. A factory reset.

17. Which type of ACL application should be implemented to control the traffic switched within a subnet with the same rules applied to all the devices in the subnet?
    A. ACLs applied to different ports of the subnet
    B. ACLs applied to the VLAN associated with the subnet
    C. ACLs applied to the subnet's VLAN interfaces
    D. ACLs applied to the LAG interfaces of the subnet

18. A network administrator sets up both 802.1X and RADIUS-based MAC-auth on a port. For which clients does the RADIUS server need accounts with their MAC addresses?

A. All clients.

B. Clients that support 802.1X.

C. Clients that support CHAP.

D. Clients that do not support 802.1X.

19. MaxNet company implements a network solution with a pair of AOS-CX switches with connections to external networks. Both switches are configured for OSPF. The administrator wants to import external routes on both switches, but assign different seed metrics to the routes, as well as import them as external Type 1 routes. How can the administrator implement this?

    A. The administrator needs to configure a route map with the route type and metrics.

    B. The administrator should define the route type and metrics using the OSPF process.

    C. The administrator needs to configure a policy to import routes with the correct route type and metrics.

    D. The administrator needs to define a class and policy map with the required route type and metrics.

20. Which of the following are benefits of the AOS-CX operating system's microservices architecture? (Select two.)

    A. Modularity allowing easier updates and maintenance

    B. Exclusively designed for unicast communication

    C. Enhanced resilience and fault isolation

    D. Slower development and deployment processes

21. What is the effect of executing the following set of commands while configuring policy-based routing on an interface?

```
policy pbr_policy_1
 10 class ip action pbr pbr_action_list_1
```

    A. Defines a list of actions to be taken on the matched traffic

    B. Defines the criteria for matching traffic

    C. Applies the PBR policy on the desired interface

    D. Creates a PBR policy from the specified action list

22. A company wants to implement access control on the Ethernet ports of AOS-CX switches. The company has a readily-available AAA solution. Which security features can the company implement along with the existing infrastructure?

    A. User-based tunneling

    B. Device fingerprinting

    C. Local user roles

    D. Port-based tunneling

23. Which network troubleshooting principle recommends reloading the last working checkpoint to revive a defective network?

    A. Divide and conquer

    B. Follow the path

    C. Spot the difference

    D. Replace configuration

24. Which of the following is the correct tunneling of user traffic between AOS-CX switches and HPE Aruba Networking mobility controllers?

    A. Using IPSec to protect the data plane traffic.

    B. Using IPSec to protect the control plane traffic.

    C. Using a GRE tunnel for the data plane traffic.

    D. Using PAPI for the data plane traffic.

25. Which type of traffic is nearly always a candidate for QoS prioritization?

    A. TCP/File Transfer

    B. UDP/Voice/Video

    C. FTP and TFTP

    D. Low priority UDP traffic

26. Which REST API method is used to create a new resource in AOS-CX switches?

    A. POST

    B. GET

    C. PUT

    D. DELETE

27. Which component of the private VLAN architecture provides isolation between devices of different secondary VLANs while providing connectivity with devices of the same secondary VLAN as well as the external world?

    A. Isolated VLAN

    B. Community VLAN

    C. Primary VLAN

    D. Promiscuous port

28. How often does an NAE agent collect data and write it to the time-series database?

    A. Every second

    B. Every five seconds

    C. Every minute

    D. Every hour

29. A multinational company has offices in several countries and uses BGP to connect its various sites over the internet. The company has recently acquired a new branch and needs to integrate its network into the existing BGP configuration. The new branch has an Autonomous System Number (ASN) that overlaps with one of the existing branches. It has a route to any subnet within this range in its IP routing table. How should the company address the issue of overlapping ASNs to successfully integrate the new branch into the existing BGP configuration?

    A. Change the ASN of the new branch to a private ASN and use BGP confederation.

    B. Disable BGP at the new branch and use static routing.

    C. Use the same ASN for the new branch as the existing branch and enable route aggregation.

    D. Implement BGP route reflectors at the new branch to avoid ASN conflicts.

30. What mechanisms enhance the resiliency protecting a VSX pair of switches from a split brain situation?

    A. Inter-Switch Link (ISL) for forwarding traffic

    B. Configuration synchronization

    C. Keepalive mechanism for peer status monitoring

    D. Active-gateway

Answers

1. Which statement correctly describes the role of IGMP in multicast group membership?

   A. IGMP is a router-to-router protocol

   B. Hosts use IGMP to join multicast groups

   C. IGMP uses IP protocol number 5

   D. Only the querier forwards multicast traffic

   Answer B is correct. Hosts use IGMP to join multicast groups.

2. A company has an existing wireless solution deployed with HPE Aruba Networking APs and mobility controllers running 8.x code. The company wants the same security and firewall policies for both wired and wireless traffic. Which solution should the company implement?

   A. 802.1X authentication

   B. Downloadable user roles using ClearPass

   C. IPSec

   D. User based tunneling

   Answer D is correct. As the company uses the HPE Aruba Networking solutions, they should implement the user based tunneling solution for the unified policy enforcement for both wired and wireless traffic.

3. Which two factors can influence the assignment of Local Priority and Color values in the prioritization stage of AOS-CX? (Select two.)

   A. Ingress packet L3 DSCP marks

   B. Egress interface settings

   C. Egress packet L2 CoS marks

   D. Classifier-based policy

   Answers A and D are correct. The ingress packet L3 DSCP marks and classifier-based policy are factors that can influence the assignment of Local Priority and Color values in the prioritization stage.

4. Why might network administrators seek alternatives to the old Simple Network Management Protocol (SNMP) and CLI in modern campus network designs? (Select two.)

A. SNMP's poll-driven approach is outdated

B. CLI is excellent for machine-to-machine interaction

C. CLI files are highly secure

D. SNMP and CLI hinder scalability

Answers A and D are correct. SNMP's poll-driven approach is outdated, while SNMP and CLI hinder scalability.

5. Which action needs to be configured at the end of an Access Control List (ACL) to forward all the traffic that does not match with any of the rules in the ACL?

   A. The implicit deny behavior of ACLs will be sufficient to achieve this.

   B. Explicitly add a rule to deny all the non-matching traffic.

   C. Create a `permit ip any any` rule.

   D. Create a `deny ip any any rule` with `log` action.

   Answer C is correct. To allow a packet without a match to be forwarded, the implicit deny rule of the ACL must be overridden by creating a "permit all" rule at the end.

6. What are two benefits of using NAE agents for network troubleshooting? (Select two.)

   A. Manual data collection

   B. Continuous real-time monitoring

   C. Automatic alerts

   D. Reduced network performance

   Answers B and C are correct. Continuous real-time monitoring and automatic alerts are two benefits of using NAE agents to troubleshoot network issues.

7. Based on the bellow OSPF configuration, how many routes will routing switches see in Area 1?

   There are two areas. There is the only one ABR for Area 1. The switch has the appropriate adjacencies to route to switches in Areas 0 and 1.

   Configuration on the ABR:

```
router ospf 1
area 0
area 1 stub no-summary
```

   Routes:

- Area 0: 5 routes (LSA Type 1 and 2)
- Area 1: 10 routes (LSA Type 1 and 2)
- External routes: 2 (LSA Type 5)

A. 15

B. 6

C. 11

D. 12

Answer C is correct. With this configuration there will be 11 routes in Area 1.

8. After performing the following configuration on an AOS-CX switch, the administrator notices that the authentication process fails, and switch ports always remain in the EAP start state. What should the administrator do to fix this problem?

Configuration:

```
Switch-1(config)# aaa authentication port-access dot1x authenticator
radius server-group cppm
Switch-1(config)#aaa authentication port-access dot1x authenticator
enable
Switch-1(config)#radius-server host cppm key plaintext aruba123 vrf mgmt
Switch-1(config)#interface 1/1/6 - 1/1/24
Switch-1(config-if)# aaa authentication port-access dot1x authenticator
Switch-1(config-if-dot1x-auth)#enable
```

A. Set the ports to client-mode.

B. Set the ports to device-mode.

C. Configure server group cppm.

D. Enable change of authorization (CoA).

Answer C is correct. The server group named cppm is missing in the configuration.

9. Which of the following are true about VSX Multi-Chassis Link Aggregation in AOS-CX switches? (Select two.)

A. VSX MC-LAG allows for active-active forwarding across two switches.

B. VSX MC-LAG is limited to only Layer 2 traffic.

C. VSX MC-LAG can reduce downtime during software upgrades

D. VSX MC-LAG requires a dedicated management switch to operate.

Answers A and C are correct. VSX MC-LAG allows for active-active forwarding across two switches and can reduce downtime during software upgrades.

10. Which command is used to configure an interface to use a specific VRF other than the default VRF?

    A. `vrf attach <VRF-NAME>`

    B. `vrf allowed <VRF-NAME>`

    C. `vrf access <VRF-NAME>`

    D. `vrf enable <VRF-NAME>`

    Answer A is correct. An interface in a VRF is configured using the `vrf attach` command.

11. What does the GET method return when used on a specific resource?

    A. A list of URIs

    B. The attributes of the resource

    C. The resource's JSON model

    D. The resource's metadata

    Answer B is correct. GET returns the attributes of the resource.

12. An ecommerce-based company wants to implement BGP for their network. The network has over 25 exit points across 20 different BGP routers. The administrator does not want to implement a fully-meshed iBGP peering between all BGP routers. Which feature should the administrator implement to reduce the number of peers the administrator needs to define?

    A. Next-hop-self

    B. BFD

    C. Peer-Groups

    D. Route reflectors

    Answer D is correct. Route reflectors can simplify the iBGP topology and prevent full-mesh iBGP peering.

13. What are the key parameters of the Unidirectional Link Detection (UDLD) protocol configuration?

    A. mode, interval, and retries

    B. aggressive, normal, and Aruba-OS

C. vlan, port, and port-type

D. mode, role, and priority

Answer A is correct. The key parameters of UDLD configuration are operating mode (mode) , packet transmission delay interval (interval), and retry count (retries).

14. What information can you obtain using the `show ip igmp` command?

   A. Only IGMP group memberships

   B. VLAN configuration details

   C. Querier state and IP address

   D. MAC address table

   Answer C is correct. The command shows the querier state and IP address.

15. Which of the following statements is true about PIM-Dense Mode?

   A. PIM-DM assumes nobody wants to receive the stream.

   B. PIM-DM is best suited for low-bandwidth environments.

   C. PIM-DM floods multicasts to all neighbours.

   D. PIM-DM requires complex configuration.

   Answer C is correct. PIM-DM floods multicasts to all neighbors.

16. What is required for an NAE script to start monitoring attributes and taking actions?

   A. An administrator login.

   B. An enabled NAE agent.

   C. CLI command execution.

   D. A factory reset.

   Answer B is correct. An enable NAE agent is required to start monitoring attributes and taking actions.

17. Which type of ACL application should be implemented to control the traffic switched within a subnet with the same rules applied to all the devices in the subnet?

   A. ACLs applied to different ports of the subnet

   B. ACLs applied to the VLAN associated with the subnet

C. ACLs applied to the subnet's VLAN interfaces

D. ACLs applied to the LAG interfaces of the subnet

Answer B is correct. When you want to control traffic switched within a subnet with the same rules applied to all the devices in the subnet, you must apply the ACL on the VLAN.

18. A network administrator sets up both 802.1X and RADIUS-based MAC-auth on a port. For which clients does the RADIUS server need accounts with their MAC addresses?

A. All clients.

B. Clients that support 802.1X.

C. Clients that support CHAP.

D. Clients that do not support 802.1X.

Answer D is correct. Clients that do not support 802.1X require MAC address for the authentication.

19. MaxNet company implements a network solution with a pair of AOS-CX switches with connections to external networks. Both switches are configured for OSPF. The administrator wants to import external routes on both switches, but assign different seed metrics to the routes, as well as import them as external Type 1 routes. How can the administrator implement this?

A. The administrator needs to configure a route map with the route type and metrics.

B. The administrator should define the route type and metrics using the OSPF process.

C. The administrator needs to configure a policy to import routes with the correct route type and metrics.

D. The administrator needs to define a class and policy map with the required route type and metrics.

Answer A is correct. A route map with a route type and metrics are required to redistribute the external routes with external Type 1.

20. Which of the following are benefits of the AOS-CX operating system's microservices architecture? (Select two.)

A. Modularity allowing easier updates and maintenance

B. Exclusively designed for unicast communication

C. Enhanced resilience and fault isolation

D. Slower development and deployment processes

Answers A and C are correct. The modularity of a microservices architecture does indeed facilitate easier updates and maintenance Also, it enhances resilience and fault isolation, which are key benefits.

21. What is the effect of executing the following set of commands while configuring policy-based routing on an interface?

```
policy pbr_policy_1
 10 class ip action pbr pbr_action_list_1
```

    A. Defines a list of actions to be taken on the matched traffic
    B. Defines the criteria for matching traffic
    C. Applies the PBR policy on the desired interface
    D. Creates a PBR policy from the specified action list

    Answer D is correct. The `policy` command is used to create a policy from the action lists in the PBR configuration.

22. A company wants to implement access control on the Ethernet ports of AOS-CX switches. The company has a readily-available AAA solution. Which security features can the company implement along with the existing infrastructure?

    A. User-based tunneling
    B. Device fingerprinting
    C. Local user roles
    D. Port-based tunneling

    Answer C is correct. As the company uses a third party AAA solution, using local user roles is the only solution in this scenario.

23. Which network troubleshooting principle recommends reloading the last working checkpoint to revive a defective network?

    A. Divide and conquer
    B. Follow the path
    C. Spot the difference
    D. Replace configuration

    Answer D is correct. The "replace configuration" principle recommends replacing a defective network configuration with the last-known working configuration.

24. Which of the following is the correct tunneling of user traffic between AOS-CX switches and HPE Aruba Networking mobility controllers?

    A. Using IPSec to protect the data plane traffic.

    B. Using IPSec to protect the control plane traffic.

    C. Using a GRE tunnel for the data plane traffic.

    D. Using PAPI for the data plane traffic.

    Answer C is correct. UBT uses a GRE tunnel to carry the user traffic between AOS-CX switch and HPE Aruba Networking mobility controller.

25. Which type of traffic is nearly always a candidate for QoS prioritization?

    A. TCP/File Transfer

    B. UDP/Voice/Video

    C. FTP and TFTP

    D. Low priority UDP traffic

    Answer B is correct. UDP, voice, and video traffic are always good candidates.

26. Which REST API method is used to create a new resource in AOS-CX switches?

    A. POST

    B. GET

    C. PUT

    D. DELETE

    Answer A is correct. POST is used to create a new resource in AOS-CX switches.

27. Which component of the private VLAN architecture provides isolation between devices of different secondary VLANs while providing connectivity with devices of the same secondary VLAN as well as the external world?

    A. Isolated VLAN

    B. Community VLAN

    C. Primary VLAN

    D. Promiscuous port

    Answer B is correct. The devices in a community VLAN can communicate with each other through L2 switching.

28. How often does an NAE agent collect data and write it to the time-series database?

A. Every second

B. Every five seconds

C. Every minute

D. Every hour

Answer B is correct. An NAE agent collects data every five seconds.

29. A multinational company has offices in several countries and uses BGP to connect its various sites over the internet. The company has recently acquired a new branch and needs to integrate its network into the existing BGP configuration. The new branch has an Autonomous System Number (ASN) that overlaps with one of the existing branches. It has a route to any subnet within this range in its IP routing table. How should the company address the issue of overlapping ASNs to successfully integrate the new branch into the existing BGP configuration?

    A. Change the ASN of the new branch to a private ASN and use BGP confederation.

    B. Disable BGP at the new branch and use static routing.

    C. Use the same ASN for the new branch as the existing branch and enable route aggregation.

    D. Implement BGP route reflectors at the new branch to avoid ASN conflicts.

    Answer A is correct. Confederation divides a large autonomous system (AS) into multiple sub-autonomus systems that can be assigned with private ASNs.

30. What mechanisms enhance the resiliency protecting a VSX pair of switches from a split brain situation?

    A. Inter-Switch Link (ISL) for forwarding traffic

    B. Configuration synchronization

    C. Keepalive mechanism for peer status monitoring

    D. Active-gateway

    Answer C is correct. The keepalive mechanism for peer status monitoring enhances VSX resiliency from a split-brain situation.

Appendix A Learning check

Overview

This appendix contains the answers to all the Learning Check sections found in the previous chapters.

Chapter 1: AOS-CX switching portfolio

This section contains the Learning Check answers for Chapter 1.

Learning Check answers

Question 1: Always-on PoE ensures that connected devices such as IP phones, wireless access points, and security cameras remain powered even during switch reboots or software upgrades, enhancing network reliability and up-time.We have more switches that also supported always on PoE, but in this case, the answer is the CX 6300 Series and CX 6400 Series.

Question 2: The AOS-CX feature that supports the time-series database to capture important statistics over a period and take actions on the captured data is NAE. NAE provides advanced analytics and automation capabilities by continuously monitoring network conditions and enabling proactive management and troubleshooting.

Question 3: The AOS-CX feature that provides dual management plane functions with zero downtime during management upgrades is VSX (HPE Aruba Networking Virtual Switching eXtension). VSX allows for high availability and enables seamless upgrades by maintaining continuous operation of network services. It ensures that one control plane remains active while the other is being upgraded, thus providing zero-downtime during the upgrade process.

Chapter 2: Virtual Switching eXtension

This section contains the Learning Check answers for Chapter 2.

Learning Check answers

Question 1: In VSX, the control plane for Layer 2 operation is unified/virtualized and up to two switches can be part of a VSX implementation.

Question 2: Smaller broadcast domains can improve performance and separate VLANs mitigate risk.

Question 3: Each VSX switch buffers and sends its new neighbor ARP/ND entries to the VSX peer every time its buffer is full (64 entries) or when it hits a timeout of five seconds.

Chapter 3: Layer 2 optimization

This section contains the Learning Check answers for Chapter 3.

Learning Check answers

Question 1: Aruba-OS modes are recommended to be used when connecting to HPE switches. When working with third-party switches, the RFC5171 modes should be used.

Question 2: Promiscuous port is a port of the primary VLAN that can communicate with all the ports of that primary VLAN including the ports of the secondary VLANs associated with that primary VLAN. This port is used to communicate outside the PVLAN domain.

Question 3: Root guard is applied to ports of the potential root switches, that is, the core layer. `admin-edge` and `admin-network` sets the edge ports for faster transitioning to forwarding state. BPDU guard is applied on edge ports to block the port in case of receipt of a BPDU, thus protecting the network against a rogue device joining at the edge port.

Question 4: AOS-CX switches support RPVST+ which can work with PVST+. In addition, RPVST+ creates a separate tree for each VLAN in the network.

Chapter 4: Advanced OSPF

This section contains the Learning Check answers for Chapter 4.

Learning Check answers

Question 1: The primary advantage of configuring multi-area OSPF in a network is that it simplifies the routing table by summarizing routes within each area.

Question 2: The routers cannot successfully become neighbors if the dead interval does not match each other.

Question 3: Intra-area routes, a default route, and inter-area routes should be learned by internal routers in a stub area.

Question 4: An ABR is responsible for summarizing routes between OSPF areas.

Question 5: The switch detects when it has lost connectivity with an OSPF neighbor more quickly.

Chapter 5: Border Gateway Protocol

This section contains the Learning Check answers for Chapter 5.

Learning Check answers

Question 1: Whether the neighbor has the same AS number as the AOS-CX switch router or not determines if there is an eBGP or iBGP relationship with a neighbor.

Question 2: Using a private ASN with BGP confederation allows the integration of the new branch without ASN conflicts.

Question 3: The `show ip bgp neighbors <NEIGHBOR-IP> advertised-routes` command specifically displays the routes being advertised to a particular BGP neighbor.

Question 4: Setting a higher weight for routes received from ISP1 will make the router prefer these routes for outbound traffic. If ISP1 fails, the routes from ISP2 will be used, as they will have a lower weight but are still available.

Question 5: The AS_PATH attribute in BGP lists the autonomous systems that a route has passed through, which is used for loop prevention and path selection.

Chapter 6: Additional Layer 3 features

This section contains the Learning Check answers for Chapter 6.

Learning Check answers

Question 1: VRF creates multiple logical routers in a single physical router, allowing customers with overlapping addresses to avoid any interference.

Question 2: The `nexthop` action overrides the routing table's longest prefix match next hop router with a next hop address that you specify.

Question 3: DHCP snooping only investigates packets received on untrusted ports. Packets received on trusted ports are forwarded.

Chapter 7: IGMP

This section contains the Learning Check answers for Chapter 7.

Learning Check answers

Question 1: By default, AOS-CX switches flood traffic throughout the VLAN in the same manner as a broadcast.

Question 2: In order for multicast traffic to work correctly on a VLAN, you must configure IGMP on L3 switches acting as gateways for the VLAN and configure IGMP snooping on every switch acting as L2 switches within the VLAN.

Question 3: IGMP and IGMP snooping are disabled by default.

Question 4: For an IGMP querier, IGMP is enabled on a Layer 3 interface.

Chapter 8: Multicast routing

This section contains the Learning Check answers for Chapter 8.

Learning Check answers

Question 1: PIM-SM routing switches must explicitly join the tree, and you must configure them to do so.

Question 2: The administrator must enable PIM and IGMP on the VLAN.

Chapter 9: Access Control Lists

This section contains the Learning Check answers for Chapter 9.

Learning Check answers

Question 1: Sequence number, action, and matching criteria are the three components of an ACL rule.

Question 2: Always specify the most specific rule with lower sequence number. Thus, denying permission to one device comes first, followed by the permit rule to allow all other devices. Permit has to be specified as there is no implicit permit. Option C has the correct specifications including the subnet masks—10 deny 10.1.10.10/255.255.255.255 any, 20 permit 10.1.10.0/255.255.255.0 any.

Question 3: The ACL is applied outbound on VLAN 10, therefore, it acts on traffic destined to Server 1 or 2. It allows traffic from all other VLANs ("any") to reach Server 1 but denies access to Server 2 ("implicit deny").

Question 4: There is no implicit deny in classifier policies.

Question 5: The ACL applied on the port takes precedence over the VLAN ACL.

Chapter 10: 802.1X authentication

This section contains the Learning Check answers for Chapter 10.

Learning Check answers

Question 1: An AOS-CX plays an authenticator role in 802.1X authentication.

Question 2: Dynamic authorization, along with NTP, should be enabled on the switch when specifying ClearPass as a RADIUS server on an AOS-CX switch.

Question 3: Digital certificates are supported by EAP-TLS for mutual client and server authentication.

Question 4: The primary function of user roles in AOS-CX switches is to provide differentiated network access and permissions for users.

Question 5: Device fingerprinting uses the MAC address, operating system, DHCP, DNS, and LLDP for finger printing.

Chapter 11: MAC authentication

This section contains the Learning Check answers for Chapter 11.

Learning Check answers

Question 1: Clients that do not support 802.1X only require MAC database for authentication.

Question 2: By default, the switch implements 802.1X authentication first when both 802.1X and MAC-auth have been enabled on a switch port.

Question 3: MACsec provides encryption and integrity for data at Layer 2 of the OSI model.

Question 4: When MACsec is enabled, the switch encrypts the 802.1Q tag of the Ethernet frame along with the payload.

Chapter 12: Dynamic Segmentation

This section contains the Learning Check answers for Chapter 12.

Learning Check answers

Question 1: Enhanced security through centralized policy enforcement is the primary advantage of using UBT.

Question 2: When deploying Dynamic Segmentation v2.0, the VSF stack consumes only one AP license.

Question 3: HPE Aruba Networking Mobility Controller is an essential component of the UBT solution when you want to ensure specific user traffic management.

Question 4: DUR is primarily used to download the user roles from ClearPass.

Question 5: The `show port-access role` command is useful to verify the communication with ClearPass.

Chapter 13: Multicast routing

This section contains the Learning Check answers for Chapter 13.

Learning Check answers

Question 1: POST should be used to create new entries.

Question 2: The REST API (Swagger) interface communicates with the switch DB. Any action, such as creating or deleting a parameter, will have an immediate effect on the system.

Chapter 14: Quality of Service

This section contains the Learning Check answers for Chapter 14.

Learning Check answers

Question 1: CoS or 802.1p are used as the classifications assigned to traffic a QoS mark in the VLAN tag.

Question 2: The port looks at the incoming DSCP and maps it to an LP and color with the DSCP map.

Question 3: Strict priority (SP) provides very low latency and jitter for high priority queues, but can start lower priority queues.

Chapter 15: NAE

This section contains the Learning Check answers for Chapter 15.

Learning Check answers

Question 1: HPE-certified scripts are scripts authored by HPE and posted to ASE and GitHub.

Question 2: You should examine the time on the switch and your desktop if you do not see expected NAE data in your web browser.

Chapter 16: Troubleshooting

This section contains the Learning Check answers for Chapter 16.

Learning Check answers

Question 1: Since the issue is with a specific application, it is best to start investigating from the application layer; that is, using the top-down approach.

Question 2: Identification is the first step of the methodology and involves understanding and documenting the problem.